CYTOCHROME P-450, BIOCHEMISTRY,
BIOPHYSICS AND INDUCTION

DEVELOPMENTS IN BIOCHEMISTRY

CYTOCHROME P-450, BIOCHEMISTRY, BIOPHYSICS AND INDUCTION

Proceedings of the 5th International Conference on
Cytochrome P-450 held in Budapest, Hungary, August 21–24, 1985

Editors:
L. VERECZKEY
K. MAGYAR

1985

ELSEVIER SCIENCE PUBLISHERS

AMSTERDAM · NEW YORK · OXFORD

Joint edition published by
Elsevier Science Publishers, Amsterdam, The Netherlands and Akadémiai Kiadó,
The Publishing House of the Hungarian Academy of Sciences, Budapest, Hungary

The distribution of this book is handled by the following publishers

for the U.S.A. and Canada
Elsevier Science Publishing Company Inc.
52 Vanderbilt Avenue
New York, N. Y. 10017

for the East European Countries, Democratic People's Republic of Korea, People's
Republic of China, People's Republic of Mongolia, Republic of Cuba and Socialist
Republic of Vietnam
Akadémiai Kiadó, The Publishing House of the
Hungarian Academy of Sciences, Budapest

for all remaining areas
Elsevier Science Publishers B. V.
P.O. Box 211
1000 AE Amsterdam, The Netherlands

ISBN 0-444-80712-8 (Vol. 27)
ISSN 0165-1714

Printed in Hungary

v

PREFACE

The 5th International Conference on Biochemistry, Biophysics and Induction of Cytochrome P-450, sponsored by the Hungarian Pharmacological Society was held in Budapest, 21-24 August, 1985. The Proceedings of the conference were published by the time of the conference to enable participants to benefit from the congress as much as possible. All papers submitted by June 1 could be included in this volume.

Cytochrome P-450 seems to have a very important role in detoxification, carcinogenesis and toxicology, therefore the study of biochemistry and biophysics of cytochrome P-450 enzymes, investigations on the mechanism of substrate and oxygen activation are very important for a better understanding of the role of this enzyme system in the living organism. The induction of this system, an important reaction of the organism in the adaptation to the environment, its toxicological and clinical relevance were specifically emphasized at the congress.

The Organizing Committee wishes to acknowledge the members of the International Scientific Advisory Board - Drs A.Archakov (USSR), M.J.Coon (USA), L.Ernster (Sweden), R.W.Estabrook (USA), G.Feuer (Canada), I.C.Gunsalus (USA), J.-Å.Gustafsson (Sweden), K.Ruckpaul (GDR) and R.Sato (Japan) - for their help in the organization of the scientific sessions, the Semmelweis University Medical School for making the conference facilities available and MOTESZ for organization in general. IUB, Fujisawa Pharmaceutical Co. (Osaka, Japan) and the Chemical Works of G.Richter, Ltd. (Budapest, Hungary) gave financial support to the conference.

We are most grateful to the staff of Akadémiai Kiadó, Publishing House of the Hungarian Academy of Sciences for their quick collaboration.

Budapest, June 3, 1985

L. Vereczkey
K. Magyar

Contents

CYTOCHROME P-450 AND ITS IMPLICATION IN ENVIRONMENTAL PROBLEMS

Biophysical Approach to the Dynamics of the Structure of Cytochrome P-450 and its Functional Implications

CYTOCHROME c AND CYTOCHROME P-450

G. Williams, G.R. Moore and R.J.P. Williams

Inorganic Chemistry Laboratory, South Parks Road, Oxford OX1 3QR. England

Now that we know a great deal about the structure, structural changes on change of oxidation state, and the dynamics of the different states of many cytochromes c and of some cytochromes b from NMR studies it is useful to review these properties in the context of the reaction sequence of cytochrome P-450. I shall take it that the important steps in cytochrome P-450 reaction are

(1) Initial state is low-spin Fe(III) but it is poised close to the high-spin state
(2) The substrate binding converts the Fe(III) to high-spin
(3) The bound molecule is reduced to high spin Fe(II)
(4) High-spin Fe(II) binds O_2 and changes to low spin
(5) Further one-electron reduction allows the production of and then the attack by the iron oxene on the substrate via H-atom and OH radical migrations
(6) The products leave and the enzyme returns to (1).

The ordered sequence requires an understanding of spin-state switches due to ligand binding $(1) \to (2)$, reduction through a protein matrix (3) and selective interaction with O_2 (4).

The Change of Spin-State

Changes of spin-state from low to high spin are brought about by weakening the bonds to the iron. Now in cytochrome P-450 the Fe(III) is bound to a thiolate, cysteine, and to a sixth ligand which is possibly a weak donor such as an hydroxyl group. It is known that the binding of substrate breaks the weak bond and not the bond to the thiolate group. It should be easy to confirm this by studies of the MCD spectrum at wavelengths around 1,500 nm. What can we learn about such reactions from the study of Fe(III) cytochrome c and other proteins?

Cytochrome c has a histidine as its strong link to the Fe(III), not a cysteine, and its weak link is through methionine 80. The low-spin to high-spin switch occurs with this protein as the temperature is raised from about 50 to 60°C and can be followed by many methods. It is also brought about by strong salt or even freeze drying. It is a slow step since it involves some displacement in the protein i.e. an internal conformation change. The nature of this change can be compared with a displacement reaction

$$\text{Fe(III)} \cdot (\text{Met } 80) + X \; \overset{\rightarrow}{\leftarrow} \; \text{Fe(III)} \; X \; + \; \text{Met } 80$$

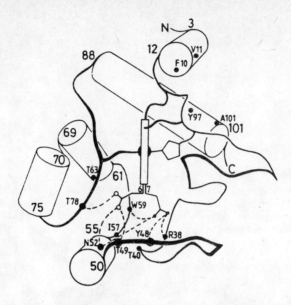

Fig.1.   The redox state change of conformation of cytochrome-c.   Note
motion of groups is indicated by black dots.   Motion is greatest around
the Met-80 side of the heme but extends to the termini.   The changes are
similar to those in the cyanide displacement reaction where breaking the
Met-80 link is accompanied by adjustment especially in the bottom and left
hand corner.

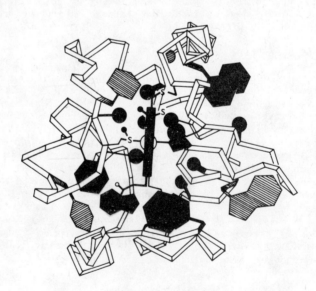

Fig.2.   Motions in cytochrome-c.   Full residues – little motion, shaded –
considerable motion.   Phe-82 is top left and shaded.

The group X can be cyanide or imidazole or Lys 79 at pH greater than 9. The kinetics of the reactions have been analysed by Sutin and coworkers and they conclude that the reaction follows a slow first order step

$$Fe(III).(Met.80) \overset{\rightarrow}{\leftarrow} Fe(III) + Met.80$$

Addition of X is then rapid. The reaction is then a first order rearrangement, Fig.1.

We have looked at the complex Fe(III)X where X is cyanide so as to compare the structure of the protein bound to X and the native protein. We use the NMR procedures described in our papers. The results show that there is no gross change in the protein structure but a variety of small movements are observed in various parts of the protein. These small movements do not only involve the groups close to the iron atom but extend throughout the protein reaching groups as far away as the termini. From a study of several proteins including calmodulin and an analysis of work by others on many proteins we conclude that the protein rearrangments are transmitted principally through changes in helix-helix contacts by rolling and sliding. In cytochrome c only minimal changes occur but in haemo-globin and P-450 we believe these changes to be equivalent to a hinge bending allowing allosteric changes and substrate binding.

Change of Redox Potential on Ligand Binding

The change from low-spin to high-spin iron (bound by substrate in P-450) causes a considerable rise in the redox potential. The change is not surprising in that the Fe(III)-ROH bond is broken and the environment is made more hydrophobic. The Fe(II) remains high-spin throughout. Notice that the change in redox potential is one way of aiding the reduction step Fe(III) to Fe(II) so that the oxygen uptake reaction can follow. However there is also a conformation change on redox change. This type of change has been followed in cytochrome c, Fig.1.

The Reduction

A reduction rate from Fe(III) to Fe(II) in the millisecond range is more than adequate to account for the reactions of cytochrome P-450. It is known that the reducing equivalents can be supplied from an external source but in vivo that source is either an Fe/S protein or a flavo-protein. Electron transfer rates of this rate have been analysed both theoretically and experimentally and it is known now that the transfer distance can be up to some 20Å. The only restriction this introduces is that the iron must not be too deeply buried away from the surface of the protein. Most of the critical experiments on this point concern cyto-chrome c. It is a feature of this protein that although it is very rigid in certain parts, Fig.2, with ring flip rates of only 10 per sec, in the region under the haem relaxation of the groups close to the iron, can be rapid e.g. of Phe-82. We see the rapid flip rate of Phe-82 as indicative of considerable ease of small motion in the iron-methionine bond geometry and in the absence of a large conformation change, this is all that is required for fast electron transfer, see Fig.2.

In cytochrome c the reduction of Fe(III) to Fe(II) causes a small conformation change but this change is important in our discussion of cytochrome P-450 since here it is likely to be much larger. The change is not just at the iron but it echoes through the protein by way of slight changes in the disposition of helices. This is the model put forward for

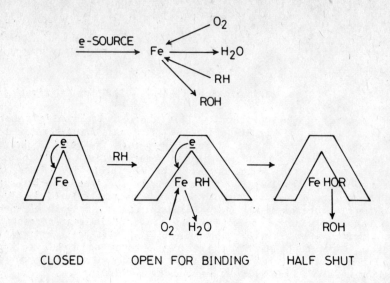

Fig.3. The proposed sequence of movements of cytochrome P-450.

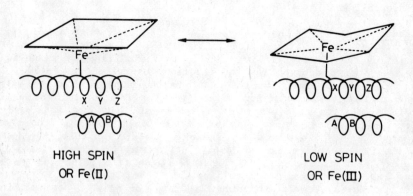

Fig.4. The trigger mehanism of haemoglobin which may be common to P-450.

the changes in haemoglobin on changes from spin state to spin state or oxidation state to oxidation state. It has been fully confirmed by the beautiful X-ray studies of Perutz, see Fig.4.

Helical Motions

Helices in proteins rest up against other objects, other helices or the haem plane via pendant hydrophobic groups such as leucine, phenylalanine or valine side chains. The contacts are then Van der Waals in nature and the constraints on change are only considerable if there is crosslinking. In most helical proteins we expect and find that the helix can move by rotation and translation like a screw drive. Another way of looking at the motion is to say that the haem can slip in the pocket. In the case of cytochrome c the helix carrying the histidine is rigidly fixed (shown by NMR) by thioether cross-links so that there is virtually no motion. Beneath the plane of the haem the protein bond angles can be adjusted and through them the loop of residues carrying methionine 80 can be rotated and translated slightly so as to generate a transmission of the conformation change to distant parts of the molecule. It is probably the helices in the 50-70 part of the sequence which allow this motion.

Cytochrome b, to which cytochrome P-450 and haemoglobin are more closely related is not constructed in the same way since both the iron-ligands are attached to helices which are not cross-linked. The b-cytochromes are made from protoporphyrin directly. We then expect that these cytochromes and therefore cytochrome P-450 will be much more readily deformed. The expectation is borne out by the known reaction cycle of P-450 but also by the changes in the spectra and redox potentials of cytochromes b in the mitochondrial and thylakoid chains.

The Solvent Cage

In redox reactions of small molecules the solvent is required to relax. One of the problems when we use X-ray structures of redox proteins to discuss reaction mechanisms is that the possible motion of groups solvating the redox centre are not known. NMR studies of proteins, especially of cytochrome c, have shown that the protein (solvent) cage around the heme is very asymmetrical. It is then the case that on one side, the histidine side of cytochrome c and possibly the thiolate side of P-450, there is little motion of amino and sidechains. On the other side there is motion which allows the entry of substrates. It is this motion which is described also by the studies on haemoglobin using Mossbauer and other techniques. We propose that as the iron - sixth ligand bond breaks in a first order uni-molecular reaction, the new internal surface which is created has on it a number of hydrophobic amino acid side chains which can rotate and move so allowing the entry of substrate and generating selectivity. The sequence of movements is shown in Fig.3 and Fig.4.

Protein-Protein Interaction

The surfaces of proteins are not adequately described by single structures since there is little constraint on the conformational mobility amongst rotamers. We should then devise methods of describing these surfaces which are covered by mobile groups, in terms of potential energy zones. Some zones will be hydrophobic, some will be positively charged and some will be negatively charged. We shall then have a mosaic of patches, representing the potential energy of a test molecule on the protein

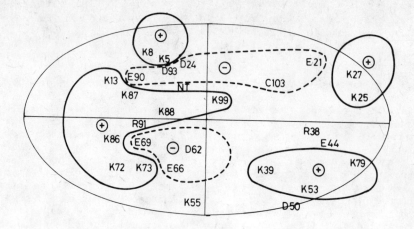

Fig.5. A projection map of cytochrome c showing the positive and negative patches.

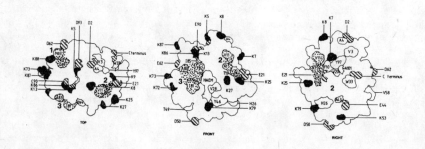

Fig.6. The surface of cytochrome c mapped by $[Cr(CN)_6]^{3-}$. The stippled areas are binding patches for anions. Note the disposition of charged groups and see Fig.5.

surface, and each patch will be dynamic, Fig.5. The interaction of proteins will be due to dynamic, mosaic-patch, matching. In other words repulsion due to space-filling as in lock and key fitting will not operate since the faces of the protein are too flat. The dynamic nature of the surfaces permits fast adjustment and fast association/dissociation even when binding is strong. This flexibility is a great advantage in a triggered set of sequential reactions such as that undergone by P-450.

Once again we have used cytochromes c as a model. We have determined experimentally the mosaic of potential energy patches by following binding of different NMR probes to the surface e.g. $[Cr(CN)_6]^{3-}$ is used to find regions of high positive potential. The method has been thoroughly described, see Fig.6. Fig.7 indicates the regions of cytochrome c which are important in two different binding reactions. A similar description will be needed for P-450.

SUMMARY

The problems to be faced in the study of large protein, as large as P-450, arise from the fact that detailed mobility may be beyond direct measurement despite good crystallographic data. This means that reaction pathways will remain conjectural. It then makes it essential for us to understand the likely contributing factors from the study of smaller proteins. In this paper we show how light may be thrown on the reactions of cytochrome P-450 through the intimate knowledge we have now of the properties of cytochrome c.

REFERENCES

A full account of the solution properties of cytochrome c, as seen by NMR will appear in J. Molec. Biol. 1985. The discussion of the spin-states and coordination chemistry of cytochrome P-450 was given by H.A.O. Hill, A. Roeder and R.J.P. Williams, Naturwissenschaften 1970, 57, 69-80.

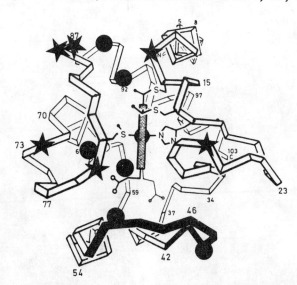

Fig.7. A view of cytochrome c showing by stars the important mobile lysines for binding to cytochrome oxidase and by filled points and zones the very different regions of high antigenic strength.

CYTOCHROME P-450 CATALYTIC MECHANISM: IMPLICATIONS FROM RESONANCE RAMAN SPECTRA

Pavel Anzenbacher

Department of Biochemistry, Charles University, Albertov 2030
128 40 Prague 2, Czechoslovakia

INTRODUCTION

Cytochromes P-450 (P-450) are hemoproteins with characteristic physical and biochemical properties conditioned by
structure of the heme active site. Strong basic axial ligand
(mercaptide sulfur) of heme iron influences electron distribution in the Fe(porphyrin)-ligand system.

Among all spectroscopic techniques, resonance Raman (RR)
is uniquely suitable for studying hemoproteins because of its
ability to get information about mode of coordination of the
heme as well as of strength of the Fe-ligand bond.

EXPERIMENTAL

Cytochrome P-450 was prepared from liver microsomes of
phenobarbital-treated rats (form P-450b) and of untreated animals (form RLM 3) by published procedures (P-450b, Anzenbacher et al.(1984a); RLM 3, Cheng and Schenkman(1982)). All
samples were highly purified with specific content of 12-18
nmol heme/mg protein. Model compound of reduced P-450.CO adduct, heme(benzylthiolate)CO was prepared by reduction of
hemin with $Na_2S_2O_4$/18-crown-6 in presence of NaH in dimethylsulfoxide.

Raman spectra were taken with Jeol JRS S1 (488,0 nm) excitation and Spex 1401 (413.1 and 457.9 nm) with Ar^+(CR2) and
Kr^+ or Ar^+(Spectra Physics 171 or 170) lasers. Samples were
either placed in capillary or in rotating NMR test tube. In
this case, a backscattering geometry was used, samples were
cooled with a flow of cold N_2; low laser power was applied
(max. 40 mW at sample) and stability of samples was checked
spectrophotometrically.

RESULTS AND DISCUSSION

An example of RR spectrum of P-450 RLM 3 is given in
Fig.1, positions of the ν_4 band and of the Fe-CO vibration
for various forms of P-450 together with data for hemoglobin
are included in Table 1.

RR band labelled ν_4 (whose origin is mainly a breathing
vibration of heme pyrroles) was chosen because its sensitivi-

ty to the electron distribution at the heme. It was shown that
at its position depends on the oxidation state of heme iron
(Yamamoto et al.(1973)). Recently it was postulated that,
more accurately, position of this band reflects the amount of
π-electron density donated from d_π orbitals of iron to π^* orbitals of porphyrin (Spiro (1983)). This effect is called π-
back donation to distinguish it from σ-donation (from ligand
to central atom) responsible for formation of metal-ligand
complexes. The π-back donation is very weak if not absent in
Fe(III) hemes; as a consequence, ν_4 occurs at 1370-1375 cm^{-1};
in spectra of Fe(II) hemes, (where this effect is present),
the ν_4 band appears at about 1355-1360 cm^{-1}.

TABLE 1
Positions of ν_4 and of νFe-CO vibrations in RR spectra of various forms of P-450 and its model complex (cm^{-1}). For comparison, data for Hb are included (taken from Spiro (1983)).

	Fe(III) RLM 3, P-450b	Fe(II) RLM 3	Fe(II).CO RLM 3	Fe(II).CO model	Fe(III) metHb	Fe(II) Hb	Fe(II).CO Hb
ν_4	1370	1341	1369	1369	1373	1356	1372
ν_{Fe-CO}	—	—	474	not de-termined	—	—	507

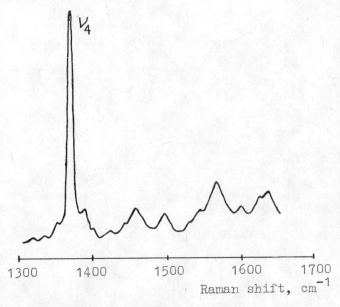

Fig. 1 RR spectrum of Fe(III) P-450 RLM 3 (excit. 413.1
nm, Kr^+ laser, rotating NMR tube, 40 mW/sample)

Ferric P-450s give the ν_4 band at 1368-1372 cm^{-1} i.e. in the lower part of interval found for Fe(III) hemes. Hence, the effect of strong both π- and σ-donating mercaptide sulfur is only weak – otherwise the π-back donation would be traced by more significant lowering of the ν_4 position. In fact, Fe(III) is rather a π-acceptor than a donor (see above). Another reason for the small effect of mercaptide sulfur to spectrum of Fe(III)P-450 may be be based on a presence of a π-electron system in close neighborhood to the heme, able to accommodate the π-electron density coming from mercaptide. As the oxygen is the most probable native ligand, tyrosine or C=O ligation should be taken into account. The tyrosine appears to be the most likely candidate (Jänig et al.(1983)), although the Fe-O-Phenyl vibrations were not until now assigned in RR spectra of P-450 (Anzenbacher et al.(1984b)).

Binding of a substrate causes strong conformational changes which alter the Fe-S interaction significantly: Champion et al.(1982) were able to detect the Fe-S vibration only in RR spectra of the substrate-bound P-450$_{CAM}$ and not in this of the substrate-free enzyme. On the other hand, position of the ν_4 shifts varies little (1372→1368 cm^{-1}, Champion et al.(1978)) which shows that the electron distribution is affected only slightly.

For the Fe(II) P-450, the π-back donation is extremely strong; it is reflected in anomalously low (in comparison with other hemoproteins) position of the ν_4 band (1340-1348 cm^{-1}). This shift of the ν_4 band was ascribed to influence of the mercaptide (Ozaki et al.(1978), Champion et al.(1978)). The formal charge of the heme iron should be (as follows from comparison with other hemoproteins) much less than +2. This "superreduced" P-450 is often an unstable species; this fact clearly reflects unusual properties of the heme iron moiety.

The complex of reduced P-450 with CO can be treated as an example of complexes with π-acceptor ligands: O_2 as native ligand, CO, NO have free or partially occupied low-lying π^* orbitals able to accept π electrons. Acceptance of π electrons by these π^* orbitals of ligand is seen by the upshift of the ν_4 band which is shifted in spectra of both reduced P-450.CO and its model complex from 1341 to 1369 cm^{-1} (Table 1). In other words, the π-electron density is no more given to π^* orbitals of porphyrin by back donation (which is visualized by the shift of the ν_4 band to position corresponding rather to ferric hemes). As a consequence, the Fe-C-O moiety becomes labilized (electron density flows to antibonding orbitals). This effect is far more important for the native, Fe-O-O structure, where the next (second) electron is supplied; the Fe-O-O structure becomes then untenable and is splitted upon production of water and active oxygen.

For the reduced P-450.CO complex, also the position of the Fe-CO vibration (shifted ~30 cm^{-1} to lower frequencies, see Table 1) indicates significant weakening of the Fe-CO bond due to lowering of the affinity of Fe to bind CO. This is again caused by effect of mercaptide sulfur which acts as a strong σ- and π-donor to the iron thus lowering its demand for σ electrons from CO or other ligand. The RR spectroscopy is thus able to monitor the overall labilization of the Fe-O-O structure which is of paramount importance for understanding of the catalytic mechanism of P-450.

REFERENCES

Anzenbacher P., Fidler V., Schenkman J.B., Kirkup R.E., Spiro T.G. (1984a) Resonance Raman study of cytochrome P-450. Proc. 9th Int. Conf. Raman Spectroscopy, Tokyo, Japan,1984, 474-475

Anzenbacher P., Šípal Z., Hodek P. (1984b) Purification of rat liver microsomal P-450 by chromatography on immobilized adamantane with a new method of detergent removal. Biomed.Biochim.Acta 43, 1343-1349

Champion P.M., Gunsalus I.C., Wagner G.C. (1978) Resonance Raman investigations of cytochrome P-450$_{CAM}$ from Pseudomonas putida. J.Am.Chem.Soc. 100, 3743-3751

Champion P.M., Stallard B.R., Wagner G.C., Gunsalus I.C. (1982) Resonance Raman detection of a Fe-S bond in cytochrome P-450$_{CAM}$. J.Am.Chem.Soc. 104, 5469-5472

Cheng K.-C., Schenkman J.B. (1982) Purification and characterization of two constitutive forms of rat liver microsomal cytochrome P-450. J.Biol.Chem. 257, 2378-2385

Jänig G.-R., Dettmer R., Usanov S.A., Ruckpaul K. (1983) Identification of the ligand trans to thiolate in cytochrome P-450 LM 2 by chemical modification. FEBS Letters 159, 58-62

Ozaki Y., Kitagawa T., Kyogoku Y., Imai Y., Hashimoto-Yutsudo C., Sato R. (1978) Resonance Raman studies of hepatic microsomal cytochromes P-450: Evidence for strong basicity of the fifth ligand in the reduced and carbonyl complex forms. Biochemistry 17, 5826-5831

Spiro T.G. (1983) The resonance Raman spectroscopy of metalloporphyrins and heme proteins. In: Physical Bioinorganic Chemistry Series, No 2, Iron Porphyrins, Part II, Chapter 3. Lever A.B.P., Gray H.B., Eds. 1st Edn, Addison-Wesley, Reading, MA, U.S.A., pp. 89-159

Yamamoto T., Palmer G., Gill D., Salmeen I, Rimai L. (1973) The valence and spin state of iron in oxyhemoglobin as inferred from resonance Raman spectroscopy. J.Biol.Chem. 248, 5211-5213

ENERGY TRANSPORT AT THE CYTOCHROME P-450

J.Vincze and Z.Vincze

Biophysical Department, 3400 Cluj, Rumania

There was a time not so long ago when the very concept of active transport was questioned by many biologists and biophysicians. Active transport of ions is considered an integral part of the function of all cells and even of many cell organels. Nevertheless from the biophysical point of view the transport through the membrane presents many tempting problems.

The architecture of membranes in relation to membrane function

Electron microscopy made possible to reveal that membranes represent the most common type of supramolecular structures in cells. Therefore this particular type of condensation of protein and lipid molecules appears to represent a basic principle of molecular arrangement that must be of decisive importance to explain basic life processes. This discovery has focused the attention to the metabolic function of membranes, while earlier, when only the plasma membrane was considered, the barrier function involving permeability control was the only aspect which was discussed and explained mainly by the passive physical properties of the membrane.

Electron microscopists proposed that membranes like mitochondrial membranes are the site of enzyme molecules and that the condensed state of enzyme in the membranes would favour organized interaction between enzyme molecules in multienzyme systems like the respiratory chain. This proposal has been found to be correct by demonstrating that for instance, the respiratory chain components are parts of the inner mitochondrial membranes. According to the concepts that have been emerging on this basis, the cell can be divided into a number of sites, each one associated with a particular function. The maintaining of this subdivision is made possible by a compartmentalization of the cell body through the enclosure of regions of cytoplasm by means of membranes.

The membranes themselves are conceived as the sites for enzymes and particularly for multienzyme systems. Depending on the type of site, the enzymic composition will vary to satisfy the requirements for the particular function of each site. The compartment enclosed by the membrane is considered to favour the maintenance of a high substrate concentration within the compartment, thus allowing high local substrate and enzyme concentration. This would account for the high efficiency of the enzymes. The ordered arrangement of enzyme molecules in the membranes would contribute to a coordination of the enzymic activity in multienzyme systems.

These two features are characteristic of the transport processes.

The transport processes

Phenomenological equations which characterize the transport phenomenon, establish the relationship between generalized forces and given fluxes. These relations, from a theoretical point of view are very important, they permit the prediction and the development of irreversible processes in thermodynamical systems, the estimation of different ways of evolution, the calculation of the final distribution of the parameters. By means of phenomenological equations, a series of problems were clarified in living systems. Parameters that occur in these equations: mass, energy, impuls, in vitro and in vivo, in experimental research, are difficult to be determined.

The black box approach is particularly useful in the initial stages of a study of the transport potentialities of an membrane. If the flux ratio for a given species differs substantially from the one predicted on the basis of potential difference and concentration gradient, one may suspect coupling between fluxes and possibly the existence of active transport. The short-circuiting approach is really only a special case of the flux ratio approach where the concentration and potential difference for all species are maintained at zero during the run. The membrane may consist of bilayers in series and the concentration and potential may vary along the path in any conceivable way. In its simpler versions it assumes, however, that the interaction with flow of solvent can be neglected.

The energy transport

As transport phenomenon, we consider the separate or simultaneous variation in space and time of some physical parameters that furnish us flux of other parameters for which the conservation laws are valid (Vincze, 1967). This definition gives us, the possibility for an ample and general treatment of transport phenomena (Vincze, 1971).

By this definition we are able to calculate the quantity of transport through a surface dS in a given interval of time dt, under the action of the generalized force grad a using the following relation:

$$W = K \int_{t_o}^{t_i} \iiint s_{/x,y,z/} \; grad \; a \; dSdt$$

where K - represents a constant.

For energy transport cytochrome P-450 is a very good example:

Reconstitution of cytochrome P-450

Cytochrome P-450 can be reconstituted into a phospholipid bilayer in the absence of added detergent by incubation of purified hemoprotein with preformed phosphatidylcholine vesicles. Salt effects demonstrated that the primary interaction between cytochrome P-450 and phospholipid vesicles is hydrophobic rather than ionic, in contrast, neither adrenodoxin reductase nor adrenodoxim will bind to phosphatidylcholine vesicles by hydrophobic interactions.

We calbulate the variation of probabilistic density of the vesicles as the function of time. The apparition of vesicles is a aleatorous event. We obtain the following equations system describing the existence probability of vesicles as the function of time:

$$\frac{dP_n t}{dt} = C_e \left[P_{n-1}/t/ - P_n/t/ \right] + C_i \left[P_{n+1}/t/ - P_n/t/ \right]$$

$$\frac{dP_1/t/}{dt} = C_e \left[P_o/t/ - P_1/t/ \right] + C_i \left[P_2/t/ - P_1/t/ \right]$$

$$\frac{dP_M/t/}{dt} = C_e \, P_{M-1}/t/ - C_i P_M/t/$$

Insertion of cytochrome P-450 into a phospholipid bilayer resulted in conversion of the optical spectrum to a low spin type, but this transition was markedly dimished if cholesterol was incorporated within the bilayer. Vesicle reconstituted cytochrome P-450 metabolized cholesterol within the bilayer, virtually 95 % of the cholesterol within the vesicle was accessible to the enzyme. The activity of the reconstituted hemoprotein was sensitive to the fatty acid component of the phospholipid. Thus, the cholesterol-binding site on vesicles-reconstitued cytochrome P-450 is in communication with the hydrophobic bilayer of the membrane.

REFERENCES

Vincze,J.: Characterization of transport processes in living systems. First European Biophysics Congress, 1971 Vienna, Vol.III, 217-222.

Vincze,J., Bicleseanu,A., Cuparencu,B., Böhm,B., Carpen,E. 1979 Effects of some autonomic Drugs on Serum Lipids Levels in Normo-lipidaemic and Triton WR-1339 treated Rats. Agressol. 20, 1, 19-50.

THERMODYNAMIC STUDIES ON DIFFERENT CYTOCHROME P-450 PROTEINS

C. Jung, P. Bendzko, O. Ristau and I.C. Gunsalus[*]

Central Institute of Molecular Biology, Academy of Sciences
of the GDR, 1115 Berlin-Buch, GDR; *Dept. of Biochemistry,
University of Illinois, Urbana, Il 61801, USA

For two bacterial cytochrome P-450 proteins, the camphor hydroxylating cytochrome P-450 from Pseudomonas putida (P-450 CAM) and the linalool hydroxylating cytochrome P-450 from Pseudomonas incognita (P-450 LIN), we have observed differences in the active site structure by optical, infrared, proton nuclear magnetic resonance and enzyme inhibition studies /Jung et al., 1985/. A relatively closed heme pocket, with rigid active site protein structure, is suggested for P-450 LIN. A more open heme pocket, with flexible active site protein structure, is concluded for P-450 CAM.

This paper represents data of thermodynamic studies which follow the idea that the different active site structures reflect different overall protein structure stabilities of the P-450 proteins. By investigating the influence of the substrate binding on the protein unfolding, conclusions about the enzyme-substrate interaction are drawn.

P-450 CAM, P-450 LIN and cytochrome LM2 isoenzyme from rabbit liver (P-450 LM2) are studied in presence and absence of their specific substrates (camphor, linalool and benzphetamine, respectively) by differential scanning calorimetry, using the Privalov-type calorimeters DASM 1M and DASM 4, and by optical temperature difference spectroscopy, using the spectrophotometer SPECORD M40 from Carl Zeiss Jena, GDR.

Tab. 1 summarizes all thermodynamic data. Following main results are obtained by the calorimetric studies: 1. The protein unfolding of all P-450 forms studied is characterized by complex thermal transitions reflecting in the thermodynamic cri-

3*

Tab.1 Comparison of thermodynamic data obtained from calorimetric and optical temperature difference spectra

sample[1]	$\varkappa$[2]	n[3]	ΔH^{cal}(kJ/mole)[4]	$T_m(^oC)$[4]	ΔH(kJ/mole)[5]	ΔS(J/deg·mole)[5]	$T_{1/2}(^oC)$[5]	$\Delta\varepsilon_{mM}$[5]
m^o_{CAM}	2.7	3	1011.0 ± 50	53.3 ± 0.3	177.7 ± 3.1	568.1 ± 10.5	39.6	82 ± 2
m^{os}_{CAM}	1.8	2	1190.5 ± 55	63.7 ± 0.5	225.3 ± 5.8	687.2 ± 18.4	54.7	57 ± 3
m^o_{LIN}	1.9	2	946.6 ± 20	53.5 ± 0.3	351.5 ± 8.4	1096.8 ± 29.3	47.3	75 ± 4
m^{os}_{LIN}	1.8	2	1065.4 ± 50	61.7 ± 0.5	380.8 ± 13.0	1184.2 ± 40.5	48.4	31 ± 3
m^o_{LM2}	2.5	3	1104.8 ± 55	55.3 ± 0.3	225.1 ± 7.9	697.2 ± 24.7	49.7	80
m^{os}_{LM2}	-	-	-	-	262.6 ± 2.1	808.8 ± 6.3	51.5	22

1) m^o and m^{os} indicate the substrate free and substrate bound proteins, respectively; 50 mM potassium phosphate buffer, pH=7; For the optical studies of P-450 LM2 20%(v/v) of glycerol had to be added to the buffer to avoid early precipitation of the protein which disturbs spectral analysis. protein concentrations: 1-2.5mg/ml for calorimetric studies, 1.4-3.5 M for optical studies;substrate concentrations: camphor 400μM, linalool 200μM, benzphetamine 1mM; 2) $\varkappa = \Delta H^{cal}/\Delta H$v.H. "thermodynamic criterion"; 3) n- number of two-state transition components determined by deconvolution of the experimental calorimetric curves according to the procedure of Potekhin and Privalov/1982/; 4) ΔH^{cal}- calorimetric transition enthalpy; ΔHv.H.- van't Hoff enthalpy; T_m- calorimetric half-transition temperature; 5) ΔH, ΔS and $\Delta\varepsilon_{mM}(mM^{-1}cm^{-1})$ were obtained by least-square fitting of the original transition curves(fig.1) derived from optical temperature difference spectra.The saturation extinction $\Delta\varepsilon_{mM}$ refers to 418nm and 387nm for the substrate free and substrate bound proteins,respectively. For P-450 LM2 the mean of the respective values of the bacterial proteins was used because precipitation of the LM2 protein above 53°C for m^o_{LM2} and above 50°C for m^{os}_{LM2} hindered the spectral analysis. For m^{os}_{LM2} $\Delta\varepsilon_{mM}$ corresponds to an amount of high-spin state of 50% at 38°C. $T_{1/2}$ - optical half-transition temperature

terion $\mathfrak{X} > 1$. Deconvolution of the experimental curves reveals 3 two-state transitions for the substrate-free P-450 CAM and P-450 LM2. This means that at least 3 cooperative units exhibit different thermostability. In contrast, the unfolding of sub-strate-free P-450 LIN takes place with high cooperativity. At least 2 cooperative units with similar thermostability are sug-gested. 2. The binding of the substrate results in an increa-se of the thermostability of the proteins. The half-transition temperature T_m is shifted to higher values by 8-10 degree. For P-450 CAM the cooperativity of the protein unfolding is increa-sed by camphor binding. In contrast, the binding of linalool to P-450 LIN does not considerably change the cooperativity of the protein unfolding. For the benzphetamine-bound P-450 LM2 the measurements were not successful until now because of protein precipitation.

The optical studies revealed following main result (Fig. 1): P-450 CAM in absence of camphor exhibits a low thermostability of the active site ($T_{1/2} \approx 40\ ^oC$) which is drastically elevated by camphor binding ($T_{1/2} \approx 55\ ^oC$). For P-450 LIN and P-450 LM2, however, the thermostability of the active site is only unessen-tially influenced by substrate binding reflecting in similar half-transition temperatures.

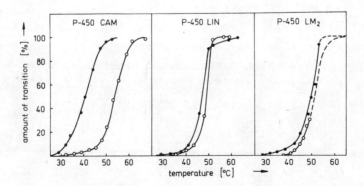

Fig. 1 Amount of protein unfolding monitored by optical temperature difference spectra, (•) substrate-free proteins
observed on the decrease of the low-spin Soret band at
418 nm, (o) substrate-bound proteins observed on the decrease of the high-spin Soret band at 387 nm.

22

Fig. 2 Temperature depen-
dence of the substra-
te dissociation con-
stants. The value at
12 °C (2.1 µM for cam-
phor; 18.4 µM for li-
nalool; 540 µM for
benzphetamine) is
set equal 100 %.
Camphor values were
replotted from Fig.4
of the paper of
Griffin and Peterson
/1972/. Benzphet-
amine values were
taken from Ristau
et al. /1979/.

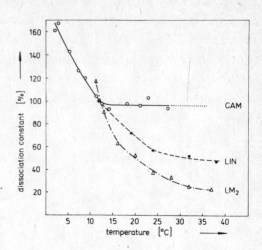

CONCLUSIONS FOR P-450-SUBSTRATE INTERACTION

1. The P-450-substrate interaction for P-450 CAM has to be
qualitatively different from P-450 LIN and P-450 LM2. This is
also indicated by the different temperature dependence of the
substrate dissociation constant for the three P-450 proteins
(Fig. 2). While for P-450 LIN and P-450 LM2 the decrease of the
dissociation constant with increasing temperature indicates
predominant hydrophobic interaction, remarkable contributions
of chemical binding for the camphor-P-450 CAM interaction is
suggested from the independence of the camphor dissociation
constant on the temperature above 13 °C.

2. Camphor binds, with high probability, in the contact region
of at least two domains and induces thereby significant changes
in the active site protein structure of P-450 CAM. This is suggested from the increase of the cooperativity of the temperature-induced transition by camphor binding.

REFERENCES

Griffin, B.W. and J.A. Peterson (1972) Biochemistry 11, 4740-
4746

Jung, C., E. Shyamsunder, S.F. Bowne, A.H.J. Ullah, I.C. Gunsalus and G.C. Wagner (1985) J. Biol. Chem. in press

Potekhin, S.A. and P.L. Privalov (1982) J. Mol. Biol. 159, 519-
535

Ristau, O., H. Rein, S. Greschner, G.-R. Jänig and K. Ruckpaul
(1979) Acta biol. med. germ. 38, 177-185

Mechanism of Oxygen and Substrate Activation
by Cytochrome P-450

INSIGHTS INTO STRUCTURE–FUNCTIONAL INTERRELATIONS IN ADRENOCORTICAL
CHOLESTEROL SIDE CHAIN CLEAVAGE SYSTEM

Usanov, S.A., Pikuleva, I.A., Turko, I.V., Chashchin, V.L., Akhrem, A.

Institute of Bioorganic Chemistry, BSSR Academy of Sciences, Minsk,
U.S.S.R.

Molecular oxygen activation is an important step in cholesterol side
chain cleavage reaction which performs the stereo- and regio-selective
insertion of the reactive oxygen intermediate by three sequential reactions
to form pregnenolone and isocapraldehyde. Oxygen activation is an electron-
dependent process and therefore, cholesterol side chain cleavage system
may be considered as a highly ordered electron transfer chain, containing
flavoprotein (adrenodoxin reductase), iron sulfur protein (adrenodoxin)
and hemeprotein (cytochrome P–450scc).

The electron transfer in the cholesterol side chain cleavage system may
be conventionally considered as a set of two processes: (i) an intercomponent electron transfer from NADPH to cytochrome P–450scc and (ii) transfer
of electrons from the heme group of cytochrome P–450scc to molecular oxygen. Studies of the selectivity and specificity of the intercomponent
electron transfer presume investigation of molecular organization of the
whole system, while to understand the second process it is necessary to
study the oxidation–reduction properties of cytochrome P–450scc which are
determined by the nature of the axial heme iron ligands. Each study requires an individual approach.

The present work represents a structure–functional study of the adrenocortical cholesterol side chain cleavage system and is devoted to investigation of the nature of axial heme iron ligands, the role of protein-
protein interaction and dynamics of steroidogenic electron transfer by
means of chemical modification with mono- and bifunctional reagents.

Identification of putative cytochrome P–450scc axial ligands in poly-peptide chain of hemeprotein

In contrast to other hemeproteins which participate in reversible oxygen binding (myoglobin, hemoglobin), electron transfer (cytochromes b-
type), dioxygen splitting (catalase, peroxidase), reduction of molecular
oxygen (cytochrome c oxidase), cytochrome P–450 catalyzes all types of
reactions and also radical attack of substrate by oxo–iron species. These
unique properties of cytochrome P–450 are determined by the axial heme
ligands which are presented with polypeptide chain amino acid radicals.
Thus, it is protein structure, both proximal and distal to the heme iron,
which dominates the unusual physico-chemical behavior.

The progress in understanding of the nature of cytochrome P–450 axial
ligands has been achieved through comparison of the physical properties of
porphyrine complexes having various axial ligands and hemeproteins with
known proximal histidine ligation and variable distal ligation. At present,

it is thought that thiolate is a proximal ligand in cytochrome P-450 hemeproteins /White, Coon, 1980/, but the sixth ligand of native cytochrome
P-450 appears to be oxygen-containing amino acid or water /Dawson et al.
1982, White, Coon, 1982/. If unique physico-chemical properties of cytochrome P-450 are explained by thiolate as a fifth ligand, the nature of the
sixth ligand of cytochrome P-450 is of key importance in controlling the
redox properties, substrate and oxygen binding and spin-spin changes. The
heme iron distance which is consistant with thiolate ligation presumably
dominates in all oxidation-reduction and spin states, however oxygen
cannot bind to ferrous cytochrome P-450 whithout dissociation of sixth
axial ligand.

Previously, we have shown that cytochrome P-450scc consists of two
domains, F(1)P-450 and F(2)P-450, respectively. The heme group appears to
be bound with F(1)P-450, representing the N-terminal sequence of cytochrome P-450scc polypeptide chain, that contains 251 amino acid residues of
481 amino acids total. Domain F(1)P-450 contains adrenodoxin-binding site.
Domain F(2)P-450 represents C-terminal sequence of cytochrome P-450scc
polypeptide chain and appears to be responsible for interaction with
phospholipid membrane.

Studies which have been done by using a number of sulfhydryl reagents
and amino acid analysis indicate that cytochrome P-450scc contains 3 or 4
sulfhydryl groups per hemeprotein molecule. Recently, we /see Akhrem et al
in this volume/ and others /Morohashi et al. 1984/ completed the cytochrome P-450scc primary structure investigations. These studies indicate that
in contrast to other cytochromes P-450, cytochrome P-450scc contains only
two cysteine residues, which can be considered as potential axial heme
iron ligands. Both cysteine residues are in domain F(2)P-450, one being
at N- (264) and an other at C-terminal sequence of this domain polypeptide
chain. The discrepancy between number of cysteine residues determined by
chemical modification methods and primary structure sequence is not clear
at present.

A set of experiments on selective chemical modification of cytochrome
P-450scc with different sulfhydryl reagents indicate that these two cysteine residues form two different pools. One cysteine is on the hemeprotein surface, is accessible for sulfhydryl reagents used. Its modification
does not lead to cytochrome P-450scc inactivation, i.e. a loss of intrinsic spectral properties inherent to native hemeprotein. This, according
to present concept, indicates to intact sulfur-heme iron bond. The other
cysteine residue is not accessible to sulfhydryl reagents. Modification of
this residue may be achieved only after preliminary cytochrome P-450scc
inactivation with subsequent heme removal. It is important, that both cysteine residues under definite conditions may react with thio-propyl-Sepharose 6B, i.e. are involved in thiol-disulfide exchange.

The question may be asked: what cysteine residue is involved in the
proximal ligand formation, and what is on the surface of the molecule? To
answer these question, the approach, the overall scheme of which is presented in Fig. 1, was used. Preliminarily, under conditions excluding cytochrome P-450scc inactivation, the exposed cysteine residue was unrever-
sibly modified with N-ethyl maleimide. Modification of this residue does
not change spectral properties of hemeprotein, while the catalytic activity in reconstituted system is even increased. Further treatment of cytochrome P-450scc with SDS and removal of heme results in regeneration of
the second sulfhydryl group. When applied to thio-propyl-Sepharose 6B,
cytochrome P-450scc interacts with the matrix by means of thiol-disulfide
exchange between sole sulfhydryl group of hemeprotein and activated groups
matrix.

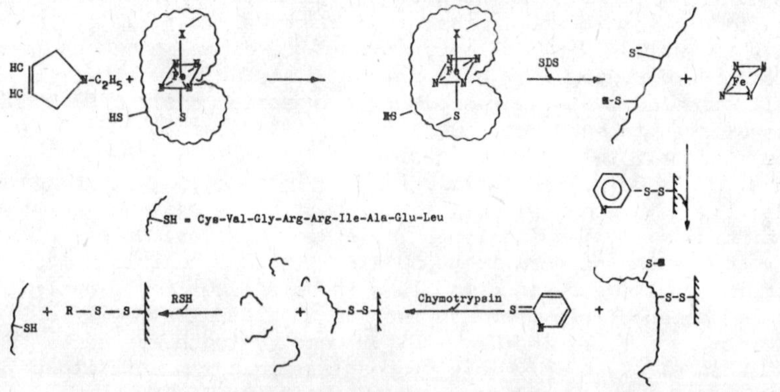

Fig. 1. The overal scheme of experiment for identification of cysteine
residue involved in heme ligation

Subsequent proteolytic modification of immobilized protein with chymotrypsin results in removal of peptide material which did not connect with
carrier. The cysteine-containing peptide involved in thiol-disulfide
exchange was eluted with 2-mercaptoethanol. Carboxymethylation, desalting
and futher sequencing allowed to identify the peptide bound with matrix
as: C V G R R I A E L.

Thus, in contrast to theoretical conclusions about the localization of
the cysteine residue serving as axial heme ligand, we were able to experimentally determine this heme-binding site, and show that if cysteine residue is indeed the fifth axial ligand in cytochrome P-450scc, this residue is cysteine at 422 position in polypeptide chain of hemeprotein.
There has been some controversy concerning the position of cysteine residue in polypeptide chain of cytochromes P-450. Owing to the similarity of
physico-chemical properties of different cytochromes P-450, the primary
structure and conformation of active site were proposed to be similar for
various forms of hemeprotein. To check this assumption, we compared the
amino acid sequences of the known cytochrome P-450 structures near the
region of cysteine residue involved in heme-binding /Fig. 2/.

P-450scc	F	G	W	G	V	R	Q	C	V	G	R	R	I	A	E	L	E	M	T	L	F		C-422
P-450b,e	F	S	T	G	K	R	I	C	L	G	E	G	I	A	R	N	E	L	F	L	F		C-436
P-450LM2	F	S	L	G	K	R	I	C	L	G	E	G	I	A	R	T	E	L	F	L	F		C-436
P-450c	F	G	L	G	K	R	K	C	I	G	E	T	I	G	R	L	E	V	F	L	F		C-461
P-450d	F	G	L	G	K	R	R	C	I	G	E	I	P	A	K	W	E	V	F	L	F		C-456
P-450LM4	F	G	L	G	K	R	R	C	I	G	E	T	L	A	R	W	E	V	F	L	F		C-456
P-4501	F	G	L	G	K	R	K	C	I	G	E	T	I	G	R	S	E	V	F	L	F		C-458
P-4503	F	G	L	G	K	R	R	C	I	G	E	I	P	A	K	W	E	V	F	L	F		C-458
P-450cam	F	G	H	G	S	H	L	C	L	G	Q	S	L	A	R	R	E	I	I	V	F		C-355

Fig. 2. Comparison of amino acid sequences of various cytochromes P-450
near heme-binding cysteine residue in C-terminal region

The cytochromes P-450 from ferredoxin-dependent monooxygenases contain
heme-binding cysteine least of all removed from N-terminal sequence. There

is some homology between the structures presented, but it is not absolute.
This may indicate that either the position of heme—binding cysteine is
different in various forms of cytochrome P—450 or the hypothesis about the
highly conserved region near heme—binding cysteine is unlikely.

Localization of cysteine residue in polypeptide chain of cytochrome
P—450scc allowed to make some conclusions: (i) cytochrome P—450 contains
cysteine residue involved in heme—binding in domain F(2)P—450, which is
according to our data, responsible for the interaction with phospholipid
membrane; (ii) cysteine residue at 264 position is on the N—terminal se—
quence of domain F(2)P—450 polypeptide chain, exposed, accessible for sulf—
hydryl reagents and appears to be outside the phospholipid membrane.

Since the cysteine residue involved in heme ligation is according to
our data in phospholipid membrane, we can conclude that heme is in phos—
pholipid membrane also; the iron—sulfur bond distance was shown to be
2.2 A /Dawson et al. 1980/. This proposal is in agreement with physico—
chemical data and hydrophobic nature of cholesterol. Since cholesterol is
a membrane component, heme, which is thought to be in the vicinity of the
cholesterol—binding site, should be embedded into interior of mitochondria
membrane.

In contrast to fifth, the nature of axial ligand trans to thiolate is
less defined. The identity of this ligand remains controversial, but can—
didates currently receiving support are water or oxygen—containing amino
acid radical. Based on selective chemical modification and physico—chemi—
cal studies we proposed that tyrosine residue may be involved in active
site formation in cytochrome P—450 LM2 /Jänig et al. 1983/ and cytochrome
P—450scc /Usanov et al. 1984/. For more complete understanding of the
functional role of this highly reactive tyrosine residue, it was necessary
to identify this residue in cytochrome P—450scc polypeptide chain. To
simplify this task, first of all it was necessary to answer the question:
in what functional domain of cytochrome P—450scc the putative tyrosine re—
sidue is located? Modification of cytochrome P—450scc essential tyrosine
residue with tetranitromethane does not effect the proteolytic modification
of cytochrome P—450scc with trypsin. Therefore, covalent chromatography on
thio—propyl—Sepharose 6B was used to separate F(1)P—450 and F(2)P—450 do—
mains of modified hemeprotein, which contained 1.6 nitrotyrosine residue
per molecule. The overall scheme of this experiment is presented in Fig. 3.

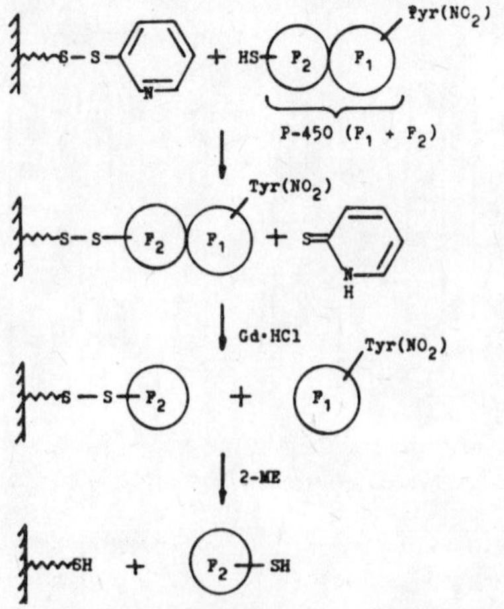

Fig. 3. Localization of nitrotyrosines in cytochrome P—450scc domains

Cytochrome P–450scc modified with tetranitromethane and proteolytically digested with trypsin interacts with thio–propyl–Sepharose 6B, indicating that cysteine residue at 264 position is not oxidized with tetranitromethane. Highly purified F(1)P–450 domain was eluted by washing the column with 6 M guanidine hydrochloride. Subsequent washing with 2–mercaptoethanol eluted F(2)P–450 together with a portion of non–digested hemeprotein. The latter was removed by preparative SDS–electrophoresis affording pure F(2)P–450 domain. The results of amino acid analysis of separated domains of cytochrome P–450scc indicate that the modification of essential tyrosine residue(s) touches upon only F(1)P–450 domain (0.55 nitrotyrosines per polypeptide chain of this domain). On the other hand, we failed to detect nitrotyrosines in F(2)P–450 domain. Moreover, all tyrosines present in this domain were recovered by the amino acid analysis. This finding substantially facilitated the further localization of nitrotyrosines in polypeptide chain of F(1)P–450 domain.

The preparation of highly purified domain F(1)P–450 from nitrated cytochrome P–450scc, proteolytic digestion with St. aureus protease, chromatography of peptides on Bio–gel P–6 and further HPLC of nitrotyrosine containing material allowed to indicate that nitrotyrosine containing peptide is following: Y D I P P W L A Y H R Y Y.

This results indicate the localization of essential tyrosine residue(s) in peptide containing four tyrosines, but did not permit to discriminate putative heme–binding residue. Although it is not clear which of this four tyrosines chelates to the heme iron, we have narrowed the choice to one of the neibouring tyrosines. No further attempt has been made to localize this residue. It follows from theoretical analyses of cytochrome P–450scc primary structure that this region of hemeprotein polypeptide chain has a poor potential to form a helix structure, but the latter could not be excluded. Therefore, this region of polypeptide chain should be located on the peripheral side of phospholipid membrane close to the matrix.

Structural approach to the localization of putative axial heme–binding ligands of cytochrome P–450scc proved to be effective in identification of amino acid residues which can potentially play a role as axial non–heme ligands and showed that this amino acid residues are distributed among structurally performed domains of cytochrome P–450scc F(1)P–450 and F(2)P–450, respectively, being differently depatured from the surface of phospholipid membrane. These data allowed to conclude that the heme group of cytochrome P–450scc is buried in phospholipid membrane and appears to be located between domains F(1)P–450 and F(2)P–450, respectively. Thus, heme group is interface between hydrophobic and hydrophylic parts of the hemeprotein. This proposal is further confirmed by our results on separation of cytochrome P–450scc domains: any attempts to separate domains of proteolytically modified cytochrome P–450scc result in hemeprotein inactivation.

Role of protein–protein complexes in electron transfer in adrenocortical mitochondria

To activate molecular oxygen, cytochrome P–450 should obtain two electrons via an electron transfer chain, by two sequential 1–electron steps. Steroidogenic electron transfer in adrenocortical mitochondria may be realized in ternary (or morehigher order) complex or through shuttle mechanism, where adrenodoxin mediates electrons from bimolecular complex with adrenodoxin reductase to a bimolecular complex with cytochrome P–450. There is experimental evidence in favor of former and latter mechanisms.

30

To study the mechanism of electron transfer between redox proteins, chemical modification with bifunctional reagents was used to stabilized protein
complexes which is active in electron transfer. Before studing proteins
interaction, it was necessary to investigate the effect of reagents used
on individual proteins.

Under conditions used, neither adrenodoxin reductase nor adrenodoxin
form significant amount of oligomers in the presence of bifunctional
reagents used. This is in accordance with the previous data that both protein are in monomeric forms when purified from mitochondrial membrane. At
the same time, cytochrome P-450scc in the presence of bifunctional imidate
gives rise to oligomeric forms up to octamer. This fact indicates a considerable cross-linking of hemeprotein, and the band pattern implies that
cytochrome P-450scc has an octameric structure, the latter being confirmed
with gel-filtration chromatography, indicating that hemeprotein molecular
weight in solution is 400 000 dalton.

The cross-linking of cytochrome P-450scc with cleavable bifunctional
imidate - dimethyl-3,3-dithiobispropionimidate, proteolytic degradation of
cross-linked hemeprotein with trypsin and subsequent analysis by two-
dimensional SDS-electrophoresis after reduction of reagent disulfide bond,
allowed us to conclude that cytochrome P-450scc exists in solution in octameric heterologous form, wherein neighbouring cytochrome P-450scc molecules
are cross-linked through F(1)P-450 and F(2)P-450 domains, respectively. A
scheme explaining the results of cross-linking studies is presented in
Fig. 4.

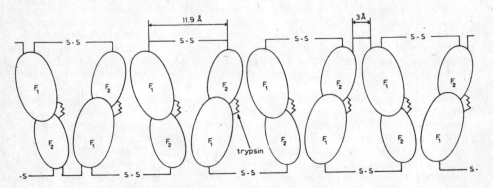

Fig. 4. Hypothetical scheme for cross-linking of cytochrome P-450scc
with dimethyl-3,3-dithiobispropionimidate

This scheme indicates the presence of intermolecular cleavable (11.9 A)
and non-cleavable (3.0 A) cross-links between F(1)P-450 and F(2)P-450 domains, respectively. Cross-linked cytochrome P-450scc retains inherent
physico-chemical properties. This means that cross-linking studies reflect
intrinsic molecular organization of hemeprotein which is only fixed with
cross-linking reagents.

Because of its central role in steroidogenic electron transfer system,
and its relatively simple structure, adrenodoxin has been the object of
the most detailed structural and functional studies. With the aim to
further clarify the problem of steroidogenic electron transfer mechanism,
we sought to stabilize complexes between adrenodoxin and other proteins by
chemical cross-linking.

Among a number of bifunctional reagents used, heterobifunctional reagent

N-succinimidyl 3-(2-pyridyl-dithio) propionate was reagent of choice to cross-link adrenodoxin reductase with adrenodoxin. Cross-linking results in appearence of a new band in SDS-electrophoresis with molecular weight 66000±2000 dalton. Since molecular weight of adrenodoxin has been estimated to be 12000±1500, but adrenodoxin reductase 52000±2000 dalton, this new species has been ascribed to stoichiometric complex of both proteins. Further evidence came from two-dimensional SDS-electrophoresis data: when disulfide bond of reagent was reduced, the protein species with molecular weight 66000±2000 dalton completely dissociated to give adrenodoxin and adrenodoxin reductase. The yield of cross-linked complex is about 42%.

Before characterization of the cross-linked complex of adrenodoxin with adrenodoxin reductase, it was necessary to remove monomeric proteins. By using gel-filtration on Sephadex G-100, ion exchange chromatography and affinity chromatography on adrenodoxin-Sepharose, a cross-linked complex has been obtained in a highly purified state as jurdge by SDS-electrophoresis. The cross-linked complex of adrenodoxin reductase with adrenodoxin is correctly orientated, since it is competent in many of the electron transfer processes which have been tested, such as cytochrome P-450scc and c reduction by NADPH. Thus, the cross-linked proteins retain the ability for intra- and intercomplex electron transfer. This allows to make some conclusions with respect to the mechanism of electron transfer: (i) the nucleotide-binding site on adrenodoxin reductase is not shielded with adrenodoxin and these sites are distinct; (ii) the complex mediates electrons from NADPH to donors without dissociation; (iii) cytochrome P-450scc can bind to adrenodoxin involved in the complex with adrenodoxin reductase.

Cleavable bifunctional imidate, dimethyl-3,3-dithiobispropionimidate, the amino group specific reagent was chosen to study the cross-linking of cytochrome P-450scc with adrenodoxin. The cross-linking leads to the appearance of a new band in SDS-electrophoresis with an apparent molecular weight 62000±3000 dalton, which is close to that expected for cross-linked product of both proteins, having molecular weight 49000±1500 and 12000±500 dalton, respectively. A direct evidence that this band correspond to a stoichiometric complex of cytochrome P-450scc with adrenodoxin, was obtained using two-dimensional electrophoresis: a cross-linked complex was dissociated with 2-mercaptoethanol to form adrenodoxin and cytochrome P-450scc, respectively.

To characterize the cross-linked complex of adrenodoxin with cytochrome P-450scc, it was necessary to separate it from monomeric proteins and cross-linked oligomers. Since cross-linking might block an adrenodoxin-binding site on cytochrome P-450scc and thus, interfere with interaction of second adrenodoxin molecule, adrenodoxin-Sepharose was chosen as affinity matrix. Adrenodoxin and small portion of cytochrome P-450scc were ran through the column, while the cross-lined complex and free hemeprotein were bound. A fraction eluted from affinity column did not contain free adrenodoxin, but were contaminated with free cytochrome P-450scc.

When stoichiometric amounts of adrenodoxin reductase and NADPH were added to cross-linked complex, cytochrome P-450scc was reduced at a high rate. In the presence of adrenodoxin reductase, NADPH and cholesterol, the cross-linked complex is active in cholesterol conversion to pregnenolone at rate similar that of in reconstituted system. Thus, the functional activity of the cytochrome P-450scc-adrenodoxin complex reflects their intrinsic molecular organization, which have been only "fixed" with bifunctional imidate, and indicate that both proteins are cross-linked at the surfaces involved in the electron transfer reactions.

The ability of cross-linked complex of cytochrome P-450scc with adreno-

doxin mediate electron transfer from NADPH to hemeprotein in the presence
of adrenodoxin reductase, allows us to summarize: (i) since the cross-
linked complex is rather active in cholesterol conversion to pregnenolone,
this means that cholesterol-binding site on cytochrome P-450scc is not
shielded by iron sulfur protein; (ii) complex of cytochrome P-450scc with
adrenodoxin may function without dissociation in electron transfer and
catalysis; (iii) functional activity of cytochrome P-450scc cross-linked
complex with adrenodoxin indicate that there are two distinct binding
sites on adrenodoxin responsible for interaction with cytochrome P-450scc
and adrenodoxin reductase; (iv) at least under experimental conditions
described the organized complex model rather than binary complex model,
is favored.

REFERENCES

1. Dawson, J.H., Anderson, L.A., Davis, I.M., Hahn, J.E. (1980) Spectroscopic investigation of the active site structure of cytochrome P-450
   with X-ray absorption spectroscopy. Dev. Biochem., 13, 565-572
2. Dawson, J.H., Anderson, L.A., Sono, M. (1982) Spectroscopic investigation of ferric cytochrome P-450cam ligand-complexes. Identification
   of the ligand trans to cysteinate in the native enzyme. J. Biol. Chem.,
   257, 3606-3617
3. Jänig, G-R., Dettmer, R., Usanov, S.A., Ruckpaul, K. (1983) Identification of the ligand trans to thiolate in cytochrome P-450LM2 by
   chemical modification. FEBS Lett., 1, 58-62
4. Morohashi, K., Fujii-Kuriyama, Y., Okada, Y., Sogawa, K., Hirose, T.,
   Inayama, S., Omura, T. (1984) Molecula cloning and nucleotide sequence
   of cDNA for mRNA of mitochondrial cytochrome P-450scc of bovine adrenal
   cortex. Proc. Natl. Acad. Sci. USA, 81, 4647-4651
5. Usanov, S.A., Pikuleva, I.A., Chashchin, V.L., Akhrem, A.A. (1984) Chemical modification of adrenocortical cytochrome P-450scc with tetranitromethane. Biochim. Biophys. Acta, 790, 259-267
6. White, R.E., Coon, M.J. (1980) Oxygen activation by cytochrome P-450.
   Ann. Rev. Biochem., 49, 315-356
7. White, R.E., Coon, M.J. (1982) Heme-ligand replacement reaction of the
   bonding atom of the axial ligand trans to thiolate as oxygen.
   J. Biol. Chem., 257, 3073-3083

SUBSTRATE DEPENDENT ACTIVITY CONTROL OF CYTOCHROME P-450 LM2

K. Ruckpaul, D. Petzold and H. Rein

Academy of Sciences of the G.D.R., Central Institute of Molecular Biology, 1115 Berlin-Buch, G.D.R.

INTRODUCTION

Cytochrome P-450 LM2 forms a thermal ferric spin equilibrium which is shifted in different amounts by different type I substrates /Rein et al. 1977/. The magnitude of the substrate induced spin shift is correlated with the catalytic activity of the enzyme /Blanck et al. 1983/. The importance of the substrate induced spin shift in the regulation of the activity of the enzyme arises the question what physico-chemical properties of the substrates determine their affinity to the enzyme and their capability to shift the spin equilibrium.
In the present paper intrinsic properties of a series of homologous tertiary amines (benzphetamine analogues) which bind to cytochrome P-450 LM2 were quantified. Molecular parameters of the substrates (^{13}C chemicals shifts, solubilities and steric parameters), their binding ($\ln K_s$) and their capability to shift the spin equilibrium ($\Delta\alpha$) were correlated to detect such peculiarities which are decisive for binding and for the magnitude of the spin shift and by this to understand what properties of the substrate are essential for control of catalytic activity of the enzyme.

MATERIALS AND METHODS

Cytochrome P-450 LM2 has been isolated from liver microsomes of phenobarbital treated male rabbits and purified according to published procedures /Coon et al. 1978; Jänig et al. 1984/. The concentrations of cytochrome P-450 were determined from the respective CO-difference spectra using $\Delta\varepsilon = 91 \ mM^{-1}cm^{-1}$ /Omura and Sato 1967/ and revealed a specific content of 16-

4 Vereczkey

17 nMol/mg protein. All experiments were carried out in 0.1 M
phosphate buffer, pH 7.4, 20 % (v/v) glycerol with a cytochrome P-450 LM2 concentration of 0.8 µM. The substrates used were
synthesized analogously to published methods by alkylation of
the respective amines /cf. Blanck et al. 1983/.
The ^{13}C NMR spectra were run on the NMR-spectrometer BP 497 FT
(Tesla, CSSR) or on PS 100 FT (Jeol, Japan). Magnetic field
strength was 2.347 T. Using proton noise decoupling conditions
each spectrum was averaged from 600 to 1500 scans. The central
signal of C^2HCl_3 (δ = 78.284 ppm related to tetramethylsilan
δ = 0 ppm) was used as internal standard. The precise assignment of the signals were performed by use of proton coupled
spectra which were obtained utilizing gated decoupling technique.
All optical spectra were performed on a spectrophotometer
UV 300 (Shimadzu, Japan) at 26 $^{\circ}C$. The optical titration experiments were carried out as usual /Ruckpaul and Rein 1984/.
The high spin shift $\Delta\alpha$ was calculated from ΔA_{∞} by use of
$\Delta\epsilon_{384\ nm/417\ nm}$ = 122 $mM^{-1}cm^{-1}$ /Petzold et al. in preparation/.
The solubility of the respective substrate (c_{satd}) was deter-
mined by means of light scattering at 330 nm.
N-Demethylation of benzphetamine was measured according to
Nash /1953/.
The molecular models of substrates were fitted with space fil-
ling Courtauld atomic models (Griffin and George, England).

RESULTS

The 12 studied tertiary alkylamines have in common that they
bind to cytochrome P-450 LM2 forming type I difference spectra
and are uniformly N-demethylated /Blanck et al. 1983/. These
compounds can be divided into 3 chemically different classes:
(1) 6 compounds without chlorine differ from benzphetamine in
length and/or degree of ramification of the alkyl chains between both phenyl rings. (2) 3 compounds contain one chlorine
atom in o- or p-position of the phenyl rings. (3) 3 compounds
contain 2 chlorine atoms in o- and/or p-position /Table 1/.

Spectral parameters K_s and

In Table 1 the K_s-values together with the ΔA_∞-values and the corresponding high spin contents α are listed in the order of increasing affinity of the substrates to cytochrome P-450 LM2. The K_s-values range from 5.2 μM $\leq K_s \leq$ 580 μM. The induced spin shift ($\Delta\alpha$) in soluble cytochrome P-450 LM2 amounts to 22 % - 72 %. In the series of substrates without chlorine substitution a high binding affinity is accompanied by an increased spin shift. Elongation of the alkyl chain or ramification does not significantly affect this relation.

Table 1

Spectral parameters of the binding of tertiary amines to cytochrome P-450 LM2.

K_s = apparent spectral dissociation constant; ΔA_∞ = final extinction in the optical substrate titration of P-450 LM2; α = spin content; ϕ = phenyl; o-Cl-ϕ=o-chlorphenyl; p-Cl-ϕ=p-chlorphenyl.

№	Substrates Formula	K_s /μM/	$\Delta_1 A_\infty$ /mM^{-1}cm^{-1}/	α /%/
1	$(\phi\text{-CH}_2\text{-CH}_2)_2\text{N-CH}_3$	580 ± 200	32 ± 6	35
2	$\phi\text{-CH}_2\text{-CH}_2\text{-N-CH}_3\text{-CH}_2\text{-}\phi$	160 ± 13	68 ± 3	65
3	$\phi\text{-CH}_2\text{-C(CH}_3)_2\text{-NCH}_3\text{-CH}_2\text{-}\phi$	140 ± 15	60 ± 3	58
4	$\phi\text{-CH}_2\text{-CHCH}_3\text{-NCH}_3\text{-CH}_2\text{-}\phi$	114 ± 9	69 ± 4	66
5	$(\phi\text{-CH}_2)_2\text{-NCH}_3$	93 ± 10	86 ± 4	80
6	$(\phi\text{-CHCH}_3)_2\text{-NCH}_3$	37 ± 2	83 ± 3	77
7	$\phi\text{-CH}_2\text{-CHCH}_3\text{-NCH}_3\text{-CH}_2\text{-p-Cl-}\phi$	30 ± 2	55 ± 3	54
8	$\text{p-Cl-}\phi\text{-CH}_2\text{-NCH}_3\text{-CH}_2\text{-}\phi$	28 ± 2	55 ± 3	54
9	$\text{o-Cl-}\phi\text{-CH}_2\text{-NCH}_3\text{-CH}_2\text{-}\phi$	14 ± 1	76 ± 5	72
10	$(\text{p-Cl-}\phi\text{-CH}_2)_2\text{-NCH}_3$	9.3 ± 0.5	25 ± 1	30
11	$\text{o-Cl-}\phi\text{-CH}_2\text{-NCH}_3\text{-CH}_2\text{-p-Cl-}\phi$	5.4 ± 0.4	66 ± 5	63
12	$(\text{o-Cl-}\phi\text{-CH}_2)_2\text{-NCH}_3$	5.2 ± 0.2	89 ± 4	82

Substitution by chlorine enhances the binding affinity but the spin shift is not significantly changed at chlorination in o-position but even remarkably diminished at chlorination in p-position as compared to the unsubstituted compounds (cpds.).

^{13}C NMR studies of the substrates

The ^{13}C signals of the 12 substrates could be completely assigned to the corresponding carbon atoms. The ^{13}C chemical shifts of the N-methyl carbon and of the quarternary phenyl carbon atom bound via one methylene bridge to the nitrogen were used to monitor electron density changes in the aliphatic and aromatic moieties of the molecules. A high field shift of the quarternary aromatic carbon atom indicates a diminished electron density at this atom. In dependence on R and Cl, resp., the chemical shift of the N-methyl carbon changes from 41.2 ppm (cpd. 1) to 39.1 ppm (cpd. 12) and that of the quarternary carbon is shifted from 130.5 ppm (cpd. 5) to 127.9 ppm (cpd. 11) (Fig. 1).

From substituent dependent chemical shifts related to benzphet amine (cpd. 4) as structural basis it can be derived that (i) chlorine binding to the phenyl ring induces a diminished electron density at the quarternary carbon atom and at the N-methyl

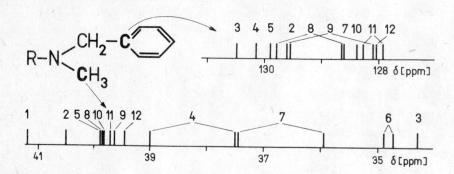

Fig. 1 Position of the ^{13}C NMR signals of two selected C-atoms of the substrates studied: the quarternary C-atom (above) and the N-methyl carbon (below).

carbon as well. Substitution in o-position of the phenyl ring induces a stronger effect than p-substitution. (ii) Besides electron density changes also steric effects on the chemical shifts of both carbon atoms are induced by substituent R. The N-methyl carbon chemical shift of substrates 3, 4, 6, 7 are rather more influenced by the γ-gauche effect induced by their C-methyl groups than by electron density changes. Therefore they are not suited to monitor electronic effects.

Correlation between molecular parameters of the substrates and K_s and the high spin amount (α)

The chemical shift (δ) of the N-methyl carbon and of the quarternary aromatic carbon as well linearly correlates with the binding affinity (ln K_s) of the substrates. Since the chemical shifts of both carbon atoms likewise correlate, this correlation indicates that also the chemical shift of the N-methyl carbon is caused by electron density changes. Obviously substituent induced electron density changes of the carbon skeleton are transmitted to the N-methyl carbon via the nitrogen which leads to a lowered basicity of the amino group. The latter is expected to result in a diminished solubility which experimentally has been evidenced. Both correlations reveal that the lowered solubility originates from a diminished basicity and that a diminished basicity as reflected in the high field shift of the δ-values enhances the binding affinity to cytochrome P-450 LM2.

Plotting ln $\left[\alpha(1-\alpha)^{-1}\right] \alpha \Delta g$ versus ln $K_s \alpha \Delta G$ significant correlations are obtained separating the 3 chemically different classes of the substrates: (i) unsubstituted; (ii) with one and (iii) with two chlorine atoms. The straight lines represent a linear dependence of the free energy Δg necessary for the stabilization of the substrate induced high spin states on the free energies ΔG of substrate binding (Fig. 2).

38

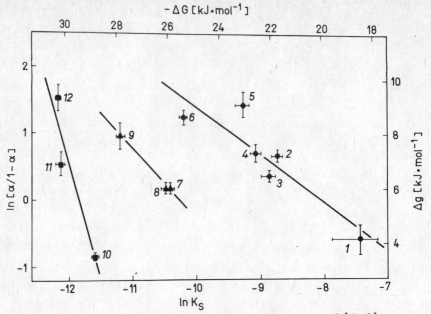

Fig. 2 Plots of the free energy $\Delta g = RT \cdot \ln \frac{\alpha(1-\beta)}{(1-\alpha)\beta} = (\beta = \alpha$ without substrate = 9,2 %) versus the binding free energy $-\Delta G = RT \ln K_s$ for each substrate. ●——● without chlorine substitution; ▲——▲ with one chlorine atom; ■——■ with two chlorine atoms.

DISCUSSION

Attempts to correlate stereochemical and electronic properties of cytochrome P-450 LM2 substrates with catalytic activity did not result in general rules about structure activity relationships /Martin and Hansch (1971)/. The presence of different cytochrome P-450 isozymes in microsomes and their capability to react with a great variety of substrates impaired or even excluded such correlations. These disadvantages have been avoided in this study by utilizing a single isozyme and a homologous series of benzphetamine derivatives. Supposition for a detailed description of the properties of a substrate to interact with the enzyme is to differentiate between its binding affinity and its capability to convert an enzyme into its biologically active conformation (intrinsic activity of the substrate /Ariens 1971/).

The K_s-values indicate the affinity of the respective substrate and its intrinsic activity can be derived from the amount of the substrate induced high spin shift due to the correlation between the conversion rate of the substrate and the corresponding high spin shift of the enzyme /Blanck et al. 1983/.
In the present study a lowered solubility of the substrates has been evidenced to be accompanied by an increased binding affinity. The molecular reason of this result is based on a diminished electron density of the quarternary aromatic carbon atom of the derivatized benzphetamines (indicated by the high field shift of their ^{13}C δ-values) which lowers the electron density of the nitrogen via inductive effects.
The decreased electron density of the nitrogen is connected with a diminished basicity which results in a lowered solubility in aqueous solution. The distinct electron distribution of each substrate determines via hydrophobicity the binding affinity.
The existence of 3 chemically different classes of substrates with class specific characteristics within this series indicates that besides the binding affinity additional factors determine the intrinsic activities of the amines. The substrates therefore additionally should be characterized by distinct peculiarities which substrate dependent induce a high spin shift of the enzyme. The 12 substrates exist in conformational states which range between a stretched and an angular form and in all transition states between. In aqueous solution the angular form should be favoured thermodynamically due to the tendency to minimize the surface. In dependence on the chemistry of each substrate, however, the angle (with the methyl bound nitrogen at the top) varies. A decreasing length of both arms of the substrate (that means a wider angle) is accompanied by an increased high spin shift indicating that a bulky molecular structure near the nitrogen obviously represents the essential property to exhibit a high intrinsic activity of the substrate.
Chlorine substitution obviously defines a decisive intrinsic property of the molecule. p- and o-Chlorination induce by the mesomeric effect about the same distribution of the pi-electrons of the phenyl ring. o-Chlorination additionally lowers the elec-

tron densitiy of the quarternary carbon atom by the inductive
effect. The diminished spin shift of the p-substituted substrates is caused by the elongation of the molecule and perturbs the
intermolecular interaction with the complementary structure of
the active center. o-Substitution, however, does not perturb
the steric requirements necessary to shift the spin state.
From the present results two different intrinsic properties of
the series of tertiary amines are discriminated. The electron
density at the nitrogen of the substrate effects the basicity
of the amino group and this way determines the binding affinity
of the substrates to cytochrome P-450 LM2. The capability of
the bound substrate to shift the spin equilibrium, however, is
strongly influenced by the compactness of the molecular structure near the nitrogen. Since the amount of the substrate induced
high spin conformation of the enzyme determines its enzymatic
activity the turnover of a substrate depends on its steric properties.

REFERENCES

Ariens, J. (1972) Drug Design, Vol. 1, Academic Press, New York,
London

Blanck, J., Rein, H., Sommer, M., Ristau, O., Smettan, G. and
Ruckpaul, K. (1983) Biochem. Pharmacol. 32, 1683-1688

Coon, M.J., Hoeven, T.A., Dahl, S.B. and Haugen, D.A. (1978)
Methods Enzymol. 52, 109-117

Jänig, G.R., Makower, A., Rabe, H., Bernhardt, R. and Ruckpaul,
K. (1984) Biochim Biophys. Acta 787, 8-18

Martin, Y.C. and Hansch, C. (1971) J. Med. Chem. 14, 777-779

Nash, T. (1953) Biochem. J. 55, 416-421

Omura, T. and Sato, R. (1967) Methods Enzymol. 10, 536-561

Rein, H., Ristau, O., Friedrich, J., Jänig, G.R. and Ruckpaul,
K. (1977) FEBS Letters 75, 19-22

Ristau, O., Rein, H., Jänig, G.R. and Ruckpaul, K. (1978) Biochim. Biophys. Acta 536, 226-234

Ruckpaul, K. and Rein, H. (1984) Cytochrome P-450, Akademie-
Verlag, Berlin

MECHANISM OF OXYGEN ACTIVATION BY CYTOCHROME P-450

H. Rein, C. Jung, O. Ristau and K. Ruckpaul

Central Institute of Molecular Biology, Academy of Sciences of
the GDR, 1115 Berlin-Buch, GDR

INTRODUCTION

The catalytic principle of cytochrome P-450 dependent monooxygenases consists in the splitting of molecular oxygen and the
insertion of one oxygen atom into a substrate. This reductive
dioxygen splitting includes the transfer of electrons from an
external donor by which the energy of the intramolecular dioxygen bond is lowered resulting in the cleavage of the molecule.
From the thermodynamic point of view the overall monooxygenase
reaction

$$SH + O_2 + XH_2 \longrightarrow SOH + X + H_2O$$

(SH, substrate; XH_2, electron donor) should proceed spontaneously because of its negative free enthalpy /Jung and Ristau,
1978/. The reaction, however, is kinetically hindered caused
by two reaction steps which require high energies, the high
enthalpy of dioxygen dissociation (460.5 kJ mol^{-1}) and that of
CH cleavage of the substrate (about 400 kJ mol^{-1}). Neglecting
its enzymatic character, a value of about 418 to 460 kJ mol^{-1}
is expected for the activation energy of the overall reaction.
For the cytochrome P-450 catalyzed reaction, however, an activation energy only of about 38 to 71 kJ mol^{-1} was found /Akhrem
et al., 1977/. That means the enzyme lowers the activation energy of the overall reaction which proceeds via several intermediate steps with low activation energy /Rein et al., 1984/.

The conditions for oxygen activation by cytochrome P-450 result from the binding of dioxygen to the heme iron. Certain
structural suppositions of the active site, however, are necessary to activate the heme bound dioxygen. In this connection,

42

the axial ligands of the heme iron are most important for functional characteristics of the hemoprotein. In contrast to most
hemoproteins in which a histidine residue occupies the 5th coordination site of the heme iron, a cysteine residue is localized at this position in cytochrome P-450 (Fig. 1).

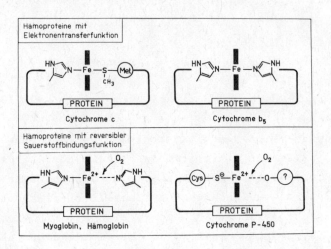

Fig. 1 Axial ligands of some hemoproteins.

Recently, a sulfur ligand has been experimentally evidenced by
Raman resonance studies on isotopically enriched $^{54}Fe/^{34}S$ and
naturally abundant samples /Champion et al., 1982/.
This paper deals with possible intermediates of the oxygen activation process, their electronic properties, derived from
quantum chemical calculations, and the relevance of the fifth
heme iron ligand for the oxygen activation. By comparison of
sulfur with imidazole as fifth ligand conclusions about necessary conditions for the oxygen activation are drawn.

THEORETICAL METHOD

The theoretical analysis is carried out in three steps:
(i) Calculation of spectral properties using a quantum chemical
method, specially developed for the calculation of electronic
transitions of organic metal complexes (π-INDO method /Jung et
al., 1983/), to verify the typical experimental optical spectra
for the carbon monoxide and dioxygen complexes of cytochrome

P-450; (ii) calculation of the electron distribution and correlation of the charges with experimental data to show that the
charge distribution is reasonably reflected by the quantum chemical method used /Jung, 1983/; (iii) postulate of possible intermediates and calculation of their electronic structure using
the quantum chemical parameters from steps (i) and (ii).
The structural models of the heme complexes base on the facts
that (i) the dπ orbitals of the iron couple the π-systems of
the axial ligands with the π-system of the porphyrin standing
perpendicularly to each other (coupled π-orbital model); (ii)
the polarization measurements of cytochrome P-450-CO crystals
/Hanson et al., 1976/ and the appearance of distinct A terms in
the MCD spectrum of the CO complex /Dawson et al., 1978/ indicate energetically degenerate electronic transitions for the
Soret and Q bands meaning that the sulfur ligand is sp hybridi-
zed; (iii) only special combinations of sulfur charge (-0.3
until -0.6) and iron-sulfur distance (1.9 - 2.3 Å) cause the
unusual optical spectrum ("hyper" spectrum) of the CO complex
of cytochrome P-450 /Jung, 1985/; (iv) the dioxygen molecule is
asymmetrically bound to the heme iron derived from X-ray structure analysis of oxymyoglobin /Phillips, 1978/ and dioxygen
picket-fence iron porphyrin complexes /Collman, 1974/.

RESULTS

The calculated spectral properties of several iron porphyrin
complexes with sulfur and imidazole as trans ligands, respectively, agree with experimental spectral data of the respective
cytochrome P-450 and hemoglobin complexes (Table 1). The calculated charge distribution can be well correlated with experimental data from Raman resonance, infrared and Mossbauer studies /Jung, 1983/. This justifies to use the quantum chemical
method for the determination of charge distributions and bond
orders of the postulated intermediates involved in the process
of oxygen activation.
In Table 2 the calculated net charges of oxygen model complexes
of cytochrome P-450 and hemoglobin are listed. The data for the
models B and D are based on the analysis of the properties of

TABLE 1   Calculated and experimental electronic transitions

band	calculation ν (f)		experiment ν	(f)
oxyhemoglobin				
Q	18.4 (0.2)	ν (o-o)	17.4	
		ν (o-1)	18.5	(0.4)
Soret	25.6 (2.9)		24.0	(2.9)
UV	30.5 (1.1)		29.2	(0.6)
oxy-cyt P-450				
Q	18.6 (0.2)		18.0	(0.3)
Soret	25.7 (3.0)		23.9	(2.4)
UV	28.2 (1.3)		27.9	(0.9)

ν: intensity-weighted averaged transition energy value of the x- and y-polarized components ($10^3 cm^{-1}$); (f): sum of the component oscillator strengths refered to the experimental value of the Soret band of oxyhemoglobin.

the lowest unoccupied molecular orbital of the porphyrin complexes A and C. The π-bond orders of iron-oxygen and dioxygen are of special interest for the activation of molecular oxygen and are therefore also listed in Table 2. According to the postulated intermediates involved in the reaction cycle of cytochrome P-450 the following model complexes have been considered: - the dioxygen heme complexes with and without a proton attached at the outer oxygen atom (models C and A, respectively), - the reduced dioxygen heme complexes with and without a proton attached at the outer oxygen atom (models D and B, respectively) which may be intermediates after the second reduction step in the monooxygenase reaction. Binding of a proton at the outer oxygen atom weakens the dioxygen bond which is further lowered in the reduced complexes. The iron-oxygen bond is also weakened by protonation, however, strengthened by reduction of these protonated complexes. Two important facts are to be noted for the cytochrome P-450 model in comparison to hemoglobin. First, reduction of the proton-attached dioxygen complex C leads to a stronger loosening of the dioxygen π-bond with antibonding character and a stronger stabilization of the iron-oxygen π-bond. Secondly, after reduction of the proton-attached dioxygen

TABLE 2   Net charges and π-bond orders of oxygen complexes

	Hemoglobin model: $N_{Im} - Fe^{II} - O_2$				P-450 model: $S^- - Fe^{II} - O_2$			
	A	B	C	D	A	B	C	D
Charges								
iron	1.080	1.043	0.907	0.667	1.124	0.984	1.362	0.922
imidazole	0.264	0.264	0.258	0.258	—	—	—	—
sulfur	—	—	—	—	−0.090	−0.110	0.051	−0.049
porphyrin	−0.761	−1.554	0.390	−0.293	−0.809	−1.544	−0.492	−0.819
inner-oxygen	−0.233	−0.300	−0.627	−0.702	−0.283	−0.317	−0.497	0.627
outer-oxygen	−0.350	−0.453	—	—	−0.442	−0.513	—	—
outer-oxygen attached a proton	—	—	0.072	0.070	—	—	0.076	0.073
bond-orders								
Fe—O	0.466	0.501	0.231	0.324	0.457	0.473	0.269	0.435
O—O	0.570	0.487	0.028	0.016	0.496	0.444	0.033	0.054

$$A\left[-Fe^{II}-O_2\right]; \quad B=\left[-Fe^{II}-O_2\right]^-; \quad C=\left[-Fe^{II}\;O_2H\right]; \quad D\left[Fe^{II}\;O_2H\right]$$

complex of the cytochrome P-450 model the supplied electron is predominantly localized (67 %) in the axial π-system (Table 3). Both facts favour the splitting of the dioxygen bond which is

TABLE 3

Percentage of distribution of the extra electron in the oxygen complexes (data derived from Table 2)

	hemoglobin model		P-450 model	
	B	D	B	D
iron	4	24	14	44
imidazole	0	0	—	—
sulfur	—	—	2	10
porphyrin	79	68	74	33
O_2	17	—	10	—
O_2H	—	8	—	13
summarized for axial system	21	32	26	67

$$B=\left[X-Fe^{II}-O_2\right]^-;$$

$$D=\left[X-Fe^{II}-O_2H^+\right]^-; \quad X=-N_{Im},\;-S^-$$

in line with an increase of the 0 - 0 distance and a decrease of the splitting energy for unbound oxygen species (Table 4).

TABLE 4   Bond distances and splitting energies for different oxygen species

	oxygen	hydrogen peroxyradical	hydrogen peroxide	hydroxyradical	water
	$O_2 \xrightarrow[H^+]{1e} $	$HO_2^{\cdot} \xrightarrow[H^+]{1e}$ $\downarrow -H^+$ O_2^{-} superoxide anion	$H_2O_2 \xrightarrow[H^+,-H_2O]{1e}$ $\downarrow -H^+$ HOO^- hydrogen peroxide anion	$OH^{\cdot} \xrightarrow[H^+]{1e} H_2O$	
binding distance	1.21 Å	1.28 Å	1.49 Å		
splitting energy ($kJ\,mol^{-1}$) ΔH	494	$276\,[O_{2(g)}^{-} \rightarrow O_{(g)}^{-} + O_{(g)}]$ $270\,[HO_{2(g)}^{-} \rightarrow HO_{(g)}^{\cdot} + O_{(g)}]$	*heterolytic* $178\,[HO_{2(aq)}^{-} \rightarrow OH_{(aq)}^{-} + O_{(g)}]$ $122\,[HO_{2(aq)}^{-}+H_{(aq)}^{+} \rightarrow H_2O_{(l)} + O_{(g)}]$ *homolytic* $208-213\,[H_2O_2 \longrightarrow 2OH^{\cdot}]$		

$O_{(g)}$, gaseous phase; $O_{(aq)}$, aqueous phase; $H_2O_{(l)}$, liquid phase

IMPORTANCE OF THE THIOLATE LIGAND

The comparative quantum chemical analysis of model complexes from cytochrome P-450 and hemoglobin, respectively, reveals the importance of a negatively charged sulfur ligand trans to the heme iron bond dioxygen for the oxygen activation process. In cytochrome P-450 the thiolate ligand specifically determines the electronic properties of the iron protoporphyrin IX complex: (i) The Soret band of the CO complex is shifted from 418 nm (hemoglobin) to 450 nm (cytochrome P-450). (ii) The electron spin resonance spectrum of ferric cytochrome P-450 exhibits g-values at 2.41, 2.25, 1.91 indicating a strong axial and rhombic distorsion of the ligand field of the heme complex with octahedral symmetry as compared to low spin complexes of methemoglobin /Rein et al., 1975/ (Table 5).

Two important electronic properties differentiate the thiolate ligand in cytochrome P-450 from the imidazole ligand in hemoglobin. The 3pπ orbitals of the sulfur show a 3-fold stronger overlapping with the iron 3dπ -orbitals than the 2pπ orbitals of the imidazole nitrogen atom. The negative charge of the sulfur ligand lowers the 3pπ orbitals ionization potential which favours charge donation to the iron. Both effects together induce a strong π-backdonation from the iron to the

TABLE 5 Experimental high field g-values of different hemoproteins and the crystal field parameters of the tetragonal (μ) and rhombic (R) distorsions of the cubic ligand field of the heme iron.

Low spin compound	High field g-values			Crystal field parameters	
	g_z	g_y*	g_x	R	μ
Acidic methemoglobin pH 6.2					
Isomer I	2.92	2.25	1.51	1.89	3.27
Isomer II	2.77	2.27	1.67	2.39	3.78
Methemoglobin-OH (pH 8.6)	2.55	2.26	1.83	3.51	4.54
Methemoglobin at (pH 10.5)	2.40	2.25	1.90	4.91	4.76
Methemoglobin-imidazole	2.93	2.29	1.46	1.83	2.91
Methemoglobin-triazole	2.93	2.28	1.47	1.84	2.99
Cytochrome P-450**	2.40	2.22	1.91	4.94	5.69

* Values corrected under the normalization condition, because these g-values are not precisely readable at zero passage of the curve.
** Prepared from rat liver microsomes from phenobarbital-treated animals.

dioxygen ligand.
Our conclusions for the oxygen activation are summarized in Fig. 2. The ferric-superoxide-like model ① is supported by the quantum chemical calculations and experimental data as well because Mossbauer data of the dioxygen cytochrome P-450 complex exhibits isomer-shifts comparable with ferric low spin hemoproteins /Lang and Marshall, 1966/. The negatively charged dioxygen ligand favours the binding of a proton ② . The insertion of an electron into the dioxygen heme complex induces a further destabilization of the dioxygen-π-bond. The calculated electron distribution in the electronic ground state supports the peroxy-like electronic structure ③ and ④ . In this scheme the cleavage of the outer oxygen atom

Fig. 2 Oxygen activation cycle of cytochrome P-450.

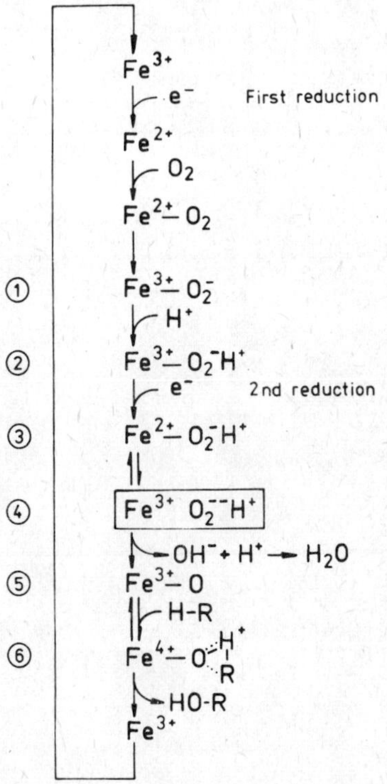

Fe^{3+}
— e^- First reduction
Fe^{2+}
— O_2
$Fe^{2+}-O_2$
① $Fe^{3+}-O_2^-$
— H^+
② $Fe^{3+}-O_2^-H^+$
— e^- 2nd reduction
③ $Fe^{2+}-O_2^-H^+$
④ $Fe^{3+}\ O_2^{--}H^+$
—OH^-+ H^+ → H_2O
⑤ $Fe^{3+}-O$
— H-R
⑥ $Fe^{4+}-O \cdots {}_R^{H^+}$
—HO-R
Fe^{3+}

accompanied with release of water may be heterolytic, meaning
ionic, and results in the formation of an oxene-iron complex.
This complex with a possible ferric-ferryl resonance may be
the substrate hydroxylating agent ((5)   and   (6)).

REFERENCES

Akhrem, A.A., D.I. Metelitza, S.N. Bielski, P.A. Kiselev, M.L.
Skurko and S.A. Usanov (1977) Croat. Chem. Acta 49, 223-235

Champion, P.M., B.R. Stallard, G.C. Wagner and I.C. Gunsalus
(1982) J. Amer. Chem. Soc. 104, 5469-5472

Collman, J.P., R.R. Gagne, C.A. Reed, W.T. Robinson and G.A.
Rodley (1974) Proc. Natl. Acad. Sci. USA 71, 1326-1329

Dawson, J.H., J.R. Trudell, R.E. Linder, G. Barth, E. Bunnenberg and C. Djerassi (1978) Biochemistry 17, 33-42

Hanson, L.K., W.A. Eaton, S.G. Sligar, I.C. Gunsalus, M. Gouter-
man and C.R. Connell (1976) J. Amer. Chem. Soc. 98, 2672-2674

Jung, C. and O. Ristau (1978) Pharmazie 33, 329-331

Jung, C. (1983) Studia Biophys. 93, 225-230

Jung, C., O. Ristau and Ch. Jung (1983) Theoret. Chim. Acta 63,
143-159

Jung, C. (1985) Chem. Phys. Letters 113, 589-597

Lang, G. and W. Marshall (1966) Proc. Phys. Soc. (London) 87,
3-34

Phillips, S.E.V. (1978) Nature 273, 247-248

Rein, H., O. Ristau and K. Ruckpaul (1975) Biochim. Biophys.
Acta 393, 373-378

Rein, H., C. Jung, O. Ristau, J. Friedrich (1984) in Cytochrome
P-450, edited by K. Ruckpaul and H. Rein Akademie-Verlag Chap.
4, pp. 163-249

STRUCTURE, FUNCTION, AND FLUCTUATIONS DURING P-450 ELECTRON TRANSFER

Stephen G. Sligar, Mark T. Fisher, and Kermit Carraway

Department of Biochemistry, University of Illinois
1209 West California Street, Urbana, Illinois   61801   USA

With the elucidation of the three dimensional x-ray of cytochrome P-450$_{cam}$ by Dr. Tom Poulos at Genex Corporation we are using this road map to define the controlling structural features in electron transfer and oxygenase chemistry events. This translates into a description of the conformational changes and functionally important motions of the cytochrome P-450$_{cam}$ macromolecule. Cytochrome P-450$_{cam}$ and its associated physiological redox transfer components provide an excellent system to study the linked dynamics of electron transfer and protein structure. Limiting in many cases is the ability to monitor internal equilibria in protein structure and, thus, deconvolute the free energy linkages operating to control physiological response. In the case of P-450$_{cam}$, as well as other 5-d-iron systems, nature has provided an observable spin-state equilibrium.

Cytochrome P-450$_{cam}$ is responsible for the hydroxylation of camphor to form 5-exo-hydroxycamphor in an overall reaction cycle involving discrete one electron transfer steps. In the normal reconstituted system, the rate limiting step appears to be the transfer of the second reducing equivalent from putidaredoxin to oxygenated cytochrome. The first electron transfer step, however, is rate limiting in many cases in bacterial P-450$_{cam}$ when other substrates are metabolized, and in many of the hepatic P-450 systems, where it has been suggested that the electron transfer rates are determined by the percent high spin ferric iron species initially present. We will discuss the first two coupled steps in the P-450$_{cam}$ reaction, substrate binding and ferric-ferrous reduction of the heme iron, in light of this spin equilibrium of the heme center.

Inasmuch as the thermal nature of this spin equilibrium and the relevant thermodynamic parameters linking it to the substrate affinities have been thoroughly documented and reviewed, we will not discuss in detail these aspects of the spin equilibrium in cytochrome P-450 and other heme proteins. Our concern is to examine the structural basis for this observed spin state equilibrium, and to examine the electron transfer processes that are linked in rate and equilibria with the ferric spin state.

A self-consistent choice for reductant of cytochrome P-450$_{cam}$ is its physiological partner, putidaredoxin. Electron transfer experiments performed between these two proteins under a number of different experimental conditions by several groups revealed that electron transfer occurs between a dienzyme complex of the two protein species. Electron transfer rates between putidaredoxin and different ferric spin states populations of cytochrome P-450$_{cam}$ have heretofore not been extensively documented. For a complete quantitative study of the regulation of electron transfer rate through substrate induced shift of the ferric spin equilibrium, it is necessary to be able to independently vary the redox, spin and substrate binding equilibrium. This can be accomplished in cytochrome P-450$_{cam}$ through the use of various substrate analogs of camphor which were found to perturb the ferric spin equilibria to varying degrees, with results indicating that the redox potential exhibits a linear free energy correlation with the ferric spin equilibria. Recent experiments have measured the first order electron transfer rates for the reaction

$$Pd^r \cdot P\text{-}450^0 \underset{k_{-1}}{\overset{k_1}{\rightleftarrows}} Pd^0 \cdot P\text{-}450^r$$

where Pd represents putidaredoxin and superscripts r and 0 refer to reduced and oxidized forms of the proteins. The activation energies for electron transfer are also linearly correlated with ferric spin state free energy with the reduction rate increasing as the percent high spin ferric species increase. The spin state relaxation rates, as determined by temperature jump spectroscopy, indicated that spin-reequilibration was not rate limiting in relation to the electron transfer event in both the substrate free and the substrate bound forms. The observed reduction kinetics are simple first-order in both cases. Hepatic cytochrome P-450, on the other hand, shows biphasic kinetics when one monitors the increase in the ferrous CO-adduct of the heme upon reduction by its physiological electron donor in reconstituted systems. One possible explanation for the apparent "biphasic" kinetics seen in hepatic cytochrome P-450 is that the rate limiting step in the reduction sequence is the formation of the high spin ferric species which then controls the reduction rate. The "fast phase" observed in the electron transport scheme might thus represent the reduction of the fraction of high spin cytochrome P-450 present in a pre-equilibrium prior to the introduction of reducing equivalents. In this hypothesis, the rate of the slow phase is postulated to reflect the rate of formation of the high spin species from the low spin ferric form with only the high spin form being reduced. Support for this hypothesis is the observation of a direct correlation between the amount of initial high spin species and the extent of the initial "burst" phase. Computer simulations using an equilibrium model were consistent with this scheme. However, it is incorrect to ignore the occurrence of a reduced low spin form of cytochrome P-450 when examining reduction kinetics. As will be discussed, significant transient population of the ferrous S = 0 state can exist.

Although many investigators using a variety of P-450 systems have reported dissociation constants for a variety of substrates, the extent of spectral change observed is also directly related to the microscopic spin state equilibrium constants. Further use of this parameter can separate the actual substrate binding free energies from the conformational free energies monitored by the spin state equilibria. As we have recently documented, ΔA_{max} and K_d are two completely independent quantities and hence we stress again that lack of a large spectral change in any P-450 upon addition of substrate does not necessarily imply that there is no binding to the protein. Rather, it simply points to the near identity of the spin state equilibrium constants in the substrate bound and free forms of the P-450.

A complete thermodynamic description of the interaction of substrates with cytochrome P-450$_{cam}$ must necessarily include documentation of the other relevant state functions such as entropy, ΔS, entholpy, ΔH, and partial specific volume ΔV. Enthalpies and entropies describing the regulation of spin state via substrate binding can be extracted by quantitating all equilibria as a function of temperature. Through data recently collected in our laboratory, one can see a profound feature of protein folding and regulatory equilibria. Strikingly obvious is the nearly identical enthalpies and widely varying entropies of the spin state equilibria. Thus, the spin equilibria of cytochrome P-450 is poised by entropy factors and not enthalpy considerations. Since entropy reflects the accessible phase space of the system, the shift in spin equilibria via substrate association is by differential "tightening" of the structure. On a broader perspective, the use of entropy to control spin state equilibria is, in some sense, necessary in view of the desired regulation of system redox potential. This follows from conservation of thermodynamic state functions and the dominant entropy factors in any adiabiatic electron transfer process. In order to explore these implications further, we will discuss the linkage of P-450 oxidation-reduction potential.

As will be shown, the free energies of redox potential and ferric spin equilibrium are indeed related, though partitioning of the total substrate binding energy into other conformer modes does occur. We will conclude our discussion by offering a model to also explain conflicting interpretations of the kinetics of electron transfer in cytochrome P-450 systems. Emerging will be a demonstration that although simplified models may be appropriate for equilibrium discussions, electron transfer in P-450 systems is kinetically controlled, and a thermodynamically complete model must be used. A major error, however, is to use any simplified three state model that neglects the Fe(2+:LS) state to describe the reduction of P-450 hemoprotein. This can easily be seen in a schematic form as indicated below, where the individual rate constants are used to calculate a family of "trajectories" of the system:

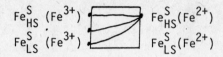

The various structural features of the cytochrome P-450$_{cam}$ molecule can interact to provide additional redistributions of redox, substrate, and spin state free energies, through alterations in dielectic constant, ionic stabilizations, and dipole-induced dipole interactions. All of these factors have a profound impact on substrate binding, spin state equilibrium, redox potential, and electron transfer rates.

ASSIGNMENT OF TYROSINE TO THE ACTIVE CENTRES OF CYTOCHROME P-450 LM2
AND CYTOCHROME P-450 CAM

Jänig, G.-R., Kraft, R., Makower, A., Rabe*, H., and Ruckpaul, K.

Central Institute of Molecular Biology, Academy of Sciences of the
GDR, 1115 Berlin-Buch, GDR
*Institute of Pathological and Clinical Biochemistry, Humboldt
University, 1040 Berlin, GDR

INTRODUCTION

An understanding of the different substrate specificity of cytochrome
P-450 (P-450) isozymes requires detailed knowledge about the active
centres including those amino acids which are functionally involved
therein and/or especially those providing the axial haem iron ligands.
There is no doubt on thiolate as 5th iron ligand. However, that amino
acid providing the assumed oxygen ligand trans to thiolate is rather
unknown. This report presents results of chemical modification
experiments indicating the location of Tyr-380 in the active centre
of P-450 LM2 and Tyr-94 in that of P-450 CAM.

MATERIALS AND METHODS

P-450 CAM was a gift from Prof.I.C. Gunsalus (University of Illinois,
Urbana, USA) which is gratefully acknowledged. Purification of P-450
LM2 from phenobarbital-treated rabbits and chemical modifications with
tetranitromethane (TNM) were performed as described previously /Jänig
et al. 1984a, 1984b/. Tryptic digestion of nitrated proteins and HPLC
separations of labelled peptides were carried out as published by
Bernhardt et al. /1984/. N-terminal amino acid sequences were determined according to the method of Chang et al. /1978/.

RESULTS AND DISCUSSION

A 30-fold molar excess of TNM over P-450 LM2 results in the nitration
of 3.4 tyrosine residues at pH 7.5. Nitration of up to 2 tyrosines
diminishes the enzymatic activity in a reconstituted system to about
20 % thus suggesting a functional importance of tyrosine(s) /Jänig et

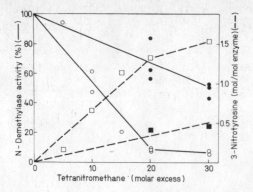

Fig. 1. Influence of nitration with tetranitromethane on the
enzymatic activity of cytochrome P-450 LM2

(– – –) Extent of nitration in the absence (□) or presence
of metyrapone (■); (———) activity in a reconstituted
system /Jänig et al. 1984a/ after nitration without (○) or
in the presence of metyrapone (●).

al. 1984a/. Modification at pH 6.0 which decreases the number of modified tyrosine residues to 1.5 inactivates P-450 LM2 to the same
extent as that modified at pH 7.5 thus implicating very reactive
tyrosine residues of functional importance (Fig. 1). Nitration in the
presence of the inhibitor metyrapone which results in a significantly
lowered extent of nitration and protects from inactivation indicates
the location of a tyrosine at the active centre of P-450 LM2. Evidence for such a location was obtained by HPLC tryptic peptide mapping of P-450 LM2 nitrated in the presence or absence of metyrapone
/Jänig et al. 1984a/. 7 fragments out of more than 65 exhibit an absorbance at 360 nm characteristic of 3-nitrotyrosine. Rechromatography
and manual micro-sequencing yield 5 labelled peptides. According to
the primary structure /Tarr et al. 1983/ they can be assigned to the
following tryptic peptides of P-450LM2 (Table 1). Thus, it can be
derived that P-450 LM2 contains 4 accessible tyrosines in positions
111, 235, 380, and 484. All of them are located in solvent-accessible
positions of the protein as evidenced by the possibility of complete
reduction of the 3-nitrotyrosines to 3-aminotyrosines by sodium dithionite. Analysis of the extent of labelling of these individual
tyrosines in P-450 LM2 nitrated at pH 6.0, 7.5, or 8.2 reveals decreasing relative reactivities in the order Tyr-380 > Tyr-235 >
Tyr-484 > Tyr-111. The high reactivity of Tyr-380 may be indicative

Table 1. Sequences of 3-nitrotyrosine-containing peptides from
cytochrome P-450 LM2

Peak	Sequence	Location in the native protein		
2a	Q I Y* R			
2b	E I Y* R	Gln-233	-	Arg-236, Tyr-235
3	E S G V G ...	Glu-474	-	Arg-487, Tyr-484
4	G Y* V I P(K)	Gly-379	-	Lys-384, Tyr-380
7	I A V V D P I F Q G Y* G ...	Ile-101	-	Arg-120, Tyr-111

for a tyrosine residue with a lowered pK value.

Sequence analyses of the rabbit isozymes P-450 LM2 and P-450 LM3b and
of rat P-450b and P-450e show a location of Tyr-380 and Tyr-484 in
homologous regions of P-450. These regions are, therefore, likely to
play an important role in the function of P-450. Out of the four
accessible tyrosine residues metyrapone protects Tyr-380 from nitration. According to the molecular parameters of metyrapone /Rossi 1983/
Tyr-380 should be located within a maximal distance of 8 Å to the haem
iron. Thus, Tyr-380 might provide the oxygen forming the axial ligand
trans to thiolate in P-450 LM2.

Whereas in P-450 LM2 about 4 tyrosines out of 12 are nitrated, in substrate-bound P-450 CAM only 1.3 nitrotyrosine residues out of totally
9 tyrosines /Haniu et al. 1982/ are obtained under comparable conditions (50-fold molar excess of TNM, pH 7.5 /Jänig et al. 1984b/).
Nitration with an 100-fold excess of TNM at pH 8.2 results in a modification of 3.5 nitrotyrosine residues per molecule. This is to be
expected because TNM reacts with negatively charged tyrosinate groups.
To prove the selectivity of nitration HPLC tryptic peptide mapping
was carried out. Manual micro-sequencing of the tryptic fragments
after rechromatography gave the following results (Table 2). It can be
derived that 6 tyrosines out of totally 9 are accessible to nitration
(positions 73, 76, 94, 177, 199, and 303). Tyr-73, Tyr-199 and Tyr-303
appear to be less reactive as resulting from incomplete modification.
Nitration in the presence of metyrapone evidences the selective protection of Tyr-94.

Table 2.   Sequences of 3-nitrotyrosine-containing peptides from
           cytochrome P-450 CAM

Peak	Sequence	Location in the native protein		
1	A Y E D Y* R	Ala-72	- Arg-77,	Tyr-76
2	A Y* E D Y* R	Ala-72	- Arg-77,	Tyr-73, Tyr-76
3	Y* L T D Q M ...	Tyr-177	- Arg-184,	Tyr-177
5	I L T S D Y* E F ...	Ile-298	- Lys-311,	Tyr-303
	A G E A Y* D F I ...	Ala-90	- Arg-107,	Tyr-94
6	A G E A Y* D F ...	Ala-90	- Arg-107,	Tyr-94
7	I L T S D Y* E ...	Ile-298	- Lys-311,	Tyr-303
8	A L Y* D Y L ...	Ala-197	- Arg-209,	Tyr-199

It is noteworthy that those tryptic peptides starting with Glu-Ala-...
(peaks 1/2, 5/6 and 8) gave only Ala instead of Glu as N-terminal
amino acid if a tyrosine present therein was nitrated. The reason for
this result is still unknown as yet. The elution of a single peptide
at different times may be due to a formylation of Ser or Thr residues.

Summarizing the data, both P-450 LM2 and P-450 CAM contain at least
one tyrosine in the active centre within a maximal distance of 8 Å
from the haem iron. Tyr-380 and Tyr-94 are 56 or 40 amino acid residues distant from Cys-436 and Cys-134, respectively, forming the 5th
iron ligands. Although there are 4 homologous regions between both
proteins these tyrosines are located in non-homologous sequences.

REFERENCES

Bernhardt, R., Makower, A., Jänig, G.-R., and Ruckpaul, K. (1984)
Biochim. Biophys. Acta 785, 186-190.

Chang, Y.-Y., Brauer, D., and Wittmann-Liebold, B. (1978) FEBS Lett.
93, 205-214.

Haniu, M., Armes, L.G., Yasunoby, K.T., Shastry, B.A., and Gunsalus,
I.C. (1982) J. Biol. Chem. 257, 12664-12671.

Jänig, G.-R., Makower, A., Rabe, H., Bernhardt, R., and Ruckpaul, K.
(1984a) Biochim. Biophys. Acta 787, 8-18.

Jänig, G.-R., Makower, A., Rabe, H., Friedrich, J., and Ruckpaul, K.
(1984b) Biomed. Biochim. Acta 43, K17-K24.

Rossi, M. (1983) J. Med. Chem. 26, 1246-1252.

Tarr, G.E., Black, S.D., Fujita, V.S., and Coon, M.J. (1983) Proc.
Natl. Acad. Sci. USA 80, 6552-6556.

MECHANISM OF HYDROPEROXIDE-DEPENDENT ALCOHOLS' OXIDATION CATALYZED BY CYTOCHROME P-450 LM-2 AND OTHER HEMOPROTEINS

METELITZA, D.I. and PLYUGACHYOVA, E.I.

Institute of Bioorganic Chemistry,
BSSR Academy of Sciences, Minsk, USSR

INTRODUCTION

The great role of hemoproteins in the metabolism and oxida tion of alyphatic alcohols is well known and many times documented (Ingelman-Sundberg,and Iohansson,1981,1984;Metelitza, 1982; Morgan et al,1982).In the oxidation of ethanol by micro somes and reconstituted systems including the varions forms of cytochrome P-450,the reductase and NADPH,the significant role of active radicals has been shown by inhibitory analysis. Free hydroxyl radicals,generated by the microsomes and recon-stituted systems from H_2O_2,that is always there present,appe-ar to mediate the slow oxidation of ethanol and other alypha-tic alcohols.Cytochrome P-450 isozyme 3a,isolated from hepa-tic microsomes of rabbit treated chronically with ethanol,was found to have a unique alcohol metabolizing ability (Morgan et al,1982).It is shown that the $HO^{\bullet}$ scavenger DMSO did inhi-bit ethanol oxidation by P-450 LM-3a.However,it is clear from the kinetic analysis that the inhibition results from competi tion between ethanol and DMSO at the active site (Morgan et al,1982).The P-450 LM-3a-catalyzed alcohol oxidation occurs at the P-450-active site and not as a result of interaction with $HO^{\bullet}$ in solution.The rat liver microsomes (Rahimtula and O'Brien,1977) and cytochrome P-450 LM-2 (Miwa et al,1977) ca-talyze the hydroperoxide-dependent oxidation of ethanol.But the mechanism of hydroperoxide-dependent alcohols' transfor-mation was not yet studied.

In the present paper we report systematic kinetic study of cumene hydroperoxide (CHP)-dependent oxidation of methanol a-nd ethanol catalyzed by cytochrome P-450 LM-2,cytochrome C, myoglobin and hemoglobin and mechanism of the inhibitory ac-tion of well-known antioxidants - 1-Naphthol,thiourea,manni-tol and DMSO-in these systems.The results show that the both alcohols are oxidized in the systems "CHP-hemoprotein" by ra-dical-chain mechanism.All the antioxidant used inhibited com-petitively alcohols' oxidation interacting with radicals $RO_2^{\bullet}$, $RO^{\bullet},HO_2^{\bullet}$ and $HO^{\bullet}$ propagating the chain.Any sings of non-radical (oxenoid) reactions are absent in CHP-dependent oxidation of both alcohols in all the systems "CHP-hemoprotein-ROH".

MATERIALS AND METHODS

Cytochrome P-450 LM-2,purified to homogeneous state,cytochrome C of horse heart and hemoglobin,four-fold recrystallized,from "Reanal" (Hungary),myoglobin of whale heart from "Serva" (FRG) were used.The concentrations of hemoproteins were spectrophotometrically determined using the molar extinction coefficients 114 (435 nm),119 (430 nm),109 (408 nm) and 110 (45o nm) $mM^{-1}cm^{-1}$ for myoglobin,hemoglobin,cytochrome C and carbonyl complex of reduced P-450 LM-2 correspondingly.Methanol,ethanol and DMSO were destilled.CHP concentration was determined by iodometric titration.The methanol oxidation occurs at 37° in phosphate buffer (pH 7,0) with participation of myoglobin,hemoglobin and cytochrome C.The ethanol oxidation was studied at 37° in the same buffer with cytochrome P-450 (pH 7,4) and hemoglobin (pH 7,0).Aldehyde formation from alcohols was spectrophotometrically measured by method (Ingelman-Sundberg,Iohansson,1982) for acetaldehyde and method (Nash,1953) for formaldehyde.In all cases,the reactions were characterized by initial rates of aldehyde formation in $M.s^{-1}$.

RESULTS AND DISCUSSION

Upon addition of CHP into buffered solutions of hemoproteins a rapid complex formation of hydroperoxide with proteins occurs.From the difference spectra of these complexes,the dissociation constants K_s were determined those equal 0,26 and o,o83 mM for myoglobin and cytochrome C (37°),0,22 and 0,12 mM for hemoglobin and cytochrome P-450 (20°).All hemoproteins formed the complexes with both alcohols also.The dissociation constants of these complexes are much higher as those for complexes "CHP-hemoproteins".The CHP and both alcohols are the competiting ligands,but in our experimental conditions,the complexes "CHP-hemoproteins" are preferencially formed.Based upon our data,the transformation of the hemoprotein complexes with CHP may be described by the following general scheme (Metelitza,1984):

$$Fe^{3+}L^{\cdot\cdot} \quad + \quad ROOH \rightleftharpoons \left[L^{\cdot\cdot}Fe^{3+}...\underset{H}{OOR} \right] \quad (1)$$

$$\left[L^{\cdot\cdot}Fe^{3+}...\underset{H}{OOR} \right] \longrightarrow L^{\cdot\cdot}Fe^{2+} + RO_2^{\cdot} + H^+ \quad (2^x)$$

$$\longrightarrow L^{\cdot}Fe^{3+} + RO^{\cdot} + HO^- \quad (2^{xx})$$

$$\longrightarrow L^{\cdot}Fe^{3+} + HO^{\cdot} + RO^- \quad (2^{xxx})$$

$$L^{\cdot\cdot}Fe^{2+} \quad + ROOH \longrightarrow L^{\cdot\cdot}Fe^{3+} + RO^{\cdot} + HO^- \quad (2^+)$$

The complexes "Fe^{3+}-ROOH" in the absence of redox-active additivities are decomposed to form radicals via reactions $2^x,2^{xx}$, 2^{xxx} and 2^+.Reactions $2^x,2^{xx},2^{xxx}$ may be rate-limiting stages for the whole process,since the rate formation of complexes "Fe^{3+}- HOOR" is very high.In correspondence with above sited scheme,the radical initiation rate in systems "Fe^{3+}- HOOR" may be described by the equation:

$$V_i = \frac{\gamma \cdot k_2 \left[Fe^{3+}\right]_o \left[ROOH\right]_o}{\left[ROOH\right]_o + \frac{k_{-1} + k_2}{k_1}} \quad ,$$

where $k_2 = (k_{2}x + k_{2}xx + k_{2}xxx)$ and $\gamma \leq 1$ determines the com
plex part,decomposed by radical way.The liberated radicals $r^{\bullet}$
($HO^{\bullet}$,$RO^{\bullet}$, $RO_2^{\bullet}$) initiate the methanol and ethanol oxidation.

The initial rates of alcohol oxidation were proportional to
the hemoprotein content and increased with the CHP and substrate concentrations,achieving the limiting values.In Table 1,
kinetic parameters of both alcohols' oxidation at 37° are sum
marized.The effective catalytic constants were obtained by relation of initial reaction rates to the hemoprotein concentration.

Table 1

The kinetic parameters of alcohols' oxidation by CHP at 37°:
CHP - 8,0 mM, MeOH - 1,0 M, EtOH - 0,8 M.

Alcohol	Hemoprotein and its concentra-tion in μM	$10^7 \cdot V$, $M \cdot s^{-1}$	$10^3 K_m^{CHP}$, M	$10^4 k_{cat}$, s^{-1}
MeOH	Myoglobin,70	0,30	4,4	4,3
	Cytochrome C,70	0,70	1,7	10,0
	Hemoglobin,50	1,00	0,9	20,0
EtOH	Hemoglobin,20	10,60	19,0	530,0
	Cyt. P-450 LM-2, 0,5	0,33	0,02	660,0

As it is shown from Table 1,the hemoproteins are poorly active biocatalysts of hydroperoxide-dependent oxidation of al-
cohols.Cytochrome P-450 LM-2 is the most active in the ethanol oxidation.

The addition of the inhibitors to systems "CHP-hemoprotein
-ROH" decreases the oxidation rates of both alcohols and their conversions.All inhibitors act by competitive manner in all
the cases.From the dependencies of reversed initial rates of
inhibited alcohol oxidation on initial inhibitors' concentrations,the inhibition constants were determined (Table 2).

Table 2.

The conditions and constants of inhibiting alcohols' oxidation.

Systems	Inhibitor and its con-centration range, mM		$10^3 K_i$, M
Cyt. P-450 LM 2 - EtOH	1-Naphthol	0,1-0,1	0,23
	Thiourea	0,5-5,0	0,12
	Mannitol	0,1-1,0	1,66
	DMSO	0,1-14,0	13,50

Hemoglobin – EtOH	1-Naphthol Thiourea DMSO	0,1–0,8 0,6–0,8 80 – 600	0,83 5,40 530
Myoglobin – MeOH	1-Naphthol Thiourea	0,1–0,8 0,4–8,0	0,12 8,0
Hemoglobin – MeOH	Thiourea DMSO	5,0–30 40–700	22 46

From all the antioxidants used, only 1-Naphthol formed the complex with cytochrome P-450 LM-2 that has $K_s = 0,4$ mM in buffer (pH 7,4) at $22°$. The inhibiting action of the antioxidants are decreased as follows: 1-Naphthol> thiourea> mannitol> DMSO.

Based upon data obtained, the following scheme of hydroperoxide-dependent oxidation of alyphatic alcohols was proposed (Plyugachyova, Metelitza, 1985):

$$r^• \ + \ RCH_2OH \longrightarrow rH \ + \ RCH^•OH \quad (3)$$
$$RCH^•OH \ + \ O_2 \longrightarrow RCH(OH)OO^• \quad (4)$$
$$RCH(OH)OO^• + RCH_2OH \longrightarrow RCH(OH)OOH \ + RCH^•OH \quad (3^x)$$
$$RCH(OH)OOH \longrightarrow CH_3CHO \ + H_2O_2 \quad (5)$$

In propagating stages of the process, the radicals $HO_2^•$ are formed from a product of reaction – H_2O_2 – by well-known Haber-Weiss mechanism. The chain termination occurs in accord with the known reactions:

$$2\ r^• \longrightarrow products \quad (6)$$
$$r^• + InH \longrightarrow rH \ + \ In^• \quad (7)$$
$$r^• + Fe^{2+} \longrightarrow r^- \ + Fe^{3+} \quad (8)$$

where InH is inhibitor, Fe^{2+} – a reduced form of hemoproteins. In the inhibitor presence, the most important chain termination way is the reaction 7. In the quasi-steady-state conditions, the initiation rate is near to the chain termination rate. Thus, the steady-state concentration of active radicals is described by the following equation:

$$[\overline{r^•}] \approx \frac{\gamma \cdot k_2 \cdot [Fe^{3+}]_o \cdot [ROOH]_o}{\left\{[ROOH]_o + (k_{-1} + k_2/k_1)\right\} f \cdot k_7 [\overline{InH}]_o}$$

where f – a stoichiometric coefficient of inhibition, meaning a number of the radicals, terminating on one antioxidant molecule. At the great CHP concentration, the inhibiting oxidation rate of alcohols is described by the equation:

$$V_o = k_3 \cdot [EtOH]_o \cdot [\overline{r^•}] \approx \frac{\gamma k_3 \cdot k_2 \cdot [Fe^{3+}]_o [EtOH]_o}{f \cdot k_7 \cdot [InH]_o}$$

The above cited equation is in full accordance with all the kinetic dependencies of inhibited oxidation of alcohols catalyzed by hemoproteins. An analysis of inhibition constants and an ability of various antioxidants to react with the different radicals show that free $HO^•$ radicals appear to mediate the slow oxidation of ethanol catalyzed by cytochrome P-450 LM-2.

In the cases of other hemoproteins,the role of HO˙ radicals
is much lesser,since the specific part of the reaction 2[XXX]
in the chain initiation process is very small as compared
with the cytochrome P-450 LM-2-system.Any signs of non-radical (two-electron or oxenoid) reactions are absent in CHP-dependent oxidation of alcohols in all the systems "CHP-hemoproteins".

REFERENCES

Ingelman-Sundberg,M.,and Iohansson (1981) The mechanism of
cytochrome P-450-dependent oxidation of ethanol in reconstitu
ted membrane vesicles. J. Biol. Chem.,256,6321-6326.
Ingelman-Sundberg,M.,and Iohansson,I. (1984) Mechanism of
hydroxyl radical formation and ethanol oxidation by ethanol-
inducible and other forms of rabbit liver microsomal cytochrome P-450. J. Biol. Chem.,259,6447-6458.
Metelitza,D.I. (1982) Oxygen activation by enzyme systems.
(Russ.) p. 167-193, Nauka, Moscow.
Metelitza,D.I. (1984)   Models of oxidoreductases.(Russ.),
p.122-137,197-223, Nauka and Tekhnika, Minsk.
Miwa,G.T.,Levin,W.,Thomas,P.E.,and Lu,A.Y.H. (1978) The
direct oxidation of ethanol by a catalase and alcohol dehyd-
rogenase-free reconstituted system containing cytochrome P-
450. Arch. Biochem. Biophys.,187,464-475.
Morgan,R.T.,Koop,D.R.,and Coon,M.J. (1982) Catalytic activity of cytochrome P-450 isozyme 3a isoluted from liver microsomes of ethanol-treated rabbits. J. Biol. Chem.,257,
13951-13957.
Nash,T. (1953) The colorimetric estimation of formaldehyde
by means of the Hantsch reaction. Biochem. J.,55,416-423.
Plyugachyova,E.I.,and Metelitza,D.I. (1985) Mechanism of
hydroperoxide-dependent methanol oxidation with participation
of hemoproteins. Biokhimiya (Russ.),in press.
Rahimtula,A.D.,and O'Brien,P.J. (1977) The role of cytochrome P-450 in the hydroperoxide-catalyzed oxidation of alcohols by rat liver microsomes. Eur. J. Biochem.,77,201-208.

AN ENTROPY-ENTHALPY COMPENSATION AMONG BACTERIAL, MICROSOMAL
AND MITOCHONDRIAL CYTOCHROME P-450'S

Yung-Yuan Huang,* Takayuki Hara,*
Stephan Sligar,** Minor J. Coon,*** and Tokuji Kimura*[†]

*Department of Chemistry, Wayne State University, Detroit, MI 48202

**Department of Biochemistry, University of Illinois, Urbana, Ill 61801

***Department of Biological Chemistry, University of Michigan
Ann Arbor, MI 48109

ABSTRACT

An optically transparent thin-layer electrode cell was used for
determination of the formal reduction potentials of bacterial, microsomal, and mitochondrial cytochrome P-450's. Temperature-dependent
experiments were carried out to obtain standard thermodynamic parameters of P-450's for the first time. All cytochromes tested have a
relatively large negative entropy, suggesting large conformational
changes in these proteins upon reduction. The extent of these changes
appears to be comparable to that found with cytochrome c. In addition,
we observed an entropy-enthalpy compensation among four P-450's. The
correlation appears to be weaker than that with cytochrome c from
various sources.

INTRODUCTION

The purpose of this study is to see the functional similarity and
dissimilarity among bacterial, microsomal, and mitochondrial cytochromes
in terms of their common properties of electron transfer reaction from
the ferric to ferrous state. We have measured here their formal reduction potentials at different temperatures to collect their standard thermodynamic parameters. As a result, we observed an enthalpy compensation effect among P-450's tested, although the extent of the effect was small
relative to that of cytochrome c from various sources.

BV and IC were obtained from K and K. BV was recrystallized twice
from cold methanol by the addition of ether. IC was recrystallized
twice from water by the addition of 2-propanol. Glucose oxidase (Type
VII) and catalase (thymol-free) were purchased from Sigma.

P-450$_{scc}$ (Hsu, 1984), P-450$_{11\beta}$ (Suhara et al., 1978), P-450$_{cam}$ (Yu
et al., 1974) and P-450$_{LM_4}$ were prepared according to the reported
methods with some modifications. The optically transparent thin layer
electrode cell was prepared as described previously (Huang and Kimura,
1983). The details of calculations were reported elsewhere (Huang and
Kimura, 1984).

---
† To whom all correspondence should be addressed
∴ This study was supported by grants from the National Institutes of
Health (AM-12713 to T.K., GM-31756 to S.S., and AM-10339 to M.J.C.).

RESULTS

In order to obtain thermodynamic parameters of the oxidation-reduction reactions, we have measured the potentials at different temperatures where P-450 is stable. P-450$_{cam}$ displayed a distinct break point at 30° C. P-450$_{LM_4}$, and P-450$_{scc}$ showed a lesser dependence on temperature, being similar to that of P-450$_{cam}$ below 30° C. Since P-450$_{11\beta}$ was only stable below 10° C, we have determined the parameters at both 10° and 2° C. These results are summarized in Table I.

When ΔH° was plotted against ΔS°, a linear line was obtained with a correlation coefficient of 0.976, suggesting a compensation effect of ΔS° against ΔH° (Figure 1).

DISCUSSION

For the first time, we could obtain the reduction potential of P-450$_{11\beta}$ and the values of ΔG°, ΔH°, and ΔS° from temperature-dependent effects of four P-450's. The large negative values of ΔS° for P-450's (-20.6 ~ -54.7 eu) may imply large conformational changes of these proteins upon reduction. Cytochrome c's from various sources have a large negative value of ΔS°, which ranges from 0 to -40 eu (Huang and Kimura, 1984). Therefore, the distribution (34 eu) of ΔS° values of P-450's from bacteria to mammalian appears to be similar to that (40 eu) of cytochrome c's, based on relatively limited amounts of data points. It is more difficult to explain the large negative value of ΔH°, namely, for P-450$_{cam}$. We can point out, however, some possibilities such as the enhancement of the π-back bonding of ferrous ion. If this is the case, the degree of π-back bonding varies considerably from P-450$_{cam}$ to P-450$_{LM_4}$. When ΔH° values were plotted against ΔS° values, we could obtain a linear relationship among P-450's. The correlation coefficient was 0.976. Similar plots for various cytochromes c gave a linear relationship with a correlation coefficient of 0.986 (Huang and Kimura, 1984), leading us to conclude an entropy-enthalpy compensation effect among cytochrome c's. For P-450's, the correlation among compensation effects appears to be weak compared with cytochromes c. The meaning of this difference is difficult to understand at present. However, one can say that in order to maintain ΔG° as constant as possible cytochrome c's strongly compensate ΔH° with ΔS°. For P-450's, the constancy of ΔG° may not be a crucial factor, and the compensation of ΔH° with ΔS° becomes, hence, weak. In the present study, availability of P-450's was limited. It is of interest to see whether yeast P-450's may have ΔH° and ΔS° values in between bacteria and mammalian.

When we calculated the compensation factor (β) or isokinetic temperature from the slope, the value of 433° K was obtained. The β-value for cytochromes was 264° K. For most entropy-enthalpy compensation processes in chemistry, the β-values lie in a relatively narrow range between 250 and 315° K (Lumry and Rajender, 1970). Our value of 433° K appears to be very high, but the literature values do not include biological oxidation-reduction reactions. The simple comparison is premature due to lack in reported values.

Table I.   Thermodynamic Parameters of P-450's at 25° C

P-450	ΔG^0	ΔH^0	ΔS^0	$E^{0'}$
	Kcal/mol	Kcal/mol	en	mV vs NHE
P-450$_{cam}$	2.75	- 13.6	- 54.7	- 128 ± 1
P-450$_{11\beta}$	6.92	- 4.08	- 36.6	- 286 ± 5
P-450$_{scc}$	6.50	0.36	- 20.6	- 285 ± 7
P-450$_{LM_4}$	8.39	- 0.13	- 28.6	- 360 ± 2

ΔG^0, ΔH^0, and ΔS^0 values were calculated as described in "MATERIALS AND METHODS". The value of P-450$_{11\beta}$ was calculated by extrapolating to 25° C, where the cytochrome is unstable and the determination was experimentally impossible.

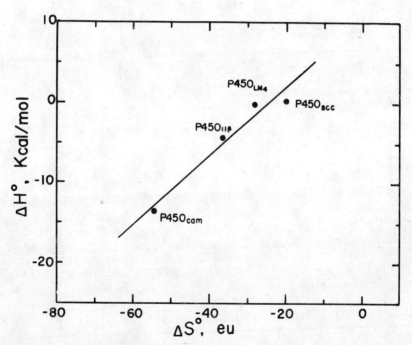

Figure 1: ΔH^0 - ΔS^0 diagrams for reduction reaction of various cytochrome P-450's. The ΔH^0 and ΔS^0 values of various P-450's calculated from Table III were plotted as ΔH^0 - ΔS^0 diagram. The linearality with a correlation coefficient of 0.9763 gave a slope of 433° K. The value for P-450$_{cam}$ in the low temperature region was used to obtain the best correction. When that, in the high temperature region was used, the correlation coefficient of 0.5714 was calculated.

6 Vereczkey

REFERENCES

Hsu, D. K. (1983), Ph.D. Thesis, Wayne State University, Detroit

Huang, Y.-Y. and Kimura, T. (1983)
    Anal. Biochem. <u>133</u>, 385-393

Huang, Y.-Y. and Kimura, T. (1984)
    Biochemistry, <u>23</u>, 2231-2236

Lumry, R. and Rajender, S. (1970)
    Biopolymers <u>9</u>, 1125-1127

Suhara, K., Gomi, T., Sato, H., Itagaki, E., Takemori, S., and
Katagiri, M. (1978)
    Arch. Biochem. Biophys. <u>190</u>, 290-299

Yu, C.-A., Gunsalus, I. C., Katagiri, M., Suhara, K., and Takemori, S.
(1974)
    J. Biol. Chem. <u>249</u>, 94-101

THE FREE RADICAL EVOKING CAPACITY OF HEPATOTOXIC AGENTS (ETHANOL AND CARBONE TETRACHLORIDE) OXIDIZED BY MICROSOMAL CYTOCHROME SYSTEM AND THE PROTECTIVE EFFECT ON THESE DAMAGES BY SCAVANGERS

Jávor,T., Horváth,Tünde, Wittmann,I., Nagy,S., Deli,J., Balogh,Eszter[+] Kádas,I.[+]

First Department of Medical University, Baranya County Hospital[+] Pécs, Hungary

We have some reason to explain the liver damaging effect of both carbontetrachloride and ethanol as chemical compounds undergone biotransformation during which highly reactive free radicals emerge giving rise to membrane destruction (5,6).

Intoxating doses of different compounds lead to liver injuries differing substancially from each other, there is no uniform way how to compare their effects. The routine liver tests are not sensitive enough (2). Supposing the free radicals as the final pathway and the products of lipid peroxidation as the pathochemical result we wanted to throw some light into the biochemical processes of injury. A series of light microscopic examination helped us to relate the biochemical findings to the pathological changes.

In our experiments on rats we wanted to substanciate the free radicals' liver damaging effect by the supposed protective effect of some chemical compounds which are well known as free radical scavengers: (+)-cyanidanol-3, allopurinol, MTDQ dimer and MTDQ monomer (1,3,4,11).

MATERIAL AND METHODS

Female Wistar strain rats with about 2oo g body weight were treated either with carbontetrachloride or ethanol. Both treated groups were divided into subgroups each of them were pretreated with either physiological saline resp. sunflower oil as controls, or with different free radical scavengers. The ethanol treatment lasted for three days and the carbontetrachloride treatment for one day. The pretreatment started 2 hours before the administration of injurious compound. 24 hours after the last treatment the animals were killed by neck dislocation. Table 1 shows the schematic diagramm of experimental procedures.

Control groups				Protected groups			

CCl$_4$ model

1.	2.	3.	4.	5.	6.	7.	8.
phys. NaCl + oil	phys. NaCl + CCl$_4$	oil + CCl$_4$	CMC + CCl$_4$	MTDQ m. + CCl$_4$ (oil)	MTDQ d. + CCl$_4$ (oil)	allopurinol + CCl$_4$ (oil)	cyanidanol + CCl$_4$ (CMC)

ethanol model

phys. NaCl	phys. NaCl + ethanol	oil + ethanol	CMC + ethanol	MTDQ m. + ethanol (oil)	MTDQ d. + ethanol (oil)	allopurinol + ethanol (oil)	cyanidanol + ethanol (CMC)

Table 1: Experimental trials and groups

For testing the injury and the protective effect we determined the serum GPT, and histological examination was carried out (lo). Studiing free radicals the following parameters were estimated: malondialdehyde, lipid peroxides and diene conjugates (7,8,9).

Some of our evaluated data were estimated by scores, the others by quantitative data we decided to make a comparative correlation by using for each our results a common nonparametric rank correlation matematical evaluation method (Mann Whitney).

The table 2 shows the upper and lower values of the measured variables. According to these the biochemical variables show no significant differences in the different groups although these deviate more in the carbontetrachloride group.

parameters	measured values	
	in groups of CCl$_4$ model	in groups of ethanol model
diene conjugates	$(6,69 \pm 1,3o)lo^{-5}$ $(22,55 \pm 25,8o)lo^{-5}$ M/g prot.	$(5,35 \pm 1,42)lo^{-5}$ $(23,4o \pm 2o,77)lo^{-5}$ M/g prot.
lipid peroxide content	$(o,76 \pm 1,o4)lo^{-6}$ $(2,82 \pm 1,24)lo^{-6}$ M/g prot.	$(o,62 \pm o,5o)lo^{-6}$ $(4,o7 \pm 1,45)lo^{-6}$ M/g prot.
MDA content	$(3,2o \pm o,53)lo^{-7}$ $(6,52 \pm 3,1o)lo^{-7}$ M/g prot.	$(2,46 \pm o,27)lo^{-7}$ $(6,15 \pm 5,9o)lo^{-7}$ M/g prot.
serum GPT	$34 \pm 7,8$ $88 \pm 5,7$ IU	$16 \pm 7,7$ $29 \pm 4,5$ IU

Table 2: Maximal and minimal values of parameters characterizing the toxic effects of free radicals measured in all of groups

Ranking the different parameters the highest protection was achieved in the carbontetrachloride treated animals with allopurinol, some protective patterns were obtained by the quinoline compounds and by cyanidanol (table 3).

In the ethanol treated groups the histology revealed only minimal changes, here the diene conjugates exhibited the most relevant alteration. The highest protective effect could be atributed in this group also to the allopurinol. In our experiments the ethanol exerted a rather mild injury, perhaps it was the reason why we were unable to demonstrate a protective effect of quinoline compounds and of cyanidanol (table 4).

GROUPS	Parameters					
	serum GPT	histology	lipid peroxide content	MDA content	diene conjugates	perfect rank
1. oil+CCl₄	6	7	8	7	6	34
2. CMC+CCl₄	7	6	7	6	4	3o
3. phys.NaCl+CCl₄	5	8	3	8	7	31
4. MTDQ m.+CCl₄	8	4	4	4	1	21
5. MTDQ d.+CCl₄	4	3	5	3	8	23
6. cyanidanol+CCl₄	3	5	1	5	3	17
7. phys.NaCl+oil	1	1	6	1	5	14
8. allopurinol+CCl₄	2	2	2	2	2	1o

GROUPS	Parameters				
	serum GPT	diene conjugates	lipid peroxide content	MDA content	perfect rank
1. oil+ethanol	8	8	6	3	25
2. CMC+ethanol	6	4	8	6	24
3. MTDQ d.+ethanol	7	6	5	2	2o
4. cyanidanol+ethanol	1	5	7	4	17
5. MTDQ m.+ethanol	5	7	2	1	15
6. phys.NaCl+ethanol	4	2	1	8	15
7. phys.NaCl	3	1	3	7	14
8. allopurinol+ethanol	2	3	4	5	14

Table 3: Result of rank order correlation between CCl₄ treated groups

Table 4: Result of rank order correlation between ethanol treated groups

CONCLUSIONS

We studied the hepatotoxic effect of carbontetrachloride and of ethanol. Using the former a well observable microscopic injury was obtained. Using ethanol only minimal changes and some shift in diene conjugation were detectable.

The highest protective effect was obtained by allopurinol. It may be understood regarding the multiple effects of this compound. It is a potent inhibitor of microsomal oxydation system therefore it is able to slow down the transformation of carbontetrachloride to the more agressive carbontrichloride. As a xanthin oxidase inhibitor it stops the free radical formation along that pathway. Finally allopurinol itself behaves as a scavanger.

LITERATURE

1.Börzsönyi,M. et al (198o) Egészségügytudomány 24, 188
   Új antioxidáns protektív hatása a morfolin és nitrit
   együttes adását követő akut májkárosodásra.

2. Clampitt,R.B.(1978) Arch.Toxicol.Suppl.1, 1 An investigation into the value of some clinical biochemical tests in the detection of minimal changes in liver morphology and function in the rat.

3. Fehér,J. et al (1982) Br.J.exp.Path. 63, 394 Biochemical markers in carbon-tetrachloride- and galactosamine--induced acute liver injuries:The effects of dihydro-quinoline-type antioxidants.

4. Flavin,D.F. Lancet (in press) Allopurinol treatment for hepatitis.

5. Glende,E.A., Recknagel,Jr.and Ro.(1969) Exp.Mol.Path. 11, 172 Biochemical basis for the in vitro pro-oxydant action of carbon tetrachloride.

6. Lewis,K.O.(1982) Lancet 11, 188 Could superoxide cause cirrhosis?

7. Mair,R.D. and Hall,R.T.(1971) in "Organic Peroxid" (Swern D.ed.) Vol II, p.535.

8. Niehaus,W.G.Jr. and Samuelsson,B.(1968) Europ.J.Biochem. 6, 126 Formation of Malonaldehyde from Phospholipid Arachidonate during Microsomal Lipid Peroxidation.

9. Recknagel,R.O. and Ghoshal,A.K.(1966) Exp.Mol.Pathol.5, 413 Quantitative Estamination of Peroxidative Degeneration of Rat Liver Microsomal and Mitochondrial Lipids after Carbon Tetrachloride Poisoning.

lo. Reitman,S. and Frankel,S.(1957) Amer.J.Clin.Path. 28, 56 A Colorimetric Method for the Determination of Serum Glutamic Oxalacetic and Glutamic Piruvic Transaminases.

11. Slater,T.F.(1981) The Royal Society of Medicine International Congress and Symposium Series:Number 47.pp.11 (+)-Cyanidanol-3 inhibition of lipid peroxidation induced by hepatotoxic chemicals in the rat.

EFFECT OF LIPIDS AND SUBSTRATES ON THE PROPERTIES OF MICROSOME LIVER FERROUS CYTOCHROME P-450 LM-2

DAVYDOV,R.M.*,*KHANINA,O.Yu.* and UVAROV,V.Yu.**

Institute of Chemical Physics, USSR Academy of Sciences*, and
Second Moscow Medical Institute, Moscow, USSR**

At the present time the properties of ferric cytochrome
P-450 and its reaction have been well investigated (Ruckpaul
1984). It is accounted for in particular the sensitivity of
the absorption, MCD and ESR spectra to changes of oxidized cytochrome P-450 induced by action of various physical and chemical factors. Unlike ferric cytochrome P-450, the spectral
characteristics of reduced protein are weakly depended on substrates, lipids and temperature. Therefore the properties of
ferrous cytochrome P-450 have been studied much less than oxidized protein. Earlier we have shown (Greschner et al.1982)
that kinetics of carbon monoxide binding with ferrocytochrome
P-450 is sensitive to the macromolecular structure and so may
be used for investigation of the reduced cytochrome P-450 properties. In this work we studied the effect of lipids, type
I substrates, Tritone N-101, serum albumin and NADPH-cytochrome P-450 reductase on the kinetics of recombination of cytochrome P-450 LM-2 from rabbit liver microsomes with CO and
their influence on circular dichroism spectra of the protein.

According to flash photolysis measurements when
$[CO]/[protein]>5$ the ferrocytochrome absorbance change at 450
during its recombination with CO can be described by three
phase kinetics. The pseudo-first-order rate constants (k_i)
and phase weights (P_i) are given in Table 1. Similar data
have been obtained by stop-flow technique. The first-order
rate constants of two slower phases are linear on CO concentration. The rate constant of the faster phase is not linear
in the same scale (Fig.1). The phase weights are independent
of CO concentration.

The rate constants changes are reversible at temperature
range 5-30°C. The phase weigths are independent of temperature.
At temperature higher than 35°C the changes become irreversible while the spectral characteristics of the protein are not
changed. The repeated decreasing of temperature from 35°C to
10°C leads to changes of the rate constants of all the phases.

Experimental data may be accounted for the presence of a
few stable conformers of ferrocytochrome P-450 LM-2 with different reactivity. Multyphase kinetics of CO binding may be
also resulted from protein aggregation.According to (Guengerich et al.1980) purified cytochrome P-450 in aqueous solution forms associates from 6 to 9 monomers. Protein aggregation

can influence both protein conformation and CO access to heme iron (II). It may well be that these factors may be varied from the relative localization of the protein molecule in the aggregate (inside or outside of the aggregate).

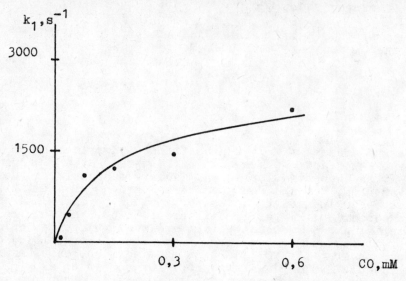

Fig.1. Dependence of the first-order rate constant of the faster phase of the reaction CO with cytochrome P-450 on CO concentration.

In the presence of type I substrates (hexobarbital, metphenetamin, benzphetamin) the recombination is described by biphasic bimolecular kinetics and relative quantum yield of carboxycytochrome P-450 photodissociation is increased by 30-40%. The change of recombination kinetics could be interpreted as action of substrates on the protein structure and or protein desaggregation. On the one hand we found that these substrates weakly influence the circular dichroism spectra of ferrous cytochrome P-450 LM-2 and its carboxycomplex in UV and Soret regions. On the other hand the increasing of the quantum yield of carboxycytochrome photodissociation at the presence of substrates indicates their effect on the stereo-chemistry of Fe-Co bound (Shimada 1979). It is supported by the results of the independent IR measurements of the carboxycytochrome P-450 (Böhm et al. 1980). In this work it was shown that type I substrates influence the CO stretching frequency of carboxycytochrome P-450. Thus all these results show the substrates effect on the stereochemistry of heme and its microenvironment.

The incorporation of the ferrous cytochrome P-450 into liposomes prepared from microsomal phospholipids results in the decreasing of the kinetic phase number to 2 and change of the rate constants (Table 1). These data show that the lipid microenvironment essentially influences the structure of protein conformers and the equilibrium between them. According to the circular dichroism data reconstruction of ferrous cytochrome P-450 into microsomal lipid vesicules leads to approximately 5% decreasing of α- spiralization. It is possible that the observed cytochrome structure and kinetics of recombination are due to protein desaggregation in liposomes. This proposal agrees with the fact that rotational mobility of reduced cytochrome P-450 in lipoproteid complexes is much more than in solution (Gut et al. 1983). The detergent effect on the recombination kinetics supports this point. In micellar solution of 0,2% Tritone N-101 where protein is partly desaggregated the kinetics of CO binding is similar to that at the presence of liposomes. It is possible that the structural and kinetic changes of cytochrome P-450 in liposomes are due to the protein microenvironment.

Table 1. Kinetic parameters of the recombination reaction of cytochrome P-450 with CO 0,37 mkM cytochrome P-450, 60 mkM CO, 15°C, pH=7,3.

Additions	$k_1(s^{-1})$ (P_1)	$k_2(s^{-1})$ (P_2)	$k_3(s^{-1})$ (P_3)
–	990 ± 250 (0,16)	170 ± 40 (0,64)	19 ± 3 (0,20)
Metphenetamin		210 ± 50 (0,64)	34 ± 4 (0,25)
Benzphetamin		175 ± 40 (0,74)	30 ± 4 (0,26)
Tritone N-101		260 ± 50 (o,67)	35 ± 4 (0,28)
Albumin	380 ± 100 (0,43)	100 ± 20 (0,48)	11 ± 3 (0.09)
Liposomes		250 ± 50 (0,51)	60 ± 6 (0,48)

Recently it has been shown that bovin albumin and NADPH-cytochrome P-450 reductase form complexes with ferric cytochrome P-450 and change its activity (Uvarov 1985). So it was of interest to study the influence of the redox state of cy-

74

tochrome P-450 on its interaction with the proteins. We found that at the presence of albumin some of the rate constants are decreased. This in particular can be accounted for the interaction with albumin. NADPH-cytochrome P-450 reductase does not essentially influence the kinetics of CO binding with cytochrome P-450.

Thus experimental data reveal that highly purified ferrous cytochrome P-450 LM-2 consists of three conformers with various reactivity. Lipid microenvironment and type I substrates markedly influence the structure of individual conformers and the equilibrium between them. We suggest that the observed effects may be important when the interaction between reduced cytochrome P-450 and molecular oxygen takes place.

REFERENCES

Böhm,S., Jänig, G.R., Fabian, H., Ruckpaul, K.(1980) "Characterization of The Active Site of Ferrous Carbonyl Complexes of Highly Purified Forms of Liver Microsomal Cytochrome P-450 by Infrared Spectroscopy" in Biochemistry, Biophysics and Regulation of Cytochrome P-450" Sweden, June 15-19, p.144.

Greshner, Z., Davydov, R.M., Jänig, G.R., Ruckpaul, K., Blumenfeld, L.A.(1982) "Substrates Effect on the Kinetics of the Reaction of Purified Microsomal Cytochrome P-450 LM-2 with CO". J.Phys.Chem.(in Russian) 56, 2884-2887.

Guengerich, F.P., Holladay, L.A., Kaminski, L.S.(1980) "Association of Cytochrome P-450 with Itself and Other Microsomal Enzymes" in "Biochemistry, Biophysics and Regulation of Cytochrome P-450" Sweden, June 15-19, 26.

Gut,J., Richter,C.,Cherry,R.J., Winterhalter, K.N, Kawato (1983). J.Biolog.Chem., 258, 8588-8594.

Ruckpaul, K., Rein,H.(1984) "Cytochrome P-450" Akademie-Verlag, Berlin.

Shimada,H., Iizuka,T.(1979) "Correlation Between the Quantum Yields of Photodissociation and C-O Stretching Frequencies of Carbon Monoxide Hemoproteins". FEBS Letters, 98, 290-294.

Archakov, A.I., Uvarov, V.Yu.(1985). "Role of Intermolecular Interaction in the Formation of Cytochrome P-450 Catalytically Active State". In present collection.

THE STOICHIOMETRY OF NADPH OXIDATION IN LIVER MICROSOMES. ELECTRON FLOW
DISTRIBUTION TO VARIOUS OXIDASE REACTIONS CATALYZED BY CYTOCHROME P-450

A.A.Zhukov, A.I.Archakov

Institute of Physico-Chemical Medicine, Moscow, USSR

Cytochrome P-450 (P-450) is classified as an external monooxygenase
which implies that the rates of consumption of the cosubstrates NAD(P)H and
dioxygen be equal to the rate of an organic substrate oxidation. With liver
microsomal P-450, however, it is difficult to gain experimental proof to
this stoichiometry: the amount of NADPH as well as dioxygen consumed exceeds
that of the substrate oxidized (Stripp et al., 1972; Cohen and Estabrook,
1971). Excess NADPH and dioxygen has been shown to be utilized for the formation of hydrogen peroxide and superoxide anion (Gillette et al., 1957;
Nordblom and Coon, 1977; Ullrich and Kuthan, 1980). It is not clear, however, if hydrogen peroxide can be formed directly via 2-electron reduction
of the enzyme-bound dioxygen in addition to that formed via O_2^- dismutation.
It is also not known if P-450 is capable of catalyzing complete 4-electron
reduction of the dioxygen molecule to yield two molecules of water. Earlier
we have demonstrated the direct H_2O_2 and water formation to occur during
NADPH free oxidation in liver microsomes (Zhukov and Archakov, 1982). This
paper deals with the further investigation on the stoichiometry of the
P-450-catalyzed oxidase reactions and its regulation.

EXPERIMENTAL PROCEDURES

Microsomes were isolated from the liver of phenobarbital-pretreated
rabbits.
NADPH oxidation was followed by the decrease in absorbance at 340 nm.
Oxygen consumption rate was determined by means of polarography using the
Clark-type electrode. The rates of benzphetamine and dimethylaniline N-demethylation and aniline p-hydroxylation were measured by standard methods
(Nash, 1953; Schenkman et al., 1967). H_2O_2 was determined by the thiocya-
nate method (Thurman et al., 1972). Oxy-P450 concentration was calculated
using the extinction coefficient $\varepsilon_{440-500} = 42 \ mM^{-1} cm^{-1}$ (Werringloer and
Kawano, 1980). To determine the rate of O_2^- formation the superoxide dismu-
tase-sensitive rate of reduction of succinylated cytochrome c was calculated. The rate of O_2^- formation was estimated using xanthine/xanthine oxidase system for calibration.
The temperature was maintained at 37 $^{\circ}C$ throughout the experiments.

RESULTS AND DISCUSSION

Table 1 presents the data on the stoichiometry of NADPH oxidation in liver microsomes. If H_2O_2 resulted entirely from the dismutation of O_2^- the

TABLE 1. The stoichiometry of NADPH oxidation in liver microsomes

substrate added	Δ NADPH (a)	ΔO_2 (b)	ΔP (c)	ΔO_2^- (d)	ΔH_2O_2 (e)	$\dfrac{a-c-e}{b-c-e}$	$\dfrac{d}{e}$
none	25	18	0	9	11	2,0	0,8
0,5 mM benzphetamine	123	108	72	29	24,3	2,3	1,2
6 mM dimethylaniline	73	52	17	–	13,3	2,0	–
0,5 mM perfluorohexane	56	–	0	10,2	10,9	–	0,9

Presented are the mean rates in nmol/min per 1 mg protein. Incubation mixture contained: 25 mM tris-HCl, pH 7,6; 150 mM KCl, 10 mM $MgCl_2$, 1 mM NaN_3, 0,6 mM EDTA, 0,3 mM NADPH, 0,5 mg/ml protein and the substrate indicated. In the experiment with no substrate added the mixture contained: 10 mM K-phosphate, pH 7,8; 150 mM KCl, 1 mM NaN_3, 0,3 mM NaCN, 0,6 mM EDTA, 0,3 mM NADPH, 0,5 mg/ml protein. Dash means that the parameter was not measured. ΔP is the rate of accumulation of the oxidized substrate.

ratio $\Delta O_2 : \Delta H_2O_2$ would be two in accordance with the stoichiometry of dismutation reaction:

$$2O_2^- + 2H^+ = H_2O_2 + O_2$$

The table shows, however, that the ratio is about unity in all experiments. Thus, another way exists for H_2O_2 formation in addition to O_2^- dismutation, namely, the direct 2-electron reduction of P-450-bound dioxygen.

The table also shows that the sum of H_2O_2 formed and substrate oxidized fail to account for NADPH and oxygen consumed indicating the reaction of water formation to occur. Water may be formed either in an oxidase reaction:

$$O_2 + 2NADPH + 2H^+ = 2NADP^+ + 2H_2O \tag{1}$$

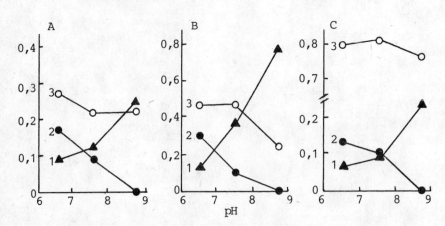

FIG. 1. pH-dependence of the ratio of P-450-catalyzed oxidase reactions. The incubation mixture contained: 25 mM tris-HCl, 10 mM $MgCl_2$, 150 mM KCl, 0,6 mM EDTA, 1 mM NaN_3, 0,3 mM NADPH, 0,5 mg/ml protein. The substrates added were: 0,5 mM benzphetamine (A), 3 mM aniline (B) and 0,5 mM perfluorohexane (C). The ordinate gives the portion of NADPH spent on the formation of O_2^- (1), H_2O_2 (2) and water (3).

or in a monooxygenase one:

$$NADPH + H^+ + O_2 + XH = NADP^+ + XOH + H_2O \qquad (2)$$

where XH is some endogenous substrate. The two mechanisms can be discriminated by calculating the stoichiometry of NADPH and O_2 consumption in water production reaction. The calculation presented in the table favours the complete 4-electron reduction of dioxygen as the way of water formation (eq. 1) rather than endogenous substrate oxidation (eq. 2).

The ratio of P-450-catalyzed oxidase reactions depends on the experimental conditions. Fig. 1 presents the influence of pH on this ratio. One can see that the ratio of 1- and 2-electron reduction reactions exerts clear pH-dependence: the 2-electron process predominates at low pH, whereas at high pH its contribution decreases as compared to that of the 1-electron reaction and reaches zero at pH 8,8. The increase in the contribution of O_2 formation reaction with increasing pH correlates with the rise in the steady state concentration of P-450 oxycomplex - the precursor of free O_2^- (Fig. 2A).

One of the objections to the possibility of H_2O_2 formation via 2-electron reduction of P-450-bound dioxygen was the failure to demonstrate synergism of NADPH and NADH in the reaction (Estabrook et al., 1979; Ullrich and Kuthan, 1980). To investigate the point in some detail we have studied the influence of pH on the interaction of NADPH- and NADH-dependent redox chains in H_2O_2 formation reaction. Fig. 2B shows that synergism is observed at low pH, but it gives place to the additive action and, moreover, competition with increasing pH. This observation is in a close agreement with the data on stoichiometry. Indeed, NADH by accelerating the 2nd reduction of the enzyme-bound dioxygen exerts synergism, when H_2O_2 is formed mainly via the

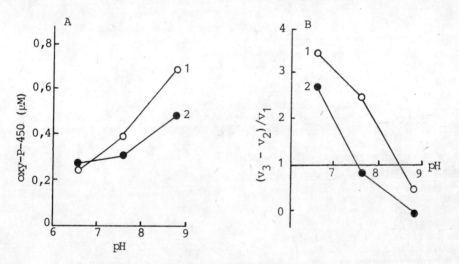

FIG. 2. pH-dependence of oxy-P-450 steady state concentration (A) and interaction of NADPH- and NADH-dependent redox chains in H_2O_2 formation reaction (B). The mixture composition is identical to that under Fig. 1 except that in measuring the oxy-P-450 concentration that of protein was 1 mg/ml. The substrates added were: 0,5 mM benzphetamine (1) and 0,5 mM perfluorohexane (2). v_1, v_2 and v_3 are the rates of H_2O_2 formation in the presence of 0,3 mM NADH, 0,3 mM NADPH and both cosubstrates together respectively.

direct 2-electron reduction of the former. On the other hand, the steady
state concentration of oxy-P-450 which decomposes to yeild O_2 has to decrease with accelerating transfer of the second electron. Hence, at high pH
when H_2O_2 is formed mainly via O_2^- dismutation the competition is observed.
The reason for the additive action of NADPH and NADH at neutral pH may be
the balance of accelerated direct and retarded indirect H_2O_2 formation under these conditions.

The contribution of water formation reaction shows no marked pH-dependence (Fig. 1). One can see, however, that the contribution rises sharply
in the presence of pseudosubstrate perfluorohexane, the portion of NADPH
spent on this reaction being about 80 per cent. In other words, the complex of P-450 and pseudosubstrate functions mainly as the 4-electron oxidase.

To summarize, the data obtained show P-450 capable of catalyzing three
types of of oxidase reactions in which it functions as the 1-, 2-, and 4-
electron oxidase. The relative contribution of each of the three reactions
to the entire process of microsomal NADPH oxidation depends on the substrate added and the composition of the incubation medium.

REFERENCES

Cohen B.S., Estabrook R.W. (1971) Microsomal electron transport reactions. 2. The use of triphosphopyridine nucleotide and/or reduced diphospho-
pyridine nucleotide for the oxidative N-demethylation of aminopyrine and
other drug substrates.- Arch. Biochem. Biophys. 143, 46-53.

Estabrook R.W., Kawano S., Werringloer J., Kuthan H., Tsuji H., Graf H.,
Ullrich V. (1979) Oxycytochrome P-450: its breakdown to superoxide for the
formation of hydrogen peroxide.- Acta Biol. Med. Germ. 38, 423-434.

Gillette J.R., Brodie B.B., La Du B.N. (1957) The oxidation of drugs by
liver microsomes: on the role of TPNH and oxygen.- J. Pharm. Exp. Therap.
119, 532-540.

Nash T. (1953) The colorimetric estimation of formaldehyde by means of
the Hantzch reaction.- Biochem. J. 55, 416-421.

Nordblom J.D., Coon M.J. (1977) Hydrogen peroxide formation and stoichiometry of hydroxylation reactions catalyzed by highly purified liver microsomal cytochrome P-450.- Arch. Biochem. Biophys. 180, 343-347.

Schenkman J.B., Remmer H., Estabrook R.W. (1967) Spectral studies of
drug interaction with hepatic microsomal cytochrome.- Mol. Pharmacol. 15,
428-438.

Thurman R.G., Ley H.G., Scholz R. (1972) Hepatic microsomal ethanol oxidation. Hydrogen peroxide formation and the role of catalyze.- Eur. J.Bio-
chem. 25, 420-430.

Ullrich V., Kuthan H. (1980) Autoxidation and uncoupling in microsomal
monooxygenations.- In: Biochemistry, Biophysics, and Regulation of Cytochrome P-450 (Gustafsson J.-A. et al., eds.), pp. 267-272, Elsevier, Amsterdam.

Werringloer·J., Kawano S. (1980) Ibid. pp. 359-362.

Zhukov A.A., Archakov A.I. (1982) Complete stoichiometry of free NADPH
oxidation in liver microsomes.- Biochem. Biophys. Res. Commun. 109, 813-
818.

A NEW APPROACH TO CYTOCHROME P-450 IMITATION: HEMOPROTEINS ENTRAPPED BY REVERSED MICELLES OF SURFACTANTS IN ORGANIC SOLVENTS

ERJOMIN, A.N. and METELITZA, D.I.

Institute of Bioorganic Chemistry,
BSSR Academy of Sciences, Minsk, USSR

INTRODUCTION

Results obtained by application of ^{31}P-NMR spectroscopy to rabbit liver microsomal membranes and a reconstituted cytochrome P-450 system show that part of the lipids exists in an inverted micellar state (so-called "lipidic particles") which may be indispensible for the structural integration of cytochrome P-450 (Stier et al, 1978). This non-lamellar phospholipid phase is presumably formed by acidic phospholipids (Stier et al, 1982). The "lipidic particles" are considered to be necessary for adaptation of cytochrome P-450 rotamers to the bilayer membrane.

Recently, a new mode of studying protein function has been introduced. It has been proposed that enzymes may be dissolved not in water but in colloidal solution of water in an organic solvent (Khmelnitzki et al, 1984). In such systems many enzymes retain their catalytic activity. The reversed micelles of surfactants with entrapped proteins simulate in vitro the conditions of enzyme action in vivo.

In this work, we have shown for the first time that the nature of the environment of the active center in model systems is distinctly reflected by its spectra. The full simulation of the spectral properties of P-450 is observed if the heme iron has the necessary axial ligands and a certain microenvironment of the active center exists. Such a microenvironment is created by the reversed micelles of the surfactants in organic solvents.

MATERIALS AND METHODS

Myoglobin (Mb) from whale heart in the oxidized form (Serva, FRG) was used. To obtain the reversed micelles in octane, sodium salt of di(2-ethylhexyl)sulfosuccinic acid (AOT) (Serva, FRG) was used. β-Mercaptoethanol (RSH) was obtained from Sigma (USA). Aniline, pyridine, N,N-dimethylaniline (DMA), and octane (Reakhim, USSR) were redistilled before use. The absolute and difference spectra of all systems in reversed micelles were recorded in thermostatically controlled cuvettes using a Specord UV VIS spectrophotometer (Carl Zeiss, GDR) at 31°C.

RESULTS AND DISCUSSION

The incorporation of Mb into the AOT micelles in octane is accompanied by the absorption maximum shift, which depends greatly on the micelles wetness. The absorption maximum is shifted from 415.5 to 407 nm with an increase of the content of either water or tris buffer in the micelles. By varying content of buffer in the AOT micelles, the equilibrium between the Mb spin states can be shifted:

$$MbOH_2 \text{ (HS, 407 nm)} \underset{+H^+}{\overset{-H^+}{\rightleftharpoons}} MbHO^- \text{ (LS, 415 nm)}.$$

In the tris buffer (pH 7.5) the dissociated form of RSH $HOCH_2CH_2S^-$ takes part in complex formation with Mb. This comp lex has the spectral parameters that are typical of hexacoordinated hemoproteins. The Mb reduction took place neither in the aqueous nor in the buffer solution. The absolute and difference absorption spectra of the Mb-RS$^-$ complexes in the AOT micelles differ greatly from those obtained in the solutions. In the absolute spectra of the complexes in AOT micelles, the absorption bands characteristic of high-spin hemoproteins - 388-390, 513-519, and 631-640 nm - appear. The Mb-RS$^-$ complexes in reversed micelles are rather stable.

In Table, the absorption bands of hemin complexes with dif ferent ligands in the reversed micelles, as well as the spectra of P-450, are summarized. The spectral characteristics of hemin complexes with RSH in the reversed micelles coincide with spectral parameters of high-spin P-450s.

TABLE. The spectral parameters of hemin and its complexes with various ligands in AOT (0.15 M) micelles. The hemin concentration is 0.34 μM and the micelles include 4% of 0.1 N NaOH solution.

Systems	Maxima of the absorption bands, nm				
	δ	γ	β	α	—
Hemin	360	398	579	–	600,693
Hemin + RSH (0.014 M)	–	388	512	–	633
Hemin + RSH (0.014 M) + aniline (0.162 M)	364	422	536	564	635
Hemin + RSH (0.014 M) + pyridine (0.123 M)	358	421	529	559	–
Hemin + RSH (0.014 M) + imidazol (0.018 M)	359	417	540	564	–
LS P-450	360	418	535	568	–
HS P-450	–	394	510	–	645

Our work showed that the spectral characteristics of Mb-RS$^-$ and hemin-RS$^-$ complexes in reversed micelles coincided with the spectral parameters of HS P-450, whereas the spectral cha racteristics of Mb-RS$^-$ complexes in solution are typical for LS forms of P-450. It may be suggested that the thiolate ligand, which is fixed on the lumenal surface of the micelles, very strongly interacts with the heme iron and displaces the iron away from the heme plane. A similar shift of the P-450 spin equilibrium in a microsomal membrane is caused by negatively charged phospholipids (Metelitza, 1982). The reversed AOT micelles as modulators of spin conversions of Mb-RS$^-$ complexes imitate one function of native membranes, the influence on the structure of a protein solubilized by them.

Upon addition of cumene hydroperoxide (CHP) to Mb solubilized by reversed AOT micelles, a rapid complex formation of hydroperoxide with protein occurs (Erjomin, Metelitza, 1983). This complex is characterized by difference spectrum with maxima 428 and 590 nm and minima 410 and 503 nm. The Mb-CHP complex is decomposed with time, accompanied by a decrease in the Soret band intensity. The transformation of Mb-CHP complex in the reversed micelles of surfactants may be described by the following scheme:

$$ROOH + \overset{..}{L}Fe^{3+} \rightleftharpoons \left[\overset{..}{L}Fe^{3+} \cdots \underset{H}{O}OR \right] \longrightarrow \begin{array}{l} \overset{..}{L}Fe^{2+} + RO_2^{\cdot} + H^+ \\ \overset{..}{L}Fe^{3+}OH + RO^{\cdot} \\ \overset{..}{L}Fe^{3+}OR + HO^{\cdot} \end{array}$$

The radicals $RO_2^{\cdot}$, $RO^{\cdot}$ and $HO^{\cdot}$ attack protoporphyrin IX and destroy it, which leads to a decrease in the Soret band intensity of hemoprotein. Thus, Mb retains its ability to react with CHP in micelles. CHP is poorly soluble in water and it is unlikely that hydroperoxide penetrates to the micelles lumen. The iron protoporphyrin of Mb is accessible to CHP. This suggests that the hydrophobic "pocket" of the heme either faces the inner surface of micelles or even that it is located in the hydrophobic layer formed by the aliphatic fragments of surfactant.

We have for the first time shown that the oxidative demethylation of tert.amines could be enhanced not only by the enzymes but also by the reversed micelles of a surfactant. The N,N-dimethylaniline demethylation in its reactions with oxygen and/or CHP in AOT (0.2 M) micelles in octane has been studied as an example of the micellar catalysis in the oxidative dealkylation of tert.amines. In the micellar system at 40°, the very rapid DMA demethylation occurs. In the absence of AOT in octane, the DMA demethylation by the dissolved oxygen was not observed at 40°. The reaction rate of the DMA transformation strongly depends from the water content in micelles and has the maximum at the water content of ~ 0.5% (v/v). With the CHP addition to the micellar system, the rate of DMA demethylation increases. In the absence of AOT, CHP does not react with DMA in octane at 40°. The oxidative demethylation of DMA exhibits the first-order kinetics with respect to the DMA concentration up to its value of 0.1 M. The initial rate of the DMA demethylation by the CHP increases with the oxidant concentration and approaches to the limiting value at [CHP] = 0.002 M.

The micelles concentrate the oxygen and DMA, thus enhancing the contact complex formation between the substrate and oxidant. We proposed the following scheme for the process in the AOT micelles:

$$PhN(CH_3)_2 \underset{O_2}{\overset{O_2}{\rightleftharpoons}} PhN(CH_3)_2 \rightleftharpoons \underset{+}{PhN}(CH_3)_2 \rightarrow \underset{+}{PhN}(CH_3)_2 + O_2^{\cdot-} \quad (1)$$

$$H^+ + O_2^{\cdot-} \rightleftharpoons HO_2^{\cdot}$$

$$\underset{+}{PhN}(CH_3)_2 + HO_2^{\cdot} \xrightarrow[-H_2O_2]{} PhN\overset{CH_2}{\underset{+}{\diagdown}}_{CH_3} \xrightarrow{+H_2O} PhN\overset{H}{\underset{CH_3}{\diagdown}} + HCOH$$

In the presence of CHP, together with reaction 1 the complex formation of DMA with CHP followed by its decomposition with the liberation of radicals occurs. The active radicals $RO_2^{\cdot}$, $RO^{\cdot}$ and $HO^{\cdot}$ oxidize the DMA in accordance with the scheme sited above. The autoinhibition of DMA demethylation might be accounted for by the radical mechanism of the process: product of the reaction – monomethylaniline – inhibits the process by its interaction with active radicals.

Assumingly, the stimulation of DMA reactions with oxygen and CHP by the reversed micelles is accounted for by the enhancing effect of the micelles on the decomposition rate of DMA complexes with an oxidant. Based upon our data, the participation of the membrane themselves in transformation of chemical compounds may be suggested as well.

REFERENCES

Erjomin, A.N., and Metelitza, D.I. (1983) Catalysis by hemoproteins and their structural organization in reversed micelles of surfactants in octane. Biochim. et Biophys. Acta, 732, 377-386.

Khmelnitzki, Yu.L., Levashov, A.V., Klyachko, N.L., Martinek, K. (1984) A microheterogeneous medium for chemical (enzymatic) reactions based upon colloidal solution of the water in organic solvents. Uspekhi khimii (Russ.) 53, 545-565.

Metelitza, D.I. (1982) Oxygen activation by enzyme systems. Nauka, Moscow, 254 p.

Stier, A., Finch, S.A.E., Bösterling, B. (1978) Non-lamellar structure in rabbit liver microsomal membranes. FEBS Letters, 91, 109-112.

Stier, A., Finch, S.A.E., Greinert, R., Höhne, M., and Müller, R. (1982) Membrane structure and function of hepatic microsomal cytochrome P-450 system. In: Liver and Aging. Ed. K. Kitane. Elsevier Biomed Press, p. 3-14.

SULFUR DEPROTONATION AS THE ORIGIN OF SIGNIFICANT ACTIVATION OF DIOXYGEN BY CYTOCHROME P-450: A QUANTUM CHEMICAL APPROACH

S.S.Stavrov, I.P.Decusar, and I.B.Bersuker

Institute of Chemistry, Academy of Sciences of
the MoSSR, Kishinev, USSR

The oxidation of substrates by cytochrome P-450 (P-450) takes place immediately after a one electron reduction of a complex of P-450 with O_2 (P-450-O_2). This reaction passes through the step of the O-O bond cleavage which is promoted by the dioxygen coordination to P-450 /Rein et al.1984/. The catalytic activity of P-450 is ordinary attributed to the coordination of the sulfur atom to the fifth coordination site of the iron atom /Rein et al.1984/.

The sulfur coordination leads to the occurence of extra absorption in optical spectra of P-450 complexes with series of ligands /Anderson et al.1983, Sono et al.1982/. Hanson /1979/ explanes the origin of the extra bands by a charge transfer transition from the sulfur lone pair $p_{\pm}(S)$ to the porphyrin $e(\pi^*)$ orbital. Jung /1985/ attributes these bands to $e(\pi_{ax}) \rightarrow b(\pi^*)$ transitions ($b(\pi^*)$ is a pure porphyrin orbital, whereas the $e(\pi_{ax})$ orbital is localized in the axial system). From our point of view both approaches are not deprived of shortcomings. In particular it follows from the paper of Hanson /1979/ that P-450-O_2 must display extra absorption bands, and this disagrees with experimental data. In the calculations by Jung /1985/ the Fe-S distances are essentially underestimated against the observed ones /Schappacher et al.1981, Ricard et al.1983/, and the effect of the cysteine group is ignored. The latter splits significantly the $e(\pi_{ax})$ orbital due to the symmetry reduction /Hanson 1979/. As a result, the $e(\pi_{ax}) \rightarrow b(\pi^*)$ has to be essentially splitted and this contradicts to the results of experiments on magnetic circular dichroism(MCD) which indicate the presence of a transition to a degenerate state in the region of 450 nm.

Up to date theoretical works run into difficulties in an attempt to find out the mechanism of a dioxygen activation by P-450, too. Thus, to explain the O-O bond cleavage after acception of an electron, it was suggested by Rein et al./1984/ that the dioxygen must form strong hydrogen bond with a certain amino acid of the apoenzyme. However, since the acception of an electron affects the order of the O-O bond slightly /Rein et al.1984/, it must be supposed that such a hydrogen bond is absent before the electron acception and takes place after that. It is clear, that in this case the hydrogen bond formation has to be connected with certain changes of the apoenzyme conformation, and it is very hard to relate such changes to the very high velocity of substrate oxidation following the electron acception immediately.

In the present paper the electronic structures of the active centers of P-450-O_2 and P-450-CO are obtained by quantum chemical calculations of their models $[FeP(SCH_3)O_2]^-$, 1, and $[FeP(SCH_3)(CO)]^-$, 2 (P=porphin). These calculations were performed in MO LCAO approximation considering additionally the energies of Coulomb repulsion of the sulfur electrons. This allows to explain both the essential dioxygen activation by P-450 and the peculiarities of the optical and MCD spectra of P-450 complexes with various ligands.

METHOD. The electronic structure calculations were performed in the MO LCAO approximation by charge iterative extended Hückel method in the form of Chumakov et al. /1982/. One of the limitations of the procedure used is a failure to consider (during the process of iterative convergence) electronic states, which are formed by totally and partially occupied MO's. Therefore, in order to find out the peculiarities of the electronic states with holes in some of the MO's, additional calculations must be performed. In the calculations we used data on Slater exponents and the geometry of the ferrous porphin from the paper of Hanson /1979/. Distances between the ferrous atom and axial ligands are given by Schappacher et al./1981/ and Ricard et al./1983/.

RESULTS AND DISCUSSION. The calculated energies of highest occupied (HO) and lowest unoccupied (LU) MO's of 1 and 2 are shown in Fig.1. The sulfur orbitals give significant contributions to the HOMO's of both complexes, the energy gaps between HOMO's and LUMO's in 1 and 2 being rather small. To investigate the structure of the ground electronic states of 1 and 2, the Coulomb repulsion energy for eletrons occupieng the sulfur HOMO $p_x(S)$ is evaluated. It is of the order of 7 eV for 1 and 8 eV for 2. It follows that the lowering of the energy of the system due to an electron transition from the $p_x(S)$ orbital to the lowest LUMO essentially exceeds the energy gap between the HOMO and LUMO. Consiquently, the states with single occupied $p_x(S)$ and $d_{xx}-\bar{\pi}_x(O_2)$ MO's for 1 and $p_x(S)$ and $e(\pi*)$ MO's for 2 are the ground ones.

To solve the problem of the ground state spin multiplicity, the configuration interaction with high energy excited states has to be considered /Kalkeren et al.1979/, and this cannot be done in the frames of the simple MO LCAO method. On the other hand the fact that the single occupied MO's are localized mainly at the sulfur atom and dioxygen in 1 or porphyrin ring in 2 allows to use the results of Anderson /1963/ to solve this problem. In his paper the exchange interaction of two paramagnetic centers through a bridge atom was considered, and it was shown that in the case of essential overlap of the orbitals of the two centers with the same orbital of the bridge atom the singlet state is the lowest one. Just this situation takes place in complexes under consideration. Indeed, the bridge orbital is formed by the ferrous d_{xz} one

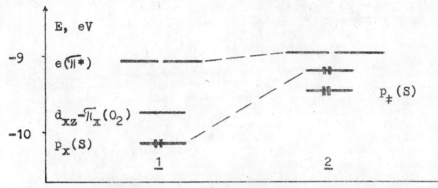

Fig.1. Calculated energies of HOMO's and LUMO's of $[FeP(SCH_3)O_2]^-$, **1**, and $[FeP(SCH_3)(CO)]^-$, **2**.

which essentially overlaps with the $p_x(S)$ and $\overline{\pi}_x(O_2)$ orbitals in **1** and with the $p_x(S)$ and $e(\overline{\pi}*)$ ones in **2**.

We conclude, that the ground states of the complexes **1** and **2** are the zero spin states with single occupied sulfur $p_x(S)$ orbital plus $d_{xz}-\overline{\pi}_x(O_2)$ or $e(\overline{\pi}*)$ MO's, respectively.

It follows from the said above that the electronic structure of the porphyrin in **2** is close to the electronic structure of porphyrin π anions. It means that the optical spectrum of **2** has to be similar to porphyrin π anion spectra. The fact that the latter are characterized by the split Soret band with a low frequency maximum near 450 nm /Lanese et al. 1972, Collman et al.1976/ explaines directly the occurence of such a feature in optical spectrum of P-450-CO. The temperature dependence of the MCD spectra of this latter /Shimizu et al.1975/ also supports the idea of a single occupied $e(\overline{\pi}*)$ MO, because it indicates the degenerate ground state of this compound. In P-450-O_2 the $e(\overline{\pi}*)$ MO is unoccupied, and the porphyrin electronic structure differs very little from that, for example, in carboxyhemoglobin, their optical spectra, hence, being similar. Analogously, the appearence of extra bands in the Soret region in complexes of P-450 with a number of ligands may be considered as due to the occupation of the $e(\overline{\pi}*)$ MO in the ground states of these complexes. The absence of such bands in other complexes can be explained by the presence of a metal-ligand LUMO which is lower in energy then the $e(\overline{\pi}*)$ MO and accepts the sulfur electron instead the latter.

The elucidation of the electronic structure of **1** allows also to elucidate the origin of the high catalytic activity of P-450. Indeed, the occupation of the $d_{xz}-\overline{\pi}_x(O_2)$ MO in P-450-O_2, even in the absence of the reducing electron, strongly activates the dioxygen due to the strong antibonding nature of $\overline{\pi}_x(O_2)$ orbital. Therefore even a short time stay of the reducing electron on the $d_{xz}-\overline{\pi}_x(O_2)$ MO results in a break of the O-O bond, because the presence of two electrons on this MO destroies the O-O bond immediately. Note that the electronic state with the double occupied $d_{xz}-\overline{\pi}_x(O_2)$ MO and single occupied $p_x(S)$ one should not necessary be the ground state of the reduced P-450-O_2. The only requirement for the process to proceed is that the electron falls in the $d_{xz}-\pi_x(O_2)$ MO during the reduction.

REFERENCES.

Anderson,P.W.(1963) Theory of magnetic exchange interactions: Exchange in insulators and semiconductors. In: Solid State Physics, 14, Academic Press, New York, 99-214

Andersson, L.A., Sono, M., and Dawson, J.H.(1983) Circular dichroism studies of low-spin ferric cytochrome P-450$_{CAM}$ ligand complexes. Biochim.Biophys Acta 748, 341-352

Chumakov, Ju.M., Dimoglo, A.S., and Bersuker, I.B.(1982) Programms for semiempirical calculations of electronic structure of molecules with taking account of their symmetry by methods of MWR and MWH. Zh.Strukt.Khim. 23, 182-183

Collman, J.P., Sorrell, T.N., Dawson, J.H., Trudell, J.R. Bunnenberg, E., and Djerassi, C.(1976) Magnetic circular dichroism of ferrous carbonyl adducts of cytochromes P-450 and P-420 and their synthetic models: Further evidence for mercaptide as the fifth ligand to iron. Proc.Natl.Acad.Sci.USA 73, 6-10

Hanson, L.K.(1979) Axial ligand effects on iron and manganese porphyrins: Extended Hückel calculations of cyt P-450 analogs and of O_2 binding to iron and manganese. Int.J.Qunt. Chem.Quant.Biol.Symp. 6, 73-87

Jung, Ch.(1985) Quantum chemical explanation of the "hyper" spectrum of the carbon monoxide complex of cytochrome P-450. Chem.Phys.Lett. 113, 589-596

Kalkeren, G.van, Schmidt, W.W., and Block, R.(1979) Superexchange in insulators: Comparison of different methods. Physica 97B, 315-337

Lanese, J.G. and Wilson, G.S.(1972) Electrochemical studies of zinc tetraphenylporphin. J.Electrochem.Soc. 119, 1039-1043

Rein, H., Jung, Ch, Ristau, O., and Friedrich, J.(1984) Biophysical properties of cytochrome P-450, analysis of the reaction mechanism – thermodynamic aspects. In: Cytochrome P-450 (Ruckpaul, K. and Rein, H., eds.) Academic-Verlag, Berlin, 163-249

Ricard, L., Schappacher, M., Weis, R., Montiel-Montoya,R. Bill, E., Gonser, U., and Trautwein, A.(1983) X-ray structure and Mössbauer properties of a synthetic analog for the active site of oxycytochrome P 450. Nouv.J.Chim. 7, 405-408

Schappacher, M., Ricard, L., and Weiss, R.(1981) Synthetic analogs for the coordination site of iron in cytochrome P-450. God.Jugosl.cent.kristalogr. 16, 21-32

Shimizu, T., Nozawa, T., Hatano, M., Imai, Y., and Sato, R.(1975) Magnetic circular dichroism studies of hepatic microsomal cytochrome P-450. Biochemistry 14, 4172-4178

Sono, H., Andersson, L.A., and Dawson, J.H.(1982) Sulfur donor ligand binding to ferric cytochrome P-450-CAM and myoglobin. J.Biol.Chem. 257, 8308-8320

XANTHOGENATES AS "SUICIDE" SUBSTRATES OF CYT-P-450

S.Yanev,H.Frank[*],H.Remmer[*],T.Stoytchev

Dep.Drug Toxicology,Inst.Physiology,Bulg.Acad.Sci.,
1113 Sofia,Bulgaria and [*]Inst.Toxicology,Univ.Tubingen,
Tubingen,FRG

Our earlier papers have shown that potassium ethylxanthoge-
nate(PEX) inhibited in vivo and in vitro some mixed-function
oxygenases: the metabolism of hexobarbital(Mitcheva et al.,
1976), aniline(Mitcheva,Stoytchev,1980), pentobarbital(Yanev
et al.,1982) and dimethylnitrosamine(Yanev et al.,1983).
The aim of present experiments was to study the metabolism
of six different alkyl derivatives of xanthogenic acid in male
rats thus to elucidate their inhibitory effect on drug metabolism.

MATERIALS AND METHODS

Male rats,weighing 180-200 g,were used.For in vivo experiments the animals were injected with different xanthogenates
(0.5 mmoles/kg i.p.) and placed in an all-glass desicator system for gas-chromatographic monitoring of volatile hydrocarbons
in the expired air by the procedure of Frank et al.,(1980). In
some experiments rats were injected with Phenobarbital(PB) 80
mg/kg i.p. for 3 days and 3-methylcholantrene(3-MC) - 20 mg/kg
i.p. and microsomes were prepared. For studying in vitro microsomal metabolism of xanthogenates the incubation system (1 ml)
contained in 0.1M Na-K phosphate buffer, pH 7.4: NADPH-generating system(NADP,0.33mM;Na-isocitrate,0.5mM,$MgSO_4$,0.1mM,iso-ci-
trate dehydrogenase,2 units),EDTA(1mM),1-2 microsomal protein
and different amount of xanthogenates. When using ^{14}C-labelled
in alkyl chain PEX the ^{14}C-binding in microsomes was determined in TCA-precipitate after washing with chloroform:methanol
/2:1/,methanol and ether, and alkaline digestion by liquid
scintilation counting.

RESULTS AND DISCUSSION

Male rats injected with different xanthogenates expired the
corresponding hydrocarbons/depending from the alkyl substituent
in the molecule and small amount of CS_2/Table 1/. Merlevede
and Peters(1965) have observed such small amount of CS_2 in the
expired air after administration of PEX in guinea pigs and human without checking the hydrocarbons production. Our further
experiments have shown that CS_2 in the expired air after xanthogenates administration is mainly due to nonenzymatic formation.

Table 1   *In vivo rates of hydrocarbons production from rats injected with different xanthogenates*

Substituent	Hydrocarbons	nmoles/kg/h
Methyl	methane	72.0
Ethyl	ethane	60.0
	ethylene	4.7
n-Propyl	propane	8.2
	propylene	2.1
i-Propyl	propane	5.3
	propylene	5.5
n-Butyl	n-butane	15.4
	n-butene	5.1
n-Amyl	n-pentane	4.0
	n-pentene	1.8

The rate of the unsaturated hydrocarbons expiration was linear up to 4h after the injection. The formation of the saturated hydrocarbons ceases after 1.5-2.0h. The exact determination of the rates of hydrocarbons production from different xanthogenates is hampered by the fact that the corresponding hydrocarbons metabolized itself with different rates.

Rat liver microsomes incubated in vitro with different xanthogenates and NADPH-generating system metabolized also to the corresponding hydrocarbons and in the same time decreased the cyt-p-450 level.There is no correlation between the rate of metabolism and extent of cyt-p-450 inhibition.PB-pretreatment increased the metabolism changed it to more oxidized-metabolites and increased also the cyt-p-450 inhibition/Table 2/. 3-MC-pretreatment did not change the rate of metabolism.

Table 2   *Production of hydrocarbons from xanthogenates incubated with rat liver microsomes*

Compound /1mM/	Untreated rats			cyt-p-450 %	PB-treated rats			cyt-p-450 %
	SH	UH	UH/SH		SH	UH	UH/SH	
Methyl	28	-	-	72	37	-	-	69
Ethyl	26	19	0.76	70	35	52	1.47	58
n-Propyl	15	33	2.13	68	27	64	2.41	54
i-Propyl	13	43	3.18	63	32	121	3.82	58
n-Butyl	13	35	2.60	53	21	85	4.03	41
n-Amyl	15	24	1.57	47	30	98	3.21	31

SH- corresponding saturated hydrocarbons(pmoles/mg/30 min)
UH-   "   unsaturated   "   "   "

The microsomal PEX metabolism seems to be cyt-p-450 dependent(Table 3): it is blocked totally by CO and partially by Metyrapon and SKF-525A and required O_2 and NADPH.Thus correlated with cyt-p-450 inhibition.The metabolism probably depends upon the oxy-radicals generating potency of the microsomes - it is stimulated by NaN_3 and blocked by addition of exogenous SOD and catalase.The same type of metabolism have been observed when incubated xanthogenates with different hydroxyl generating systems:$Fe/EDTA/H_2O_2$ or hemoglobin.It is interesting that such kind of chemical cleavage of xanthogenate molecule has been

Table 3   Microsomal PEX metabolism

Conditions	Ethane	Ethylene	Cyt-p-450
	(pmoles/mg/30 min)		%
Air	24.8	14.8	74.1
N_2	2.0	-	97.7
O_2	18.0	28.8	54.0
$CO/O_2(9:1)$	1.2	-	95.0
Metyrapon(1mM)	17.3	9.6	82.1
SKF-525A(1mM)	21.8	10.5	79.4
NaN_3(1mM)	28.6	135.4	61.6
Catalase(2000U)	1.7	-	93.5
SOD(12U)	1.9	-	98.2

observed when subjected different salts of xanthic acids to pyrolysis at very high temperature.In the presence of NaN_3 the
V_{max} value for ethylene formation from PEX increased drasticaly
(from 17 to 255 pmoles/mg/30 min).In the same time the K_m value
(0.025mM) is very closed to the I_{50} value of PEX for inhibition
of dimethylnitrosamine metabolism(0.045mM) by rat hepatocytes
(Yanev et al.,1983).Microsomal ^{14}C-PEX metabolism leads to covalent binding of ^{14}C-derived species which is NADPH-dependent
and it is stimulated by NaN_3(Table 4).

Table 4   Binding of ^{14}C-ethylxanthogenate to rat
liver microsomes in vitro

Treatment	Microsomal binding (^{14}C-PEX nmoles/mg/30)
^{14}-PEX(1mM)	2.1±0.05
^{14}-PEX(1mM)+NADPH	9.6±0.07
^{14}-PEX(1mM)+NADPH +NaN_3	15.4±0.09

Metabolism of xanthogenates seems to be a radical-induced
process, induced by oxy-radicals/O_2^-,OH·/(Fig 1).In vivo these
radicals may originate from different sources(cyt-p-450,hemog-
lobin).That leads to the formation of two basic types of reactive metabolites - reactive sulphur species and highly reactive
unsaturated alkyl radicals.Both metabolites led to irreversible
inhibition of cyt-p-450 by different places of covalent binding
- sulphur to the apoprotein/-SH groups/ of cyt-p-450(Neal,Hal-
pert,1982) and alkyl radicals to the cyt-p-450 heme(Ortiz de
Montellano,Correia,1983).Thus xanthogenates behaved as typical
oxy-radical scavengers/like methional and DMSO/ and unusual
"double suicide" substrates of cyt-p-450 system.

REFERENCES

1.H.Frank,T.Hintze,D.Bimboes,H.Remmer(1980) - Monitoring lipid peroxidation by breath analysis:Endogenous hydrocarbons
   and their metabolic elimination - Tox.appl.Pharmacol.,56,
   337-344
2.E.Merlevede,J.Peters(1965) - Metabolisme des xanthates

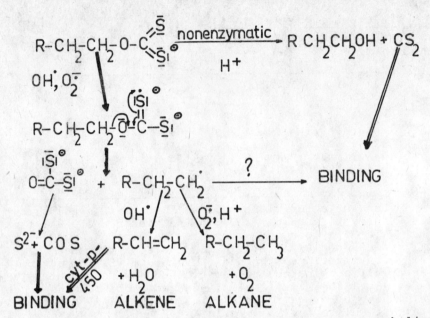

Fig. 1   *Proposal pathways of xanthogenates metabolism*

chez l'homme et chez le cobaye - Arch.Belges de Med.Soc.Hyg.
Med.Travail et Med.Legale,23,513-551

3. M.Mitcheva,T.Stoytchev(1980) - Influence of some thiol compounds on the aniline toxicity and aniline metabolism - Exp.
Med.Morph.,19,36-40

4. M.Mitcheva,S.Yanev,T.Stoytchev(1976) - Influence of potassium
ethylxanthogenate on hexobarbital metabolism,Farmacia,26,
45-48

5. R.Neal,J.Halpert(1982) - Toxicology of thiono-sulphur compounds - Ann.Rev.Pharmacol.Toxicology,22,321-339

6. P.R.Ortiz de Montellano,M.A.Correia(1983) - Suicidal destruction of cytochrome P-450 during oxidative drug metabolism -
Ann.Rev.Pharmacol.Toxicology,23,481-503

7. I.Tyden(1966) - Gas chromatographic study of the pyrolysis
of potassium salts of xanthic acids - Talanta,13,1353-1360

8. S.Yanev,G.Hauber,M.Schwenk,H.Remmer(1983) - Inhibitory effects of some thiol compounds on the metabolic activation of
dimethylnitrosamine and dimethylhydrazine in guinea pig hepatocytes and enterocytes - In:Extrahepatic drug metabolism
and chemical carcinogenesis,J.Rydström,J.Montelius,M.Bengts-
son,eds.,Elsevier Science Publishers,N.Y.,509-600

9. S.Yanev,I.Janku,T.Stoytchev,I.Havlik,V.Krebs(1982) - Effects
of potassium ethylxanthogenate and 2,3-dimercaptopropane
sulphonate sodium on the pentobarbital pharmacokinetics and
metabolism in male mice - Eur.J.Drug Metab.Pharmac.,7,21-29

EVIDENCE FOR A NOVEL CARBON-CENTERED FREE RADICAL FROM CCl_4 METABOLISM

R.G. Thurman, M.D. Galizi (Dept. of Pharmacology, Univ. of North Carolina, Chapel Hill, NC, USA), H.D. Connor (Dept. of Chemistry, Kentucky Wesleyan College, Owensboro, KY, USA) and R.P. Mason (Lab. Mol. Biophysics, NIEHS, Research Triangle Park, NC, USA)

SUMMARY

Electron spin resonance (ESR) spectroscopy has been used to monitor free radicals formed during CCl_4 metabolism by perfused livers from phenobarbital-treated rats. Radical metabolites were trapped with phenyl N-tert-butylnitrone (PBN). In addition to the PBN/$CCl_3^{\cdot}$ adduct, spectroscopic evidence for a PBN adduct of a carbon dioxide anion radical ($CO_2^{\overline{\cdot}}$) about 2 orders of magnitude more intense than the $CCl_3^{\cdot}$ adduct in effluent perfusate and rat urine was obtained.

INTRODUCTION

It is well known that CCl_4 is reductively dehalogenated to the $CCl_3^{\cdot}$ radical by cytochrome P-450 (Noguchi, A., et al., 1982). Spectroscopic evidence for this radical comes from a number of studies with isolated microsomes as well as from studies in vivo (Poyer, J.L., et al., 1980; Albano, E., et al., 1982). Previous studies examined the ESR spectra of membrane preparations or organic extracts of livers or membranes after the addition of an appropriate spin trap. Little information exists, however, on possible aqueous adducts. Therefore, we have examined perfusate and urine for PBN adducts formed during CCl_4 metabolism.

MATERIALS AND METHODS

Livers from phenobarbital-treated female Sprague-Dawley rats were perfused as described elsewhere (Scholz, R., et al., 1973). Effluent perfusate from the liver flowed past a Teflon-shielded Clark-type O_2 electrode and was collected in polyethylene bottles for ESR analysis. PBN (10 mM) was dissolved in the perfusate, and CCl_4 was bound to albumin (22.5%) prior to infusion. ESR spectra of aqueous perfusate and urine were measured employing a Varian E109 spectrometer equipped with a TM_{110} cavity. $CHCl_3(2)$: MeOH(1) extracts of perfusate were degassed in vacuo and analyzed using a TE_{102} cavity.

RESULTS AND DISCUSSION

When PBN was infused, an increase in O_2 uptake of about 15% was observed, possibly due to the partial mixed-function oxidation of the spin trap (Fig. 1). The subsequent infusion of CCl_4 resulted in a slight increase in O_2 uptake followed by a progressive decrease in O_2 uptake for the next 30 min of perfusion (Fig. 1).

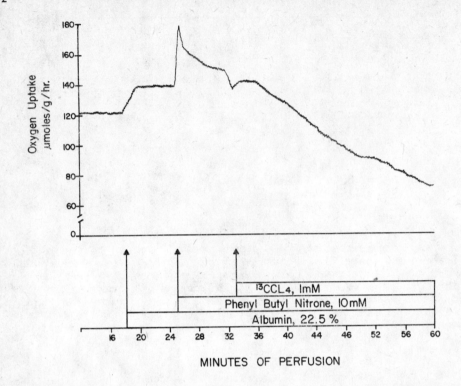

Fig. 1: Effect of PBN and CCl₄ on O₂ uptake by the isolated, perfused liver.
Liver from a fed, phenobarbital-treated rat was perfused with Krebs-Henseleit
bicarbonate buffer for time indicated. [O₂] was monitored continuously with a
Clark-type O₂ electrode and values were converted into rates employing the
influent-effluent concentration difference, the flow rate and the liver wet
weight.

ESR analysis of the aqueous perfusate produced a six-line spectrum (Fig.
2A). In organic extracts of perfusate, the PBN/CCl₃· spectrum was obtained (Fig.
2B). The absence of the former species in the extract suggested an ionic species.
Moreover, the species in the aqueous perfusate was between 1 and 2 orders of
magnitude more intense than the PBN/CCl₃· adduct. An identical spectrum was
generated in vitro with a Fenton hydroxyl radical generating system containing
formate (100 mM), a well-known hydroxyl radical scavenger (Table; Kalyanaraman,
B., et al., 1984). We conclude, therefore, that this is the PBN/CO₂⁻ adduct.
Similar hyperfine coupling constants have been observed for this adduct in
chemical systems (Aurian-Blajeni, B., et al., 1982). Further evidence that this is
indeed the PBN/CO₂⁻ radical adduct was obtained from studies of the effect of
pH on the coupling constants. The pKₐ was 2.85 for this radical adduct generated
either by the perfused liver or the Fenton system.
    Why then has this adduct been overlooked in the past in biological systems?
Most previous studies, as mentioned above, used organic extracts of tissue or
hydrophobic membrane preparations. Since the PBN/CO₂⁻ adduct is highly ionic,
it does not appear in organic extracts but is observed in aqueous perfusate (Fig.
2B).

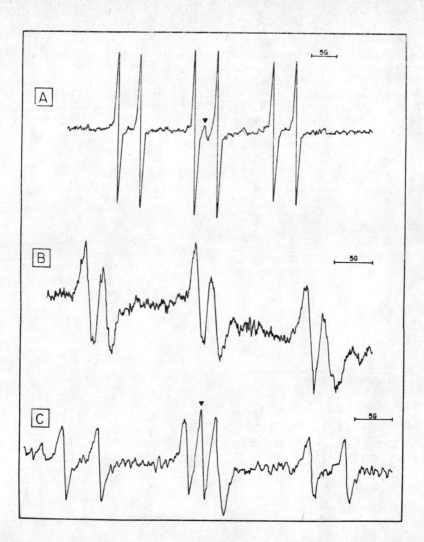

Fig. 2: <u>ESR spectra of perfusate and urine</u> A. Spectrum of aqueous perfusate after perfusion with PBN and CCl_4. Sample was treated with O_2 for 10 min followed by N_2 for 5 min. ▼ = ascorbate radical. Settings: 20 mW microwave power; 0.53 G modulation amplitude; 8 sec time constant; 0.67 G/min scan rate. B. Spectrum of organic extract of 8 parts perfusate to 1 part $CHCl_3$-MeOH. Settings: 20 mW; 2 G; 1 sec; 0.87 G/min. C. Spectrum of rat urine 1 h following gavage with PBN and CCl_4 solution. Settings: 20 mW; 0.53 G; 16 sec; 0.42 G/min.

By analogy, we predicted that this species should be filtered by the kidney and appear in the urine. Indeed, after treatment of aerated urine with ascorbate oxidase and catalase to decrease the ascorbate (▼) signal, a six-line spectrum with identical coupling constants to the aqueous perfusate and the Fenton system was observed in urine (Fig. 2C).

Metabolically, CCl_4 is converted into $CCl_3^{\bullet}$ by cytochrome P-450 by an anaerobic dehalogenation (Tomasi, A., et al., 1980; Poyer, J.L., et al., 1980). Indeed, in the perfused liver, adequate $PBN/CCl_3^{\bullet}$ spectra were only obtained when the O_2 level was reduced (Fig. 2B). It is probable that the next step involves the formation of the peroxy radical ($CCl_3O_2^{\bullet}$) which is converted into $^{\bullet}COCl$ which is trapped by PBN (Packer, J.E., et al., 1978). This radical adduct hydrolyzes to the carboxylic acid form which is observed spectroscopically (PBN/CO_2^{-}). Further experiments are necessary to elucidate the pathway(s) responsible for the generation of this interesting, novel metabolite from CCl_4.

TABLE: ESR Hyperfine Coupling Constants

			GAUSS	
Source	Presumed Structure	pH	a_{β}^{H}	a^{N}
Effluent Perfusate	PBN/CO_2^{-}	7.0	4.6	15.8
Fenton System	PBN/CO_2^{-}	6.9	4.6	15.8
Urine	PBN/CO_2^{-}	6.0	4.5	16.0
Extract of Perfusate	$PBN/CCl_3^{\bullet}$		1.85^{*}	14.45^{*}

*computer-simulated value.

REFERENCES

Albano, E., Lott, K.A.K., Slater, T.F., Stier, A., Symons, M.C.R., and Tomasi, A. (1982) Biochem. J., 204: 593-603.

Aurian-Blajeni, B., Halmann, M. and Manassen, J. (1982). Photochem. & Photobiol. 35: 157-162.

Kalyanaraman, B., Mottley, C. and Mason, R.P. (1984). J. Biochem. Biophys. Methods 9: 27-31.

Noguchi, T., Fong, K.-L., Lai, E.K., Alexander, S.S., King, M.M., Olsen, L., Poyer, J.L. and McCay, P.B. (1982). Biochem. Pharmacol. 31: 615-624.

Packer, J.E., Slater, T.S. and Willson, R.L. (1978). Life Sci. 23: 2617-2621.

Poyer, J.L., McCay, P.B., Lai, E.K., Janzen, E.G. and Davis, E.R. (1980). Biochem. Biophys. Res. Commun. 94: 1154-1160.

Scholz, R., Hansen, W. and Thurman, R.G. (1973). Eur. J. Biochem. 38: 64-73.

Tomasi, A., Albano, E., Lott, K.A.K. and Slater, T.F. (1980). FEBS Lett. 122: 303-306.

STEREOSPECIFICITY AND ISOTOPE EFFECTS OF CYTOCHROME P-450-DEPENDENT ETHANOL OXIDATION

G. Ekström, C. Norsten, T. Cronholm, and M. Ingelman-Sundberg, Department of Physiological Chemistry, Karolinska institute, Stockholm, Sweden.

INTRODUCTION

Recently, an ethanol-inducible form of cytochrome P-450 has been found (Koop et al. 1982, Ingelman-Sundberg and Hagbjörk 1982), which is effective in the oxidation of the alcohol (Morgan et al. 1983, Ingelman-Sundberg and Johansson 1984). Immunochemical analysis have revealed that almost all microsomal cytochrome P-450-dependent oxidation of ethanol can be accounted for by the presence of this type of cytochrome P-450 (Ingelman-Sundberg and Jörnvall 1984, Koop et al. 1984). However, the turnover for ethanol of this enzyme in intact microsomes appears to be about 4 times higher, than is reached under iron-free conditions in reconstituted membranes with limiting amounts of NADPH-cytochrome P-450 reductase (Ingelman-Sundberg and Jörnvall 1984). Interestingly, addition of small amounts of chelated iron to reconsituted membranes specifically causes a 4-fold enhancement of the rate of ethanol oxidation catalyzed by the ethanol-inducible form of P-450, whereas such an inrease of ethanol oxidation is not observed using other types of P-450 (Ingelman-Sundberg and Johansson 1984). Furthermore, the increase of ethanol oxidation by the addition of chelated iron is accompanied by a stoichiometric decrease in the rate of hydrogen peroxide production, and the reaction is inhibited, both in the presence and in the absence of chelated iron, by scavengers of hydroxyl radicals (Ingelman-Sundberg and Johansson 1984). These findings indicate that the reaction in the reconstituted membranes, to a great extent is dependent on the generation of hydroxyl radicals in a Fenton-like reaction. In order to further evaluate the mechanism of cytochrome P-450-dependent ethanol oxidation, we have examined the stereo-specificity and D(V/K) isotope effects of ethanol oxidation in i) liver microsomes, ii) reconstituted membranous systems containing phenobarbital-inducible and ethanol-inducible forms of cytochrome P-450 and iii) the xanthine-xanthine oxidase system, known to oxidize ethanol entirely in a hydroxyl radical-dependent reaction. We here present preliminary data of these studies.

EXPERIMENTAL PROCEDURES

Sources of enzymes and chemicals, methods for membrane reconstitution and some analytical methods have previously been described (Ingelman-Sundberg and Johansson 1984). Liver microsomes were prepared from imidazole-treated male rabbits in 0.25 M sucrose, washed with 1.14

% KCl and suspended in 50 mM potassium phosphate buffer, pH 7.4.

Incubations with ethanol were carried out using 25 ml stoppered tubes containing 5 ml tubes with reagent for derivatization of the acetaldehyde formed. The acetaldehyde was determined either spectrophotometrically as described elsewhere (Ingelman-Sundberg and Johansson 1981) or by capillary GC/MS. In the latter case, the 5 ml inner tube contained 5 mM 2,4-dinitrophenylhydrazine in 1 ml of dimethylformamide. The dinitrophenylhydrazones were extracted twice with hexane, washed with water and dried with sodium sulphate before GC/MS analysis. The kinetic isotope effects of the ethanol oxidation was measured by a noncompetitive and a competitive method (Northrop 1977). In the noncompetitive method, the Vmax and Km values of the cytochrome P-450 catalyzed ethanol oxidation were determined using either CH_3-CH_2-OH or CH_3-CD_2-OH as substrates, with spectrophotometric analysis of the product. In the competitive method, the rate of product formation was determined by GC/MS in a 50 mM 1:1 mixture of either $(1-^{13}C)-CH_3-CH_2-OH$ and CD_3-CD_2-OD or CH_3-CD_2-OH and CD_3-CH_2-OH. Butyraldehyde was used as an internal standard. Under the working up conditions used, no exchange of hydrogen atoms occurred.

RESULTS AND DISCUSSION

The cytochrome P-450-catalyzed oxidation of ethanol exhibited no stereospecificity neither when liver microsomes were used, nor when the reaction was reconstituted in artificial membranes containing purified forms of cytochrome P-450 LM_2 or ethanol-inducible P-450 (cytochrome P-450 LMeb) with small amounts of Fe-EDTA (Table I).

Table I. Absence of stereospecificity in cytochrome P-450-dependent ethanol oxidation. Incubations were carried out with liver microsomes from imidazole-treated animals corresponding to 4 mg of protein or with membrane vesicles corresponding to 0.5 nmol of P-450, for 30 or 60 min at $37^\circ C$. The incubation mixtures contained 1 mM NADPH and 50 mM of either $(1-D_R)$-ethanol or $(1-D_S)$-ethanol. Incubations with the microsomes were performed in the presence of 1 mM sodium azide. The ratio of non-deuterated to deuterated products formed is shown in the table.

System used	$(1-D_R)$-ethanol $CH_3CHO:CH_3CDO$	$(1-D_S)$-ethanol $CH_3CHO:CH_3CDO$
Microsomes	0.57 : 0.43	0.48 : 0.52
Ethanol-inducible P-450 + 2 μM Fe-EDTA	0.64 : 0.36	0.54 : 0.46
P-450 LM_2 + 10 μM Fe-EDTA	0.68 : 0.32	0.56 : 0.44

The D(V/K) isotope effects of ethanol oxidation was examined by two
different methods (see Experimental Procedures). As seen from table II,
isotope effects of about 4 were reached when liver microsomes from
imidazole-treated rabbits were incubated with C-1-deuterated and
C-1-nondeuterated ethanol. This result was obtained both when the
noncompetitive kinetic method and the competition method, based
on analysis on capillary GC/MS, were used. Much smaller isotope effects
were registered using the xanthine-xanthine oxidase system. Here ethanol
oxidation is carried out with entirely hydroxyl radicals as the active
oxidizing species, and the isotope effects were about 1.5, according to
results reached using both types of analytical methods. Similar values
were also observed using reconstituted membrane vesicles containing
cytochrome P-450 LM_2 in the presence of EDTA-chelated iron. The isotope
effects reached using membranes containing the ethanol-inducible form of
cytochrome P-450 were intermediary between the value obtained in
microsomes and in the xanthine-xanthine oxidase system.

Table II. D(V/K) isotope effects of ethanol oxidation in liver
microsomes, in reconstituted membranous systems containing cytochrome
P-450 and in the xanthine-xanthine oxidase system. Vmax and Km values
were determined using CH_3CH_2OH or CH_3CD_2OH as substrates at eight
different concentrations. In a second set of experiments, the enzyme
containing systems were incubated with a 50 mM 1:1 mixture of either
CH_3CD_2OH and CD_3CH_2OH or $(1-^{13}C)-CH_3CH_2OH$ and CD_3CD_2OD. The products
were analyzed by capillary GC/MS as outlined under Experimental
Procedures.

System used	$\dfrac{CD_3CHO}{CH_3CDO}$	$\dfrac{CH_3{}^{13}CHO}{CD_3CDO}$	H $\dfrac{Vmax}{Km}$ / D $\dfrac{Vmax}{Km}$	v*/
Cytochrome P-450 LMeb	2.86	3.10	n.d.	4.5
Cytochrome P-450 LMeb + 2 µM Fe-EDTA	2.06	1.74	n.d.	22
Cytochrome P-450 LM_2 + 10 µM Fe-EDTA	1.51	1.43	2.35	5.5
Microsomes	4.54	3.47	3.71	4.1
Xanthine-Xanthine oxidase	1.32	1.45	1.87	-

*/ the rate is given as nmol of ethanol converted per min and nmol of
P-450.

There was a shift in the isotope effect of the ethanol oxidation,
catalyzed by cytochrome P-450 LMeb, upon addition of a small amount of
chelated iron. The rate of the reaction was enhanced 5-fold by the
addition of Fe-EDTA (2 µM), whereas the isotope effect became more
similar to that of the xanthine-xanthine oxidase system. This may
indicate that the reaction mechanism in the presence of chelated iron is
not identical with the mechanism in its absence. Thus, in the presence

of Fe-EDTA, the results favour a mechanism to the major extent being
hydroxyl radical-dependent, whereas in its absence, also additional
mechanisms might be operating. According to the isotope effects, ethanol
oxidation by cytochrome P-450 LM$_2$ in the presence of Fe-EDTA appears to
be entirely dependent upon hydroxyl radicals, a finding in line with
previous findings (Ingelman-Sundberg and Johansson 1981). Using our
reconstituted systems, it was not possible for us to reach similar
isotope effects as those registered in intact microsomes. This finding
might indicate that there are additional factors of importance that have
to be added to the reconstituted system, in order to reach the
environment giving the more native-like mechanism of cytochrome
P-450-dependent ethanol oxidation.

The oxidation of ethanol by alcohol dehydrogenase (ADH) and catalase has
previously been shown to be stereospecific and exert D(V/K) isotope
effects of 3.0 and 1.9, respectively (Damgaard, 1982). Our results
indicate that cytochrome P-450 differs from these enzymes in this
reaction by having no stereospecificity.

ACKNOWLEDGEMENTS

This work was supported by grants from Magn. Bergvalls Stiftelse and
from the Swedish Medical Research Council.

REFERENCES

Damgaard, S.E., 1982, The D(V/K) isotope effect of the cytochrome
P-450-mediated oxidation of ethanol and its biological applications.
Eur. J. Biochem. 125, 593-603
Ingelman-Sundberg, M. and Johansson, I., 1981, The mechanism of
cytochrome P-450-dependent oxidation of ethanol in reconstituted
membrane vesicles. J. Biol. Chem., 256, 6321-6326
Ingelman-Sundberg, M. and Johansson, I., 1984, Mechanisms of hydroxyl
radical formation and ethanol oxidation by ethanol-inducible and other
forms of cytochrome P-450. J. Biol. Chem. 259, 6447-6458
Ingelman-Sundberg, M. and Jörnvall, H., 1984, Induction of the
ethanol-inducible form of rabbit liver microsomal cytochrome P-450 by
inhibitors of alcohol dehydrogenase. Biochem. Biophys. Res. Commun.
124, 375-382
Koop, D. R., Morgan, E. T., Tarr, G. E. and Coon, M. J. Purification
and characterization of a unique isozyme of cytochrome P-450 from
liver microsomes of ethanol-treated rabbits. J. Biol. Chem. 257,
8472-8480
Koop, E. T., Nordblom, G. D. and Coon, M. J., 1984, Immunochemical
evidence for a role of cytochrome P-450 in liver microsomal ethanol
oxidation. Arch. Biochem. Biophys. 235, 228-238
Morgan, E. T., Koop, D. R. and Coon, M. J., 1982, Catalytic activity
of cytochrome P-450 isozyme 3a isolated from liver microsomes of
ethanol-treated rabbits. J. Biol. Chem. 257, 13951-13957
Northrop, D. B., 1977, Isotope effects on enzyme-catalyzed reactions;
Determining the absolute magnitude of hydrogen isotope effects. Eds.
Cleland, W. W., O'Leary, M. H. and Northrop, D. B., University Park
Press, London, Baltimore, Tokyo, pp. 122-152

ETHANOL AND ACETONE-INDUCIBLE CYTOCHROME P-450-DEPENDENT OXIDATION OF BENZENE AND ACETONE

Johansson, I. and Ingelman-Sundberg, M., Department of Physiological
        Chemistry, Karolinska institute, Stockholm, Sweden

Ethanol-treatment in vivo rapidly enhances the metabolism of several
organic solvents in mouse, rat and rabbit liver microsomes (cf. Sato
et al. 1981, Johansson and Ingelman-Sundberg, 1985). Benzene, carbon
tetrachloride, chloroform, toluene, trichloroethylene and
1,2-dichloroethane are among these solvents (Sato et al. 1981). A
similar induction of the sovent metabolism is also seen following
starvation or acetone treatment of the animals. The molecular basis of
this process of induction remains essentially unknown. The finding of
an ethanol-inducible species of rabbit liver microsomal cytochrome
P-450 (Koop et al. 1982, Ingelman-Sundberg and Hagbjörk 1982)
suggested that the induction of this form of cytochrome P-450 might be
the mechanism for the ethanol-dependent elevation of the solvent
metabolism. This appears also to be true concerning the cytochrome
P-450-dependent transformation of carbon tetrachloride to chloroform
and to products being able to initiate lipid peroxidation (Johansson
and Ingelman-Sundberg 1985). Using e.g. antibodies towards the
ethanol-inducible P-450-form, it could be shown that this form of
cytochrome P-450 apparently was responsible for almost all microsomal
metabolism of carbon tetrachloride. However other solvents, among them
benzene, were not substrates for the ethanol-inducible cytochrome
P-450 form, when the enzyme was incorporated into artificial membranes
together with NADPH-cytochrome P-450 reductase. One can identify three
possible reasons for this unexpected finding: i) ethanol induces also
additional forms of cytochrome P-450, ii) other factors are present in
the microsomes that have to be added in order to be able to
reconstitute the activity and iii) the enzyme is modified during
solubilization and purification. Here we present evidence for the
existence of a high affinity cytochrome P-450-dependent metabolic site,
in microsomes from ethanol or acetone-induced animals, that oxidizes
acetone and benzene at high turnover, with very similar properties
to the ethanol-inducible form of cytochrome P-450.

EXPERIMENTAL PROCEDURES

Male Sprague-Dawley rats (175 g) were starved for 24 hours and
subsequently treated with acetone (5 ml/kg) for two subsequent days.
The acetone was administrated intragastrically in a 33 % (v/v) saline
solution. The rats received water ad libitum but no food. Other groups
of rats were treated with ethanol according to DeCarli and Lieber
(1967). The livers of the animals were homogenized in three volumes of

10 mM sodium/potassium phosphate buffer, pH 7.4, containing 1.14 %
KCl. The microsomes were prepared by ultracentrifugation and washed
once before suspension in 50 mM potassium phosphate buffer, pH 7.4.
Benzene metabolism was determined using the head space technique
described by Sato and Nakajima (1979). Water soluble benzene
metabolites were also quantified as described previously (Johansson
and Ingelman-Sundberg, 1983) using ^{14}C-benzene as substrate. Acetone
metabolism was studied using ^{14}C-labelled substrate. The incubations,
performed in sealed tubes, were stopped by the addition of 0.1 ml 40%
trichloroacetic acid. The water phases were treated with a stream of
nitrogen at 56°C for 15 min, the residues were allowed to stand
at room temperature for 48 h, and 100 μl aliquots were counted in a
scintillation counter after addition of water to the original volume.

RESULTS AND DISCUSSION

The oxidation of benzene and acetone in liver microsomes from
acetone-treated rats were inhibited by typical cytochrome P-450
inhibitors as well as by agents known to interact with the
ethanol-inducible cytochrome P-450 form in rabbits (Table I).

Table I. Effect of Fe(III)-EDTA, mannitol, inhibitors of cytochome
P-450 and compounds known to interact with the ethanol-inducible
P-450-form on the metabolism of acetone and benzene in liver
microsomes isolated from acetone-treated rats. Incubations were
carried out in a total volume of 1 ml 50 mM potassium phosphate
buffer, pH 7.4, with microsomes corresponding to 1 mg of protein in
the presence of 50 μM benzene or 1.36 mM ^{14}C-acetone at 37°C for
15 or 20 min. Substrates were added in water and the incubations were
started by the addition of NADPH to 0.5 mM.

Compund added	benzene oxidation		acetone oxidation	
	nmol/ mg protein/ min			
		% inh.		% inh.
None	1.00		1.75	
Carbon monoxide	0.34	67	n.d.	
Mannitol, 100 mM	1.00	0	1.54	12
-"- , 500 mM	0.82	18	1.17	33
Fe-EDTA, 10 μM	0.03	97	0.65	63
Ethanol, 200 mM	0	100	0	100
Aniline, 1 mM	0	100	0.07	96
Imidazole, 0.5 mM	0.55	45	0.35	80
-"- , 5 mM	0.21	79	0	100
Acetone, 5 mM	0.29	71	-	-
-"- , 50 mM	0	100	-	-
Benzene, 0.1 mM	-	-	0.21	78
-"- , 1 mM	-	-	0.04	98
SKF-525A, 200 μM	0.25	75	0.82	53
Metyrapone, 1 mM	1.1	0	1.38	21

However, mannitol, a scavenger of hydroxyl radicals, did not inhibit the reaction and, furthermore, chelated iron, known to specifically enhance the rate of ethanol oxidation catalyzed by the ethanol-inducible cytochrome P-450 form, did inhibit the rate of acetone or benzene oxidation under these conditions. This indicates that the mechanism of these oxidations is different from that of ethanol oxidation catalyzed by the ethanol-inducible form of cytochrome P-450 in the presence of small amounts of chelated iron (Johansson and Ingelman-Sundberg, 1984).

The apparent kinetics of benzene and acetone oxidation was studied in liver microsomes from control and acetone-treated rats (Table II). Concerning benzene oxidation, two Km-values were determined in both types of microsomes, one high affinity (20 µM) and one low affinity (1 mM). However, the Vmax values of both the high affinity and the low affinity reaction were more than 10-fold higher in microsomes from acetone-treated animals. A similar difference of the Vmax of acetone oxidation was evident between the two groups of animals. In this case, only one Km value, about 1 mM, could be identified in both types of microsomes.

Table II. Kinetics of benzene and acetone oxidation in liver microsomes from control and acetone-treated rats. Benzene oxidation was determined by measuring the amount of water-soluble metabolites formed. Incubations with microsomes from control rats were carried out using material corresponding to 2 mg of protein. The Km and Vmax values were obtained from results of incubations at 10 different substrate concentrations. Other conditions as described in Table I.

Substrate	Type of microsomes	Km mM	Vmax nmol/mg, min
Acetone	Control	0.88	0.26
	Acetone	1.26	3.09
Benzene	Control	0.020/1.0	0.02/0.25
	Acetone	0.016/1.0	0.33/2.5

Similar kinetics of benzene oxidation, as obtained using microsomes from acetone-treated animals, were also observed using liver microsomes from ethanol-treated rats. The results are consistent with the existence of an ethanol and acetone-inducible form of cytochrome P-450 in rat liver microsomes, participating in the metabolism of acetone and benzene, having properties very similar to the ethanol-inducible form of cytochrome P-450 described in the rabbit. It appears that such a form of cytochrome P-450 might be of a great physiological significance in the transformation of acetone to products acting as precursors of glucose (cf. Casazza et al. 1984). The Km-value determined for acetone falls within the physiological concentrations of acetone seen during starvation. However, further work is needed before the molecular basis of this type of cytochrome

P-450-catalyzed reaction is understood.

ACKNOWLEDGEMENTS

This work was supported by grants from Magn. Bergvalls Stiftelse and from the Swedish Medical Research Council. The skilful technical assistance of Miss Elisabeth Wiersma is gratefully acknowledged.

REFERENCES

DeCarli, L. M. and Lieber, C. S., 1969, Fatty liver in the rat after prolonged intake of ethanol with a nutritionally adequate liquid diet. J. Nutr. 91, 331-336

Casazza, J. P., Felver, M. E. and Veech, R. L., 1984, The metabolism of acetone in the rat. J. Biol. Chem. 259, 231-236

Ingelman-Sundberg, M. and Hagbjörk, A.-L., 1982, On the significance of the cytochrome P-450-dependent hydroxyl radical-mediated oxygenation mechanism. Xenobiotica 12, 673-686

Johansson, I. and Ingelman-Sundberg, M., 1983, Hydroxyl radical-mediated cytochrome P-450-dependent metabolic activation of benzene in microsomes and reconstituted enzyme systems from rabbit liver. J. Biol. Chem. 258, 7311-7316.

Johansson, I. and Ingelman-Sundberg, M., 1984, Mechanisms of hydroxyl radical formation and ethanol oxidation by ethanol-inducible and other forms of rabbit liver microsomal cytochrome P-450. J. Biol. Chem. 259, 6447-6458

Johansson, I. and Ingelman-Sundberg, M., 1985, Carbon tetrachloride-induced lipid peroxidation dependent on an ethanol-inducible form of rabbit liver microsomal cytochrome P-450. FEBS lett. 183, 265-269

Koop, D. R., Morgan, E. T., Tarr, G. E. and Coon, M. J., 1982, Purification and characterization of a unique isozyme of cytochrome P-450 from liver microsomes of ethanol-treated rabbits. J. Biol. Chem. 257, 8472-8480.

Sato, A. and Nakajima, T., 1979, A vial-equilibration method to evaluate the drug-metabolizing enzyme activity for volatile hydrocarbons. Toxicol. Appl. Pharmacol. 47, 41-46

Sato, A., Nakajima, T. and Koyama, Y., 1981, Dose-related effects of a single dose of ethanol on the metabolism in rat liver of some aromatic and chlorinated hydrocarbons. Toxicol. Appl. Pharmacol. 60, 8-15.

Component Interactions in Cytochrome P-450 Dependent Monooxygenatic Systems

ROLE OF INTERMOLECULAR INTERACTION IN THE FORMATION OF CYTOCHROME P-450 CATALYTICALLY ACTIVE STATE

Archakov A.I., Uvarov V.Yu.

Research Institute of Physico-Chemical Medicine, Malaya Pirogovskaya 1a, 119828, Moscow, USSR

Cytochrome P-450 is a typical membrane hydrophobic protein which needs phospholipids to provide its functioning/Ruckpaul and Rein. 1983/, /Ingelman-Sundberg and Johansson. 1980/. The incorporation of isolated cytochrome P-450 into the phospholipid bilayer or the addition of phospholipids causes a 2-3 times increase in the haemoprotein catalytic activity with organic peroxide used as a cosubstrate /Archakov et al. 1980/. Cytochrome b_5 is also able to increase cytochrome P-450 catalytic activity in hydroxylase reactions /Uvarov et al. 1983/. It was interesting to investigate the mechanism of phospholipids and cytochrome b_5 reactivation action on cytochrome P-450. Thus, a study was undertaken to investigate the influence of phospholipid charge, fatty acid composition and also the effect of other proteins on the catalytic activity and structure of isolated cytochrome P-450 (LM_2). It was assumed that the phospholipid and protein activation action on cytochrome P-450 results from the influence of intermolecular interactions upon its conformational rigidity.

MATERIALS AND METHODS

Rabbit liver microsomal cytochrome P-450 was isolated by the method of Karuzina et al /1979/. Cytochrome b_5 was isolated by the method of Spatz and Strittmatter /1971/. Purified cytochrome b_5 contained 60 nmoles of haemoprotein per 1 mg protein. Liposomes and proteoliposomes were obtained as described by Archakov et al /1980/. Cytochrome P-450 was incorporated into 1.2-di-O-palmitoylphosphatidylcholine (DPPhC) and 1.2-di-O-hexadecylphosphatidylcholine (DHPhC) at $T=43^{\circ}C$.
The haemoprotein conformational state and thermal stability were estimated from the circular dichroism spectra (CD). CD spectra were registered as described by Archakov et al /1980/. The rate of p-nitroanisole (p-NA) O-demethylation in the presence of cumole hydroperoxide (CHP) was registered by the accumulation of p-nitrophenol, the rate of dimethylaniline (DMA) N-demethylation and aniline (AN) p-hydroxylation in the presence of CHP was estimated as demonstrated by Karuzina et al /1979/.

RESULTS AND DISCUSSION

Shown in Table 1 are the data on the influence of different phospholipids on the structure and catalytic activity of cytochrome P-450. The addition of various phospholipids increases

TABLE 1. Phospholipid Influence on Cytochrome P-450 Catalytic Activity and Confomational State

System	CHP-dependent reaction $(k_{cat} \cdot 10^3, c^{-1})$			α-helix content %
	p-NA	DMA	AN	
Soluble cytochrome P-450	70	640	70	60
Microsomes	120	1600	130	-
Proteoliposomes				
MPL	200	2000	220	56
PhC	200	2200	200	55
AS	220	-	-	55
PhC/PhA	220	-	-	55
PhC/CTAB	220	-	-	55
DPPhC	70	-	-	54
DHPhC	70	-	-	54

Footnotes: MPL - mixture of microsomal phospholipids; PhC - - phosphatidylcholine; AS - asolectin; PhC/PhA - liposomes from the mixture of PhC and 10 mol % phosphatidic acid; PhC/CTAB - - liposomes from the mixture of PhC and 10 mol % CTAB (cetyltrimethylammonium bromide).

cytochrome P-450 catalytic activity without much effecting the α-helix content. Liposomes from MPL containing about 10% phosphatidylserine and phosphatidylinositol, liposomes from AS containing 15% phosphatidylinositol and 10% PhA as well as PhC liposomes containing 10% PhA were shown to produce the same effect as PhC. It means that the negative charge does not effect the catalytic activity and confromational state of cytochrome P-450. No effect was either observed after the addition of CTAB to PhC; therefore, a positive charge does not influence cytochrome P-450 catalytic activity and conformational state as well. However, synthetic phospholipids containing palmitic acid residues were not shown to activate cytochrome P-450 at a temperature lower than the point of phase transition. Thus, cytochrome P-450 activation depends upon the phase state of the phospholipid bilayer.

Shown in Table 2 are the thermodynamic parameters of p-nitroanisole O-demethylation reaction catalized by isolated and membrane-bound cytochrome P-450. Cytochrome P-450 activation after its incorporation into the liquid phospholipids can be explained by the decrease in the change of the reaction entro-

TABLE 2. <u>The Influence of Different Phospholipids on the</u>
<u>Thermodynamic Parameters of p-NA Oxidation Reaction</u>

Experimental Conditions	k_{cat}, s^{-1}	$\Delta H^{\#}$ kcal/mole	$-\Delta S^{\#}$ e.u.
Soluble LM_2	0.07	14.4	16.7
LM_2 incorporated into liposomes from:			
PhC	0.2	18.0	3.7
PhC/PhA	0.2	17.5	4.5
PhC/CTAB	0.2	17.3	4.2
MPL	0.2	18.0	3.1
AS	0.2	17.6	4.6
DPPhC	0.07	8.3	36.8
DHPhC	0.07	8.2	36.7

pic factor. In the case of solid phospholipids the absence of
their activation action despite the decrease in the activation
enthalpy can be accounted for by the increase in the entropic
expenditure. The data show that a liquid phospholipid bilayer is likely to provide the enzyme optimal conformational rigidity necessary for the catalytic act. A hydroxylase reaction
catalized by the enzyme incorporated into a solid phospholipid
bilayer as well as by a solubilized enzyme has a higher entropic barrier. It should be noted that changes in the enthalpy
factor do not correlate with the rate of catalysis.

As has been shown previously, cytochrome P-450 incorporation into the phospholipid bilayer of different composition is
accompanied by the increase in the enzyme stability and conformational rigidity /Archakov et al. 1980/. It should be
pointed out that this effect is independent of the phospholipid bilayer composition. In view of the fact that cytochrome
P-450 interaction with the phospholipid bilayer can be registered as spectral changes, similar to those of type 1 substrates binding /Fig.1A/, the kinetic curves of the haemoprotein interaction with the lipid bilayer were obtained /Fig.1B/.
The data presented undoubtedly prove the high affinity of cytochrome P-450 to the negatively charged phospholipids.

Therefore, the phospholipid negative charge considerably
increases the rate of cytochrome P-450 interaction with liposomes without, however, affecting its catalytic activity.

The obtained results imply that the haemoprotein interaction with the negatively charged residue of phospholipids
phosphoric acid does not influence the catalytic activity,
while the intermolecular hydrophobic, unspecific interactions
considerably effect the enzyme catalytic activity. It can be
assumed that protein-protein interactions will affect the
structure and catalytic activity in the same way.

Table 3 shows the possibility of isolated cytochrome P-450
activation as resulting from its complex formation with other
proteins. No specific influence of the proteins has been shown.

108

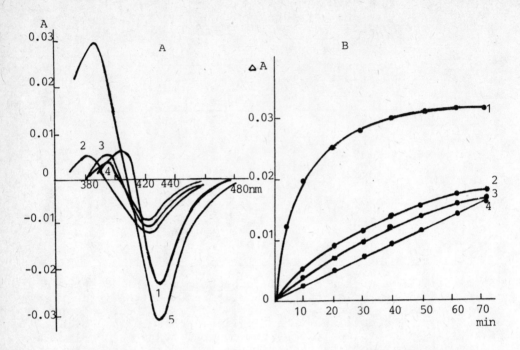

/Fig. 1A,B/. Spectral changes (A) and kinetics (B) of cytochrome P-450 incorporation into the liposomal lipid bilayer.
Incubation mixture of 2.5 ml contained 0.83 µM cytochrome P-450
and 250 µM phospholipid. 1- liposomes from PhC/PhA; 2 - liposomes from MPL; 3 - liposomes from PhC/CTAB, 4 - liposomes from
PhC; 5 - 1mM hexabarbital. Apparent K_D for liposomes from
PhC/PhA - 25 µM; MPL - 80 µM, PhC/CTAB - 290 µM.

TABLE 3. <u>Protein Influence on the Catalytic Properties and
Structure of LM_2</u>

Experimental Conditions	k_{cat}, s^{-1}	$\Delta H^{\#}$ kcal/mole	$-\Delta S^{\#}$ e.u.
Soluble LM_2	0.06	14.4	17
+ cyt d-b_5 (1:4)	0.09	11.6	27
+ FP (1:4)	0.09	10.5	30.3
+ BSA(1:30)	0.10	7.1	40.5
+ cyt t-b_5 (1:10)	0.06	–	–

<u>Footnotes</u>: FP - NADPH-dependent reductase; BSA - bovin serum
albumine.

All the three proteins investigated - d-cytochrome b_5, FP, BSA
have produced the same activating effect. This effect was not

observed in the case of cytochrome b_5 trypsin fragment which does not bind to cytochrome P-450. The proteins activation action on cytochrome P-450 results from the decrease in $\Delta H^{\neq}$ change which makes the effect different from that produced by the phospholipids - their activation action results from the decrease in $\Delta S^{\neq}$ change.

At the same time, proteins as well as phospholipids stabilize cytochrome P-450 structure and increase its conformational rigidity /Table 4/. Thus, both proteins and phospholi-

TABLE 4. Induced Conformational Mobility and Thermal Stability of Cytochrome P-450 in Water Solution, Proteoliposomes and in Protein-Protein Complexes.

System	$T_{1/2}^{d}$, 0 C	α-helix content %		
		without additions	diaminooctan (8.4 mM)	$Na_2S_2O_4$ (1.8 mM)
Soluble cytochrome P-450	45	60	65	70
+MPL	72	56	57	61
+b_5 (1:4)	62	37	37	37
+FP (1:1)	57	27	30	29
+BSA (1:1)	57	35	34	34

Footnotes: $T_{1/2}^{d}$- temperature causing 50% decrease in ellipticity at 220 nm.

pids facilitate the reactions catalized by cytochrome P-450. Cytochrome P-450 molecule in both cases becomes more stable and rigid.

It should be noted that cytochrome b_5 and FP binding to cytochrome P-450 is characterized by the same spectral changes as in the case of type I substrates or phospholipids /Fig.2/. Cytochrome b_5 and FP affinity to cytochrome P-450 is very high: K_D is two orders lower than that for type I substrates. A question arises whether lipids and proteins can produce an additive effect on isolated cytochrome P-450. The results in Table 5 show that the increase of cytochrome b_5 content in proteoliposomes causes a decrease in the rate of p-NA O-demethylation reaction which occurs due to the increase in the reaction entropic barrier. It can be accounted for by the occurrence of new cytochrome P-450-cytochrome b_5 interactions in proteoliposomes alongside with lipid-protein interactions already existing. It causes a further increase in cytochrome P-450 conformational rigidity which is accompanied by the decrease in k_{cat} as in the case of solid phospholipids.

Therefore, cytochrome b_5 which functions as cytochrome P-450 activator in the absence of phospholipids turns into the inhibitor of a hydroxylase reaction in proteoliposomes. An isokinetic curve gives the best illustration of the data discussed here. The location of all points on one line shows that

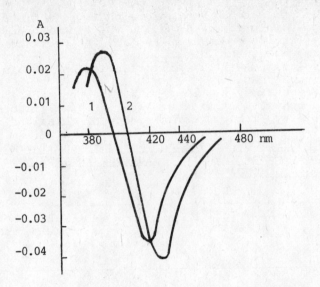

/Fig.2/. Spectral changes observed during cytochrome P-450 binding to other proteins. Incubation mixture of 2.5 ml contained 1.7µM cytochrome P-450 and 3.5µM FP; 2 - 3µM cytochrome P-450 and 15µM cytochrome b_5. Apparent K_D in water solution for cytochrome b_5 - 2.4µM with FP - 3.3µM, in proteoliposomes from MPL - 1.3µM and 1.2µM respectively.

TABLE 5. The Influence of Cytochrome P-450/Cytochrome b_5 Ratio in·Proteoliposomes on the Thermodynamic Properties of p-NA O-demethylation.

System	$\Delta H^{\#}$ kcal/mol	$-\Delta S^{\#}$ e.u.	$k_{cat} 10^{-3}$ s^{-1}
PhC + cyt P-450	18.0±2.7	3.7±0.6	200±30
PhC + complex cyt P-450:b_5=			
1:2	12.1±1.8	24.4±3.7	100±15
1:4	9.6+1.4	30.7±4.6	90±14
1:6	6.5+1.0	43.0±6.4	70±10

the change in both the thermodynamic parameters occurs due to the same factor.

The obtained results once again confirm the assumption that the activation action of phospholipids occurs due to the influence of lipid-protein interaction on the entropic factor of a hydroxylase reaction, while the activation action of proteins results from the decrease in the enthalpy factor of the reaction catalized by cytochrome P-450.

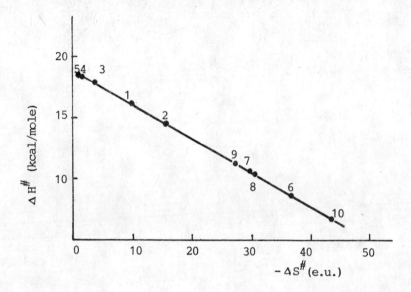

/Fig.3/.Isokinetic dependence between $\Delta H^{\#}$ and $\Delta S^{\#}$ in p-NA oxidation reaction catalized by LM_2 and its complexes with proteins and phospholipids. 1 - microsomal LM_2; 2 - soluble LM_2; 3 - +asolectin; 4 - +PhC; 5 - +MPL; 6 - +DPPhC; 7 - +FP; 8 - +cyt b_5; 9 - +cyt b_5+MPL; 10 - +cyt b_5+DPPhC.

REFERENCES

1. Archakov A.I., Bachmanova G.I., and Uvarov V.Yu.(1980), Interaction of cytochrome P-450 with phospholipid bilayer, Biochemistry, Biophysics and Regulation of Cytochrome P-450 (J-A Gustafsson et al. eds), Elsevier/North-Holland Biomedical Press, 551-558.
2. Ingelman-Sundberg M., Johansson I. (1980), Catalytic properties of purified forms of rabbit liver microsomal cytochrome P-450 in reconstituted phospholipid vesicles, Biochemistry 19, 4004-4011.
3. Karuzina I.I., Bachmanova G.I. et al (1979), Cytochrome P-450 properties and its isolation from rabbit liver microsomes, Biokhimia 44, 1149-1159.
4. Ruckpaul K., and Rein H. (1983)Cytochrome P-450, Akademie - - Verlag, Berlin
5. Spatz L. and Strittmatter P. (1971), A form of cytochrome b_5 that contains an additional hydrophobic sequence of 40 amino acid residues.Pros.Nat.Acad.Sci.USA, 68, 1042-1046.
6. Uvarov V.Yu., Wernicke D., Bachmanova G.I., and Archakov A.I. (1983) Influence of cytochrome b_5 on the functional activity and conformational state of cytochrome P-450. Biokhimia 48, 1542-1547.

A MODEL FOR MOLECULAR ORGANIZATION OF THE CHOLESTEROL SIDE CHAIN CLEAVAGE CYTOCHROME P-450 FROM ADRENAL CORTEX MITOCHONDRIA

A.A.Akhrem, T.B.Adamovich, V.N.Lapko, S.A.Sherman, S.A.Usanov, V.M.Shkumatov, V.L.Chashchin

Institute of Bioorganic Chemistry Byelorussian SSR Academy of Sciences, Minsk, U.S.S.R.

Among cytochromes P-450 an exceptional place is occupied by the cholesterol side chain cleavage cytochrome P-450 /Cytochrome P-450scc/ in the key reaction of steroid biosynthesis in adrenal cortex mitochondria. Its most peculiar feature is a three-step cholesterol transformation to pregnenolone without any release of intermediates. Numerous investigations allowed of a rather detailed specification of cytochrome P-450scc and of the cholesterol hydroxylation system as a whole. However, at a molecular level the mechanism of action of this unique hemoprotein remains obscure. The study of such mechanism requires investigations of the relationships between structure and function of cytochrome P-450scc. Such investigations should involve the study of correlation between the functionally important sites and groups in the hemoprotein molecule and its primary structure. The aim of the present work is to solve this problem.

Our previous investigations and literature data allow to ascribe the main functional properties of cytochrome P-450scc to the presence in the hemoprotein molecule of specific sites: i) a catalytic centre containing a prosthetic group, protoporphirin IX; ii) a site responsible for biospecific interactions with the electron transport protein, adrenodoxin; iii) hydrophobic site responsible for the interaction with phospholipid membrane; iv) regulatory site reversibly interacting with pregnenolone to result in high-to-low spin transition and in inhibiting the electron transport.

Such functional variety and thereby stipulated presence of specific sites is, as a rule, manifested by a peculiar topography of protein molecules as individual domains with independent functional significance. Indeed, as it is shown in our earlier publications /Akhrem et.al.1980/ using the limited trypsinolysis, the native molecule of cytochrome P-450scc consists of two domains, F(1) with Mr 29 800 and F(2) with Mr 26 600, linked by a small loop of polypeptide chain whose cleavage does not result in any essential hemoprotein inactivation (not above 20%). The limited trypsinolysis does not significantly affect Kd for adrenodoxin (from 0,14 to 0,37 μM), pregnenolone (from 1,7 to 3,3 μM), for Tween 20 (from 6,7 to 7, 1 μM). The activity of cytochrome P-450scc after its cleavage to F(1) and F(2) is 90% of the starting value. Upon the prolongation of the trypsinolysis a further cleavage of domain F(2) to F(3) with Mr 16 800 has been observed, with retention of F(1); a change in hemoprotein properties is in this case expressed less as compared with limited cleavage to F(1) and F(2). Based on intrinsic spectral and catalytic properties, 70% of cytochrome P-450scc is native after proteolytic modification.

It turned out that the domains do not dissociate after the linking loop of polypeptide chain is cleaved; their interaction is noncovalent, it is disturbed in the presence of either SDS or guanidine hydrochloride.

The separation of fragments in the polypeptide chain of the corresponding domains F(1) and F(2) in the pure form for the structural analysis has
been achieved using chromatography on activated thiol-Sepharose, since only
F(2) is covalently bound with SH-groups of Sepharose. After immobilization
of F(2) on thiol-Sepharose, associated with it F(1) was eluted by the buffer
containing guanidine hydrochloride and the remaining F(2) was eluted upon
the addition to the buffer of 2-mercaptoethanol.

The structural analysis of individual domains showed that F(1) correlates with N-terminal- and F(2) - with the C-terminal part of the polypeptide chain of cytochrome P-450scc. The loop of polypeptide chain connecting
F(1) and F(2) in the native protein molecule is actually of an insignificant length, and is limited by R(250) F(1) and N(257) F(2), whereas trypsin
acts on the peptide bonds R(250)-R(251), R(251)-K(252), K(252)-T(253) and
R(256)-N(257) to form F(2) heterogeneous in N-terminal amino acids. The formation upon prolonged trypsinolysis of F(3) is stipulated by the cleavage
of R(399)-N(400) peptide bond in F(2).

Our further investigations were aimed at elucidation of functional importance of domains assuming the localization therein of functionally important sites.

To achieve dissociation of domains F(1) and F(2) without any loss of
hemoprotein functional properties, a search of conditions for their disagg-
regation has been undertaken, that turned out to be unsuccessful. Even in
the buffer containing 1M KCl and 0,3% sodium cholate where cytochrome P-450
is disaggregated to monomers with Mr 58 000, F(1) and F(2) are present in
the form of aggregates with Mr 400 000 and their continuos storage under
such conditions, as different from the native cytochrome, is accompanied by
spectral changes characteristic of inactivation and by a loss of biospecific affinity to adrenodoxin. However, gel-chromatography of F(1) and F(2)
showed that resulting from continuous incubation of F(1) and F(2) in the a-
bove buffer, in parallel with inactivation process, a fraction with Mr
180 000 emerges besides the fraction with Mr 400 000; a more effective separation is achieved upon the addition of Tween 80 (up to 0,3%) to the buffer containing 1M KCl and 0,3% sodium cholate. An undertaken analysis shows
that the fraction with Mr 400 000 contains 80-85% of F(2) and 15-20% of F(1)
whereas the second fraction has the reverse ratio of domains, and is a he-
mecontaining one, (A(418)/A(280) for the first fraction is 0,3, for the
second - 1,12). Based on the above data a conclusion has been drawn that
the heme is localized in F(1). A nonstability of cytochrome P-450scc modified by trypsin to F(1) and F(2) in a buffer with 1M KCl, 0,3% sodium cholate and 0,3% Tween 80 proves an important role of the polypeptide loop connecting F(1) and F(2), that does not permit the domains to dissociate in
the native protein in the presence of detergents.

To localize the adrenodoxin-binding domain, F(1) and F(2) resulting
from the limited trypsinolysis were placed on a column with adrenodoxin immobilized on Sepharose. However, a stable interaction between F(1) and F(2)
did not allow to selectively isolate these by biospecific chromatography of
the adrenodoxin-binding domain, so that these were sorbed on adrenodoxin-
Sepharose in associated state.

To identify the domain interacting with immobilized adrenodoxin, the
complex (immobilized adrenodoxin- adrenodoxin-binding domain) has been fixed using the cleavable bifunctional reagent. To exclude undesirable steric
effects, immobilization of adrenodoxin on Sepharose has been effectuated
via a long space group. A general scheme of the experiment is given in
Fig.1.

First, the modification of a SH-group of cytochrome P-450scc has been

performed by N-ethylmaleimide. A subsequent treatment of hemoprotein by heterobifunctional reagent, N-succinimidyl-3-(2-pyridyldithio)propionate, pe-
rmitted to introduce into the cytochrome molecule of three activated SH-group. Adrenodoxin immobilized on Sepharose has been preliminarily thiolated by

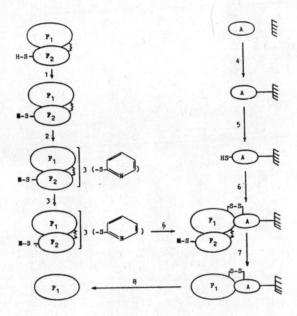

Fig.1. Scheme of experiment on identification of adrenodoxin-binding domain
    of cytochrome P-450scc.

methyl-4-mercaptobutyrimidate to introduce additional SH-groups. At a final
stage the modified cytochrome P-450scc dissociated to domains F(1) and F(2)
was sorbed on thiolated adrenodoxin-Sepharose. Simultaneously, the formation
of a covalent bond between immobilized adrenodoxin and the adrenodoxin-binding domain of cytochrome P-450scc occured. That the complex is biospecifi-
cally bound is proved by the retention under the given conditions of the capacity to enzymatically reduce cytochrome P-450scc in the presence of stoichiometric amounts of adrenodoxin reductase. To remove the domain not interacting with adrenodoxin, the sorbent has been washed by the buffer containing guanidine hydrochloride. The domain interacting with adrenodoxin and
covalently bound with it, was eluted by 5% 2-mercaptoethanol. The electrophoretic analysis showed that the sample eluted with adrenodoxin-Sepharose,
contains above 92% of domain F(1). The obtained data allowed to conclude
that the site of interaction with adrenodoxin is situated in domain F(1).
    To investigate the orientation of cytochrome P-450scc in mitochondrial
membrane and to localize the domain interacting with membrane, the trypsin
effect on the membrane-binding hemoprotein has been studied in submitochondrial particles obtained after distruction of mitochondria by ultrasonica-
tion. Electrophoretic analysis of the trypsin-treated particles showed that
in the process of hydrolysis the protein band intensity is decreased as characteristic of cytochrome P-450scc. Simultaneously, an accumulation of the
material corresponding to domains F(1) and F(2) occured.
    The above data enable us to conclude that the membrane-binding cytoch-

rome P–450scc as well as the isolated hemoprotein consists of two domains, F(1) and F(2), and in the case of membrane–binding cytochrome P–450scc a loop of polypeptide chain binding the domains is not buried into the membrane but remains exposed into solution. We assume, that not all the molecule of cytochrome P–450scc "pearces" the mitochondrial membrane, but it is bound with the latter via one of the domains whereas the other domain is partially or completely exposed into aqueous phase. Since adrenodoxin is a typical soluble protein not bond with membrane, its interaction with membrane–bound cytochrome P–450scc in submitochondrial particle is realised via the site of hemoprotein localized above the membrane. Based on the fact the adrenodoxin–binding site of cytochrome P–450scc is localized in domain F(1), we have assumed that F(1) is exposed into aqueous phase, and F(2) is buried into membrane. To prove the proposed orientation of cytochrome P–450scc in mitochondrial membrane, the accessibility of the surface tyrosine residues of individual domains has been studied under conditions of lactoperoxidase radioiodination. With this aim the enzymatic iodination of isolated and bound with submitochondrial particles hemoprotein has been investigated. The radioactively labeled cytochrome P–450scc was isolated by the standard procedure, then subjected to the limited trypsinolysis up to domains F(1) and F(2), and upon their separation by gel–electrophoresis, the label distribution into fragments has been examined. In the case of membrane–bound hemoprotein the radioactivity ratio amounted to 80% for F(1) and 20% for F(2) whereas for isolated hemoprotein the ratio was 90% and 10%, respectively. It appears that although exposed into aqueous phase F(1) partially interacts with membrane, so that some of its surface tyrosine residues are less accessible to radioiodination as compared with isolated protein.

Based on investigations undertaken here, we have concluded that F(1) is a catalytic domain revealing the enzymatic properties of cytochrome P–450scc whereas F(2) is responsible for the protein interaction with mito chondrial membrane.

The high spin form of cytochrome P–450scc is known to undergo spectral changes characteristic of high–to–low spin transition, in the presence of pregnenolone, Tween 20, Tween 80 and under elevated temperature (up to 35 C; pH – to 8). These alterations are reversible upon the removal of the low spin form inductors (pregnenolone, Tween 20, and Tween 80) decreasing temperature and pH as well as under the action of adrenodoxin and cholesterole. We have concluded that adrenodoxin stabilises the high spin conformation in a complex with cytochrome P–450scc, thus hindering the high–to–low spin transition. Since the spectral changes reveal the alterations in the environment of heme group localized in F(1) which is adrenodoxin–binding as well, it can be assumed that the effect of low spin form inductors is directed at F(1). This is also in accord with the fact that the cleavage of F(1) and F(2) to F(1) and F(3) does not seriously affect Kd of mentioned inductors. In view of the data obtained the effect of temperature seems to be of interest, since, as it is shown /Akhrem et.al.1979/ for the membrane–bound cytochrome P–450scc, the temperature increase causes the reverse spectral changes to those with isolated hemoprotein. Althogh the role of membrane–bound domain is assigned by us to F(2), which allows to assume the temperature induced conformational changes in this domain, a direct stabilizing effect of mitochondrial membrane on F(1) seems most likely, as this domain, although exposed into aqueous phase, partially interacts with membrane. It is also likely that the heme containing site of F(1) is in contact with membrane.

To elucidate the structural basis of functioning of cytochrome P–450scc we have established its primary structure /Fig. 2/.

Recently established is the nucleotide sequence of the site of complementary DNA /Morohashi et.al.1984/ coding cytochrome P-450scc, wherein the results published in 1980 /Akhrem et.al./ have been used.

```
                                                  30
I S T K T P R P Y S E I P S P G D D G W L N L Y H F W R E K
                                                  60
G S Q R I H F R H I E N F Q K Y G P I Y R E K L G N L E S V
                                                  90
Y I I H P E N V A H L F K F E G S Y P E R Y D I P P W L A Y
                                                  120
H R Y Y Q K P I G V L F K K S G T W K K D R V V L N T E V M
                                                  150
A P E A I K N F I P L L N P V S Q D F V S L L H K R I K Q Q
                                                  180
G S G K F V G N I K E D L F H F A F E S I T N V M F G E R L
                                                  210
G M L E E T V N P E A Q K F I D A V Y K M F H T S V P L L N
                                                  240
V P P E L Y R L F R T K T W R D H V A A W D T I F N K A E K
                                                  270
Y T E I F Y Q D L R R K T E F R N Y P G I L Y C L L K S E K
                                                  300
M L L E D V K A N I T E M L A G G V N T T S M T L Q W H L Y
                                                  330
E M A R S L N V Q E M L R E E V L N A R R Q A E G D I S K M
                                                  360
L Q M V P L L K A S I K E T L R L H P I S V T L Q R Y P E S
                                                  390
D L V L Q D Y L I P A K T L V Q V A I Y A M G R D P A F F S
                                                  420
S P D K F D P T R W L S K D K D L I H F R N L G F G W G V R
                                                  450
Q C V G R R I A E L E M T L F L I H I L E N F K V E M Q H I
                                                  480
G D V D T I F N L I L T P D K P I F L V F R P F N Q D P P Q A
```

Fig.2. Complete amino acid sequence of cholesterol side chain cleavage
       cytochrome P-450.

The results from determination of the primary structure on sequenation of DNA and protein, fully coincide, which is very important for evaluating the adequacy of the two methods, since cytochrome P-450scc is one of a few examples of bulky membrane proteins whose primary structure has been established in paralled by two independent methods.
The obtained data allow the following conclusions to be made:
a)Determination of the primary structure of cytochrome P-450scc by the protein chemistry methods unambiguosly proves the existence of hemoprotein in the form of one functional subunit.
b)In case thiolate is the fifth axial ligand, then the catalytic centre of hemoprotein is formed by both domains, since F(1) is heme-containing, and both cystein residues pretending to the role of axial ligand, are located in F(2). Such organization of the molecule of cytochrome P-450scc does, in principle, explain our failure to separate the domains with the retention of the native enzyme properties.
c)From the two cystein residues in domain F(2) (positions 264 and 422 in polypeptide chain) and based on great homology of the site in the region of Cys(422) with other cytochromes P-450, it can be concluded that Cys(422)

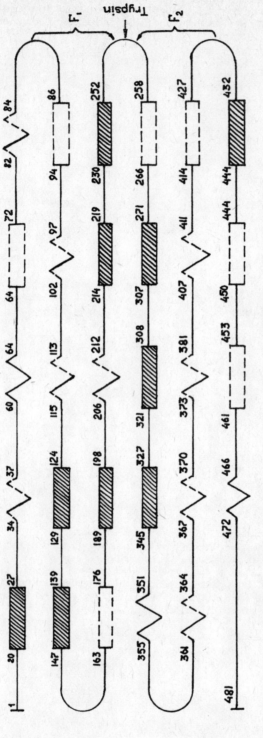

Fig. 3. The elements of the secondary structure of adrenocortical cytochrome P-450scc.

is an axial ligand. Along with that, accoding to the data from prolonged trypsinolysis of F(1) and F(2) to F(1) and F(3) the loss of the native properties is insignificant, although the peptide bond R(399)–W(400) i.e. the one preceding Cys(422) is hydrolyzed.

d)The polarity values for F(1) and F(2) calculated according to Capaldi /1972/ are close and equal to 42,6 and 42,2, respectively. These data do not allow to explain the predominance of interaction of F(2) with mitochondrial membrane. We have assumed that the interaction of F(2) with membrane is stipulated by the presence in domain of the necessary for such contact elements of secondary structure.

The calculation of elements of the cytochrome P–450scc secondary structure performed by the method /Ptitsyn, Finkelstein.1983/ allowed to detect the possible α–helix and β–sheets segments in the polypeptide chain of this hemoprotein. It is evident that extended α–helix (having more than three coils) which is thought to be more preferable for interaction with phospholipid membrane, are located in central– and C–terminal sequence of the cytochrome P–450scc polypeptide chain; this does not contradict to the above mentioned proposal that F(1) P–450scc domain is to a large extent exposed into aqueos phase /Fig. 3/.

REFERENCES

1. Akhrem A.A., Lapko V.N., Lapko A.G., Shkumatov V.M., Chashchin V.L.(1979) Isolation, structural organization and mechanism of action of mitochondrial steroid hydroxylating systems. Acta biol. med. germ., 38,p.257–273.
2. Akhrem A.A., Vasilevsky V.I., Shkumatov V.M., Chashchin V.L.(1980) Structural organization of cytochrome P–450scc from adrenal cortex. In: Microsomes, Drug Oxidation and Chemical Carcinogenesis. N–Y: Acad. Press, 1, p.77–84.
3. Akhrem A.A., Vasilevsky V.I., Adamovich T.B., Lapko A.G., Shkumatov V.M., Chashchin V.L.(1980) Chemical characteristics of the cholesterol side chain cleavage cytochrome P–450 (P–450scc) from bovine adrenal cortex mitochondria. Dev. Biochem., 13, p.57–64.
4. Chashchin V.L., Vasilevsky V.I., Shkumatov V.M., Akhrem A.A.(1984) The domain structure of the cholesterol side–chain cleavage cytochrome P–450 from bovine adrenocortical mitochondria. Biochim. Biophys. Acta, 787, p.27–38.
5. Chashchin V.L., Vasilevsky V.I., Shkumatov V.M., Lapko V.N., Adamovich T.B., Berikbaeva T.M., Akhrem A.A.(1984) The domain structure of the cholesterol side–chain cleavage cytochrome P–450 from bovine adrenocortical mitochondria. Localization of the haem group and domains in the polypeptide chain. Biochim. Biophys. Acta, 791, p.375–383.
6. Capaldi R.A., Vanderkooi G.(1972) The low polarity of many membrane proteins. Proc. Natl. Acad. Sci. USA, 69, p.930–932.
7. Morohashi K., Fujii–Kuriyama Y., Okada Y., Sogawa K., Hirose T., Inayama S., Omura T.(1984) Molecular cloning and nucleotide sequence of cDNA for mRNA of mitochondrial cytochrome P–450(SCC) of bovine adrenal cortex. Proc. Natl. Acad. Sci. USA, 81, p.4647–4651.
8. Ptitsyn O.B., Finkelstein V.(1983) Theory of protein secondary structure and algorithm of its prediction. Biopolymers, 22, p.15–26.

MECHANISMS OF COMPONENT INTERACTIONS IN THE CYTOCHROME P-450 LM2
REDUCTION

Blanck, J., Smettan, G., Ziegler, M. and Ruckpaul, K.

Central Institute of Molecular Biology, Academy of Sciences of
the GDR, 1115 Berlin-Buch, GDR

INTRODUCTION

The reduction reaction, which reflects first the three-compo-
nent interaction in the catalytic cycle of cytochrome P-450 dependent monooxygenases between NADPH-dependent cytochrome P-450
reductase, cytochrome P-450, and lipid, has been investigated
with respect to analyse interactions between the components and
by this to understand which way the system is regulated on a
molecular level.
The data so far obtained evidence that the anaerobic P-450 reduction proceeds inhomogeneously in microsomal and reconstituted systems, as well. Two main partial reactions are discriminated and discussed to be due to
1) randomly distributed proteins in the membrane matrix and regulation of the electron transfer by diffusion control and
   differently favoured P-450 molecules /Taniguchi et al., 1979/,
2) P-450 and reductase association in functional clusters and
   randomly distributed P-450 /Peterson et al., 1976/,
3) different redox states of the reductase in the interaction
   with P-450 /Oprian et al., 1979/,
4) a spin state dependence of the electron transfer, the low spin
   component being unfavourably reduced by a rate limiting spin
   state relaxation or by a following iron-coordination change
   in the transfer reaction sequence /Tamburini et al., 1984/.
In order to understand the component interaction and to discriminate between the different models, the anaerobic P-450 reduction in reconstituted systems was comparably treated by variation of the protein stoichiometry and of the lipid composition,

as well, by means of stopped-flow- and laser temperature-jump
kinetics. The experiments favour the cluster approach and evidence distinct regulation mechanisms with respect to P-450, reductase, and lipid interaction and its modulation by substrates.

MODEL TREATMENT OF THE NADPH-DEPENDENT P-450 REDUCTION

According to the stoichiometry of the electron exchanging proteins (reductase (R), P-450 (P)) in native systems of $R/P \cong$
1/20, the reductase must cycle in order to reduce the cytochrome completely. Therefore the
reaction scheme considers complex formation between both
proteins, electron transfer in
the donor/acceptor complex, followed by the dissociation of
that complex and recycling of
the reductase. The "substrate"
and "product" equilibria are
assumed to be rapid within the
liposomes, and the reductase is
approximated to be always in the
reduced state, due to the rapid
electron transfer between NADPH
and reductase. The mathematical

$$R + P \xrightarrow{K_{RP}} RP \xrightarrow{k_1} RP' \xleftarrow{K_{RP'}} R + P'$$

$$K_{RP} = \frac{R \cdot P}{RP} \; ; \; K_{RP'} = \frac{R \cdot P'}{RP'} \; ; \; K_{RP} = K_{RP'} = K$$

$$R_0 = R + RP + RP'$$

$$P_0 = P + P' + RP + RP'$$

$$P_r = P' + RP'$$

$$[RP] = \frac{R}{K + R} [P_0 - P_r]$$

$$V_r = k_1 [RP] = k_1 \frac{R}{K + R} [P_0 - P_r] = k_{app} [P_0 - P_r]$$

treatment results in a rate equation which is determined by the
rate constant of the electron transfer k_1, a free R dependent
saturation term R/K+R (R is constant throughout the reaction,
and depends on the total protein concentrations R_0, P_0), and by
the concentration of unreduced P-450, $(P_0 - P_r)$. The observed exponential time function of the P-450 reduction is this way refered to the competition of the "product" equilibrium, and the
saturation function $k_{app}(R_0, P_0)$ is a measure of the amount of
RP and by this ofthe dissociation constant K of that complex.
By means of this approach the electron donor/acceptor complex
RP has been kinetically titrated, a direct spectroscopic titration proved unsuccesful.

PHASE CONTROL BY STRUCTURAL IMPLICATIONS

The experimental data of reconstituted vesicle systems as obtained by the cholate gel filtration technique /Ingelman-Sundberg and Glaumann, 1977/ were treated by a two-exponential time function $y=b_0+\sum_i \varphi_i \cdot e^{-k_i t}$ and the respective phase contributions φ_i and rate constants k_i determined by means of nonlinear regression analysis. A distinct dependence of both parameters on the R/P ratio was observed. With relatively increasing R the amount of the physiologically most important rapid partial reaction increases and exhibits saturation behaviour (Fig. 1).

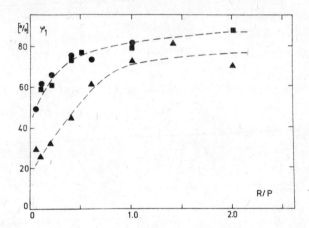

Fig. 1. Anaerobic P-450 LM2 reduction in reconstituted vesicles; amount of the fast partial reaction (φ_1) in dependence on the molar ratio reductase/P-450 (R/P). 0.05 M phosphate buffer, pH 7.4, 20 $^\circ$C. [Benzphetamine] = 1 mM. ▲—▲ dioleoylphosphatidylcholine(DOPC),●—● dioleoylethanolamine (DOPE): phosphatidylserine (PS) (3:1, w/w), ■—■ microsomal lipid mixture

As has been proved by sedimentation analysis previously /Blanck et al., 1979/, the rapid process is due to associated P-450 and reductase, too. Disintegration of such functional clusters by detergents results in the dissappearence of the rapid reaction, the time function becomes monophasic exponential, the rate constant of which corresponds to the slow reaction parameter in mixed kinetics. After disintegration the total amount of P-450 is reduced according to the above scheme /Rohde et al., 1983/.

   In the reconstituted liposomal systems reductase obviously

favours the cluster reaction. That result would either imply
reductase association to reductase-unsupported P-450 clusters,
or a reductase induced formation of reductase-supported clusters. The cluster reaction is further favoured by lipids of negatively charged head groups. It is tempting to directly attri-
bute the surface charge of the membrane lipids a regulating capability in cluster formation, possibly via lipid phase separation. Phosphatidic acid could be evidenced to selectively sur-
round P-450 in reconstituted systems /Bösterling et al., 1981/.
Charge and the lamellophobic characteristics favour phosphatidylethanolamine to selectively interact with P-450 and reductase as evidenced by NMR-investigations by use of a phosphonic
acid ester as marker molecule /Bayerl et al., 1985/. At low
R/P stoichiometry ($<$0.1), as given in microsomes, the cluster
reaction refers to more than 50 % of the P-450 amount. That implies a high degree of P-450 association in reductase-supported
clusters. The existence of P-450 clusters within microsomes and
liposomes of a high degree of association have been evidenced
/Schwarz et al., 1982 a,b/, the association of reductase, however, is open till now for experimental proof. At equimolar
stoichiometry, on the other hand, a P-450/reductase association
could be assured /Gut et al., 1982/. Thus we are left with
functional clusters of P-450 which could include reductase as
the structural background for the interpretation of the mechanism of component interactions in the P-450 reduction process.

RATE CONTROL BY FUNCTIONAL INTERACTION OF REDUCTASE AND P-450

The electron transfer between the NADPH-dependent reductase
and P-450 significantly differs in rate by more than one order
of magnitude either in the cluster or in the random process.
The reaction mechanism, however, could be proved to obey the
above model treatment in both cases. In the disintegrated systems only random reduction occurs, this facilitating the kinetic analysis.
The reconstituted liposomal systems, on the other hand, exhibit phase and rate specificities in dependence on the component ratio. This peculiarity has been overcome in the analysis
of the physiologically relevant rapid cluster process by phase

corrected protein ratios $R/P_{eff} = R/(P \cdot \gamma_1)$ /Blanck et al., 1984/. Fig. 2 shows the dependence of the rate constant k_{app} of the cluster reaction on the protein ratio R/P_{eff} . k_{app} increases with increasing protein ratio and exhibits saturation characteristics. According to the reaction scheme the drawn dependence represents a kinetic titration of a 1:1 reductase/P-450 complex, as evidenced by the calculated dissociation curve of

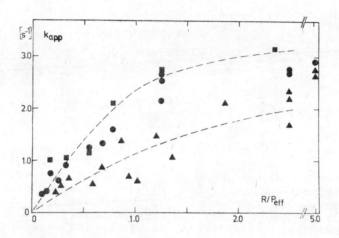

Fig. 2. Anaerobic P-450 LM2 reduction in reconstituted vesicles; dependence of the rate constant k_{app} of the fast partial reaction on the phase corrected protein ratio R/P_{eff}. Conditions and denotation as in Fig. 1.

the RP complex. A lipid specificity, again in favour of negatively charged species, is significant. Tab. 1 shows the respective reaction parameters. An extended view of the importance of negatively charged lipids presents Fig. 3, which exhibits an additional substrate dependence.

Table 1. Reaction parameters of the P-450 LM2 reduction

System	$k_1/s^{-1}/$	K /μM/
Microsomal mixture MIC	3.50 ± 0.38	0.048 ± 0.034
DOPE/PS (3:1, w/w)	3.46 ± 0.41	0.051 ± 0.026
DOPC	3.64 ± 0.43	0.470 ± 0.140

A charge dependence of the reductase/P-450 interaction in the electron transfer could be observed after chemical modification of the N-terminal amino group and of a functionally linked lysine of P-450 LM2, which resulted in a decreased function /Bernhardt et al., 1983, 1984/. Possibly the lipid charge contributes to that interaction, the mechanism of which, however, is rather unknown.

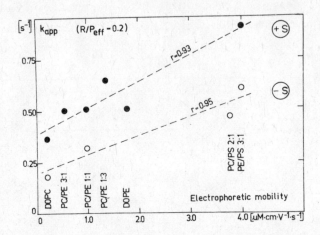

Fig. 3. Anaerobic P-450 LM2 reduction in reconstituted vesicles; dependence of the rate constant k_{app} of the fast partial reaction at constant R/P_{eff} on the electrophoretic mobility of the vesicles /Ingelman-Sundberg et al., 1983/. Conditions and abbreviations as in Fig. 1.

ELECTRON TRANSFER CONTROL BY SUBSTRATE INTERACTION WITH P-450

The interaction of the electron transfer components is further modified by substrate binding to P-450. As evidenced by French et al. /1980/ benzphetamine decreases the dissociation constant of the reductase/P-450 complex in reconstituted LM2 systems in DLPC by about twofold. This way with respect to reductase unsaturated native and reconstituted systems a substrate control via K and thus by the amount of the reactive reductase/P-450 complex could be exerted. The order of magnitude of that change, however, is low as compared to reductase and lipid control. The substrate control, instead of, is rather due to the induced high spin shift of the P-450 spin equilibrium, as supported by Tamburini et al. /1984/ in saturated systems.

RATE LIMITATION BY SPIN STATE RELAXATION

The interpretation of the biphasicity of the electron transfer reaction as being due to a rate limiting spin state relaxation

which would originate the slow partial reaction is rather unlikely. By means of laser temperature-jump investigations the
relaxation of substrate-free and substrate-bound P-450 LM2 could
be proved to proceed in the ns-time scale (Tab. 2) /Ziegler et
al., 1982/.

Table 2. Kinetic parameters of the spin state relaxation of
P-450 LM2, 0.1 M phosphate buffer, pH 7.4, 28 $^{\circ}$C

Species	$\mathcal{T}$ /ns/	$k_{1(ls/hs)}$ /s^{-1}/	$k_{2(hs/ls)}$ /s^{-1}/
P-450 LM2	182 $\pm$ 28	5.9 . 10^5	4.9 . 10^6
P-450 LM2 + benzphetamine	122 $\pm$ 24	3.0 . 10^6	5.2 . 10^6

CONCLUSIONS

The analysis of the component interaction in the P-450 LM2 reduction may be summarized as follows

1) In disintegrated and in cluster systems, as well, the intermolecular interaction of P-450 with reductase proceeds via
formation of a reactive 1:1 complex, which by amount and intrinsic electron transfer capability determines the P-450 reduction rate. The dissociation constant K of that complex significantly depends on lipid in favour of negatively charged
species.

2) The lipid species controls the formation of P-450 clusters
which may further associate reductase. Lipid phase separation
and specific cluster boundary formation is indicated. The mobility of the proteins inside a P-450/reductase cluster must
be rather high in order to fit the observed kinetic mechanism.

3) Substrates are capable of modifying the dissociation constant
K of the P-450/reductase complex in unsaturated systems. That
control is rather less effective as compared to the spin state
control of the P-450 reduction.

4) The outlined component interactions in the first electron
transfer reaction establish the 1-electron reduced reductase
associated ·ternary complex, which in turn decomposes by introduction of the second electron and subsequent substrate conversion. These interactions therefore regulate besides the
substrate induced spin control the steady-state amount of that
complex and this way the catalytic turnover of the substrate.
The observed correlations of first electron transfer and sub-

strate conversion for a series of benzphetamine analogues in
microsomal and reconstituted systems, as well /Blanck et al.,
1983; Schwarze et al., 1985/, support this view.

REFERENCES

Bayerl, T., Klose, G., Ruckpaul, K. and Schwarze, W. (1985)
Biochim. Biophys. Acta 812, 437-446.

Bernhardt, R., Ngoc Dao, N.T., Stiel, H., Schwarze, W., Friedrich, J., Jänig, G.-R. and Ruckpaul, K. (1983) Biochim. Biophys.
Acta 745, 140-148.

Bernhardt, R., Makower, A., Jänig, G.-R. and Ruckpaul, K. (1984)
Biochim. Biophys. Acta 785, 186-190.

Blanck, J., Behlke, J., Jänig, G.-R., Pfeil, D. and Ruckpaul, K.
(1979) Acta Biol. Med. Germ. 38, 11-21.

Blanck, J., Rein, H., Sommer, M., Ristau, O., Smettan, G. and
Ruckpaul, K. (1983) Biochem. Pharmacol. 32, 1683-1688.

Blanck, J., Smettan, G., Ristau, O., Ingelman-Sundberg, M. and
Ruckpaul, K. (1984) Eur. J. Biochem. 144, 509-513.

Bösterling, B., Trudell, J.R. and Galla, H.J. (1981) Biochim.
Biophys. Acta 643, 547-556.

French, J.S., Guengerich, F.P. and Coon, M.J. (1980) J. Biol.
Chem. 255, 4112-4119.

Gut, J., Richter, C., Cherry, R.J., Winterhalter, K. and Kawato,
S. (1982) J. Biol. Chem. 257, 7030-7036.

Ingelman-Sundberg, M. and Glaumann, H. (1977) Biochim. Biophys.
Acta 599, 417-435.

Ingelman-Sundberg, M., Blanck, J., Smettan, G. and Ruckpaul, K.
(1983) Eur. J. Biochem. 134, 157-162.

Oprian, D.D., Vatsis, K.P. and Coon, M.J. (1979) J. Biol. Chem.
254, 8895-8902.

Peterson, J.A., Ebel, R.E., O'Keeffe, D.H., Matsubara, T. and
Estabrook, R.W. (1976) J. Biol. Chem. 251, 4010-4016.

Rohde, K., Blanck, J. and Ruckpaul, K. (1983) Biomed. Biochim.
Acta 42, 651-662.

Schwarz, D., Pirrwitz, J. and Ruckpaul, K. (1982 a) Arch. Biochem. Biophys. 216, 322-328.

Schwarz, D., Pirrwitz, J., Coon, M.J. and Ruckpaul, K. (1982 b)
Acta Biol. Med. Germ. 41, 425-430.

Schwarze, W., Blanck, J., Ristau, O., Jänig, G.-R., Pommerening,
K., Rein, H. and Ruckpaul, K. (1985) Chemico-Biol. Interact.
in press.

Tamburini, P.P., Gibson, G.G., Backes, W.L., Sligar, S.G. and
Schenkman, J.B. (1984) Biochemistry 23, 4526-4533.

Taniguchi, H., Imai, Y., Iyanagi, T. and Sato, R. (1979) Biochim. Biophys. Acta 550, 341-356.

Ziegler, M., Blanck, J. and Ruckpaul, K. (1982) FEBS Lett. 150,
219-222.

LIPID POLYMORPHISM AND THE MICROSOMAL MEMBRANE STRUCTURE

Gert van Duijn, Arie J. Verkleij and Ben de Kruijff

Institute of Molecular Biology, State University of Utrecht, Padualaan 8, 3584 CH Utrecht, The Netherlands

INTRODUCTION

The endoplasmic reticulum of eukaryotic cells is a complex intracellular membrane network and can be considered to be the main metabolic factory of the cell. It is the organelle in which the synthesis and processing of the cellular proteins, lipids and carbohydrates occurs.

By using ^{31}P nuclear magnetic resonance (NMR) techniques it has been shown that at 37°C approximately 60% of the rat liver microsomal phospholipids undergo isotropic motion at the NMR time scale, while at 4°C nearly all the lipids give rise to a signal which is typical for a bilayer organization (De Kruijff et al., 1980b). Furthermore, these authors showed that fully hydrated microsomal phosphatidylethanolamine (PE), which amounts to 20-25% of the total lipids, prefers the hexagonal (H_{II}) phase above 5°C. Therefore it has been suggested that non-bilayer phospholipid phases might occur in this membrane and that these are possibly related to functional properties of the microsomal membrane.

With respect to membrane fusion it could be argued that this is a very common event in the highly dynamic endoplasmic reticulum, e.g. during sugar translocation and/or protein glycosylation (Paiement et al., 1980). Furthermore, during isolation of this cell organelle the membrane is disrupted resulting in the formation of small vesicles, which process involves membrane fusion. In addition, Zilversmit and Hughes (1977) found an extensive exchange of rat liver microsomal phosphatidylcholine (PC), while Hutson and Higgins (1982) observed the rapid transmembrane movement of newly asymmetrically synthesized PE in rat liver endoplasmic reticulum. Van den Besselaar et al. (1978) reported that this process is temperature dependent. These authors demonstrated a rapid exchangeability of at least 90% of the PC at 25°C and 37°C, suggesting a rapid transbilayer movement, while below 10°C the microsomal PC was not exchanged as a single pool. A dependence on temperature, which is very interesting in view of the polymorphism of PE, has also been recently reported for the general susceptibility of microsomal phospholipids towards phospholipase A_2 (Sundler, 1984).

We attempted to gain insight into the role of PE in the appearance of non-bilayer phospholipid structures in biological membranes such as of the microsome, and into the relationship between lipid polymorphism and physiological events. One approach is to study the structural and functional aspects of PE headgroup modifications in model and biological membranes. Since under appropriate conditions trinitrobenzenesulfonic acid (TNBS) can label most of the PE in the endoplasmic reticulum membrane, this molecule appears to be a suitable candidate to study the effect of PE modifications on membrane structure and function. As a first step in this approach we investigated the effects of TNBS labelling on fluidity, hydration and polymorphism of model membranes composed of PE and mixtures of PC and PE, respectively, using ^{31}P NMR, differential scanning calorimetry, small-angle X-ray scattering and freeze-fracture electron microscopy.

Furthermore, the possibility that transiently occurring non-bilayer structures, e.g. inverted micelles, arise from the influence of proteins such as cytochrome P-450 (Stier et al., 1978) cannot be excluded. This specialized microsomal cytochrome is involved in hydroxylation of many different kinds of substrates and drugs. During electron transport to molecular oxygen in various hydroxylation and oxygenation reactions, toxic partial reduction products of oxygen may be formed. These extremely reactive products are possible initiators of lipid peroxidation in biological membranes.

Despite many studies concerning: (1) the oxygen radical initiation and the mechanism of peroxidative reactions (Frankel, 1980; Porter et al., 1980), (2) the rate of lipid peroxidation (Corliss and Dugan, 1970), and (3) the effects of lipid peroxidation on the functional properties of the membrane (Shaw and Thompson, 1982; Barsukov et al., 1980; Gast et al., 1982), virtually nothing is known about the influence of lipid peroxidation on the macroscopic organization of membrane phospholipids. This question becomes even more pertinent in relation to lipid polymorphism and the effects of peroxidation on the functional aspects of membranes. Therefore, the second purpose of this study was to investigate the influence of oxygen-induced phospholipid peroxidation on the phase behavior of aqueous dispersions of both PC and PE. The structural organization of the phospholipids was monitored by ^{31}P NMR and small-angle X-ray scattering. Peroxidation was followed via the rate of formation of malondialdehyde (MDA). The effect of autoxidation on the chemical structure of hydrated PE was monitored by ^{13}C NMR and a determination of the percentage of free amino groups.

RESULTS

DEPE/DETNPPE mixtures

TNBS reacts with the amino group of PE under the formation of trinitrophenylphosphatidylethanolamine (TNPPE) (Fig. 1). ^{31}P NMR and ^{2}H NMR results indicate that TNPPE by itself is anhydrous in character (data not shown). Even in excess of water, in contrast to "normal" phospholipids, the phosphate region of the fluid TNPPE molecules cannot undergo long-axis rotation, presumably due to intermolecular ring interactions.

As revealed by ^{31}P NMR experiments (data not shown), incorporation of up to 10 mol% dielaidoyl-trinitrophenylphosphatidylethanolamine (DETNPPE) in fully hydrated mixtures with dielaidoylphosphatidylethanolamine (DEPE) results in a decrease of the bilayer to hexagonal (H_{II}) phase transition temperature, while no effect is detectable for the gel (L_β) to liquid-crystalline (L_α) transition temperature. Small-angle X-ray diffraction experiments confirm the ^{31}P NMR results and are

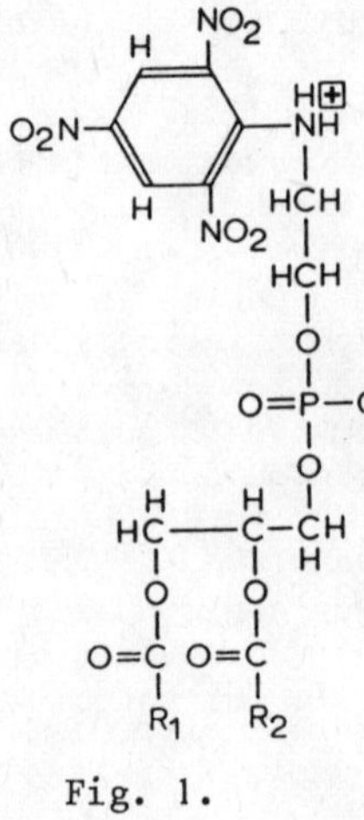

Fig. 1.

Table 1. Repeat distances (Å) in L_β and L_α phases and between brackets tube diameters (Å) in hexagonal (H_{II}) phases of hydrated DEPE-DETNPPE mixtures as a function of temperature[a].

mol% DETNPPE in DEPE	25°C	35°C	40°C	45°C	50°C	55°C	60°C	65°C
0	65.1	65.1	55.0	53.8	53.5	53.1	52.4 (75.8)	(73.9)
1	65.1	64.0	55.0	53.8	53.5	52.8	51.7 (75.1)	(72.7)
2.5	65.1	63.0	55.0	53.8	53.5	52.4 (75.8)	51.0 (73.9)	(72.1)
10	65.1	58.3	53.8	52.8 (73.9)	52.1 (72.7)	52.8 (70.5)	52.4 (68.3)	52.8 (67.3)
30	65.1	57.4	53.5	56.6 (72.7)	(71.6)	(68.3)	(65.8)	(61.8)
50	65.1	(67.3)	(70.5)	(67.8)	(66.3)	(63.5)	(58.9)	(56.4)

[a]The tube diameter was calculated from the first order repeat distance X $\frac{2}{\sqrt{3}}$

shown in Table 1. Between 35°C and 40°C the interbilayer repeat distance for hydrated pure DEPE reduces from 65.1 Å to 55.0 Å due to the transition from the gel state to the liquid-crystalline state. At 65°C the tube to tube distance in the hexagonal (H_{II}) phase is 73.9 Å. Up till 10 mol%, DETNPPE clearly decreases the bilayer to hexagonal (H_{II}) phase transition temperature of DEPE, while the gel to liquid-crystalline transition temperature appears to be unaffected. However, above 10 mol% the interbilayer repeat distance suggests that the lipids are already in the liquid-crystalline state at 35°C, suggesting a slight fluidizing effect of DETNPPE in agreement with DSC data (not shown).

DETNPPE does not influence the lamellar repeat distance both in the gel- and the liquid-crystalline state (Table 1). On the contrary, the tube diameter in the hexagonal (H_{II}) phase decreases as a result of DETNPPE incorporation. As an example, at 60°C the tube to tube distance decreases from 75.8 Å in pure DEPE to 58.9 Å in a sample comprised of DEPE and DETNPPE in a 1:1 molar ratio. A considerable part of this effect is due to the strong decrease in tube diameter of hexagonally organized PE's with increasing temperature (Seddon et al., 1984). However, to a limited extent the repeat distance also appears to decrease as a result of DETNPPE incorporation. This can be seen from the tube diameter observed just above the relevant bilayer to hexagonal (H_{II}) phase transition temperature.

DETNPPE incorporation strongly influences the enthalpy (ΔH) of the gel to liquid-crystalline transition. The ΔH values decrease linearly from 7.0 kcal/mol DEPE in pure DEPE to about 4.0 kcal/mol DEPE after incorporation of up till 10 mol% DETNPPE (Fig. 2).

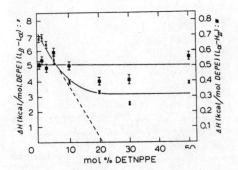

Fig. 2. DETNPPE concentration-dependent enthalpy changes (ΔH) of DEPE for the gel to liquid-crystalline transition (x) and the bilayer to hexagonal (H_{II}) phase transition (●).

In fully hydrated samples containing more than 20 mol% DETNPPE, the enthalpy of the main transition reaches a level of 3.2 kcal/mol DEPE. Extrapolation of the first linear part of the enthalpy versus concentration curve

(Fig. 2) indicates a 1 to 4 molar interaction between DETNPPE and DEPE. The
enthalpy of the bilayer to hexagonal (H_{II}) phase transition of DEPE is hardly affected by the incorporation of DETNPPE.

DOPC/DOTNPPE mixtures

From the former results presented in this paper it can be concluded that
the trinitrophenyl derivative of PE destabilizes the bilayer organization in
hydrated mixtures with PE. However, one could question whether this effect
is restricted to systems containing PE. This question is especially relevant
in relation to PC-PE ratios found in biological membranes. Since in rat liver microsomes PC and PE represent 60% and 25%, respectively, of the phospholipids (Depierre and Dallner, 1975), we investigated the temperature-dependent phase behavior of a hydrated sample comprised of dioleoylphosphatidyl-
choline (DOPC)-dioleoyl TNPPE (7:3) by using ^{31}P NMR and freeze-fracture
electron microscopy.

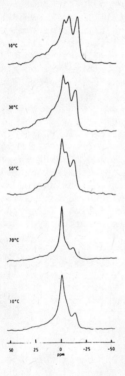

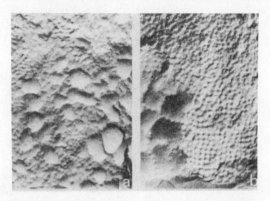

Fig. 4. Freeze-fracture micrographs of an a-
queous dispersion of DOPC-DOTNPPE (7:3), quenched from room temperature after heating till
70°C. Magnification x 55,000.

Fig. 3. Proton-decoupled 36.4 MHz ^{31}P NMR spectra obtained from a fully hydrated mixture of
DOPC-DOTNPPE (7:3), recorded as a function of
temperature.

As registered by ^{31}P NMR (Fig. 3), up till 70°C no hexagonal (H_{II}) phase
can be observed. However, the spectra reveal a sharp isotropic component,
which is superimposed on a bilayer type of spectrum and becomes more dominant at higher temperatures. Also after cooling the aqueous lipid sample
from 70°C to 10°C, the isotropic peak is still present in the spectrum. This
"isotropic phase", which shows a profound hysteresis, possibly represents
stable changes in the macroscopic organization of the lipids. Indeed, as illustrated in Fig. 4, freeze-fracture electron microscopy confirms this suggestion since not only bilayers but also both randomly arranged and more closely packed lipidic particles are visible in the hydrated DOPC-DOTNPPE (7:3)

mixtures when quenched at room temperature after heating till 70°C.

Phospholipid peroxidation measurements

   Oxygen-initiated peroxidation of phospholipids was measured via the formation of thiobarbituric acid reactive material. Both egg-PC and egg-PE showed
a considerable amount of thiobarbituric acid reactive material immediately
after dispersing the lipids in buffer. This is most likely caused by the
presence of endogenous hydro-peroxides in the phospholipids. In control experiments (phospholipid dispersions stored under nitrogen at 40°C), there
was no increase in MDA formation during 24 h (Fig. 5).

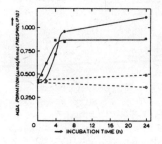

Fig. 5. Time-dependent MDA formation in aqueous
dispersions of egg-PC and egg-PE at 40°C. Egg-PC
under O_2 (■); egg-PC under N_2 (□); egg-PE under
O_2 (●); egg-PE under N_2 (o).

In contrast, there was a substantial increase in MDA formation for both phospholipids in the presence of oxygen. This reached maximal levels after 6-8 h
of incubation (Fig. 5). PE is apparently more sensitive to peroxidation than
PC, as demonstrated by the difference in maximal MDA formation.

Effect of lipid peroxidation on headgroup structure in PE

   The ethanolamine headgroup of PE has also been shown to be very sensitive
towards lipid peroxidation (Corliss and Dugan, 1970). Figure 6 shows the
time-dependent decrease of TNBS-reactive headgroups in PE model membranes
during lipid peroxidation.

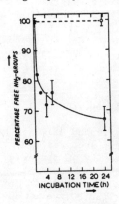

Fig. 6. Time-dependent reactivity of egg-PE towards TNBS during incubation at 40°C under an
O_2 (●) or a N_2 (o) atmosphere.

After 24 h of incubation under oxygen at 40°C, about 33% of the ethanolamine
headgroups were modified in such a way that they were no longer reactive towards the amino-indicating probe. However, the most dramatic disappearance
of free amino groups was monitored during the first 2 h of lipid peroxidation. In control experiments (PE stored under nitrogen at 40°C for 24 h),

134

all of the PE headgroups remained TNBS positive.

This result, together with data obtained from ^{13}C NMR experiments (data not shown), strongly suggests the formation of Schiff bases between about 30% of the PE molecules and aldehydes (e.g. MDA) derived from the peroxidized acyl chains.

Phospholipid organization

Fully hydrated egg-PE molecules show a polymorphic phase behavior, as monitored by ^{31}P NMR (Fig. 7) and small-angle X-ray diffraction (not shown). Figure 7 shows the temperature-dependent shift from the ^{31}P NMR characteristics of a lamellar system (25°C) to a line shape with a reversed asymmetry and a reduced line width (40°C), typical of phospholipid molecules organized in a hexagonal (H_{II}) phase (Cullis and De Kruijff, 1976).

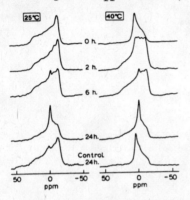

Fig. 7. Temperature-dependent proton-decoupled 36.4 MHz ^{31}P NMR spectra obtained from aqueous dispersions of egg-PE as a function of the time the lipid dispersion was incubated under an oxygen atmosphere. The sample for the control measurements was stored under nitrogen for 24 h at 40°C.

In PE dispersions undergoing peroxidation, rather dramatic changes were monitored during phospholipid peroxidation by ^{31}P NMR. After 2 h of peroxidation, a ^{31}P NMR signal was observed at 40°C, which is typical of a mixture of a lamellar and a hexagonal phase (Fig. 7). The bilayer-stabilizing effect of lipid peroxidation in PE was complete after 6 h of incubation. ^{31}P NMR spectra typical of a hexagonal (H_{II}) organization were not observed at temperatures lower than 70°C (data not shown). In control experiments the hexagonal (H_{II}) phase was still present at 40°C after 24 h of incubation under nitrogen (Fig. 7). Both at 25°C and at 40°C, the ^{31}P NMR spectra, typical of phospholipid molecules organized in a lamellar system, contained a small isotropic peak. This isotropic signal became more dominant after 24 h of peroxidation.

In the small-angle X-ray diffraction profile of the (24 h) peroxidized PE no defined diffraction bands could be observed at 40°C but, instead, a broad diffuse scattering profile was observed (not shown).

Freshly prepared egg-PC liposomes are organized in extended bilayers over a large temperature range, as evidenced by the low-field shoulder and high-field peak (Cullis et al., 1976) separated by approximately 40 ppm (Fig. 8A) in the ^{31}P NMR spectrum. Figure 8B shows the ^{31}P NMR spectra of PC dispersion after 24 h of lipid peroxidation. As shown in Fig. 8, the peroxidation did not affect the ^{31}P NMR characteristics of the lamellar PC system during the 24 h incubation. Only a very small isotropic peak was visible in the ^{31}P NMR spectra obtained from the peroxidized PC liposomes. In addition, the effective residual chemical shift anisotropy, which is the distance between the low-field shoulder and the high-field peak and which is a measure of the local order of the phosphate region (Seelig, 1978), was not affected by the peroxidation process.

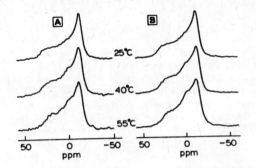

Fig. 8. Proton-decoupled 36.4 MHz ^{31}P NMR spectra of an aqueous dispersion of egg-PC before (A) and after (B) 24 h of lipid peroxidation.

DISCUSSION

PE-headgroup trinitrophenylation and lipid polymorphism

The first purpose of the present study was to investigate the effects of trinitrophenylation on fluidity, hydration and polymorphism of PE-containing model membranes.

The influence of TNPPE on the gel to liquid-crystalline phase transition of PE was studied by means of DSC. Incorporation of more than 5 mol% TNPPE slightly decreased the temperature of the main transition in PE bilayers. However, this effect was predominantly due to a broadening of the transition temperature range. In contrast, TNPPE clearly reduced the enthalpy (ΔH) of the gel to liquid-crystalline phase transition of PE. This could be the result of an interference by the adjacent bulky TNPPE molecules on the condensation of the PE-acyl chains to the gel state. Apparently, the trinitrophenyl derivatives are solubilized in PE bilayers to a limited extent since raising the proportion of TNPPE above 20 mol% did not further decrease the ΔH. Extrapolation to zero of the linear part of the enthalpy versus TNPPE concentration curve leads to the conclusion that there is a 1:4 stoichiometry between TNPPE and PE, and that TNPPE molecules exceeding 20 mol% are segregated in PE-poor domains.

TNPPE molecules also showed a strong influence on the polymorphic phase behavior of hydrated mixtures with PE. As measured by ^{31}P NMR and small-angle X-ray diffraction, the incorporation of up to 10 mol% TNPPE in hydrated mixtures with PE, resulted in an almost linear decrease in the bilayer to hexagonal (H_{II}) phase transition temperature. This linearity indicates that in the concentration range between 0 and 10 mol%, the TNPPE molecules are distributed homogeneously in aqueous aggregates of PE. Furthermore, just above the relevant bilayer to hexagonal (H_{II}) phase transition temperature, TNPPE incorporation also reduced the tube to tube distance to a limited extent.

Apart from hexagonal (H_{II}) phase preferring PE systems, TNPPE also strongly influenced the macroscopic organization of hydrated samples containing PC. PC is by itself a typical bilayer-forming phospholipid. Although biologically relevant TNPPE concentrations did not induce a hexagonal (H_{II}) phase in mixtures with PC, the lipids were organized in closely packed lipidic particles, as indicated by isotropic ^{31}P NMR signals and as visualized by freeze-fracture electron microscopy. This phase behavior is commonly observed in mixtures of bilayer and hexagonal (H_{II}) phase preferring lipids (De Kruijff et al., 1980a). The lipidic particles may be interpreted as inverted micelles, possibly representing an intermediate stage between bilayer and hexagonal (H_{II}) phase (Verkleij, 1984).

In order to gain insight in the molecular origin of the bilayer destabilizing capacity of TNPPE, it is useful to consider various factors determining the polymorphic phase behavior of PE's, e.g. (i) the molecular shape concept and (ii) interbilayer attractive forces. (i) The shape-structure model of Cullis and De Kruijff (1979) relates the overall dynamic molecular shape of a phospholipid to its macroscopic organization in aqueous dispersions. It should be noted that this shape concept is inclusive of the following factors: (1) geometrical space occupied by the isolated molecule, (2) intermolecular interactions, and (3) hydration properties of the lipid headgroups.

In comparison to PE, the TNPPE headgroup occupies a larger area. In addition, incorporation of the bulky TNPPE headgroup will disrupt the formation of hydrogen bonds between PE molecules. With regard to the apolar part of the molecules, it can be concluded that incorporation of TNPPE in bilayers consisting of PE and/or PC with corresponding acyl chains will not influence the mean cross-sectional area of the acyl chains. These properties of the trinitrophenyl derivative would suggest a more cylindrical shape and consequently a bilayer-stabilizing effect of the molecule. Therefore, the dehydration property of TNPPE must be the most important factor, responsible for the pronounced bilayer-destabilizing property of the molecule. This suggestion is reinforced by the finding that TNPPE incorporation in PE decreased the repeat distance in the H_{II} phase, probably reflecting a reduced water content inside the tubes. This is in agreement with the observations of Seddon et al. (1984) that the long spacings in hexagonally organized PE molecules decrease at lower water concentrations.

(ii) The formation of hexagonal (H_{II}) phases has been suggested to be an interbilayer fusion process (Cullis et al., 1980). With respect to this phenomenon, the orientation of the TNPPE-rings in hydrated mixtures with PC and/or PE is of importance. In relation to the hydrophobic character of the TNPPE molecule, two possibilities can be considered. First, the trinitrophenyl rings could be oriented along the interface, and second, the trinitrophenyl groups of TNPPE could be located in the aqueous phase, possibly oriented perpendicular with respect to the bilayer surface. In combination with the tendency for interbilayer ring-ring interaction, such orientations would result in attractive forces between adjacent bilayers. Consequently, in both cases the interbilayer distance would decrease locally, resulting in interbilayer fusion and H_{II} phase formation. The required local dehydration of the bilayers emphasizes the exclusive role of the anhydrous TNPPE molecules in the formation of non-bilayer structures.

Effect of peroxidation on PC and PE model membranes

The second aim of this study was to establish whether peroxidation of aqueous dispersions of egg-PC and egg-PE leads to changes in macroscopic structure of the phospholipids.

Quantitative measurements of peroxidation in phospholipid dispersions revealed that egg-PC is more sensitive to peroxidation than egg-PC. The higher level of MDA in PE dispersions during lipid peroxidation is most likely due to the higher degree of unsaturation in the acyl chains. In addition, egg-PE showed a decreased reactivity towards the amino-indicating probe (TNBS) during lipid peroxidation. This result is in agreement with previously published data (Corliss and Dugan, 1970) and indicates structural modifications in the ethanolamine headgroup, most likely due to the formation of a Schiff base. Quantitative comparison between the formation of MDA and the loss of reactivity of PE toward the amino-indicating probe, as measured after 24 h of incubation under oxygen, strongly suggests cross-linking between PE headgroups via the bifunctional MDA.

In contrast to the results obtained with egg-PC, phospholipid peroxidation strongly influenced the polymorphic phase behavior of egg-PE in aqueous dispersions. ^{31}P NMR measurements revealed the transition from a lamellar organization at 25°C to a hexagonal (H_{II}) phase at 40°C in a freshly prepared egg-PE dispersion. After 2 h of incubation under oxygen about 75% of the PE molecules were organized in bilayers at 40°C. At this level of peroxidation 20% of the ethanolamine headgroups have been modified or cross-linked with MDA. After 6 h of peroxidation, the bilayer stabilization in the PE system belwo 70°C was complete, while 74% of all PE molecules still contained a free ethanolamine headgroup.

The structural modifications can be related to the polymorphic phase behavior of egg-PE in terms of the shape-structure model of Cullis and De Kruijff (1979). Peroxidation of PE will result in a decreased amount of polyunsaturated acyl chains and will give rise to the appearance of trans-cis and trans-trans configurations in the acyl chains (Porter and Lehman, 1982), thereby decreasing the cross-sectional area of the apolar part of the molecule. However, this effect alone cannot explain the more dramatic bilayer stabilization of egg-PE to 70°C after 6 h of incubation. Most likely the structural modifications in the ethanolamine headgroup play a quantitatively more important role with respect to the changed phase behavior. The formation of Schiff bases or cross-links in the PE headgroup region will increase the molecular area of the lipid-water interface and will disrupt the possible intermolecular hydrogen bonds, giving rise to a more cylindrical shape of the molecule. Apparently, the presence of 25% headgroup-modified molecules in the PE dispersions is sufficient to stabilize all phospholipid molecules in the lamellar system after 6 h of peroxidation.

Finally, the results presented in this paper suggest that both PE headgroup modification - and phospholipid peroxidation - experiments can contribute to studies concerning the possible relation between the presence of non-bilayer structure preferring phospholipids in the microsomal membrane and functions of the endoplasmic reticulum.

REFERENCES

Barsukov, L.J., Victorov, A.V., Vasilenko, J.A., Evstigneeva, R.P. and Bergelson, L.D. (1980) Biochim. Biophys. Acta 598, 153-168.
Corliss, G.A. and Dugan, L.R. (1970) Lipids 5, 846-853.
Cullis, P.R. and De Kruijff, B. (1976) Biochim. Biophys. Acta 436, 523-540.
Cullis, P.R. and De Kruijff, B. (1979) Biochim. Biophys. Acta 559, 399-420.
Cullis, P.R., De Kruijff, B. and Richards, R.E. (1976) Biochim. Biophys. Acta 426, 433-446.
Cullis, P.R., De Kruijff, B., Hope, M.J., Nayar, R. and Schmid, S.L. (1980) Can. J. Biochem. 58, 1092-1100.
De Kruijff, B., Cullis, P.R. and Verkleij, A.J. (1980a) Trends Biochem. Sci. 5, 79-81.
De Kruijff, B., Rietveld, A. and Cullis, P.R. (1980b) Biochim. Biophys. Acta 600, 343-357.
Depierre, J.W. and Dallner, G. (1975) Biochim. Biophys. Acta 415, 411-472.
Frankel, E.N. (1980) Prog. Lipid Res. 19, 1-22.
Gast, K., Zirwer, D., Ladhoff, A.M., Schreiber, J., Koelsch, R., Kretschmer, K. and Lasch, J. (1982) Biochim. Biophys. Acta 686, 99-109.
Hutson, J.L. and Higgins, J.A. (1982) Biochim. Biophys. Acta 687, 247-256.
Paiement, J., Beaufay, H. and Godelaine, D. (1980) J. Cell Biol. 86, 29-37.
Porter, N.A. and Lehman, L.S. (1982) J. Am. Chem. Soc. 104, 4731-4732.
Porter, N.A., Weber, B.A., Weener, H. and Khan, J.A. (1980) J. Am. Chem. Soc. 102, 5597-5601.

138

Seddon, J.M., Cevc, G., Kaye, R.D. and Marsh, D. (1984) Biochemistry 23, 2634-2644.
Seelig, J. (1978) Biochim. Biophys. Acta 515, 105-140.
Shaw, J.M. and Thompson, T.E. (1982) Biochemistry 21, 920-927.
Stier, A., Finch, S.A.E. and Bösterling, B. (1978) FEBS Lett. 91, 109-112.
Sundler, R. (1984) Chem. Phys. Lipids 34, 153-161.
Van den Besselaar, A.M.H.P., De Kruijff, B., Van den Bosch, H. and Van Deenen, L.L.M. (1978) Biochim. Biophys. Acta 510, 242-255.
Verkleij, A.J. (1984) Biochim. Biophys. Acta 779, 43-63.
Zilversmit, D.B. and Hughes, M.E. (1977) Biochim. Biophys. Acta 469, 99-110.

MEMBRANE PROTEIN INTERACTIONS IN BIOTRANSFORMATION

Stier, A., Finch, S. A. E., Greinert, R., Taniguchi, H.

Max-Planck-Institut für biophysikalische Chemie,
D-3400 Göttingen, FRG

INTRODUCTION

Homologous and heterologous protein-protein interactions of components
of the drug biotransformation system and their dependence on membrane lipid
structure and composition may be important for the (1) transfer of the
first and second electron to cytochrome P-450, (2) substrate gating
of this electron transfer, (3) the rate limiting step of mixed function
oxygenation (MFO), (4) transfer of products of phase I enzymes (P-450's)
to phase II enzymes like epoxide hydrolase (EH) or UDP-glucuronyl transferase (GT) by a handshaking mechanism and (5) recycling oxygenation to
name some cooperative functions of these membrane bound enzyme systems.

Here we report on interactions of cytochrome P-450 LM2 (LM2) and
cytochrome P-450 LM4 (LM4) with cytochrome P-450 reductase (RED), cytochrome b_5 (b_5) and EH as determined by measurements of rotational diffusion.
The object was to investigate the possibility that long-lived complexes
(heterooligomers) exist, by comparison of rotational diffusion of the
proteins when reconstituted alone or in pairs. Also modulation of rotational behaviour of substrates, reduction of P-450 and lipid composition
was investigated.

ROTATIONAL DIFFUSION MEASURED BY PHOTOSELECTION TECHNIQUES

Photoselection exploits motional modulation of a long-lived - > μs -
polarized emission (in our case delayed fluorescence) of a chromophore
attached to the protein. To this end highly purified LM2, LM4 or EH from
rabbit liver microsomes (Imai et al., 1980) was covalently labelled with
diiodofluorescein iodoacetamide (Greinert et al., 1982). These, with other
unlabelled proteins, were incorporated into vesicles of various mixtures
of lipids by detergent-dialysis (Boesterling et al., 1979).

For photoselection a linearly polarized laser pulse was used to excite
those chromophores in the population with their absorption dipole moments
parallel to the polarization of the exciting light. Thereafter the time-
dependent spatial reorientation of the selected chromophores was measured
by recording the intensity of delayed fluorescence emission parallel
$[I_{||}(t)]$ and perpendicular $[I_{\perp}(t)]$ to the plane of the exciting light
(Greinert et al., 1982a). From the two observed decay curves we calculate
an anisotropy decay curve r(t) according to Eq. (1) (see Fig. 1)

$$r(t) = \frac{I_{||}(t) - I_{\perp}(t)}{I_{||}(t) + 2I_{\perp}(t)} = \frac{d(t)}{s(t)} \qquad (1)$$

Fitting exponential functions to this curve yields kinetic information on the motion of the molecules. The general case of complex rotational behaviour of an asymmetric molecule is described by rather complex functions derived from theory (Kawato and Kinoshita, 1981). Uniaxial rotation is a limiting case in which description simplifies to Eq. (2)

$$r(t)/r_0 = A_1 \exp(-t/\phi_1) + A_2 \exp(-t/\phi_2) + A_3 \qquad (2)$$

with a ratio of the relaxation time constants $\phi_1/\phi_2 = 4$. This model is applicable to membrane proteins where rotational motion is restricted by the 2-dimensional matrix to a single axis usually normal to the plane of the membrane. The factors A_1, A_2 and A_3 are trigonometric functions of the angle between the dye-label and the axis of rotation of the protein – any change in their magnitude is indicative of a change of conformation. This angle is not known a priori in our case. Theoretically it can be visible for limiting conditions by a special kinetics, e.g. for 90° by a monoexponential anisotropy decay (vide infra). The value of the independent residual anisotropy A_3 may contain in addition information on a fraction of proteins which rotate slowly on the experimental time-scale ($\phi_{rot} > 10$ ms) due to formation of large clusters. Whereas this case of strict *uniaxial rotation* allows detailed structural information (shape, size, depth of immersion in the membrane) to be retrieved; a wobbling motion superimposed on rotation (*wobbling rotation*) destroys the simple relationship described and restricts information to detection of relative changes of conformation and rotation between two experimental states without clearcut distinction between these two parameters.

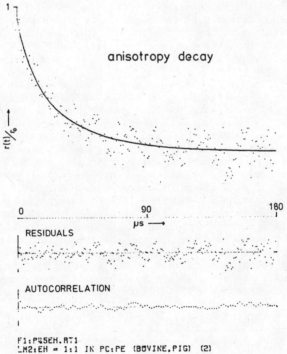

F1:P45EH.RT1
LM2:EH = 1:1 IN PC:PE (BOVINE,PIG) (2)

Fig. 1.   Typical anisotropy decay curve of a LM2-EH reconstituted system.

UNIAXIAL ROTATION OF CYTOCHROME P-450 LM2

The anisotropy decay of reconstituted LM2 follows the kinetics which theory predicts for uniaxial rotation (see Table I). Several arguments point to the structure of a disc-like rotamer of five or six spherical LM2 molecules more or less deeply immersed in the membrane. Dean and Gray (1980) determined detergent solubilized monodisperse LM2 to have a spherical structure by hydrodynamic measurements. The most abundant oligomer

TABLE I
Effects of benzphetamine (1 mM) and dithionite reduction (0.1 mM) on the anisotropy decay of LM2 in reconstituted membranes (egg PC : PE : PA, protein : lipid = 1 : 10 (w/w), cholate dialysis, 37 °C, abbrev. see Table II).

Anisotropy decay (compare Eq. (1))

	A_1	A_2	A_3	ϕ_1[1) [µs]	ϕ_2 [µs]
control	0.13	0.36	0.57	129	31
benzphetamine	0.01	0.34	0.66	204	38
benzphetamine + reduced	0.18	0.15	0.71	51	13

[1) $\phi_1 = \phi_{rot}$ = rotational relaxation time

size produced by solubilization with a gradual decrease of detergent concentration was a pentamer/hexamer (Dean and Gray, 1982). The assumption that LM2 is located in the endoplasmic reticulum membrane with its polar surface facing the cytoplasm appears to be reasonable with regard to its function and explains experiments using polar probes (see e.g. Depierre and Dallner, 1974). Furthermore, for this kind of a rotamer one can calculate reasonable relations between the observed rotational relaxation time and its diameter and depth of immersion into the membrane (Greinert et al., 1982b). An alternative model to explain the observed rotational behaviour would be an aggregate composed of three cylindrical, membrane-spanning LM2 molecules.

Binding of benzphetamine, a type I substrate prototype, greatly changes the conformation of LM2, as evidenced by a reorientation of the axis of the fluorescent chromophore by 45° to an angle of 90° (vide supra) and slows down rotation. This means for the 6-membered rotamer an increase in diameter and depth of immersion by a few Å (vertical shift); for the model of an aggregate of 3 cylinders a considerable lateral expansion or an increase of the number of subunits. Reduction of LM2 by dithionite in the presence of substrate greatly enhances rotation. This may be explained by an upward vertical shift with a strong lateral compression of the rotamer or a dissociative rearrangement of the oligomer.

The existence of LM2 as a hexamer reduces the number of collision partners of reductase by a factor of 6, bringing the ratio of LM2 to RED molecules from around 20 down to a ratio of LM2 oligomers to RED molecules of 3 to 4. The substrate-induced vertical shift of the rotamer may be a mechanism gating electron transfer between the flavine of the reductase and the heme of the cytochrome as the active sites of these proteins become set to the same level in the plane of the membrane.

WOBBLING ROTATION OF CYTOCHROME P-450 LM4

LM4 behaves quite differently from LM2 with regard to the reconstitution procedure and its rotation, in part due to a considerably stronger tendency to aggregate. Only the detergent CHAPS effected solubilization of LM4 which is a precondition for successful reconstitution. Experimental approaches using a great variety of conditions mainly regarding phospholipid composition and choice of detergents resulted in quite different rotational behaviour of the P-450's and provides strong suggestion that conditions of reconstitution of P-450 are critical for the biochemical function, particularly for LM4. Under many conditions suitable for LM2 reconstitution, LM4 appeared to form large aggregates which were immobile in the ms time range. However, LM4 in vesicle membranes prepared from total microsomal lipid extracts exhibits fast wobbling rotation ($\phi_1/\phi_2 \neq 4$) (see Table II). As already stated above from this kind of rotational

TABLE II
Effects of α-naphthoflavone (20 μM) and reduction by dithionite on the anisotropy decay of LM4 in reconstituted membranes (total microsomal lipid, protein : lipid = 1 : 10 (w/w), CHAPS dialysis, 30 °C).

	Anisotropy decay (compare Eq. (1))				
	A_1	A_2	A_3	ϕ_1 [μs]	ϕ_2 [μs]
control	0.58	0.21	0.19	53.2	5.6
+ α-naphthoflavone	0.60	0.30	0.03	181	16.1
+ dithionite	0.50	0.45	0.03	187	18.0
+ α-naphthoflavone + dithionite	0.47	0.33	0.18	218	25.2

behaviour no conclusions on shape, size and quaternary structure can be drawn. Binding of the substrate α-naphthoflavone greatly reduces the rate of rotation with minor effects on conformation (compare factors A_1 - A_3 in Table II). The effects of reduction by dithionite on rotation of LM4 were small in contrast to the effects seen by reduction of LM2.

INTERACTION OF P-450 LM2 WITH P-450 REDUCTASE AND EPOXIDE HYDROLASE

When RED is reconstituted with LM2 in different molar ratios it has a distinct influence on LM2 rotation at a molar ratio of 1 : 1 only (see Table III). In this case LM2 performs a wobbling rotation. Interestingly, this becomes faster with decreasing unsaturation of the membrane lipids, while that of LM2 alone becomes slower. This strongly suggests the formation of a complex with a 1 : 1 stoichiometry between RED and LM2. Enzymic reduction of this complex by NADPH causes its mode of rotation to revert to uniaxial (see Table IV) with a rate faster than the rate of the control. However, enhancement of rotational rate is not as pronounced as in a reconstituted system containing RED 17 mol % of the total protein. In this case the degree of enhancement is similar to that observed when LM2 reconstituted in the absence of RED is reduced by dithionite. Thus the molecular processes associated with substrate binding (data not shown) and reduction - vertical shift or oligomer rearrangement - are not appreciably hindered by the presence of RED. Presumably reduction destabilizes the heterooligomer and results in a partial dissociation. This would be

TABLE III
Interaction of P-450 LM2 with P-450 reductase reconstituted with
different lipid mixtures by cholate dialysis (30 °C).

PC : PE : PA (unsaturated)[1]		PC : PE : PA (polyunsaturated)[2]	
molar ratio LM2 : RED	ϕ_{rot} = rotational relaxation time [μs]	molar ratio LM2 : RED	ϕ_{rot} = rotational relaxation time [μs]
1 : 0 (control)	114	1 : 0 (control)	51
1 : 1	73[3]	1 : 1	131[3]
5 : 1	107	5 : 1	68
		10 : 1	53

[1] phosphatidylcholine (PC) : phosphatidylethanolamine (PE) :
phosphatidic acid (PA) = 2 : 1 : 0.12 (w/w, from egg)
lipid : protein = 10 : 1 (w/w)
[2] phosphatidylcholine (PC) : phosphatidylethanolamine (PE) :
phosphatidic acid (PA) = 2 : 1 : 0.12 (w/w, from pig liver and
bovine brain)
lipid : protein = 10 : 1 (w/w)
[3] wobbling rotation, all other cases uniaxial rotation

consistent with the view that reductase may form only transient complexes
with LM2 during the MFO cycle.

TABLE IV
Effect of enzymatic reduction on interaction of P-450 LM2 and P-450
reductase in PC : PE : PA (2 : 1 : 0.12, w/w) vesicles (protein :
lipid = 1 : 10 (w/w)) at 30 °C.

	ϕ_1 [μs]	ϕ_2 [μs]	(compare Eq. (1))
1. control	114	27	uniaxial
2. LM2 : RED 1 : 1	73	14	–
3. 2. + 100 μM NADPH	84	22	uniaxial
4. LM2 : RED 5 : 1	107	25	uniaxial
5. 4. + 100 μM NADPH	65	16	uniaxial

The rotational diffusion of EH is not influenced by the presence of
an equimolar amount of LM2 nor is the rotation of LM2 effected by an
equimolar amount of EH (see Table V). This lack of mutual binding affinity
– at least for the oxidized LM2 and in the absence of substrates – con-
trasts with the behaviour of LM4.

TABLE V

Interaction of epoxide hydrolase with P-450 LM2 in PC : PE (2 : 1, w/w, from pig liver and bovine brain) vesicles (protein : lipid = 1 : 10 (w/w)) at 30 °C.

	ϕ_1 [µs]	ϕ_2 [µs]	(compare Eq. (1))
1. epoxide hydrolase alone	46	11	uniaxial
2. EH* : LM2 1 : 1	41	10	uniaxial
3. LM2 alone	43	10	uniaxial
4. LM2* : EH 1 : 1	40	10	uniaxial

* protein labelled with DJFJA

INTERACTION OF P-450 LM4 WITH REDUCTASE, b_5, EPOXIDE HYDROLASE

RED or EH when reconstituted with LM4 greatly retard wobbling rotation of LM4 to a similar extent without changing the mode of rotation, while b5 has no appreciable effect (see Table VI). Further experiments under conditions of the various steps of MFO should show whether the observed tendency of LM4 to form aggregates is one aspect of its ability to form well-defined supramolecular structures.

TABLE VI

Interaction of P-450 LM4 with P-450 reductase, cytochrome b5 and epoxide hydrolase (total microsomal lipid, protein : lipid = 1 : 10 (w/w), CHAPS dialysis, 30 °C).

	molar ratio	ϕ_1 [µs]	ϕ_2 [µs]	(compare Eq. (1))
LM4 (control)		53.2	5.6	
LM4 : RED	1 : 1	107	21.2	
LM4 : b5	1 : 1	55.4	10.8	
LM4 : EH	1 : 1	107	13.8	

CONCLUSIONS

The differences of motional behaviour observed between LM2 and LM4 and their implications for protein-protein interactions are most remarkable. LM2 performs uniaxial rotation, LM4 wobbling rotation. The interpretation of kinetic data within the frame of a general bilayer membrane model allows us to obtain a picture of the quaternary structures of the functional states of MFO which is more detailed for LM2 than for LM4. A model of a disc-like oligomer which is shifted vertically in the membrane (in opposite directions) by substrate binding and reduction can be inferred from our results on LM2. When equimolar with RED in the membrane P-450s form complexes with it which have rather different mobilities. Reduction dissociates the LM2-RED complex. Oxidized unliganded LM2 does not interact with EH, but LM4 does. Membrane lipid structure has a strong impact on the motion of the P-450s which is particularly marked in the case of LM4. This, and the method of reconstitution are probably most important for the

enzymic functions of reconstituted systems. There are a few indications
that conclusions from our experiments on reconstituted systems may also
apply to the microsomal membrane. The reduced carbon monoxide complex in
microsomes isolated from phenobarbital-induced male rats exhibited
rotational behaviour similar to that of reduced LM2 in our reconstituted
systems. This was measured by detecting the decay of absorption anisotropy
photolysis of the heme-CO complex (Kawato et al., 1982). LM2 reconstituted
with cholate solubilized liver microsomes was found to exhibit the same
rotational behaviour as in reconstituted systems (Stier et al., 1984). On
the other hand, Schwarz et al. (1982) found evidence for the formation of
large clusters of P-450 in liver microsomes from phenobarbital-induced
rabbits from saturation transfer EPR spectroscopic studies.

ACKNOWLEDGEMENTS

    This work was supported by the National Foundation for Cancer Research,
Washington D. C.

REFERENCES

1. Bösterling, B., Stier, A., Hildebrandt, A. G., Dawson, J. H. and Trudell,
   J. R. (1979) Reconstitution of cytochrome P-450 and cytochrome P-450
   reductase into phosphatidylcholine-phosphatidylethanolamine bilayers:
   characterization of structure and metabolic activity. Molec. Pharmacol.
   16, 332 - 342.
2. Dean, W. L. and Gray, R. D. (1982a) Hydrodynamic properties of monomeric
   cytochromes P-450 LM2. Biochem. Biophys. Res. Comm. 107, 265 - 271.
3. Dean, W. L. and Gray, R. D. (1982b) Relationship between state of
   aggregation and catalytic activity for cytochrome P-450 LM2 and NADPH
   cytochrome P-450 reductase. J. Biol. Chem. 257, 14679 - 14685.
4. Depierre, J. W. and Dallner, G. (1975) Structural aspects of the membrane of the endoplasmic reticulum. Biochim. Biophys. Acta 415, 411 -
   472.
5. Greinert, R., Finch, S. A. E. and Stier, A. (1982a) Cytochrome P-450
   rotamers control mixed-function oxygenation in reconstituted membranes.
   Rotational diffusion studied by delayed fluorescence depolarization.
   Xenobiotica 12, 717 - 726.
6. Greinert, R., Finch, S. A. E. and Stier, A. (1982b) Conformation and
   rotational diffusion of cytochrome P-450 changed by substrate binding.
   Biosci. Rep. 2, 991 - 994.
7. Imai, Y., Hashimoto-Yutsudo, C., Satake, H., Girardin, A. and Sato, R.
   (1980) Multiple forms of cytochrome P-450 purified from liver microsomes of phenobarbital- and 3-methylcholanthrene-pretreated rabbits.
   J. Biochem. 88, 489 - 503.
8. Kawato, S., Gut, J., Cherry, R. J., Winterhalter, K. H. and Richter, C.
   (1982) Rotation of cytochrome P-450. I. Investigations of protein-
   protein interactions of cytochrome P-450 in phospholipid vesicles and
   liver microsomes. J. Biol. Chem. 257, 7023 - 7029.
9. Kawato, S. and Kinosita JR, K. (1981) Time-dependent absorption anisotropy and rotational diffusion of proteins in membranes. Biophys. J.
   36, 277 - 296.
10. Schwarz, D., Pirrwitz, J. and Ruckpaul, K. (1982) Rotational diffusion
    of cytochrome P-450 in the microsomal membrane - evidence for a cluster-
    like organization from saturation transfer electron paramagnetic
    resonance spectroscopy. Arch. Biochem. Biophys. 216, 322 - 328.

11. Stier, A., Finch, S. A. E., Greinert, R., Müller, R. and Taniguchi, H.
    (1984) Structure of endoplasmic reticulum membrane and regulation of
    drug metabolism, in: Pharmacological, morphological and physiological
    aspects of liver aging. Benzooijen, C. F. A., ed., EURAGE, Rijswijk,
    133 - 139.

STUDIES ON LOCATION AND INTERACTION OF ELECTRON-CARRYING ENZYMES IN MICROSOMAL MEMBRANES AND IN MODEL SYSTEMS

V.Lyakhovich, A.Grishanova, O.Gromova, D.Mitrofanov, S.Eremenko, O.Skalchinsky, A.Krainev, L.Weiner

Institute of clinical and experimental medicine, AMS USSR
Institute of chemical kinetics and combustion, AS USSR
630091 Novosibirsk, USSR

The functional activity of the monooxygenase system of liver microsomes is determined by the interaction between its main components, namely, membrane-bound proteins of cytochrome P-450 and NADPH-cytochrome P-450 reductase (below refferred to as reductase).

The kinetic studies of cytochrome P-450 , ameters in the reconstituted systems of the metabolism of xenobiotics (Miwa et al.1979, French et al.1980) have shown that the active complex containing one molecule of cytochrome P-450 and one molecule of reductase is formed during catalysis. In this case, the system shows the highest oxidating activity. The interaction of cytochrome P-450 with the reductase in the soluble reconstituted system of the metabolism of xenobiotics is interpreted as a model of their interaction in microsomal membranes.

However, monooxygenases are membrane-bound enzymes and, most probably, their organization and function in biological membranes obey the laws different from those in a soluble system reconstituted from purified enzymes. Therefore, of paramount importance is the development of methical approaches that would allow of the quantitative characterization of the microsomal system as a whole, on the one hand, and to obtain information about the characteristics of microsomal electron carriers in artificial membranes, on the other.

The objective of this work was to determine an optimal molar ratio between the electron carriers in microsomal monooxygenase systems, at which the oxidating activity of the system would be maximum. To examine the effect of the ratio between cytochrome P-448 and reductase on the efficiency of benz(a)pyrene (BP) hydroxylation, we used also reconstituted microsomal membranes with a variable content of cytochrome P-448. Fluorescence microphotolysis and EPR methods were employed to study the interaction of the reductase with artificial membranes.

MATERIALS AND METHODS

Microsomes were prepared from the livers of male Wistar rats by conventional differential centrifugation. The induction of microsomal enzymes was performed by intraperitoneal injections of 3-methylcholanthrene (MC)(25 mg/kg body wt) for 3 days and phenobarbital (PB)(80 mg/kg body wt) for 4 days.

Cytochromes P-448 and P-450 were purified from microsomes as in (Guengerich and Martin, 1980). The preparations of cytochromes used in the experiments contained 16-18 nmoles per mg protein.

Antibodies to cytochrome P-448 were obtained from the sera of adult rabbits after immunization by purified electrophoretically homogeneous cytochrome P-448 as recommended (Kamataki et al.1976).

Reconstituted microsomal membranes were obtained by self-assembly of liver microsomes solubilized in 4% sodium cholate during gel filtration through Sephadex LH-20 (Mishin et al. 1979). The isolated cytochrome P-448 was incorporated into the reconstituted membranes by adding certain amounts of the enzyme to solubilized untreated or PB-treated microsomes with subsequent incubation for 15 min at room temperature before chromatography on Sephadex LH-20 (Mishin et al.1984). The membranes were also reconstituted from a mixture of the solubilized untreated and MC-treated microsomes.

The total content of cytochrome P-450 was determined according to (Omura, Sato, 1964). The specific contents of cytochromes P-448 and P-450 were determined by rocket immunoelectrophoresis (Pickett et al.1981).

Reductase prepared from liver microsomes of PB-induced rats was purified by Dignam and Strobel's method (Dignam, Strobel, 1975) including solubilization of microsomal suspension by Renex-690 and chromatography on DEAE-Sephadex A-25. Reductase was then additionally purified on a 2',5'-ADP Sepharose 4B (Yasukochi, Masters, 1976). The activity of the reductase was determined at 22°C, as described by Phillips and Langdon (Phillips, Langdon, 1962). The enzyme purified to 20-22 units per 1 mg of the protein was used in experiments.

BP hydroxylase activity was determined from the rates of 3-OH BP accumulation (Lyakhovich et al.1978). The metabolism of BP in the presence of antibodies to cytochrome P-448 was determined after preliminary incubation of microsomal preparations with the antibodies for 10 min at 37°C (Masuda-Mikawa et al.1979).

Lateral mobilities of lipid and the reductase built-in the multibilayers from the natural and synthetic lipid were measured using the system described in (Eremenko et al.1985). In experiments, we used reductase and phosphatidylethanolamine labeled by fluorescein isothiocyanate (FITC), as well as self-fluorescence of reductase prostetic groups, FAD and FMN.

Insertion of the reductase into multibilayers from egg lecithin and dimyristoylphosphatidylcholine was perfomed as described in (Wu, Yang, 1984). Special experiments have proved this procedure to be equivalent to cholate-dialysis method (Halpert et al.1979).

The treatment of experimental curves of the fluorescence recovery after photobleaching and the calculation of diffusion coefficients were made by the known method (Yguerable et al.1982).

Liposomes from egg lecithin containing cholesterol (10% w/w) and built-in probes (I-IV) and reductase were prepared

as in (Krainev et al.1985). To study the interaction of the
reductase with liposome membranes, we used spin-labeled probes of the following structure:

All enzymatic reactions of the reduction were carried out
in 0.1 M K-phosphate buffer (pH 7.6) and 0.1 mM EDTA under
anaerobic conditions (Weiner et al.1982). EPR spectra and the
kinetics of the reduction of probes I-IV were registered using
an ER-200 D spectrometer. The size of liposomes was tested by
quasi-elastic light scattering method using the system described in (Eremenko et al.1981).

RESULTS AND DISCUSSION

1. Catalytic activity of membrane-bound cytochrome P-450.

We have previously shown that the main properties of the
reconstituted membrane-bound monooxygenase system and those of
the native one are similar, including the rates of oxidation
of xenobiotics. Thus, the activity of BP metabolism in reconstituted membranes was 85-100% of the native microsomes.

Preparations of native and reconstituted microsomal membranes were obtained with molar ratios of cytochrome P-448 and
reductase ranging from 2:1 to 32:1. To determine the catalytic
activity of membrane-bound cytochrome P-448, antibody inhibition studies of BP metabolism were performed. The difference
between the rate of inhibition by an excess of antibodies to
cytochrome P-448 and the uninhibited rates was attributed to
cytochrome P-448. Cytochrome P-448 activity was calculated
from the amount of metabolism inhibited by antibodies to cytochrome P-448 divided by the cytochrome P-448 content.

Fig.1 shows that the highest BP hydroxylase activity of cytochrome P-448 was achieved when the molar ratio of cytochrome P-448 to reductase in microsomes did not exceed 6:1. There
was no enhancement of BP hydroxylase activity in preparations
with higher ratios due to increased cytochrome P-448.

The results obtained allow us to make the following conclusions. First, the catalytic activity of cytochrome P-448 in

150

BP hydroxylation in membrane microsomal systems is the highest when the ratio of reductase and cytochrome ranges 1:3 to 1:6, which is essentially different from the equimolar ratio of electron carriers necessary to achieve the maximum activity of cytochrome P-448 in nonmembrane monooxygenase systems.

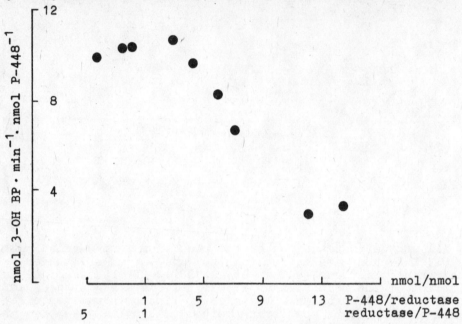

Fig.1   BP hydroxylase activity of cytochrome P-448 in microsomal membranes reconstituted from PB-treated microsomes with incorporated cytochrome P-448 with different molar ratios of cytochrome and reductase

Second, the activity of cytochrome P-448 drastically decreases as the ratio exceeds 1:8.

The interaction of cytochrome P-450 molecules with the reductase is commonly interpreted in two ways. Either this is the interaction at random collisions at their lateral diffusion in the microsomal membrane or the existence of complexes, which assumes a certain rigidity of molecular bonding.

Experimental results for reconstituted nonmembrane monooxygenase systems, liposomes with built-in carriers and microsomes "diluted" with lipids were interpreted by the authors in the framework of the model of lateral diffusion of monomeric carrier molecules. On the other hand, evidence for the functional integration of the monooxygenase components came from the experiments with microsomal membranes (Schwarz et al.1982, Baskin, Yang, 1980). However, Baskin, Yang (1982) showed also the formation of reductase-cytochrome P-450 complexes at their ratio 1:2 in artificial monooxygenase systems.

2. Interaction of the reductase with liposome membranes

To study this reaction and to determine the localization of reductase active centers relative to the membrane surface we

used spin-labeled analogs to fatty acids and phosphatidylethanolamine. EPR spectra of probes I-IV included in the liposomes are reproduced in Fig. 2. The order parameters (S) of these spectra and the temperature dependence of S for probes II-
IV are listed in Table 1. These results indicate that the nitroxyl fragment III is localized in a more ordered region of
the lipid bilayer than IV, which is in agreement with well-
known the bilayer "flexibility profile" (McConnell, 1976).
A rather high mobility of probe II can be accounted for by an
additional averaging of anisotropic motion of the nitroxyl
fragment owing to its rotation around chemical bonds that isolate it from the membrane "crystalline" surface.

Table 1   Order parameters (S) of EPR spectra of II-IV inserted into liposomes

Label	Temperature				
	2°C	12°C	22°C	32°C	42°C
II	0.31	0.26	0.23	0.18	0.18
III	0.72	0.67	0.60	0.56	0.55
IV	0.30	0.26	0.23	0.17	0.16

Lipid concentration 10 mg/ml, II-IV concentration $3 \cdot 10^{-4}$M,
0.1 M K-phosphate buffer (pH 7.6). For probe I, the parameter S was 0.22 at 22°C.

We measured the kinetics of the reduction of probes II-IV
built-in liposomes with ascorbate-ion over the temperature
range from 2 to 42°C. The data on the mobility of probes
(Table 1) and on the reduction with ascorbate-ion allow us to
conclude that nitroxyl fragments II-IV are localized in the
lipid bilayer at the distances relative to the membrane surface that correspond to their chemical structures.
The reductase can reduce spin probes to diamagnetic hydro-
xylamines in the presence of NADPH, while NADPH does not reduce nitroxyl radicals (Weiner et al.1982). The sizes of liposomes remainded unchanged upon insertion of reductase (350±50 Å).
The reduction rate of spin probes is described by

$$\frac{d[R]}{dt} = K_{app} [R] [Red],$$

where [R] is the radical concentration, [Red] is the enzyme
concentration reduced to the same reaction volume. The reduction rate is characterized by the constant K_{app}.
Temperature dependences of K_{app} of the reduction of probes
with the reductase built-in liposomes are shown in Table 2.
As is seen, whereas probes I and II are reduced over the
whole temperature range. III and IV are reduced only at 32°C.
The higher reduction rate of I compared to II may be explained
by that the nitroxyl fragment I is more mobile because of a great number of chemical bonds that isolate it from the membrane
surface and, thus, can more effectively interact with the built-in reductase. The fact that III and IV are not reduced at
low temperatures evidences that their nitroxyl fragments are
inaccessible to the reductase active centres.
Thus, proceeding from the chemical structures of probes I-
IV and from the results obtained, we may conclude that the active centre of the reductase, which reduces I-IV, is located

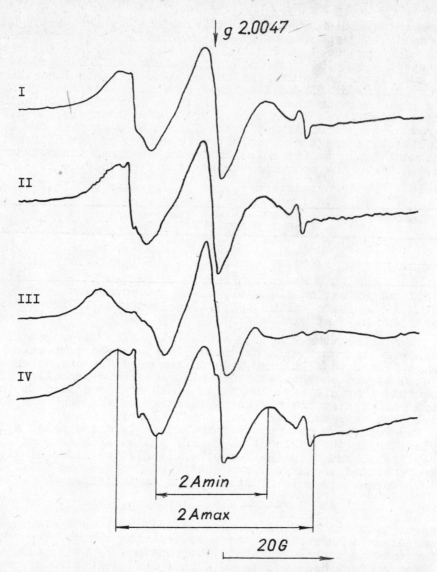

Fig.2 EPR spectra of I-IV insreted into sonicated liposomes
from egg lecithin. Concentrations of I-IV are $3 \div 5 \cdot 10^{-5}$
M. Modulation is 0,5 Hz, microwave power is 12,5 mW.
The order parameter is described as

$$S = \frac{Amax - Amin}{27,55}$$

nearby the membrane surface and is at the depth 5 Å.
    The reduction rate of I and II,calculated on the assumpti-
on that the electron transfer from the reductase into probe is
diffusion-controlled, was $2.7 \cdot 10^{-4}$M s^{-1}(Hardt,1979), which is
10^{-5} times higher than the experimental rate.This is evidence
that the rate-limiting step of the reduction can be either

steric restrictions during the interaction of nitroxyl fragments of I-IV with the enzyme active centre, or a slow occurence of the chemical act of reduction of nitroxyl groups.

Table 2 Apparent constants (K_{app}, $M^{-1}s^{-1}$) of the reduction
of I-IV by reductase built-in liposomes

Label	Temperature			
	2°C	12°C	22°C	32°C
I	$8.7 \cdot 10^2$	$1.5 \cdot 10^3$	$2.0 \cdot 10^3$	$2.4 \cdot 10^3$
II	$3.5 \cdot 10^2$	$4.5 \cdot 10^2$	$7.3 \cdot 10^2$	$1.9 \cdot 10^3$
III	0	0	0	$1.5 \cdot 10^3$
IV	0	0	0	$6.5 \cdot 10^2$

Concentrations: $\overline{[NADPH]}$ = 10^{-3}M, $\overline{[lipid]}$ = $2 \div 5 \cdot 10^{-3}$M, $\overline{[reductase]}$ = $5 \div 20 \cdot 10^{-9}$M, $\overline{[I-IV]}$ = $5 \cdot 10^{-5}$M, buffer(see Materials and Methods)

3. Lateral mobility of the reductase in lecithin multibilayers

Investigation of lateral mobility of microsomal carriers in artificial and natural membranes will permit us to clarify whether their molecular organization is different and what stoichiometry of the carriers is in these cases.

The results obtained by us as follows. 1. Lateral mobility of labeled FITC reductase built-in multibilayers is twice as low as that of the native reductase. 2. The diffusion coefficient for nonlabeled protein changed from $7.5 \cdot 10^{-8}$ to $9.7 \cdot 10^{-8}$ cm^2/s, while for FITC-labeled protein - from $3.7 \cdot 10^{-8}$ to $6.8 \cdot 10^{-8}$ cm^2/s as the temperature was varied from 13°C to 38°C. 3. The diffusion activation energy was about 2.0 kcal/mol and about 4.6 kcal/mol for the native and FITC-labeled protein, respectively. 4. The "jump" (3.5 times) of the diffusion coefficient at the point of phase transition (23°C) was observed for the FITC-labeled reductase built-in the multibilayers from dimyristoylphosphatidylcholine, while for FITC-labeled phosphatidylethanolamine the diffusion coefficient changed about 200 times.

All these experimental data taken together indicate that in lipid multibilayers the reductase moves freely across the membrane, and its movement does not limit the electron transfer into cytochrome P-450.

Method co-insertion of the reductase and cytochrome P-450 into artificial membrane systems and investigation of the diffusion of fluorescence-labeled proteins in microsomes seem to be promising for solving the problems under consideration.

The results reported here evidence that reductase -cytochrome P-450 complexes with a stoichiometry different from 1:1 can exist in microsomal systems. However, in model systems (multibilayers and liposomes) the reductase moves quite freely and does not diffuse into the depth of the lipid matrix. We believe that method of co-insertion of microsomal carriers will make it possible to elucidate the nature of their interactions and stoichiometry in artificial matrices, as well as to determine whether they are adequate to the microsomal system.

REFERENCES

Baskin L.S., Yang C.S. (1980) Biochemistry 19, 2260-2264
Baskin L.S., Yang C.S. (1982) Biochim. Biophys. Acta 684, 263-271
Dignam J.D., Strobel H.W. (1975) Biochem. Biophys. Res. Commun. 63, 845-852
Eremenko S.I. (1981) Ph.D Thesis, Novosibirsk
Eremenko S.I., Skalchinsky O.B., Weiner L.M., Gromova O.A., Lyakhovich V.V. (1985) In: Biological Membranes (USSR) in press
French J.S., Guengerich F.P., Coon M.J. (1980) J.Biol. Chem. 255, 4112-4119
Guengerich F.P., Martin M.V. (1980) Arch. Biochem. Biophys. 205, 365-379
Halpert J., Glaumann H., Ingelman-Sundberg M.(1979)J.Biol. Chem. 254, 7436-7441
Hardt S.L. (1979) Biophys. Chem. 10, 239-243
Kamataki T., Belcher D.H., Neal R.A. (1976) Mol. Pharmacol. 12, 921-932
Krainev A.G., Mitrofanov D.V., Weiner L.M., Lyakhovich V.V. (1985) In: Biological Membranes (USSR) in press
Lyakhovich V.V., Tsyrlov I.B., Gromova O.A., Rivkind N.B. (1978) Biochem. Biophys. Res. Commun. 81, 1329-1335
Masuda-Mikawa R., Fujii-Kuriyama Y., Negishi M., Tashiro Y. (1979) J. Biochem. 86,
McConnell H.M. (1976) In: Spin Labeling, (Berliner L.J., ed) Academic Press, chap. 13
Mishin V.M., Grishanova A.Yu., Lyakhovich V.V. (1979) FEBS Lett. 104, 300-302
Mishin V.M., Grishanova A. Yu., Lyakhovich V.V. (1984) Biokhimya (USSR) 49, 686-691
Miwa G.T., West S.B., Huang M.-T., Lu A.Y.H.(1979) J. Biol. Chem. 254, 5695-5700
Omura T., Sato R. (1964) J. Biol. Chem. 239, 2379-2385
Phillips A.H., Langdon R.G. (1962) J. Biol. Chem., 237, 2652-2660
Pickett C.B., Jeter R.L., Morin J., Lu A.Y.H. (1981) J. Biol. Chem. 256, 8815-8820
Schwarz D., Pirrwitz J., Ruckpaul K. (1982) Arch. Biochem. Biophys. 216, 322-328
Weiner L.M., Gritzan N.P., Bazin N.M., Lyakhovich V.V. (1982) Biochim. Biophys. Acta 714, 234-242
Weiner L.M., Woldman Ya. Yu., Lyakhovich V.V. (1982) In: Cytochrome P-450: Biochemistry, Biophysics and Environmental Implication.(Hietanen E., Laitinen M., Hänninen O., eds), Elsevier, Amsterdam/New York, 501-508
Wu E.S., Yang C.S. (1984) Biochemistry 23, 28-33
Yasukochi Y., Masters B.S.S. (1976) J. Biol. Chem. 251, 5337-5344
Yguerable J., Shmidt J., Yguerable E.E.(1982)Biophys. J. 40, 69-74

MAXIMUM P-450-DEPENDENT MONOOXYGENASE ACTIVITY
IN THE ABSENCE OF PHOSPHOLIPID

D. Müller-Enoch, P. Churchill**, S. Fleischer**
and F.P. Guengerich*.

Department of Physiological Chemistry, University
of Ulm, Oberer Eselsberg, D-7900 Ulm, F.R.G. and
*Department of Biochemistry and **Molecular Biology, Vanderbilt University, Nashville, TN, USA

I. INTRODUCTION

In numerous studies phospholipid has been reported to be
necessary for optimal catalytic activity of a number of mammalian cytochrome P-450 reconstituted systems, composed of
NADPH-P-450 reductase and any of a number of the isozymes of
P-450 (Strobel et al., 1970, Lu and West, 1978). In this study
we find the requirement of phospholipid in reconstituted P-450
systems could be eliminated when complexes of P-450 and NADPH-
P-450 reductase are prepared in high concentrations of each
enzyme prior to measuring monooxygenase activity.

II. MATERIALS AND METHODS

Phospholipids: PC $(12:0)_2$ was purchased from Serdary Research Laboratories Inc. (London, Ontario, Canada). PC $(18:1)_2$
was purchased from Avanti Biochemicals Inc. Other individual
molecular species of mixed fatty acyl phospholipids used in
this study were synthesized according to Eibel (1980).
Enyzmes: $P-450_{PB-B}$, $P-450_{BNF-B}$, $P-450_{BNF/ISF-G}$, and NADPH-
P-450 reductase were purified to electrophoretic homogeneity
using procedures described elsewhere (Guengerich et al., 1982).
The procedures used in the preparation of these enzymes have
been shown to reduce the levels of the non-ionic detergent
Lubrol PX to about 30 µg/mg of protein (2 monomers/P-450 mon-
omer) and the levels of PC and cholate <1 nmol/nmol of enzyme
(Guengerich and Davidson, 1982).
General Assays: 7-Ethoxycoumarin O-deethylase (Ullrich and
Weber, 1972), 7-ethoxyresorufin O-deethylase (Prough et al.,
1978), and scoparone O-demethylase (Müller-Enoch et al., 1981).

III. RESULTS AND DISCUSSION

A number of phospholipids and phospholipid mixtures, used as
soluble molecules or as vesicles, are able to increase the 7-
ethoxycoumarin O-deethylase activity by reconstituted enzyme
systems composed of rat liver $P-450_{PB-B}$ and NADPH-P-450 reductase. The effect of PC $(12:0)_2$ on the $P-450_{PB-B}$ supported reaction is shown in Fig. 1. The presence of PC $(12:0)_2$, even

below the critical micellar concentration increases the activity of $P-450_{PB-B}$ when the P-450 and the NADPH-P-450 reductase are present in low concentrations. However, the degree of apparent stimulation by phospholipid decreased as the concentration of the enzymes increased.

By preincubating $P-450_{PB-B}$ and NADPH-P-450 reductase at high concentrations (10.6 and 19.2 µM, respectively) for 1 h at 25°C we could obtain optimal catalytic activity of 39 nmol product formed min^{-1} (nmol $P-450^{-1}$). This rate was higher than that observed at lower concentrations of the two enzymes, even in the presence of PC $(12:0)_2$ (Fig. 1.).

The effect of phospholipid vesicles on the $P-450_{PB-B}$-dependent 7-ethoxycoumarin O-deethylase activity was increased to a similar extent by a single species of PC, a binary mixture of PC and PE, or a ternary mixture of PC, PE, and PP. The nature of the fatty acyl substituents which were altered here does not appear to markedly influence the stimulation of activity.

When $P-450_{PB-B}$ and NADPH-P-450 reductase were incorporated into phospholipid vesicles by the cholate dialysis method, dilution of the protein-containing vesicles did not decrease the specific activity. This result is due to the fact that dilution of a phospholipid vesicle does not result in any real dilution of the P-450 and NADPH-P-450 reductase. The volume of a phospholipid vesicle remains constant, and is independent of its concentration in solution. The specific activities obtained were 34 and 36 nmol 7-hydroxycoumarin formed min^{-1} (nmol $P-450)^{-1}$ with PC $(18:1)_2$ and with a mixture of microsomal phospholipids, respectively.

The 7-ethoxycoumarin O-deethylase activity of $P-450_{PB-B}$ was found to be dependent upon the concentration of $P-450_{PB-B}$ and NADPH-P-450 reductase preincubated in the absence or presence of PC (12:0), as shown in Fig. 2. When the enzymes were preincubated at a molar ratio of P-450 to reductase of 1:2 at a concentration P-450 greater than 4 µM, there was no activating effect by PC $(12:0)_2$.

The data are consistent with the view, that a complex of $P-450_{PB-B}$ and NADPH-P-450 reductase is catalytically active. Assuming an equimolar ratio of the two enzymes in the complex,

$$P-450_{PB-B} + NADPH-P-450 \text{ reductase} \rightleftharpoons P-450_{PB-B}:NADPH-P-450 \text{ reductase}$$

the K_D of the complex can be calculated using a quadratic equation (Miwa et al., 1979). The K_D values calculated from concentration dependent studies as shown in Fig. 2 of the $P-450$: NADPH-P-450 reductase complex for both $P-450_{PB-B}$ and $P-450_{\beta NF-B}$ was decreased severalfold in the presence of soluble PC $(12:0)_2$. Therefore, at low enzyme concentrations, the addition of PC increases the activity by increasing the steady state concentration of the catalytically active complex.

The mixed function oxidations of 7-ethoxycoumarin, 7-ethoxyresorufin, and scoparone were also examined with $P-450_{PB-B}$, the other rat liver P-450 isozymes $P-450_{\beta NF-B}$ and $P-450_{\beta NF/ISF-G}$; and rabbit liver P-450 isozyme $P-450_{LM-2}$. In no case examined did PC $(12:0)_2$ increase the catalytic activity when the P-450s were incubated with NADPH-P-450 reductase at a 1:2

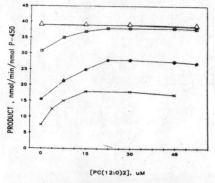

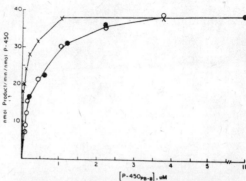

Fig. 1: Effect of PC (12:0)$_2$ on the 7-ethoxycoumarin
O-deethylase activity of P-450$_{PB-B}$ and NADPH-P-450 reductase. Varying concentrations of P-450$_{PB-B}$ and NADPH-
P-450 reductase were mixed in 0.2 ml of 0.1 M Tris-HCl
buffer (pH 7.6) containing 5 mM MgCl$_2$, 0.66 mM 7-ethoxy-
coumarin, and the indicated concentration of PC (12:0)$_2$.
Such mixtures were incubated at 35 °C for 1 min and
reactions were initiated by the addition of 0.5 mM
NADPH. The final concentrations of P-450$_{PB-B}$ and NADPH-
P-450 reductase, respectively, in the cuvette were
0.048 and 0.080 μM (x), 0.19 and 0.35 μM (●), and 1.06
and 1.93 μM (□). The activity represented by (△) was
obtained by preincubating P-450 and the reductase at
concentrations of 10.6 and 19.3 μM, respectively, in
a tube in a final volume of 30 μl for 1 h at 25 °C.
Following this incubation, an aliquot of this preformed complex was added to the cuvette containing
substrate and NADPH in the absence or presence of the
indicated amount of phospholipid to start the reaction.

Fig. 2: 7-Ethoxycoumarin O-deethylase activity of mixtures
of P-450$_{PB-B}$ and NADPH-P-450 reductase in the absence or
presence of PC (12:0)$_2$ as a function of enzyme concentration. At lower concentrations of P-450$_{PB-B}$ (O), the cytochrome was mixed with a 2-fold molar excess of NADPH-P-450
reductase in 0.1 M Tris HCl buffer (pH 7.6) containing
5 mM MgCl$_2$ and 0.66 mM 7-ethoxycoumarin for 60 s at 35 °C
in the presence (x) or absence (O) of 30 μM PC (12:0)$_2$. At
these lower concentrations, the reaction was started by the
addition of NADPH to a final concentration of 0.5 μM. At
the higher concentrations of P-450$_{PB-B}$ (●), the enzymes
were mixed in a small volume (30 μl) of 0.1 M Tris HCl
buffer (pH 7.6) containing 20 % (v/v) glycerol for 60 min
at 23 °C. The reaction was started by the addition of an
aliquot (1-5 μl) to the cuvette containing the assay media
described above including 0.5 mM NADPH.

TABLE I. METABOLISM OF VARIOUS SUBSTRATES BY PREFORMED COMPLEXES OF NADPH-P-450 REDUCTASE
AND VARIOUS P-450s IN THE ABSENCE AND PRESENCE OF PC (12:0)$_2$

PREFORMED P-450:NADPH-P-450 REDUCTASE COMPLEXES WERE MADE IN SMALL TEST TUBES BY MIXING
THE TWO ENZYMES IN A 1:2 MOLAR RATIO, WITH A P-450 CONCENTRATION OF > 5 μM. AFTER SUCH MIX-
TURES WERE MAINTAINED AT 23 °C FOR 60 MIN, ALIQUOTS WERE ADDED TO CUVETTES CONTAINING ALL
OTHER ASSAY COMPONENTS FOR THE DETERMINATION OF ENZYMATIC ACTIVITY.

P-450 Isozyme		Rate (nmol product min^{-1}(nmol P-450)$^{-1}$)		
		Substrate		
		7-Ethoxy-coumarin	7-Ethoxy-resorufin	Scoparone
Rat P-450$_{PB-B}$	- PC (12:0)$_2$	39	<0.01	12.5
	+ PC (12:0)$_2$	39	<0.01	12.6
Rat P-450$_{BNF-B}$	- PC (12:0)$_2$	92	34	0.5
	+ PC (12:0)$_2$	92	33	0.5
Rat P-450$_{BNF/ISF-G}$	- PC (12:0)$_2$	3.6	11.2	2.0
	+ PC (12:0)$_2$	3.3	11.1	2.1
Rabbit P-450$_{LM-2}$	- PC (12:0)$_2$	14.8	0.01	1.4
	+ PC (12:0)$_2$	14.5	0.01	1.3

molar ratio at concentrations higher than 5 µM prior to utilization for activity measurements. The data are shown in Table I.

REFERENCES

1. Eibel, H. (1980): Synthesis of Glycerophospholipids. Chem. Phys. Lipids 26, 405-429.
2. Guengerich, F.P., Dannan, G.A., Wright, S.T., Martin, M.V., and Kaminsky, L.S. (1982): Purification and Characterisation of Liver Microsomal Cytochromes P-450: Electrophoretic, Spectral, Catalytic, and Immunochemical Properties and Inducibility of Eight Isozymes Isolated from Rats Treated with Phenobarbital or ß-Naphthoflavone. Biochemistry 21, 6019-6030.
3. Guengerich, F.P. and Davidson, N.K. (1982): Interaction of Epoxide Hydrolase with Itself and Other Microsomal Proteins. Arch. Biochem. Biophys. 215, 462-477.
4. Lu, A.Y.H. and West, S.B. (1978): Reconstituted Mammalian Mixed-Function Oxidase: Requirements, Specificities, and other Properties. Pharmacol. Ther. Part A, Chemother. Toxicol. Metab. Inhibitors 2, 337-358.
5. Miwa, G.T., West, S.B., Huang, M.-T., and Lu, A.Y.H. (1979): Studies on the Association of Cytochrome P-450 and NADPH-Cytochrome c Reductase during Catalysis in a Reconstituted Hydroxylating System. J. Biol. Chem. 254, 5695-5700.
6. Miwa, G.T. and Lu, A.Y.H. (1981): Studies on the Stimulation of Cytochrome P-450-Dependent Monooxygenase Activity by Dilauroylphosphatidylcholine. Arch. Biochem. Biophys. 211, 454-458.
7. Müller-Enoch, D., Sato, N., and Thomas, H. (1981): O-Demethylation of Scoparone and Studies on the Scoparone-induced Spectral Change of Cytochrome P-450 in Rat Liver Microsomes. Hoppe-Seyler's Z. Physiol. Chem. 362, 1091-1099.
8. Prough, R.A., Burke, M.D., and Mayer, R.T. (1978): Direct Fluorometric Methods for Measuring Mixed-Function Oxidase Activity. In: Methods Enzymol., eds. S. Fleischer, L. Packer, Vol. 52 pp. 372-377, Academic Press, New York-San Francisco-London.
9. Strobel, H.W., Lu, A.Y.H., Heidema, J., and Coon, M.J. (1970): Phosphatidylcholine Requirement in the Enzymatic Reduction of Hemoprotein P-450 and in Fatty Acid, Hydrocarbon, and Drug Hydroxylation. J. Biol. Chem. 245, 4851-4854.
10. Ullrich, V. and Weber, P. (1972): The O-Dealkylation of 7-Ethoxycoumarin by Liver Microsomes. Hoppe-Seyler's Z. Physiol. Chem. 353, 1171-1177.

HETEROLOGOUS RECONSTITUTION OF CYTOCHROME P-450LM2 ACTIVITY WITH BACTERIAL ELECTRON TRANSFER SYSTEMS

Bernhardt, R. and Gunsalus, I.C.[*]

Central Institute of Molecular Biology, Academy of Sciences, 1115 Berlin-Buch, GDR, and *University of Illinois, Department of Biochemistry, Urbana-Champaign, USA

INTRODUCTION

Recognition and specific orientation between P-450 and its electron donor is a necessary prerequisite for the electron transfer. Previously it has been shown that two amino groups of P-450 LM2 are involved in its interaction with microsomal reductase and a charge-pair mechanism of recognition between both proteins has been proposed /Bernhardt et al. 1983, 1984/. Unlike the mammalian system the bacterial camphor- and linalool-hydroxylating monooxygenases are soluble and consist of 3 components. Preliminary results show also the involvement of one amino group each of P-450CAM and P-450LIN in the interactions with the respective redoxins /Bernhardt and Gunsalus, in preparation/. From these data the question arises, if there is a common mechanism of recognizing the electron donor by P-450 or even a common electron donor binding domain on these evolutionary distinct P-450's (P-450LM2 - P-450CAM, P-450LIN).

MATERIALS AND METHODS

The binding constant of putidaredoxin (Pd) and linredoxin (Ld) for P-450LM2 and FITC-modified P-450LM2 was determined by optical difference spectroscopy and by fluorescence measurements. Pd (3-55 μM) and Ld (2-26 μM) induced difference spectra were recorded with a Varian Cary 219 split beam UV/visible spectrophotometer using tandem cuvettes. The P-450LM2 concentration was 14 μM in 50 mM K^+PO_4 buffer, pH 7.0. ΔA_{max} was determined from $\Delta A_{386} - \Delta A_{420}$. For the fluores-

cence measurements 0.6 mol FITC/mol P-450LM2 were bound
(λ_{Exc} = 480 nm and λ_{Em} = 490 - 620 nm). The final P-450LM2
concentration was 1.4 μM. The data were corrected for the inner
filter effect of the redoxins. The free redoxin concentration
has been taken into account for calculation of K_D in both
experiments.
P-450LM2 activity was estimated from the rate of NADH oxidation or HCHO formation /Nash 1953/ in the reconstituted system
with 0.58 μM P-450LM2, 42.2-105.6 μM Pd and 0.56 μM Pd-reductase (PdR) or 0.58 μM P-450LM2, 7.2-34 μM Ld and 0.72 μM Ld-
reductase (LdR). The incubations were performed at pH 7.4 in
0.1 M K^+PO_4 buffer with 10 % glycerol.

RESULTS AND DISCUSSION
By optical difference spectroscopy and fluorescence measurements with a labeled P-450LM2 (0.6 mol FITC/mol P-450) it can
be shown that the bacterial redoxins, Pd and Ld, bind to
P-450LM2 with dissociation constants of about 500 μM and
100 μM, respectively. The K_D-values for the homologous systems
are 0.115 μM for a 1:1 reconstituted complex of P-450LM2 and
NADPH-cytochrome P-450 reductase (in presence of dilauroylphosphatidylcholine) /French et al. 1980/; 17 μM for P-450CAM/
Pd and 10 μM for P-450LIN/Ld (Table 1).

Table 1   Binding constants of cytochrome P-450LM2, P-450CAM
          and P-450LIN for different electron donors

	K_D(μM)
P-450LM2 + microsomal reductase (+DLPC)	0.115[*]
P-450CAM + Pd	17
P-450LIN + Ld	10
P-450LM2 + Pd	~ 500
P-450LM2 + Ld	~ 100

[*]data from French et al. /1980/

Like microsomal reductase the bacterial redoxins induce a
high spin shift of the heme iron of P-450LM2, whereby the
decrease at 420 nm is much more pronounced than the increase

at 386 nm. When bound to P-450CAM Pd induces a type II spectrum.
Ld induces an atypical spectrum with P-450LIN leading only to a
decrease in the absorption around 400 nm without any peak. The
origin of this behavior is not understood as yet.

Since the bacterial electron donors obviously bind rather specifically to P-450LM2 the catalytic activities of the heterologous systems have been proved. Heterologous reconstitution
of P-450LM2 with Pd and PdR as well as with Ld and LdR in the
presence of benzphetamine lead to NADH-oxidation. This was
rather unexpected since no substrate consumption has been observed at heterologously combining P-450CAM with Ld and LdR
and P-450LIN with Pd and PdR, respectively /Ullah et al. 1983/.
To exclude that the NADH-consumption was due to uncoupling leading to the formation of only H_2O_2 the product formation was
followed by determination of HCHO. During the first 5 minutes
of incubation a linear HCHO-formation was observed, the rate
of which declined during the next 5 minutes (Fig. 1).

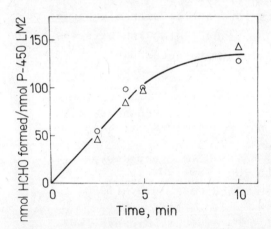

Fig. 1   Time-dependence
of HCHO-formation in re-
constituted systems of
P-450LM2 with bacterial
Pd and Pd-reductase
(0-58 µM: 105.6 µM :
0.56 µM) and Ld and Ld-
reductase (0.58 µM :
21 µM : 0.7 µM), respec-
tively

▽——▽ = P-450LM2/Ld/LdR

o——o = P-450LM2/Pd/PdR

Thus, the efficiency of the product formation in the homologous reconstituted microsomal and heterologous microsomal/
bacterial systems can be compared by changing the electron
donor concentration and extrapolating V_{max}. Table 2 shows
that under the conditions used V_{max} in the heterologous
systems is in the same order of magnitude as in the homologous
one.

Table 2 Turnover numbers of homologously and heterologously
         reconstituted P-450LM2

$$V_{max}(min)$$

P-450LM2 + microsomal reductase + DLPC   41[*]

P-450LM2 + Pd + PdR (0.58 µM:42.2-105.6µM:0.56 µM)   42

P-450LM2 + Ld + LdR (0.58 µM:7.15-34 µM:0.72 µM)   52

[*] data from Koop and Coon /1979/ (P-450LM2:reductase = 1:1.1)

The mechanism of the heterologous hydroxylation is not known
yet. The results are consistent with a model where P-450LM2
ist not very selective with respect to its electron donor.
There could, on the other hand, be a rather conserved recognition domain on P-450 for the electron donor, so that the
latter one recognizes also a "strange" P-450. This idea is
supported by the fact that modification of P-450LM2 by FITC,
which is known to disturb its interaction with microsomal reductase /Bernhardt et al. 1983, 1984/, also inhibits the NADH-
oxidation via bacterial redoxins (data not shown).

REFERENCES

Bernhardt, R., Ngoc Dao, N.T., Stiel, H., Schwarze, W.,
Friedrich, J., Jänig, G.-R. and Ruckpaul, K. (1983)
Biochim. Biophys. Acta 745, 140-148

Bernhardt, R., Makower, A., Jänig, G.-R., and Ruckpaul,K.
(1984) Biochim. Biophys. Acta 785, 186-190

French, J.S., Guengerich, F.P., and Coon, M.J. (1980)
J. Biol. Chem. 255, 4112-4119

Koop, D.R., and Coon, M.J. (1979) Biochem. Biophys. Res.
Commun. 91, 1075-1081

Nash, J. (1953) Biochem. J. 55, 416-421

Ullah, A.H.J., Bhattacharryya, P.K., Bakthavachalam, J.,
Wagner, G.C., and Gunsalus, I.C. (1983) Fed. Proc. 42, 1897

SUICIDE INACTIVATION OF RAT LIVER CYTOCHROMES P-450 BY CHLORAMPHENICOL: MECHANISM, ISOZYME-SELECTIVITY, AND STRUCTURAL REQUIREMENTS

James R. Halpert[1], Natalie E. Miller[1], Lee D. Gorsky[2], and Laurence S. Kaminsky[3]

1. Department of Pharmacology and Toxicology, College of Pharmacy, University of Arizona, Tucson, Arizona, 85721, USA
2. Department of Biological Chemistry, Medical School, University of Michigan, Ann Arbor, Michigan, 48109, USA
3. Wadsworth Center for Laboratories and Research, New York State Department of Health, Albany, New York, 12201, USA

INTRODUCTION

Despite the considerable effort devoted in recent years to elucidating the details of the catalytic mechanism of substrate oxidation by cytochromes P-450, relatively little information is available regarding the role of particular amino acid residues in the enzymes in regulating such essential functions as substrate binding or interaction with NADPH-cytochrome P-450 reductase. A major reason for the slow progress in this regard has been the lack of specific irreversible inhibitors which could be used to map the various binding regions on the P-450s.

Several years ago we reported that the antibiotic chloramphenicol acts as a suicide substrate of the major phenobarbital-inducible isozyme of rat liver cytochrome P-450 (P-450 PB-B2). In contrast to the vast majority of suicide substrates of P-450, chloramphenicol was found to inactivate the cytochrome by virtue of the covalent modification of the protein and not the heme moiety. This fact, together with the rather specific stoichiometry of 1.5 nmol chloramphenicol metabolites bound/nmol P-450 inactivated, suggested that chloramphenicol might prove to be a valuable tool for identifying amino acid residues at or near the active site of the enzyme (Halpert and Neal 1980). Subsequently, the major covalently-bound species of chloramphenicol was identified as an adduct of chloramphenicol oxamic acid and the ε-amino-group of one or more lysine residues in the cytochrome P-450, and evidence was presented that modification of lysine residues is primarily responsible for the inactivation of the P-450 by chloramphenicol (Halpert 1981). Later studies suggested that chloramphenicol might be inactivating the cytochrome by impairing its ability to accept electrons from the P-450 reductase (Halpert et al. 1983).

In the present investigation we have sought 1) to determine whether the inhibition of the enzymatic reduction of cytochrome P-450 PB-B2 resulting from its covalent modification by metabolites of chloramphenicol is sufficient to account for the loss of monooxygenase activity 2) to delineate the defect in the modified cytochrome responsible for its impaired reduction 3) to determine the isozyme-selectivity of chloramphenicol as an inhibitor of rat liver cytochromes P-450 in vivo 4) to determine which structural features of the chloramphenicol molecule are responsible for its effectiveness and isozyme-selectivity as a suicide substrate of rat liver cytochromes P-450.

RESULTS

Mechanism of the inactivation. Preparations of cytochrome P-450 PB-B2 from rats treated in vivo with 300 mg/kg chloramphenicol (CAP PB-B2) were isolated, and their catalytic, spectral, and physical properties were compared to those of the native PB-B2 (Halpert et al. 1985). The CAP PB-B2 exhibited: 1) a

12*

RESULTS (Continued)

60-70% loss in the rate of NADPH-supported monooxygenase activity with the substrates benzphetamine, 7-ethoxycoumarin, and p-nitroanisole; 2) a 60% decrease in the extent of enzymatic P-450 reduction catalyzed by NADPH-cytochrome P-450 reductase under both aerobic and anaerobic conditions; 3) a 60% decrease in the steady-state level of the ferrous dioxygen complex in the presence of substrates; 4) a 60% decrease in the magnitude of the type I spectral change induced by benzphetamine; and 5) a shift in the wavelength maximum for the chemically-reduced ferrous carbonyl complex from 450 to 451.5 nm. On the other hand, the ability of the CAP PB-B2 to catalyze the iodosobenzene-supported metabolism of 7-ethoxycoumarin and p-nitroanisole was unchanged. Furthermore, the results of steady-state kinetic, gel filtration, and spectral titration studies all suggested that complex formation with the reductase is unaltered by covalent binding of chloramphenicol. The results are consistent with the hypothesis that the binding of chloramphenicol metabolites to amino acid residues in the PB-B2 in close vicinity to the heme moiety blocks electron transport from NADPH-cytochrome P-450 reductase, thereby leading to a loss of monooxygenase activity. Whether inhibition of electron transfer results from 1) modification of amino acid residues in the cytochrome P-450 involved in complementary charge interactions with the reductase or 2) steric hindrance by the bulky bound chloramphenicol metabolites, remains to be elucidated.

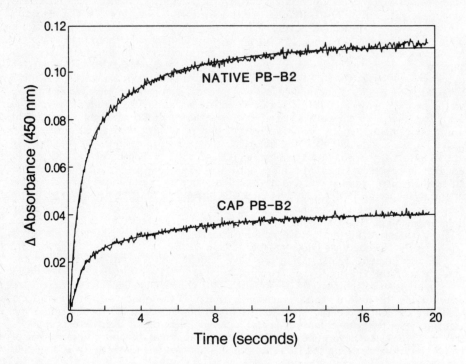

Fig. 1. Time course of anaerobic reduction of CAP and native PB-B2 by NADPH cytochrome P-450 reductase as assayed by stopped-flow spectrophotometry.

RESULTS (Continued)

Isozyme-selectivity of chloramphenicol. Untreated rats and rats treated with the inducers phenobarbital, β-naphthoflavone, pregnenolone-16α-carbonitrile, and clofibrate were injected intraperitoneally with 100 mg or 300 mg/kg chloramphenicol, and inhibition of specific cytochrome P-450 isozymes was assessed by monitoring the metabolism of warfarin (isozymes UT-A, PB-B, PB-C, PCN-E, and BNF-B), testosterone (isozymes UT-A and UT-F), isosafrole (isozymes BNF-B and ISF-G) or lauric acid (clofibrate-induced isozyme) in subsequently prepared hepatic microsomal preparations. Of the eight major cytochrome P-450 isozymes which could be monitored in this fashion, three were inhibited by more than 50% by the higher dose of chloramphenicol, whereas no evidence of inhibition of the remaining isozymes was obtained. Isozyme PB-C, which is present in significant amounts in untreated rats and which is induced approximately 2-fold by phenobarbital (Guengerich et al. 1982), was the most susceptible cytochrome P-450 to inhibition by chloramphenicol in vitro as well as in vivo. Isozyme PB-B, the major phenobarbital-inducible isozyme, and isozyme UT-A, a male-specific testosterone 2α-and 16α-hydroxylase, were intermediate in their susceptibility to chloramphenicol. In contrast, the major isozymes induced by β-naphthoflavone, pregnenolone -16α-carbonitrile, and clofibrate, as well as a constitutive testosterone 7 α-hydroxylase (UT-F), were not inhibited by chloramphenicol.

Structural requirements of the suicide inactivation of rat liver cytochromes P-450 by chloramphenicol. The importance of three different structural features of the chloramphenicol molecule in regulating its effectiveness and selectivity as a suicide substrate of rat liver cytochromes P-450 was investigated. These structural features are 1) the propanediol side chain 2) the p-nitro group and 3) the dichloromethyl moiety. The analog 1-p-nitrophenyl-2-dichloroacetamidoethane (pNO$_2$DCAE) was synthesized first and tested as a suicide substrate of the major phenobarbital-inducible (PB-B) and β-naphthoflavone-inducible (BNF-B) isozymes of rat liver cytochrome P-450 in vitro. The maximal rate of inactivation of the PB-B by the analog was the same as with chloramphenicol itself ($k_{inactivation}$ = 0.5 min $^{-1}$) but the inhibitor concentration required for half-maximal inactivation (K_I) was 0.7 μM for pNO$_2$DCAE as opposed to 15 μM for chloramphenicol. Furthermore, whereas chloramphenicol does not inactivate isozyme BNF-B, this isozyme was inactivated by pNO$_2$DCAE at a rate approximately half that of PB-B. Similarly, the analog 1-phenyl-2-dichloroacetamidoethane was found to inactivate both isozymes PB-B and BNF-B at rates comparable to those attained with the p-nitrophenyl analog. In contrast, analogs containing a difluoromethyl or monochloromethyl instead of a dichloromethyl group were inactive, whereas substitution of a dibromomethyl group caused a slight enhancement of the rate of inactivation. Increasing or decreasing the number of methylene groups between the benzene ring and the amide nitrogen was found to decrease the rate of inactivation of PB-B. Thus, at present, 1-phenyl-2-dichloroacetamidoethane represents the minimal structure required for highly efficient inactivation of isozymes PB-B and BNF-B.

$$NO_2 - \bigcirc - CH_2 - CH_2 - NH - \overset{\overset{\displaystyle O}{\|}}{C} - CHCl_2$$

1-p-Nitrophenyl-2-dichloroacetamidoethane

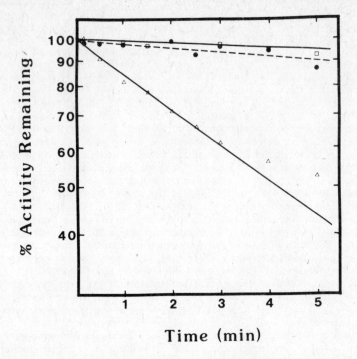

Fig. 2. Effect of preincubation with chloramphenicol and 1-p-nitrophenyl-2-dichloroacetamidoethane on the ethoxycoumarin deethylase activity of purified P-450 BNF-B in a reconstituted system.
- ● no inhibitor
- ▫ 100 μM chloramphenicol
- ▲ 5 μM 1-p-nitrophenyl-2-dichloroacetamidoethane

REFERENCES

Guengerich, F.P., Dannan, G.A., Wright, S.T., Martin, M.V., and Kaminsky, L.S. (1982). Purification and characterization of liver microsomal cytochromes P-450: Electrophoretic, spectral, catalytic, and immunochemical properties and inducibility of eight isozymes isolated from rats treated with phenobarbital or β-naphthoflavone. Biochemistry 21:6019-6030.

Halpert, J. (1981). Covalent modification of lysine during the suicide inactivation of rat liver cytochrome P-450 by chloramphenicol. Biochem. Pharmacol. 30:875-881.

Halpert, J. and Neal, R.A. (1980). Inactivation of purified rat liver cytochrome P-450 by chloramphenicol. Mol. Pharmacol. 17:427-431.

Halpert, J., Naslund, B., and Betner, I. (1983). Suicide inactivation of rat liver cytochrome P-450 by chloramphenicol in vivo and in vitro. Mol. Pharmacol. 23:445-452.

Halpert, J., Miller, N., and Gorsky, L. (1985). On the mechanism of the inactivation of the major phenobarbital-inducible isozyme of rat liver cytochrome P-450 by chloramphenicol. J. Biol. Chem., in press.

ON THE REGULATIVE FUNCTION OF CYTOCHROME b5 IN CYTOCHROME P-450-DEPENDENT DRUG OXIDATIONS

Peter Hlavica and Ines Golly

Institut für Pharmakologie und Toxikologie der Universität, Nussbaumstrasse 26, D-8000 München 2, F.R.G.

INTRODUCTION

The role of cytochrome b_5 as a potential electron donor to oxygenated ferrocytochrome P-450 is now well established (White and Coon, 1980). Recent findings suggest that cytochrome b_5 might also act as an effector in cytochrome P-450-catalyzed drug oxidations, modulating the interaction of substrates with the oxidase (Tamburini and Gibson, 1983; Hlavica, 1984; Morgan and Coon, 1984).

The present study concerns the influence of cytochrome b_5 and some of its analogues on the kinetics of rabbit liver microsomal N-oxidation of 4-chloroaniline. It will be shown that, dependent on the type of reductase coupled to the hemoprotein, the adduct thus formed affects the rate of product release from the ternary cytochrome P-450 complex.

MATERIALS AND METHODS

Microsomal fractions from the livers of phenobarbital-treated rabbits were prepared by the method of Jagow et al. (1965). Cytochrome b_5 (holo-b_5), apocytochrome b_5 (apo-b_5) and manganese-protoporphyrin IX-substituted cytochrome b_5 (Mn(III)-b_5) were prepared as described by Morgan and Coon (1984). Incorporation of the proteins into rabbit liver microsomal fractions was achieved by incubating appropriate amounts of each component in 0.15 M KH_2PO_4/Na_2HPO_4, pH 7.4, for 30 min at 37°C. Subsequently, microsomes were collected by centrifugation at 105,000 x g for 60 min, washed with ice-cold 0.15 M KCl and resuspended in phosphate buffer, pH 7.4. Reduced pyridine nucleotide oxidase, cytochrome P-450(P-450) and cytochrome c reductase activities were measured as described previously (Hrycay and Estabrook, 1974). Reoxidation of b_5 was assayed by monitoring the decrease in absorbance at 424 nm relative to 448 nm at 25°C using an Aminco-Morrow stopped-flow apparatus attached to an Aminco Dasar/DW-2 spectrophotometer system. The final concentrations of the reactants in the observation cell were as follows: 0.15 M KH_2PO_4/Na_2HPO_4, pH 7.4; microsomal fraction equivalent to 1 µM P-450 and 1.18 µM b_5; NADH or NADPH, 1.2 mM each; 4-chloroaniline (4-CA), 1 mM; gas phase as indicated in the legend to the figure. The reaction mixtures for measuring N-oxidation of 4-CA contained: 0.15 M KH_2PO_4/Na_2HPO_4, pH 7.4; microsomal fraction with or without extra-bound b_5, equivalent to 0.2 or 1.0 µM P-450; NADPH or NADH, 1.2 mM each; 4-CA, 1 mM. Reactions were carried out for 3 and 20 min, respectively, at 37°C.

RESULTS AND DISCUSSION

Incorporation into rabbit liver microsomal fractions of detergent-solu-

TABLE 1

INFLUENCE OF CYTOCHROME b_5 ON VARIOUS ENZYME ACTIVITIES IN RABBIT LIVER MICROSOMAL FRACTIONS

Parameter measured	Microsomal fraction used	
	Native	Supplemented with holo-b_5
[b_5]/[P-450]	0.43	1.33
NADH oxidase[a]	0.66 $\pm$ 0.03	0.85 $\pm$ 0.07
NADPH oxidase[a]	8.02 $\pm$ 0.33	3.03 $\pm$ 0.18
NADH-P-450 reductase[b]	0.93 $\pm$ 0.01	1.06 $\pm$ 0.04
NADPH-P-450 reductase[b]	2.21 $\pm$ 0.17	1.15 $\pm$ 0.15
NADPH-cyt.c reductase[c]	24.80 $\pm$ 5.40	22.80 $\pm$ 2.40
NADH-dependent N-oxidase[d]	1.18 $\pm$ 0.04	0.81 $\pm$ 0.34
NADPH-dependent N-oxidase[d]	7.31 $\pm$ 0.35	5.56 $\pm$ 0.69
NADH/NADPH-dependent N-oxidase[d]	7.57 $\pm$ 0.44	9.15 $\pm$ 0.05

[a]Activity expressed as nmol reduced nucleotide oxidized/min/ nmol P-450

[b]Activity expressed as nmol P-450 reduced/min/nmol P-450

[c]Activity expressed as nmol cyt.c reduced/min/mg protein

[d]Activity expressed as nmol 1-chloro-4-nitrosobenzene formed/ min/nmol P-450

bilized b_5 results in considerable decrease in NADPH-P-450 reductase activity (Table 1). It could be argued that b_5 inhibits the reaction by diverting electrons away from P-450. However, two lines of evidence raise doubts concerning the validity of this hypothesis. Firstly, b_5 fails to affect NADPH-cytochrome c reductase activity in the enriched microsomal fractions, although both hemoproteins compete for the common electron donor (Table 1). Finally, NADPH oxidation is not stimulated, but severely blocked (Table 1). These findings suggest that extra-bound b_5 acts to impair transfer of the first electron to P-450, and this accounts for the observed decrease in the initial and maximum rate of N-oxidation of 4-CA (Tables 1 and 2). Such an explanation evidently does not hold, when NADH is the sole source of electrons. Under these conditions, incorporated b_5 slightly stimulates NADH-dependent reduction of P-450, but the rate of NADH-driven N-oxidation of 4-CA is none the less decreased (Tables 1 and 2). Moreover, there is a striking difference in the fortified microsomes in the rates of NADH- and NADPH-dependent N-oxidation of 4-CA, while the rate of electron flow to ferricytochrome P-450 in the presence of NADPH is equal to that in the presence of NADH (Table 1). It was, therefore, of interest to examine, whether the ob-

TABLE 2

INFLUENCE OF CYTOCHROME b_5 AND SOME OF ITS DERIVATIVES ON THE KINETIC PARA-
METERS OF NADH- AND NADPH-DRIVEN N-OXIDATION OF 4-CHLOROANILINE

K_m is expressed in μmol/l; V_{max} indicates nmol 1-chloro-4-nitrosobenzene
formed/min per nmol of cytochrome P-450.

Type of microsomal fraction used	Molar ratio of b5 or derivative to P-450	Electrons transferred from NADH		Electrons transferred from NADPH	
		K_m	V_{max}	K_m	V_{max}
Native	0.36	78.0+3.1	0.83+0.03	51.0+1.5	10.5+0.3
Supplemented with holo-b5	1.46	91.0+5.5	0.57+0.03[a]	51.0+7.0	5.6+0.8[b]
Supplemented with holo-b5 and treated with antiserum[c]	1.49	66.7+3.0	0.58+0.02	-	-
Supplemented with Mn(III)-b5	1.78	41.0+2.0	0.93+0.05	41.7+6.1	12.8+1.9
Supplemented with apo-b5	2.19	78.0+11.5	0.85+0.12	56.0+7.0	10.5+1.3

[a]$p < 0.001$; [b]$p < 0.005$; [c]microsomal fraction incubated with antiserum to
NADPH-cytochrome c(P-450) reductase.

served discrepancy might be due to differences in the rate of transfer of
the second electron to oxyferrous P-450. As shown in Table 1, this step is
likely to be mediated via b5 in the fortified microsomal preparations,
since there is a marked NADH/NADPH synergism in the N-oxidation of the amine substrate. In these preparations, the steady-state level of reduced b5
was determined as 1.50 μM in the presence of NADH and 1.36 μM in the presence of NADPH. Arylamine-dependent reoxidation of the hemoprotein proceeds at a rate of 2.52/min and 1.58/min, respectively, with NADH and
NADPH as the electron donor (Fig. 1). These findings establish evidence
that, in the enriched microsomes, the observed differences in the V_{max}
values for NADH- and NADPH-driven N-oxidation of 4-CA (Table 2) are not the
consequence of b5-induced differences in the rates of transfer of the first
and second electron to P-450. Neither does b5 modify the affinity of 4-CA
for P-450 (Table 2). Nevertheless, the phnomenon is closely related to the
ability of b5 to undergo redox changes, since Mn(III)-b5 or apo-b5 fail to
produce analogous effects (Table 2).

Since $V_{max} = k \cdot E_t$, we measured the amount of P-450 participating in the
NADH- and NADPH-dependent N-oxidation reaction in the fortified microsomal
fractions. With NADH, 11.2% of the total amount of P-450 present were found
to be active in this process as compared to 19.6% being operative in the
presence of NADPH. The drastic decrease in the V_{max} value for N-oxidation
of 4-CA found in microsomes containing extra-bound b5 and NADH as the electron donor(Table 2) must, therefore, be due to a severe decrease in the
rate of breakdown of the ternary P-450 complex to release product.

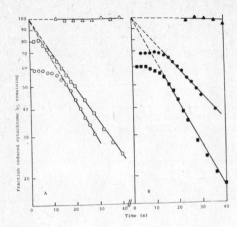

Fig. 1. Reoxidation of cytochrome b5 in enriched microsomes containing either NADH (A) or NADPH (B) as the electron donor. Conditions:☐,■, aerobic, 4-CA absent; o,●, aerobic, 4-CA present;△,▲, CO in the atmosphere (CO:O$_2$ = 4:1), 4-CA present.

In summary, the capacity of enriched microsomes for the N-oxidation of 4-CA differs markedly with the type of electron donor used. Under the specific conditions adopted throughout these experiments, this is not due to differences in the rates of electron flow to P-450 or changes in the affinity of 4-CA for the oxidase. The results are interpreted to mean that coupling of b5 to P-450 via the action either of NADH-cytochrome b5 reductase or NADPH-cytochrome c(P-450) reductase yields distinct functional protein aggregates characterized by differing conformational states. Such conformational differences might account for the observed variance in the catalytic capacity of the P-450 system towards certain types of substrates. In our case, switching of the NADH-dependent electron transport system to P-450 serves to stabilize the arylamine-oxygen-P-450 complex.

REFERENCES

Hlavica, P. (1984) On the Function of Cytochrome b5 in the Cytochrome P-450-Dependent Oxygenase System, Arch. Biochem. Biophys. *228*, 600-608.

Hrycay, E.G. & Estabrook, R.W. (1974) The Effect of Extra Bound b5 on Cytochrome P-450-Dependent Enzyme Activities in Liver Microsomes, Biochem. Biophys. Res. Commun. *60*, 771-778.

Jagow, R., Kampffmeyer, H. & Kiese, M. (1965) The Preparation of Microsomes, Naunyn-Schmiedeberg's Arch. Exp. Pathol. Pharmakol. *251*, 73-87.

Morgan, E.T. & Coon, M.J. (1984) Effects of Cytochrome b5 on Cytochrome P-450-Catalyzed Reactions. Studies with Manganese-Substituted Cytochrome b5, Drug Metab. Dispos. *12*, 358-364.

Tamburini, P.P. & Gibson, G.G. (1983) Thermodynamic Studies of the Protein-Protein Interactions between Cytochrome P-450 and Cytochrome b5. Evidence for a Central Role of the Cytochrome P-450 Spin State in the Coupling of Substrate and Cytochrome b5 Binding to the Terminal Hemoprotein, J. Biol. Chem. *258*, 13444-13452.

White, R.E. & Coon, M.J. (1980) Oxygen Activation by Cytochrome P-450, Annu. Rev. Biochem. *49*, 315-356.

INTERACTION OF CYTOCHROME P-450 WITH PHOSPHOLIPIDS IN RABBIT
LIVER MICROSOMES AS STUDIED BY PHOSPHORUS NUCLEAR MAGNETIC
RESONANCE USING PHOSPHONIC ACID DIETHYL ESTER
T. Bayerl, G. Klose, K. Ruckpaul, W. Schwarze

Dept. of Physics, Karl-Marx-University, 7010 Leipzig Linnestr. 5
and Central Institute of Molecular Biology, Acad. of Sciences
GDR, 1115 Berlin-Buch, Lindenberger Weg 70 (GDR)

INTRODUCTION

31P-NMR is a well established method in studying membrane structures.
However, in biological membranes the application of this method is
restricted mainly by its sensitivity. Moreover, the signals of membrane
phospholipids are broad and complex due to the large chemical shift
anisotropies of the different lipids and to various dynamical processes
such as chemical exchange and immobilization. Thus the low number of
phospholipids interacting with proteins in biological membranes or the
interaction of biological active substances with distinct phospholipid
species cannot detected by 31P-NMR. By use of hexane phosphonic acid
diethyl ester (PAE) as a phosphorus probe molecule and its incorporation
in natural membranes part of these difficulties can be overcome. The
31P-NMR signals of this small and hydrophobic molecule do not interfere
whith those of the phospholipids and the chemical shift is sensitive to
its environments. The behaviour of PAE in artificial membranes and the
physical properties of PAE treated membranes have been studied by NMR
and other methods (Klose et al. 1984, Hentschel and Klose 1985). Moreover,
it has been shown (Bayerl et al. 1985a) that PAE can be incorporated at
low concentrations in rabbit liver microsomes without any alteration of
the bilayer structure.
The interaction of cytochrome P-450 (P-450) with lipids in hepatic microsomes (particularly phosphatidylethanolamine (PE)), recently studied by
fluorescence methods (Kawato et al. 1982) and ERP (Bösterling and Stier
1983), is still a subject of controversial discussions. The postulated
function of PE as a boundary, lipid of P-450 in microsomes could not be
proved by NMR, mainly due to the limitations of this method as mentioned
above.
The aim of this work is to demonstrate the applicability of PAE to the
analysis of P-450 - lipid interactions in microsomes and reconstituted
liposomes by 31P-NMR. The results indicate a specifically interaction of
PE with P-450 in microsomes.

RESULTS AND DISCUSSION

The 31P-NMR spectra of smooth rabbit liver microsomes with PAE incorporated
show a significant temperature dependence between 4 - 37°C (Fig. 1). The
changes in the 31P-NMR lineshape of the broad and asymmetric microsomal
phospholipid signal with increasing temperature are caused by the
temperature dependence of the lateral diffusion of the lipids as discussed

elsewhere (Bayerl et al. 1984). PAE gives rise to two narrow and symmetric signals at -37.7 ppm (denoted as signal A) and -35.0 ppm (denoted as signal B) respectively. The latter signal grows at temperatures above 15°C (Figs. 1 B-D). All these changes were found to be completely reversible. Low concentrations of the paramagnetic shift reagent europium chloride cause a shift of signal B toward higher field whereas signal A remained unchanged (Fig. 1 E). Further characteristics of both PAE signals are as follows: signal B exhibit a saturation point which is reached at a molar ratio of microsomal lipids / PAE = 8. Above this ratio its integral intensity remained unchanged. Moreover, signal B was observed only in intact microsomes but not in trypsin treated microsomes or in liposomes of total microsomal lipids.

The chemical shift difference between signals A and B (about 2.5 ppm) indicates two different sites of PAE in microsomes. The shift of signal B on addition of europium chloride suggests that site B is localized near the preferential binding site of paramagnetic ions.

By measurements of the chemical shift of PAE in different organic and anorganic solvents it was proved that PAE form hydrogen bonds via the phosphoryl group to corresponding protondonors, its binding strength determine the chemical shift of PAE in a range of about 5 ppm. It was found that the chemical shift of signals A and B corresponds to PAE hydrogen bonded to water and amine groups respectively. The latter was demonstrated by use of hexyl amine hydrochloride which posses the same protondonor group as the headgroup of PE which represents one of the main phospholipid constituents of microsomes. Moreover, it was excluded that one of the both signals arises from PAE dispersed as macroscopic oil droplets in water. Therefore it is reasonable to assume that the two sites are caused by the formation of hydrogen bonds of different strength with water as protondonor in site A and molecules such as PE as protondonor in site B respectively. As the chemical shift of signal A is identical to that of PAE incorporated in pure multilayers of vesicles of phosphatidylcholine (PC) one can conclude that signal A represents the part of PAE interacting with PC in microsomes.

The lateral diffusion of PAE can be assumed to be faster than that of the lipids, which follows from the smaller molecular order paramter of PAE compared with that of the lipid in model systems and which is supported by Monte Carlo calculations in a model system. This faster diffusion averages the phosphorus chemical shift anisotropies of PAE to zero. As we did not observe signal B in trypsin treated microsomes nor in multilayers of liposomes of microsomal lipids under identical conditions signal B must be originate from mutual interactions between proteins, lipids and PAE. This is supported by NMR measurements in reconstituted systems containing P-450 and P-450 reductase. When both proteins were reconstituted in liposomes of microsomal lipids signal B was observed at 37°C. However, the reconstitution of only one of these proteins gave no rise for signal B. Obviously the interaction of the two proteins is essential for the formation of this signal. By reconstitution of both proteins in liposomes with different lipid composition it was proved that the presence of PE is essential also for the occurence of signal B whereas PC and the acidic lipids had no influence (Bayerl et al. 1985a) (Fig.2). The role of PE was stressed also by the addition of PE rich liposomes to PAE treated microsomes at 37°C which caused an increase in intensity of signal B (possibly due to an enrichment of PE in microsomes by fusion) whereas the addition of pure PC liposomes had no any effect. The conclusion seems reasonable that signal B is caused by a mutual interaction of both proteins and PE with PAE via a hydrogen bond to the PE

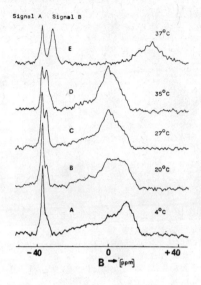

Fig.1:

81MHz 31P-NMR spectra of smooth
rabbit liver microsomes with PAE
(molar ratio of phospholipid/PAE =
3.6) at the temperatures as
indicated (A-D) and on addition
of 5mM europiumchloride (E).

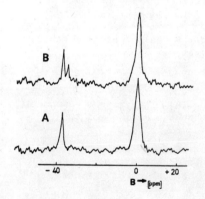

Fig.2:

81MHz 31P-NMR spectra of cytochrome P-450 and cytochrome
P-450 reductase reconstituted
in liposomes of PC/PA = 15:0.3
(mol) (A) and of PC/PE/PA =
10:5:0.3 (mol) (B) with PAE at
$37^{\circ}C$.

headgroup. The PE - PAE interaction is supported also by the opposite
effective molecular shapes of both molecules. Such a mutual interaction
can be explained on the basis of the esthablished existence of clusters
of P-450 and P-450 reductase which are surrounded by a certain lipid.
In connection with the results reported above and the experimentally
esthablished phase transition of PE in microsomes in the 20 -30°C
temperature range it seems reasonable to conclude that PE represents this
boundary lipid. At the PE phase transition to the gel state PAE is
excluded from the PE rich regions. Therefore at low temperatures no signal
B can be observed in microsomes. Thus the incorporation of PAE in
microsomes enables the observation of two different environments in microsomes, one in the PC rich region and the other in the PE rich region
neighboured to P-450 by 31P-NMR. This advantage has been used to study
the effect of some biological active substances with respect to signal A
and signal B. Benzphetamine as a typical P-450 substrate had no effect at
concentrations up to 10 mM. The addition of Na dithionite caused a
decrease of signal B indicating that the state of P-450 (oxidized or
reduced) is of importance in the PE/PAE interaction at site B.
Furthermore the effect of two widely used nonionic detergents (Triton
X-100 and octylglycoside) on signals A and B has been examined at sub
solubilizing detergent concentrations. The titration of such detergents to
microsomes with PAE at 37°C causes a shift of signal A toward the
position of signal B with increasing concentration. This is caused by a
fast exchange of PAE and membrane constituents with mixed micelles formed
by the detergent. The dependence of this shift on the detergent
concentration is a measure of its solubilization power. The concentration
at wich signal A begins to shift can be related to the maximum number of
detergent molecules incorporated in the microsomal membrane without the
formation of mixed micelles and changes in membrane structure. By this
method the effective saturation concentrations of detergent molecules in
microsomes were determined to be as a molar ratio total microsomal
phospholipids / detergent = 10 (Triton X-100) and 2.5 (octylglycoside)
respectively. Above this concentrations the formation of mixed micelles
coexisting with the membrane takes place. A more detailed study of these
processes will be published elsewhere (Bayerl et al. 1985b).

REFERENCES

Bayerl,T., Klose,G., Ruckpaul,K., Gast,K., Mops,A. "Temperature dependent
    lateral diffusion of phospholipids in hepatic microsomes as studied
    by 31P-NMR" (1984) Biochim.Biophys.Acta 769, 399-403.

Bayerl,T., Klose,G., Ruckpaul,K., Schwarze,W. (1985a) "Interaction of
    hexane phosphonic acid diethyl ester with phospholipids in hepatic
    microsomes and reconstituted liposomes as studied by 31P-NMR"
    Biochim.Biophys.Acta 812, 437-446

Bayerl,T., Klose,G., Blanck,J., Ruckpaul,K. (1985b)"The interaction of
    sub solubilizing concentrations of triton X-100 and octylglycoside
    with hepatic microsomes studied by 31P-NMR" Biochim.Biophys.Acta
    submitted

Bösterling,B., Stier,A. (1983) "Specifity in the interaction of phospho-
    lipids and fatty acids with vesicle reconstituted Cytochrome P-450"
    Biochim. Biophys.Acta 729,258-266

Hentschel,M., Klose,G. (1985) "Effect of the concentration and the molecular structure of phosphonic acid dialkyl esters on the kinetics of praseodymium permeation through egg PC small unilamellar vesicles" Biochim.Biophys.Acta 812, 447-452

Kawato,S., Gut,J., Cherry,R.J., Winterhalter,K.H., Richter,C. (1982) "Rotation of Cytochrome P-450" J.Biol.Chem. 257, 7023-7029

Klose,G., Hentschel,M., Bayerl,T. (1985) Proceedings of the 7th School on Biophysics of Membrane Transport, Zakopane (Poland), Wroclaw University Press, in the press

REGULATIVE FUNCTION OF CYTOCHROME b5 IN THE ELECTRON TRANSFER FROM NADPH-CYTOCHROME P-450 REDUCTASE TO CYTOCHROME P-450

Ines Golly and Peter Hlavica

Institut für Pharmakologie und Toxikologie der Universität, Nussbaumstrasse 26, D-8000 München 2, F.R.G.

INTRODUCTION

Cytochrome b_5 is well-known to be able to donate the second electron to oxygenated ferrocytochrome P-450 in certain types of monooxygenations (White and Coon, 1980). Recent findings by various laboratories suggest that cytochrome b_5 might modulate the catalytic capacity of the cytochrome P-450 system through protein-protein interaction (Tambutini and Gibson, 1983; Hlavica, 1984; Morgan and Coon, 1984).

The present report centres on the regulative function of cytochrome b5 in the process of electron transfer from NADPH-cytochrome c(P-450) reductase to cytochrome P-450.

MATERIALS AND METHODS

Microsomal fractions from the livers of phenobarbital-treated rabbits were prepared by the method of Jagow et al. (1965). Cytochrome b_5 (b_5) and manganese-protoporphyrin IX-substituted cytochrome b_5 (Mn(III)-b_5) were prepared as described by Morgan and Coon (1984). Incorporation of the proteins into rabbit liver microsomal fractions was achieved by incubating appropriate amounts of each component in 0.15 M KH_2PO_4/Na_2HPO_4, pH 7.4, for 30 min at 37°C. The washed microsomal pellets were resuspended in phosphate buffer, pH 7.4. Purification of cytochrome P-450 LM2 (P-450) and incorporation of the hemoprotein into liposomes constituted of dimyristoylphosphatidylcholine was performed as described previously (Hlavica, 1984). NADPH-oxidase activity was measured by recording the decrease in absorbance at 340 nm in reaction mixtures containing microsomal fraction (MF) equivalent to 2 μM P-450, 1 mM hexobarbital (HB), and 10-80 μM NADPH in 0.15 M phosphate buffer, pH 7.4; in some experiments the MF contained extra-bound b5 or Mn(III)-b5. NADPH-P-450 reductase activity was measured under anaerobic conditions by monitoring the increase in absorbance at 450 nm relative to 500 nm in reaction mixtures containing MF equivalent to 2 μM P-450, 1 mM HB, glucose-oxidase/catalase system, and 1-8 μM NADPH in 0.15 M phosphate buffer, pH 7.4; the sample cell was gassed with CO for 3 min. In some experiments, the MF was fortified either with b5 or Mn(III)-b5. Microsomal NADPH-cytochrome c reductase was assayed as indicated elsewhere (Hlavica and Hülsmann, 1979); for some experiments, the reductase was prepared to apparent homogeneity. Binding of the enzyme to liposomal cytochrome P-450 in the presence of reducing equivalents was studied by measuring the increase in absorbance at 390 nm relative to 418 nm in tandem cells containing 2 μM P-450 and 0.1 mM $Na_2S_2O_4$ in 0.15 M phosphate buffer, pH 7.4. Reductase (0.125-1.0 μM) was added to the contents of the sample cuvette; optical path length was 1 cm.

RESULTS AND DISCUSSION

Incorporation into rabbit liver microsomal fractions of detergent-solubilized b_5 causes a remarkable decrease both in the maximum velocity of P-450 reduction and the affinity of NADPH for the mixed-function oxidase system (Fig. 1B). Decreased P-450 reduction could be assumed to result from the ability of b_5 to serve as an alternate electron acceptor shunting electrons away from P-450. This explanation evidently does not hold, since extra-bound b_5 diminishes the extent of HB-triggered NADPH utilization (Fig. 1A), although stimulation of the NADPH-oxidase activity would be expected. Further, there is no decrease in the enriched MF in the rate of electron flow to cytochrome c (Fig. 2), albeit cytochromes b_5 and c compete for the same electron transfer system. The experiments with cytochrome c also demonstrate that the observed deceleration of P-450 reduction (Fig. 1B) is not due to a general alteration in the electron transferring capacity of NADPH-cytochrome c(P-450) reductase by b_5, but is strongly dependent on the type of terminal electron acceptor used. The phemomenon is, therefore, likely to arise from selective interaction of b_5 with P-450 and is closely associated with the ability of the former component to undergo redox changes, since Mn(III)-b_5 fails to elicit an analogous effect (Fig. 1B).

Cytochrome b_5 is thus proposed to exert its inhibitory action on the NADPH-cytochrome P-450 reductase activity either by interferring with the process of coupling of NADPH-cytochrome c(P-450) reductase to P-450 or by impairing the step of electron transfer in some other way. The latter conclusion is compatible with the "mixed-type" inhibition pattern depicted in

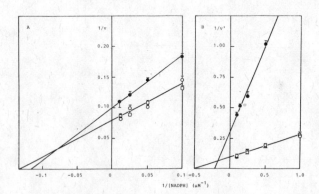

Fig. 1. *Influence of extra-bound cytochrome b_5 and Mn(III)-b_5 on the kinetics of NADPH binding, in the presence of hexobarbital, to the microsomal mixed-function oxidase system, as measured in terms of NADPH oxidation (A) or cytochrome P-450 reduction (B). o, controls; □, MF fortified with Mn(III)-b_5 to yield a molar ratio to P-450 of 0.91:1; ●, MF fortified with b_5 to yield a molar ratio to P-450 of 1.66:1; v indicates nmol NADPH oxidized/min/nmol P-450; v' indicates nmol P-450 reduced/min/nmol P-450.*

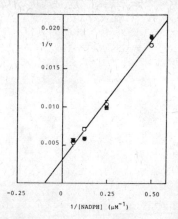

Fig. 2. Influence of extra-bound cytochrome b5 on the kinetics of NADPH binding, in the presence of hexobarbital, to microsomal NADPH-cytochrome c reductase. o, control; •, MF fortified with b5 to yield a molar ratio to P-450 of 1.66:1; v indicates nmol cytochrome c reduced/min/mg protein.

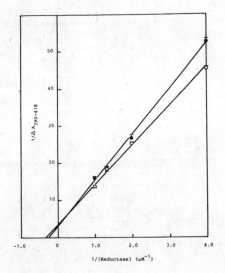

Fig. 3. Influence of extra-bound cytochrome b5 on the binding of solubilized NADPH-cytochrome c(P-450) reductase to liposomal cytochrome P-450LM2. o, control; •, reconstituted system supplemented with b5 to yield a molar ratio to P-450LM2 of 1.8:1.

13*

Fig. 1B. Such kinetics can arise, when the rate constant (k) for breakdown of the reductase/P-450 complex to release reduced hemoprotein makes an important contribution to K_m, and b_5 acts to modify k. Whatever, b_5 does not appear to change the affinity of NADPH-cytochrome c(P-450) reductase for reduced P-450LM$_2$ (Fig. 3). Further work is in progress to elucidate the mechanism, by which b_5 alters electron transfer from the reductase to the terminal oxidase.

REFERENCES

Hlavica, P. (1984) On the Function of Cytochrome b_5 in the Cytochrome P-450-Dependent Oxygenase System, *Arch. Biochem. Biophys. 228,* 600-608.

Hlavica, P. & Hülsmann, S. (1979) Studies on the Mechanism of Hepatic Microsomal N-Oxide Formation, *Biochem. J. 182,* 109-116.

Jagow, R., Kampffmeyer, H. & Kiese, M. (1965) The Preparation of Microsomes, *Naunyn-Schmiedeberg's Arch. Exp. Pathol. Pharmakol. 251,* 73-87.

Morgan, E.T. & Coon, M.J. (1984) Effects of Cytochrome b_5 on Cytochrome P-450-Catalyzed Reactions. Studies with Manganese-Substituted Cytochrome b_5, *Drug Metab. Dispos. 12,* 358-364.

Tamburini, P.P. & Gibson, G.G. (1983) Thermodynamic Studies of the Protein-Protein Interactions between Cytochrome P-450 and Cytochrome b_5. Evidence for a Central Role of the Cytochrome P-450 Spin State in the Coupling of Substrate and Cytochrome b_5 Binding to the Terminal Hemoprotein, *J. Biol. Chem. 258,* 13444-13452.

White, R.E. & Coon, M.J. (1980) Oxygen Activation by Cytochrome P-450, *Annu. Rev. Biochem. 49,* 315-356.

EFFECT OF GEL-LIQUID CRYSTAL PHASE TRANSITION ON CONFORMATIONAL REARRANGEMENT OF CYTOCHROME P-450 AND ITS REDUCTION KINETICS

Kisel M.A., Romanov S.L., Andreyuk G.M., *Smettan G. and Kisselyov P.A.

Institute of Bioorganic Chemistry, Academy of Sciences of the BSSR, Minsk, USSR; * Central Institute of Molecular Biology, Academy of Sciences, 1115 Berlin-Buch, GDR

INTRODUCTION

The data available at present indicate to a relationship between catalytic activity of the cytochrome P-450-containing enzyme system and physical state of the liver microsomal phospholipid matrix. It is assumed that resulting from phase transition of phospholipid bilayer into liquid crystal state, a drastic change in the lateral protein mobility occurs, which is revealed by breaks on the Arrhenius curves characterizing functional behaviour of the monooxygenase system /Duppel and Ullrich, 1976; Peterson et al. 1976/. On the other hand, the alteration of activation energy of the reactions catalysed by cytochrome P-450 may be stimulated by thermal rearrangements of the latter. The model membranes composed of protein components of the enzyme system and phospholipids allow of gradual studies on the effect of lipid physical state both on individual proteins and catalytic activity of the system as a whole. A similar approach has been applied to our present investigation. A model membrane contained dimyristoylphosphatidylcholine as lipid component, since the gel-liquid crystal phase transition in such system occurs about 20°C, which allows to examine the protein properties before and after the phase transition without their thermal inactivation. The temperature-induced rearrangements of cytochrome P-450 were traced by the alterations in the hemoprotein spin equilibrium as well as measuring its intrinsic fluorescence. Functional behaviour of the reconstituted system is characterized by the reduction of cytochrome P-450 by cytochrome P-450-reductase.

MATERIALS AND METHODS

The bilayer proteoliposomes containing cytochrome P-450 and cytochrome P-450-reductase were obtained by removal of pre-added sodium cholate on Sephadex G-25 column and then characterized as described earlier /Kiselev et al. 1984/. The rate of reduction of cytochrome P-450 was measured by the "stop-flow" method on Durrum D110 (USA) instrument under unaerobic conditions. The equilibrium constants for the low and high spin form were calculated from the difference absorption spectra /Ristau et al. 1978/. Fluorescence spectra were registered on Jobin Ivon 3CS instrument (France).

RESULTS

Effect of the proteoliposomes phase transition on spin equilibrium of cytochrome P-450. We have studied the effect of physical state of phospholipid bilayer on spin equilibrium of cytochrome P-450 incorporated into liposomes from dimyristoylphosphatidylcholine (DMPC) and egg phosphatidylcholine (PC), the former ones being characterized by distinct gel-liquid crystal transition at 23°C. Under different temperature provided in experimental and control cuvettes containing cytochrome P-450 incorporated into liposomes a difference spectrum emerges with a maximum at 388 nm and minimum at 418nm. The above observation testifies to the increase in high spin hemoprotein in experimental cuvette. In Fig.1 dependencies of equilibrium constants K of spin forms of cytochrome P-450 incorporated into liposomes from DMPC and PC have been presented in Arrhenius coordinates. In the case of proteoliposomes containing PC a linear dependence is observed in the temperature range investigated. The proteoliposomes from DMPC are characterized by breaks at 20°C and 25°C whose positions coincide with the initial and final points of the proteoliposomes phase transition as registered using the difference scanning microcalorimetry /Akhrem et al. 1982/.

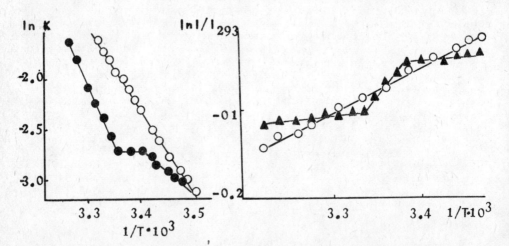

Fig.1. Dependence of cytochrome P-450 spin equilibrium constant on temperature: a) O-O in proteoliposomes from PC; b) ●-● in proteoliposomes from DMPC. 500:1 lipid/protein ratio.
Fig.2. Dependence of intrinsic fluorescence intensity of cytochrome P-450 on temperature: a) O-O in proteoliposomes from PC; b) ▲-▲ in proteoliposomes from DMPC. 500:1 lipid/protein ratio.

Thermal quenching of intrinsic cytochrome P-450 fluorescence in proteoliposomes. Incorporation of cytochrome P-450 into liposomes is accompanied by a shift of maximum protein fluorescence from 325 nm to 312 nm. Simultaneously, a growth of a re-

lative contribution of tyrosine component into total spectrum
is observed. A change of spectral shape of intrinsic fluorescence in liposomes as compared to protein in solution, testifies
in favour of conformational rearrangements in hemoprotein under
interactions with lipids. In Fig.2 the results from investigation of thermal quenching of fluorescence of cytochrome P-450
inserted into liposomes from PC are presented as fluorescence
intensity vs inverse temperature. The Arrhenius dependence for
proteoliposomes composed of PC is linear over the temperature
interval investigated whereas for proteoliposomes from DMPC the
breaks at 23°C and 26°C clearly seen in the plot, which indicates to a correlation between protein conformational mobility
and physical state of lipid bilayer.

NADPH-dependent reduction of cytochrome P-450 in proteoliposomes. In Fig.3 the temperature dependence of the cytochrome
P-450 reduction by NADPH-cytochrome P-450-reductase in proteoliposomes from DMPC is presented. The process of enzymatic reduction is characterized by two phases, slow and fast. A break
in the Arrhenius curve is observed at 20°C for both phases. The
activation energy values (ΔE) in the high temperature region
are~0 and 5.2 kcal/mol for a fast and slow phases, respectively
Below the break point ΔE increases up to 12.5 kcal/mol for the
fast phase, and up to 11.5 kcal/mol for the slow. A portion of
cytochrome P-450 reduced in the fast phase of reaction increases from 20% at 11°C to 50% at 20°C. The temperature elevation
above 25°C leads to a linear quantitative decrease of hemoprotein in this phase. A break in the Arrhenius dependence characterizing a slow phase may be ascribed to a decrease of bilayer
viscosity as a result of gel- liquid crystal transition with an
ultimate increase in lateral mobility of components of the monooxygenase system./Peterson et al. 1976/

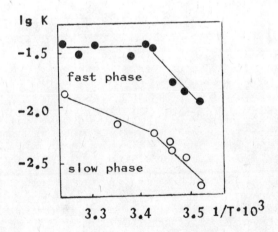

Fig.3. Dependence of cytochrome P-450 reduction rate constant
on temperature in proteoliposomes from DMPC. Cytochrome P-450/
cytochrome P-450-reductase/lipid ratio is 1.0:0.2:500.0.

184

A comparison of temperature dependence of the cytochrome
P-450 reduction rate in microsomes and in proteoliposomes reveals considerable differences for both cases of the activation
energy values and the shape of these dependences. Note as an
example that during reduction of cytochrome P-450 in microsomes a break is observed in the slow phase only. According to
the conclusions of /Peterson et al. 1976/, the fast phase of
reaction is determined by the reduction of cytochrome P-450
arranged in clasters. However, the presence of a break on the
Arrhenius curve calls for a more accurate definition of the
concept in the case of clasters surrounded by phospholipids
undergoing phase transition.

CONCLUSIONS

1. A comparison of temperature dependences of the spin
equilibrium constant and quenching of intrinsic fluorescence of
cytochrome P-450 inserted in liposomes from DMPC and PC, reveals significant conformational rearrangements of hemoprotein in
the region of the lipid bilayer phase transition.
2. When interpreting the nature of a break in the fast
phase of cytochrome P-450 reduction, we proceeded from the fact
that the given phase is stipulated by the existence of a functionally active complex, either polycomponent (claster) /Peter-
son et al. 1976/ or two-component (cytochrome P-450 and its reductase) /Gut et al. 1982/. Then, according to our results, a
drastic change in thermodynamic parameters of the reaction in
the region of phase transition is due to : a) conformational
rearrangements in cytochrome P-450; b) different concentration
of functionally active complex in gel-like and liquid crystal
states of lipid bilayer.

REFERENCES

Akhrem A.A., Andrianov V.T., Bokut S.B., Luka Z.A., Kisel M.A.,
Skornyakova T.G.,Kiselev P.A., 1982. Thermotropic behaviour of
phospholipid vesicles reconstituted with rat liver microsomal
cytochrome P-450. Biochim. Biophys. Acta, 682, 287-295.
Duppel W., Ullrich V., 1976. Membrane effects on drug monooxygenation activity in hepatic microsomes. Biochim. Biophys.
Acta, 426, 399-407.
Gut J., Richter C., Cherry R.J., Winterhalter K.H., Kavato S.J.
1982. Rotation of cytochrome P-450. J. Biol. Chem., 257, 7023-
7029.
Kiselev P.A., Smettan G., Kisel M.A., Elbe B., Zirwer D., Gast
K., Ruckpaul K., Akhrem A.A., 1984. Reconstitution of the liver
microsomal monooxygenase system in liposomes from dimyristoylphosphatidylcholine. Biomed. Biochim. Acta, 43, 281-293.
Peterson J.A., Ebel R.E., O'Keeffe D.H., Matsubara T., Estabrook R. W., 1976. Temperature dependence of cytochrome P-450
reduction. J. Biol. Chem., 251, 4010-4016.
Ristau O., Rein H., Jänig G.-R., Ruckpaul K., 1978. Quantitative analysis of the spin equilibrium of cytochrome P-450 LM_2
Biochim. Biophys. Acta, 576, 226-232.

A COMPARATIVE STUDY OF C_{27}-STEROID HYDROXYLASE COMPONENTS FROM BOVINE LIVER MITOCHONDRIA

Gilevich, S.N., Shkumatov, V.M., Chashchin, V.L.

Institute of Bioorganic Chemistry, Byelorussian SSR Academy
of Sciences, Minsk, USSR

INTRODUCTION

ω-Hydroxylation of 5β-cholestane-$3\alpha,7\alpha,12\alpha$-triol(ChT) and several other C_{27}-steroids, an important stage in bile acids biosynthesis, is catalyzed in liver mitochondria by an electron transfer chain consisting of a NADPH-dependent flavoprotein (hepatoredoxin reductase, HR), an iron-sulphur protein(hepatoredoxin, Hd) and specific cytochrome P-450($P-450_{27}$). The purification and characterization of these components was hindered by their low stability and content as well as other difficulties(Atsuta, Okuda, 1982; Oftebro et al., 1981). Our investigation of structural and functional interrelation between HR, Hd and corresponding adrenocortical proteins involved in cholesterol side chain cleavage included the following stages: i) isolation of highly purified HR, Hd and $P-450_{27}$ using original chromatographic approaches; ii) elucidation of main physical and chemical characteristics of HR and Hd in comparison with analogous properties of adrenodoxin reductase(AR) and adrenodoxin(Ad); iii) reconstitution of both steroid hydroxylases with reciprocal substitution of the reductase components; iv) estimation of the functional role of Lys, dicarboxylic acids and His residues in the ferredoxins by means of chemical modification.

MATERIALS AND METHODS

AR, Ad, cytochrome P-450scc, HR, and Hd were purified according to the methods developed in our laboratory(Akhrem et al., 1979; Shkumatov et al., 1983; Gilevich et al., 1985). $P-450_{27}$ was solubilized from submitochondrial particles with sodium cholate and fractionated with polyethylene glycol (Oftebro et al., 1981) followed by chromatography on cholate-Sepharose column equilibrated with 0.1 M sodium phosphate buffer, pH 7.4, containing 0.5 mM EDTA and 20% glycerol. $P-450_{27}$ eluted in the presence of 0.5% sodium cholate and 0.5 M NaCl was further purified on Ad-Sepharose column in analogy with P-450scc.

NADPH-cytochrome c(cyt.c) reductase, NADPH-P-450scc reductase and cholesterol side chain cleavage activities were assayed as described(Radyuk et al., 1982; Shkumatov et al., 1983). ω-Hydroxylation of ChT by $P-450_{27}$ was performed in the same manner using 50 μM ChT and 0.4 μCi of ^{3}H-labelled ChT. The molar ratio of AR(HR):Ad(Hd):P-450scc($P-450_{27}$) = 1:2:1 was used throughout the reconstitution experiments with the exception of HR and Hd specific activities measured in the presence of saturating amounts of other components.

The chemical modification of Hd was performed in sodium phosphate buffer, pH 7.0, containing: a) 12.5 μM Hd and 90 mM succinic anhydride; b) 25 μM Hd, 0.25 M ethylene diamine and 5 mM 1-ethyl-3-dimethylaminopropylcarbodiimide(EDC); c) 10 μM Hd and 5 mM diethylpyrocarbonate(DEPC). The number of modified residues was assayed spectrophotometrically by conventional methods.

RESULTS AND DISCUSSION

The isolation of Hd extracted from mitochondria as a result of crude mechanical homogenization includes DEAE-cellulose DE-32 ion-exchange chromatography, ammonium sulphate fractionation, Sephadex G-75 gel filtration as well as most effective purification steps, 1-amino-2-hydroxypropyl-Sepharose ion-exchange chromatography, followed by cyt.c-Sepharose affinity chromatography. To prevent the lysine-containing binding domain of cyt.c from modification, the synthesis of the latter adsorbent was performed by coupling to bromoacetyl-Sepharose of the protein via additional imidazole moieties introduced by modification of free carboxyl groups with histamine:

$$SEPHAROSE-OCH_2CH(OH)CH_2NH(CH_2)_2NHCOCH_2N\text{---}C(CH_2)_2NHCO-CYT.C$$

The adsorbent obtained posesses a 2 - 3 times higher affinity to Ad and Hd than cyt.c coupled to BrCN-activated Sepharose, and can be successfully used for both the purification of several electron transfer proteins and for the study of protein complexing participated by cyt.c(Akhrem et al., 1984). The procedure described resulted in electrophoretically homogeneous Hd(index of purity, A_{415}/A_{280} = 0.84; N-terminal amino acid - Ser) with high yield.

In the course of HR purification we have more extensively used the possibilities of Ad-Sepharose biospecific chromatography. To introduce a spacer between the ligand and insoluble matrix, Ad was modified with methyl-4-mercaptobutyrimidate and covalently attached to bromoacetyl-Sepharose due to the additional thiol groups:

$$SEPHAROSE-OCH_2CH(OH)CH_2NH(CH_2)_2NHCOCH_2S(CH_2)_3CNH-Ad$$

The preparation of HR extracted from intact mitochondria by sonication was twice chromatographed on Ad-Sepharose column to electrophoretic homogeneity (A_{450}/A_{270} = 0.13; N-terminal amino acid - Gly).

The isolation of P-450$_{27}$ has been complicated by low stability and content, and hydrophobicity of the enzyme as well as by large amounts of contaminating species of cytochrome P-450 and other hemoproteins(Atsuta, Okuda, 1982). The purification of the cholate extract on ω-aminooctyl-Sepharose column, in spite of a single success reported(Wikvall, 1984), is troublesome because of an irreversible adsorption in the course of hydrophobic-ionic chromatography, application of non-ionic detergents, and unsufficient separation of contaminating cytochromes. To overcome the difficulties mentioned, we have synthesized and tested several alkyl-, ω-aminoalkyl-, aryl- and steroid-substituted Sepharoses; among these cholic acid attached to the Sepharose amino derivative has proved to be a most suitable affinity ligand:

$$SEPHAROSE-OCH_2CH(OH)CH_2NH(CH_2)_2NHCOR \quad (RCOOH, cholic acid)$$

Cholate-Sepharose chromatography of P-450$_{27}$ led to significant purification

of the enzyme which was almost quantitatively eluted from the column under
mild conditions. After Ad–Sepharose chromatography, the preparation was
purified up to the heme content of 4 nmol/mg protein and it did not contain
hemoprotein contaminants.

Properties of the purified reductase components

The main physical and chemical properties of purified HR and Hd have been
found to coincide or closely resemble the analogous parameters of AR and
Ad, respectively(Table 1). We have also revealed the similarity in the
amino acid compositions, CD spectra and antigenic properties of the corres-

Properties	HR	Hd
Molecular weight	50000(51000)	12500(13000)
λ_{max}, nm	270, 375, 450 (272, 377, 450)	277, 320, 415, 455 (276, 320, 414, 455)
ε_{max}, $M^{-1} cm^{-1}$	ε_{450}=11300(–"–)	ε_{415}=9800(–"–)
Number of amino acids	428 – 430(433)	115(114)
Isoelectric point	7.2(5.4)	4.2(4.0)
Specific activity:		
[a]NADPH–cyt.c reductase	12(8.8)	52(60)
[a]NADPH–P–450scc reductase	–	8.3(7.5)
[b]Cholesterol side chain cleavage	0.14(0.15)	–

Table 1. Properties of purified HR and Hd. Corresponding values for AR and
Ad are given in parentheses. Specific activity is expressed in μmol of hemoprotein reduced/min/mg enzyme(a) or μmol of pregnenolone formed/min/mg
enzyme(b).

ponding reductase components(data not shown).

Comparison of the amino acid compositions of Hd and Ad has revealed the
common structural features of the mitochondrial steroidogenic ferredoxins:
high content of dicarboxylic and low of aromatic amino acids, the lack of
tryptophane. There are few differences in the Ad and Hd amino acid compositions, the latter lacking of the fifth Cys residue. Quite similar N–terminal amino acid sequence(residues 1 to 12) has also confirmed the close phy-
logenetic relationship between the ferredoxins.

The corresponding reductase components from liver and adrenal cortex mitochondria posess approximately the same catalytic efficiency towards enzymatic reduction of hemoproteins and cholesterol side chain cleavage reac-
tion(Table 1). A reciprocal substitution of AR and HR, as well as Ad and Hd
has been already used for the purification(see above) and reconstitution
(Wikvall, 1984) of the mitochondrial monooxygenases. The reconstitution of
both steroid–hydroxylating systems on the basis of highly purified components has demonstrated the reciprocal catalytic competency of AR and HR,
Ad and Hd, respectively, either combination posessing approximately the
same efficiency(Table 2). Obviously, the binding and catalytic sites of AR
and HR, and also of Ad and Hd, are formed by identical or similar in pro-

Reductase components	Product formation, nmol/nmol cytochrome P-450/min	
	[a]P-450scc	[b]P-450$_{27}$
None	0	0
AR + Ad	6.1	15.2
HR + Ad	6.9	14.6
AR + Hd	7.2	14.3
HR + Hd	6.7	15.3

Table 2. Reciprocal substitution of the reductase components in the biosynthesis of pregnenolone(a) and 5β-cholestane-3α,7α,12α,27-tetrol(b) by the reconstituted steroid hydroxylases.

perties amino acid residues, suggesting the unified mechanism of the protein-protein interactions and the electron transfer. Nevertheless, the terminal components of the systems have revealed very low(P-450$_{27}$) or zero (P-450scc) activity towards reciprocal substrates(cholesterol and ChT, respectively).

Chemical modification of Hd

Both NADPH-cyt.c reductase and cholesterol side chain cleavage activities of Hd have proved to be most sensitive towards modification of the protein COOH-groups involved in the complex formation with HR and other electron transfer proteins(Table 3). In all the cases studied chemical modification

Amino acid residues	Reagent	Mol of modified residues/mol Hd	Activity, %	
			a	b
Lys	Succinic anhydride	4	94	63
Asp, Glu	Ethylene di-amine + EDC	2 - 3	38	18
His	DEPC	3	79	60

Table 3. The influence of chemical modification on NADPH-cyt.c reductase(a) and cholesterol side chain cleavage(b) activities of Hd.

caused conformational changes of the polypeptide chain but did not influence either optical absorption or chirality of the [2Fe-2S]chromophore. The results obtained suggest a similar organization of the protein globules of the ferredoxins: polar amino acid residues are essential for stabilization of the native conformation, and dicarboxylic ones form the electron transfer proteins binding domain.

Thus, the mitochondrial monooxygenases function in accordance with the general principle suggesting the presence of both universal electron transfer components with complementary interactions between them, and the terminal part of the chain responsible for the reaction specificity.

REFERENCES

1. Akhrem, A.A., Lapko, V.N., Lapko, A.G., Shkumatov, V.M., and Chashchin, V.L.(1979) Isolation, structural organization and mechanism of action of mitochondrial steroid hydroxylating systems. Acta biol. med. germ., 38, 257 - 273.
2. Akhrem, A.A., Gilevich, S.N., Shkumatov, V.M., and Chashchin, V.L.(1984) Selectively immobilized cytochrome c as an effective affinity ligand for electron transfer proteins. Biomed. Biochim. Acta, 43, 165 - 177.
3. Atsuta, Y., and Okuda, K.(1982) Partial purification and characterization of 5β-cholestane-3α,7α,12α-triol and 5β-cholestane-3α,7α-diol 27-monooxygenase. J. Lipid Res., 23, 345 - 351.
4. Gilevich, S.N., Shkumatov, V.M., Chashchin, V.L., and Akhrem, A.A.(1985) Purification and properties of hepatoredoxin reductase from bovine liver mitochondria. Dokl. Akad. Nauk BSSR(Russian), 29, 181 - 184.
5. Oftebro, H., Saarem, K., Björkhem, I., and Pedersen, J.I.(1981) Side chain hydroxylation of C_{27}-steroids and vitamin D_3 by a cytochrome P-450 enzyme system isolated from human liver mitochondria. J. Lipid Res., 22, 1254 - 1264.
6. Radyuk, V.G., Shkumatov, V.M., Chashchin, V.L., and Akhrem, A.A.(1982) Study of protein-protein interactions in an enzymatically complete cholesterol side chain cleavage system. Biokhimiya, 47, 1792 - 1801.
7. Shkumatov, V.M., Gilevich, S.N., Chashchin, V.L., and Akhrem, A.A.(1983) C_{27}-steroid hydroxylating system from bovine liver mitochondria: isolation of ferredoxin(hepatoredoxin) by affinity chromatography on cytochrome c-Sepharose. Bioorg. Khim., 9, 1231 - 1236.
8. Wikvall, K.(1984) Hydroxylations in biosynthesis of bile acids. Isolation of a cytochrome P-450 from rabbit liver mitochondria catalyzing 26-hydroxylation of C_{27}-steroids. J. Biol. Chem., 259, 3800 - 3804.

DEPENDENCE OF CATALYTIC PROPERTIES OF THREE FORMS OF CYTOCHROME P-450 FROM MICE LIVER MICROSOMES ON PHOSPHOLIPID COMPOSITION OF THE MODEL MEMBRANE

P.A.Kisselyov[1], P.Keipanen[2], R.Yuvonen[2], M.Lang[3], O.Hänninen[2], and A.A.Akhrem[1]
1. Institute of Bioorganic Chemistry, BSSR Academy of Sciences, Minsk, USSR. 2. Physiology department, University of Kuopio, Finland. 3. "Labsystem", Helsinki, Finland

INTRODUCTION

Oxidation of a wide spectrum of organic compounds from endogenic and exogenic sources by the monooxygenase enzyme system from mammalian liver microsomes is achieved owing to the presence in microsomes of isoenzymes of cytochrome P-450 with overlapping substrate specificity /1/. On the other hand, it is known that the optimal catalytic activity of the membrane bound enzymes is realized in the presence of definite type phospholipids /2,3/. It is right to ask, to what extent the protein/lipid interaction affects the catalytic activity of similar by their structure and function of isoenzymes of cytochrome P-450. In the present work a comparative study of catalytic properties of the three forms of cytochrome P-450 from DBA/2J and C57BL/6N lines mice liver microsomes depending on phospholipid composition of the model membrane. Employed as substrate were 7-ethoxycoumarin, coumarin, ethoxyresorufin and ethylmorphine. The two former compounds are similar by their structure. Moreover, their enzymatic conversion by the given system affords one and the same product, 7-hydroxycoumarin. However, for their preparation two different types of reactions are realized, namely, O-dealkylation and 7-hydroxylation.

MATERIALS AND METHODS

Egg phosphatidylcholine (PC), egg phosphatidylethanolamine (PEA), phosphatidylserine from cattle brain (PS), phosphatidylglycerol (PG), and synthetic dilauroylphosphatidylcholine were from "Sigma". Cytochrome P-450 was isolated from C57/BL/6N line mice liver microsomes (Form B6) induced by phenobarbital as well as DBA/2J line (form D2 and P-450$_{pyr}$) induced by phenobarbital and pyrazol, respectively /4/. The cytochrome P-450-reductase from Wistar line mice liver microsomes, 30,5 units specific activity, was employed. The bilayer proteoliposomes containing cytochrome P-450 and cytochrome P-450-reductase with 1/10 and 1/1000 protein/protein and protein/lipid molar ratio, respectively, were obtained by dialysis with sodium cholate as detergent at 4°C. The reconstitution of catalytic activity of the monooxygenase system in the presence of DLPC was achieved by incubation of protein/lipid mixture for 15 hours at 4°C. The process of ethylmorphine demethylation was monitored by

formaldehyde accumulation. The products of oxidation of 7-etho-
xycoumarin, coumarin and ethoxyresorufin were analyzed fluomet-
rically. The system was incubated at 30°C for 20 min in Tris-
-HCL buffer, pH 7,4. The process was characterized by the initial rate of product accumulation in nmol of product per nmol of
cytochrome P-450 in minute. The values of maximal reaction rates (V_{max}) and Michaelis constant (K_m) are calculated from
Lineweaver-Burk plot.

RESULTS

The data presented in Table 1 reveal the effect of phospholipid composition of liposomes on the oxidation of 7-ethoxy-
coumarin, coumarin, ethoxyresorufin and ethylmorphine by the
reconstituted system containing cytochrome P-450 D2 as terminal
oxidase.

Table 1

Substrate specificity of cytochrome P-450 D2 in
liposomes with different phospholipid composition

Compo-sition of li-posomes	Initial oxidation rate in nmol of product per nmol of cytochrome P-450 in min			
	7-ethoxycoumarin ($1 \cdot 10^{-4}$ mol/l)	coumarin ($1 \cdot 10^{-4}$ mol/l)	ethoxyresorufin ($5 \cdot 10^{-7}$ mol/l)	ethylmorphine ($1 \cdot 10^{-2}$ mol/l)
DLPC	0,6	0,05	0,010	8,0
PC	1,0	0,13	0,004	6,5
PS	0,9	0,08	-	-
PC/PEA	0,8	0,18	0,001	6,0
PC/PS	2,1	0,25	0,012	7,5
PC/PG	2,0	0,18	0,007	6,0

Phospholipid mixtures are composed of equimolar amounts of
lipid components.

By their effect on the rate of O-dealkylation of 7-ethoxy-
coumarin, the phospholipids investigated may be devided into
two groups. DLPC, PC, PS and an equimolar mixture of PC and PEA
belong to the first group. Within this lipid group the reaction
rates differ by up to 40%. To the second group belong the PC
equimolar mixtures with negatively charged phospholipids, PS
and PG. In this case a 3-3,5 times rate acceleration is observed in the reaction relative to proteoliposomes containing DLPC.
A substitution of PC by PG in the mixture does not practically
affect the rate of O-dealkylation of 7-ethoxycoumarin. The reaction of 7-hydroxylation of coumarin proceeds the fastest in
the presence of liposomes from PC/PS. Peculiar features of the
process are: i) liposomes involving another negatively charged
lipid, PG, are considerably less effective and may be compared
(by their activating effect) to a mixture of neutral lipids,
PC/PEA, and ii) the application of individual PS as a lipid
component of a system leads to a three-fold rate decrease of
hydroxylation reaction, as compared to its mixture with PC.
Ethoxyresorufin is used as substrate to characterize the monooxygenase systems after the animals were treated by polyaromatic compounds. The rates of its oxidation by liver microsomal
fraction of either intact or phenobarbital-induced animals are,
as a rule, low, which is reflected in the reconstitution of catalytic activity of cytochrome P-450 D2 in liposomes (Table 1).
As different from other investigated here reaction, a relative-

ly high oxidation rate of ethoxyresorufin by cytochrome P-450
D2 in the presence of DLPC has been observed. As seen from
Table 1 the rate of oxidative N-demethylation of ethylmorphine
insignificantly vary in different by their composition liposomes, which may reflect the peculiarities of the given reaction. However, to our opinion, it is an artifact and is stipulated by and insufficient precision and sensitivity of the method employed for the reaction product analysis. A dependence
of the rate of oxidation of 7-ethoxycoumarin vs its concentration by the system involving cytochrome P-450 B6 in the presence of PC is described by the sigmoidal curve (not presented
data). A cooperativity degree of the process is sufficiently
high and is characterized by the Hill coefficient of 1,8. Dilution of PC by the equimolar amount of PEA leads to a decrease
in Hill coefficient up to 1,3. In the presence of other investigated phospholipids or their binary mixtures with PC, the
cooperativity of enzymatic process is not revealed. Kinetic
parameters of oxidation of 7-ethoxycoumarin and coumarin by
cytochrome P-450 B6 are given in Table 2.

Table 2

Kinetic parameters of oxidation of 7-ethoxycoumarin and
coumarin by cytochrome P-450 B6 in liposomes with
different phospholipid composition

Composition of liposomes	7-ethoxycoumarin deethylation		Coumarin 7-hydroxylation	
	V_{max}, nmol of product/nmol of cyt.P-450 in min	$K_m \cdot 10^6$, mol/l	V_{max}, nmol of product/nmol of cyt.P-450 in min	$K_m \cdot 10^6$, mol/l
DLPC	0,06	10,0	0,02	16,0
PC	-	-	0,03	20,0
PS	0,11	14,2	0,04	16,0
PC/PEA	0,18	11,8	0,03	18,0
PC/PS	0,41	10,0	0,09	17,0
PC/PG	0,39	10,0	0,08	17,0

Phospholipid mixtures are composed of equimolar amounts of
lipid components.

In accordance with these data the reaction of O-dealkylation of 7-ethoxycoumarin is most effective in the presence of
binary phospholipid mixture, with higher values of V_{max} being
observed for liposomes containing PC along with PS and PG. Similar tendence is retained in the reaction of 7-hydroxylation
of coumarin by the same form of cytochrome P-450 (Table 2).
Cytochrome P-450 is known to possess high catalytic activity towards 7-ethoxycoumarin and coumarin /4/. The activity
is also fully revealed upon its incorporation into liposomes,
which follows from the comparison of the values presented in
Tables 1-3. However, the maximal reaction rates of 7-hydroxyl-
ation of coumarin by cytochrome P-450$_{pyr}$ in liposomes from PS,
PC as well as equimolar mixture of the latter with PEA and PS
are close. A sertain increase of V_{max} is only observed in the
presence of PG. The effect of PG on cytochrome P-450 is more
vividly shown in the case of reaction of O-dealkylation of
7-ethoxycoumarin. Moreover, the reaction features for a low
rate when individual PS is used as a lipid component.

Table 3

Kinetic parameters of oxidation of 7-ethoxycoumarin and coumarin by cytochrome P-450$_{ayr}$ in liposomes with different phospholipid composition

Composition of liposomes	7-ethoxycoumarin deethylation		Coumarin 7-hydroxylation	
	V_{max}, nmol of product/nmol of cyt.P-450 in min	$K_m \cdot 10^6$, mol/l	V_{max}, nmol of product/nmol of cyt.P-450 in min	$K_m \cdot 10^6$, mol/l
DLPC	2,85	3,0	0,22	9,5
PC	3,45	3,0	0,40	10,0
PS	1,65	3,0	0,30	10,0
PC/PEA	2,80	3,4	0,56	10,0
PC/PS	3,75	2,9	0,56	11,0
PC/PG	6,25	2,9	0,74	11,0

Phospholipid mixtures are composed of equimolar amounts of lipid components.

CONCLUSIONS

1. Since those same phospholipids or their binary mixture exert different effects on kinetics of oxidation reactions of 7-ethoxycoumarin, coumarin, ethoxyresorufin, and ethylmorphine by cytochrome P-450 D2, correlation between the reaction type and the phospholipid composition-dependent alterations of catalytic activity of monooxygenase systems incorporated into liposomes may be assumed.

2. In the present work we have used three forms of cytochrome P-450 in the presence of one and the same electron transport protein, cytochrome P-450-reductase. Therefore, the obtained differences in the effect of liposomal phospholipid composition on the rate of oxidation of coumarin and 7-ethoxycoumarin by different isoenzymes, should be primarily ascribed to differences in interaction of phospholipids with cytochrome P-450.

3. From the data in Tables 1-3 it follows that the activating effect of the negatively charged phospholipids, PS and PG, is more vivid in equimolar mixture with PC. A substitution of the one negatively charged lipid by the other is in a number of cases nonequivalent. It indicates on the complex relation between the phospholipids charge and the catalytic activity of monooxygenase.

REFERENCES

1. D.A.Haugen, T.A.Van der Hoeven and M.J.Coon 1975. Purified liver microsomal cytochrome P-450. J. Biol. Chem., v.250, 3567-3570.
2. G.Lenaz 1979. The role of lipids in the structure and function of membranes. Subcellul. Biochem., v.6, 233-343.
3. P.J.Quinn, D.Chapman 1980. The dynamic of membrane structure. CRC Critic. Rev. Chem., v.8, 1-117.
4. P.Kaipainen, D.W.Nebert and M.J.Lang 1984. Purification and characterization of mouse liver microsomal cytochrome P-450 with high coumarin 7-hydroxylase activity. Eur. J. Biochem., v.144, 425-431.

DEPENDENCE OF PROGESTIN BINDING TO MICROSOMAL CYTOCHROME P-450 ON MEMBRANE LIPID AND DETERGENT CONCENTRATIONS

Nikolaus Kühn-Velten and Wolfgang Staib

Abt.f.Physiologische Chemie II, Universität,
D-4000 Düsseldorf, FRG

INTRODUCTION

Hydroxylation of progesterone at the C17-position is one important step within androgen and estrogen biosynthesis in testes and other steroidogenic tissues. The reaction is catalyzed by the microsomal steroid-17 -monooxygenase system (EC 1.14.99.9) which includes cytochrome P-450 as the substrate acceptor site (Menard & Purvis 1973; Barbieri et al 1981; Nakajin et al 1981). Binding of the physiological ligand, progesterone, and its analogues, progestins, to this cytochrome results in generation of type I difference spectra. It is well known that the apolar substrates of cytochrome P-450-dependent monooxygenases may accumulate in the lipid environment of these enzymes and that the composition of this lipid phase may perhaps affect the catalytic activity of such enzymes (Parry et al 1976; Sato et al 1979; Greenfield et al 1981; Backes et al 1982). This phenomenon may be responsible for the marked differences reported for spectral dissociation constants K_S or Michaelis constants K_m of the steroid-17 -monooxygenase (ranging from 0.04 to 100 μM). Therefore we considered it useful to investigate the effect of changing the concentrations of microsomes from rat testes or of detergent extracts from these microsomes upon the characteristics of progestin binding to the cytochrome P-450 of the steroid-17 -monooxygenase and to calculate progesterone accumulation in microsomal lipids in a systematic manner.

MATERIALS AND METHODS

Testes from male Han:Wistar rats aged 3 months were homogenized in 0.25 M sucrose, 0.01 M Tris, pH 7.4. Microsomes were prepared by differential centrifugation and finally resuspended in 0.15 M KCl, 0.01 M Tris, pH 7.4. Microsomal proteins were solubilized either in presence of Triton CF-54 (TCF-54, Rohm & Haas) or in presence of cholate (Kühn-Velten et al 1984a,b). Induction of type I difference spectra in microsomes or detergent extracts was routinely determined using progesterone (4-pregnene-3,20-dione, Calbiochem) as the ligand (Kühn-Velten et al 1984b); in some cases, medrogestone (6,17α-dimethyl-4,6-pregnadiene-3,20-dione, Kali-Chemie), hydroxyprogesterone (17 -hydroxy-4-pregnene-3,20-dione, Sigma), and pregnadienone (3,5-pregnadiene-20-one, Steraloids) were also tested. To investigate the distribution of progesterone between lipid and water phases of microsomal suspensions, 800 pmol progesterone and 500 Bq ^{3}H-progesterone were added to 1 ml microsomal suspensions. Radioactivity and therewith progesterone concentrations were determined in the aqueous supernatants and resuspended pellets after centrifugation at 150000 g for 45 minutes at 1°C. Lipids were extracted according to Folch et al (1957), and protein contents were measured in presence of lauryl sulfate sodium salt (Markwell et al 1981). It was assumed that the amount of progesterone in the pellets was either bound to the cytochrome or nonspecifically associated with membrane lipids.

RESULTS AND DISCUSSION

The apparent spectral dissociation constants K_S for the interaction of several progestins with cytochrome P-450 in crude testis microsomes depended strongly on the concentrations of microsomal suspension in the cuvettes (fig 1). As has already been reported for ligand binding to cytochrome P-450 from liver microsomes, this observation can be explained by the fact that with increasing membrane concentrations, a greater portion of a given amount of ligand is dissolved in lipid phase (see fig 2). Using the equations set up by Parry et al (1976) and Backes et al (1982), it was possible to calculate provisional K_S values (from the ordinate intercepts in fig 1) that are corrected for ligand accumulation in the lipid phase (Barbieri et al 1981): 12 nM for progesterone, 70 nM for medrogestone, 120 nM for hydroxyprogesterone, and 270 nM for pregnadienone.

The calculation of exact values for progesterone distribution between water and microsomal lipids is demonstrated in fig 2. The dependence of nonspecific progesterone accumulation in membrane lipids on microsome concentration is shown in fig 2C. The actual ligand concentrations in the lipid (fig 2D) and water (fig 2E) phases decreased with increasing microsome concentration

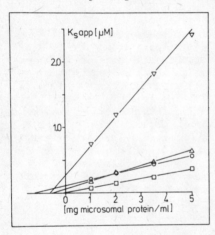

Figure 1. Dependence of apparent spectral dissociation constants ($K_S{}^{app}$) of testicular cytochrome P-450 on microsomal concentrations for progesterone (□), medrogestone (△), hydroxyprogesterone (○), and pregnadienone (▽) as the ligands.

Figure 2. Distribution of progesterone within different compartments of microsomal suspensions: amounts of progesterone bound to microsomes (A); amounts of progesterone bound specifically to cytochrome P-450 as determined from spectral measurements (B); amounts of progesterone bound in microsomal lipids calculated as the difference A-B (C); partition coefficients K_p (F) calculated from the ratio of progesterone concentration in microsomal lipids (D) and of progesterone concentration in the aqueous phase (E).

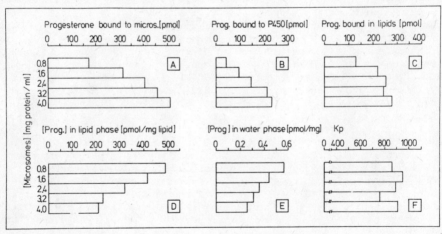

in the tubes, but the partition coefficient K_p remained rather constant at about 850. From this and parallel studies, we calculated K_p values of 80 for hydroxyprogesterone, 150 for pregnadienone, and 180 for medrogestone. By this systematic calculation, we confirmed previous results showing high progesterone accumulation in testis microsomes (Sato et al 1979).

Several investigations have provided evidence that the ligand-binding sites of microsomal cytochromes P-450 are buried within the lipid membrane (Sato et al 1979; Backes et al 1982; Taniguchi et al 1984). In this case, the provisional K_s values estimated from the ordinate intercepts in fig 1 must be multiplied by K_p to obtain preliminary K_s values for the cytochrome P-450 itself (Backes et al 1982): 10.2 μM for progesterone, 9.6 μM for hydroxyprogesterone, 12.6 μM for medrogestone, and 40.5 μM for pregnadienone. Our study therefore allows the very interesting conclusion that the progestin-binding cytochrome P-450 from rat testis appears to have similar affinities for progesterone and hydroxyprogesterone. This result seems to be important in view of the idea that both the 17 -hydroxylation of progesterone and the C17,20-cleavage of hydroxyprogesterone are catalyzed by a single cytochrome P-450 in the testis (Nakajin et al 1981).

Finally, we have compared the effects of cholate and TCF-54 on progesterone-induced spectral shifts of testicular cytochrome P-450 since both detergents have been proved to be useful for solubilization of microsomal steroid-17 -monooxygenase (Nakajin et al 1981; Chasalow et al 1982; Kühn-Velten et al 1984a,b). In our hands, cholate or 0.3% TCF-54 were less effective in solubilization than 2.0% TCF-54 (fig 3A). If extracts were diluted with the respective detergent-free buffers, the apparent K_s values decreased as was expected (fig 3B). Since those plots (as recommended by Backes et al 1982), however, gave different abscissa intercepts and therefore different relative K_p values for different TCF-54 concentrations, we changed the abscissa scale to obtain K_p values independent from lipid concentration (fig 3C) in accordance with the generally accepted definition for K_p (Parry et al 1976). The provisional K_s values (ordinate intercepts in fig 3B,C) were 60 nM for microsomes, 80 nM for cholate extracts and 250 nM for TCF-54 extracts. The relative K_p values (abscissa intercepts in fig 3C) were 25 for TCF-54 and

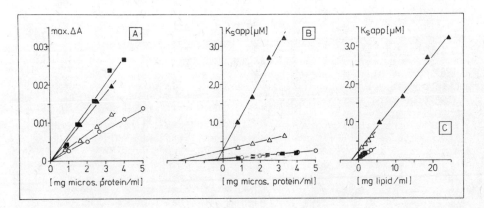

Figure 3. Maximal progesterone-induced difference spectra (A) and apparent spectral dissociation constants (B,C) in microsomal suspensions (■, 0.4 mg lipid/mg protein) or in detergent-solubilized extracts (○, cholate, 0.6 mg lipid/mg protein; △, 0.32% TCF-54 extracts, 1.0 mg lipid/mg protein; ▲, 2.0% TCF-54 extracts, 6.6 mg lipid/mg protein).

70 for cholate assuming 100 for microsomal lipids. If, again, provisional K_S values were multiplied with the (relative) K_p values, it became obvious that the preliminary K_S of the cytochrome itself was not crucially affected by cholate or TCF-54.

SUMMARY

In conclusion, this study illustrates the critical role of steroid accumulation in lipids for estimation of spectral dissociation constants K_S for cytochromes P-450 of steroid monooxygenases. This phenomenon was found to be highly important for the apparent affinity of steroids to those enzymes. The amounts and properties of lipids (including detergents) were found to change apparent K_S chiefly by determining steroid partition between water and lipid and much less by influencing the preliminary K_S value of the cytochrome P-450 itself. These problems may be responsible for the wide range of kinetic constants for steroid monooxygenases reported so far.

REFERENCES

Backes WL, Hogaboom M, Canady WJ (1982): The true hydrophobicity of microsomal cytochrome P-450 in the rat. J.Biol.Chem. 257, 4063-4070.

Barbieri RL, Canick JA, Ryan KJ (1981): High-affinity steroid binding to rat testis 17-hydroxylase and human placental aromatase. J.Steroid Biochem. 14, 387-393.

Chasalow FI, Marr H, Taylor G (1982): A new assay and solubilization procedure for steroid 17,20-lyase from rat testes. Steroids 39, 617-630.

Folch J, Lees M, Sloane-Stanley GH (1957): A simple method for the isolation and purification of total lipids from animal tissues. J.Biol.Chem. 226, 497-509.

Greenfield NJ, Gerolimatos B, Szwergold BS, Wolfson AJ, Prasad VVK, Lieberman S (1981): Effects of phospholipid and detergent on the substrate specificity of adrenal cytochrome P-450. J.Biol.Chem. 256, 4407-4417.

Kühn-Velten N, Bos D, Staib W (1984a): Solubilisierung von mikrosomalem und mitochondrialem Cytochrom P-450 des Rattenhodens mit Hilfe vin Triton CF-54. Hoppe-Seylers Z.Physiol.Chem. 365, 1021-1022.

Kühn-Velten N, Bunse T, Schürer N, Staib W (1984b): Direct effects of androgens on progesterone binding and metabolism in rat testis microsomes. Hoppe-Seylers Z.Physiol.Chem. 365, 773-779.

Markwell MAK, Haas SM, Tolbert NE, Bieber LL (1981): Protein determination in membrane and lipoprotein samples. In: Methods in Enzymology Vol 72 (Lowenstein JM, ed.), Academic Press, New York, pp. 296-303.

Menard RH, Purvis JL (1973): Studies of cytochrome P-450 in testis microsomes. Archs.Biochem.Biophys. 154, 8-18.

Nakajin S, Hall PF, Onoda M (1981): Testicular microsomal cytochrome P-450 for C21 steroid side chain cleavage. J.Biol.Chem. 256, 6134-6139.

Parry G, Palmer DN, Williams DJ (1976): Ligand partitioning into membranes: its significance in determining Km and Ks values for cytochrome P-450 and other membrane bound receptors and enzymes. FEBS Lett. 67, 123-129.

Sato B, Huseby RA, Matsumoto K, Samuels LT (1979): Molecular nature of interaction of steroids with biomembranes related to androgen biosynthesis. J.Steroid Biochem. 11, 1353-1359.

Taniguchi H, Imai Y, Sato R (1984): Substrate binding site of microsomal cytochrome P-450 directly faces membrane lipids. Biochem.Biophys.Res.Commun. 118, 916-922.

RECONSTITUTION OF LIVER MONOOXYGENASE SYSTEM IN THE BILAYER OF GIANT LIPOSOMES AND IN SOLUTION

Bachmanova G.I., Voznesensky A.I., Kanaeva I.P., Karyakin A.V., Scotzely-as E.D., Archakov A.I.
Research Institute of Physico-Chemical Medicine, Moscow, USSR

Two methodic approaches are used in the reconstitution of liver monooxygenase system: i) the incorporation of enzymes into the bilayer of liposomes composed of different phospholipids /PhLs/ and ii) the reconstitution of hydroxylase systems in solution containing isolated purified enzymes in the presence of PhLs or detergents /Trudell and Bösterling, 1983/. These methods are included in the study of the role of protein-protein and protein-lipid interactions in the functioning of monooxygenase system components and also the monooxygenase system molecular organization. The use of giant liposomes /GL/ for the monooxygenase system reconstitution is the most adequate method for the study of the monooxygenase system molecular organization due to the fact that the structural properties of GL are similar to those of the initial microsomal membranes /Bachmanova et al.,1984; Schwarz et al.,1984/. The reconstitution of hydroxylase activities in solution allows to model the reactivating action of PhLs and to investigate its mechanism. In this study we investigated the incorporation of microsomal monooxygenase system components into the bilayer of GL and aggregation state of NADPH-specific flavoprotein and cytochrome P-450 in solutionin the presence of nonpolar detergent Emulgen 913.

MATERIALS AND METHODS

Microsomes /MC/ were obtained from liver of male rabbits induced by phenobarbital /Karuzina et al.,1979/. Cytochrome P-450 /LM2/ and NADPH-cytochrome P-450 reductase /NADPH-R/ were obtained from rabbit liver microsomes induced by phenobarbital /Karuzina et al.,1979; Dignam and Strobel, 1977/. Cytochrome b5 /b5/ and NADH-cytochrome b5 reductase /NADH-R/ were obtained from rabbit liver microsomes /Spatz and Strittmatter,1971; Schater and Hulquist,1980/. GL were obtained from the mixture of microsomal PhLs by the method of reversed-phase evaporation /Szoka and Papahadjopoulos,1978/. LM2 and NADPH-R molecular weight was estimated in the presence or absence of Emulgen 913 by gel filtration method in the column with gel Toyo Pearl HW-55 and HW-60.

RESULTS AND DISCUSSION

Reconstitution of Monooxygenase System in the Bilayer of GL

The incorporation of NADH-R, NADPH-R, b5 and LM2 into the bilayer of GL was investigated by gel filtration method. The incubation mixture contained 10 nmoles of each enzyme and 5 mg of PhL of GL in 1 ml of 100 mM Tris-HCl buffer, pH 7.6, containing 0.1 mM EDTA and 1 mM dithiothreitol. The mixture was incubated for 40 min at 37°C and then placed into an ice bath for 30 min. To separate proteoliposomes /PL/ from the proteins that failed to incorporate, the mixture was put in the column with Sephacryl C-1000. It was shown that some amount of protein was eluted together with PhL. The proteins not incorporated into GL are eluted separately from PhL. GL and proteins were put into column separately under similar conditions as control. GL were shown to be eluted in the same fractions as PL.The mixture of four carriers was eluted in the following fractions.

The incorporation of the microsomal monooxygenase system four carriers does not affect the size of GL, their structure and the structure of bilayer. PL diameter was 200-270 nm, which corresponds to the mean diameter of microsomal vesicles. The absence of IMP's in PL may be accounted for the lower protein:lipid ratio in PL membranes (1:500) as compared to MC (1:50) /Borovyagin et al.,1984/.

The data on the chemical composition and activities of reconstituted and original microsomal monooxygenase system are given in Table 1. All four carriers are incorporated into GL. The incorporated enzymes possess functional activity. b5 interacts with its reductase and reduces cytochrome c in the presence of NADH, the rate constants of these reaction being identical in PL and MC. LM2 catalyzes dimethylaniline /DMA/ oxidation in the presence of cumole hydroperoxide /CHP/ at the rate similar to that of DMA oxidation in MC. The enzyme stoichiometry in PL differs greatly from that in microsomal system /Table 1/.

In PL as well as in MC both haemoproteins are reduced under anaerobic conditions in the presence of NADH or NADPH, the second order rate constants being identical /Table 1/. The first order rate constants of NADH- and NADPH-dependent DMA oxidation are higher in PL than in MC. As b5 is responsible for electron transfer to LM2 and this stage may be the limiting one, arbitrary second order rate constants of DMA oxidation were calculated taking into consideration the concentration of both haemoprotein in PL and MC. The constants were identical for both systems.

Thus, by means of NADH-R, NADPH-R, b5 and LM2 incorporation into GL bilayer we managed to reconstitute the monooxygenase system, the stoichiometry of its components differing greatly from that of original microsomal one. However, the second order rate constants of haemoproteins reduction and DMA oxidation in these systems proved to be identical. The equality of the rate constants of these reactions despite the difference in stoichiometry of carriers in MC and PL implies the absence of rigid nondissociating clusters of proteins in the membrane and confirm the assumption that proteins diffuse freely along the surface of PhL bilayer.

Reconstitution of Monooxygenase System in Solution

It is well known that the microsomal monooxygenase system may be reconstituted from isolated NADPH-R and cytochrome P-450 in the presence of PhL or detergents /Trudell and Bösterling, 1983/. To clarify the mechanism of reactivating action of detergents on hydroxylase activities we investigated DMA oxidation rate and aggregate state of NADPH-R and LM2. The maximal DMA oxidation rate in the presence of NADPH was observed with Emulgen 913 concentration being 0.05 g/l in system containing 0.33 µM NADPH-R, 0.33 µM LM2, 100 mM K-phosphate buffer, pH 7.5, 300 µM NADPH and 1 mM DMA (T=37°C). The estimation of the enzymes molecular weight in the presence of the detergent at given concentration helped to identify the formation of mixed LM2 and NADPH-R complexes /Table 2/. No protein complexes were detected in the absence of detergent.

Table 1. <u>Composition and Kinetic Properties of Reconstituted and Micro-somal Monooxygenase Systems</u>

Indeces	PL	MC
NADH-R (nmole·mg^{-1} PhL)	0.044	0.14
NADPH-R (nmole·mg^{-1} PhL)	0.58	0.14
Cytochrome b5 (nmole·mg^{-1} PhL)	1.2	2.6
LM2 (nmole·mg^{-1} PhL)	0.62	4.2
Stoichiometry NADH-R:NADPH-R:b5:LM2	1:15:30:15	1:1:20:30
Rate constant of NADPH-dependent cytochrome c reduction, min^{-1}	1.2	1.2
CHP-dependent DMA oxidation, min^{-1}	10.2	9.1

NADH-dependent reactions:

Rate constants of cytochromes reduction

		PL	MC
b5	s^{-1}	1.0	4.3
	M^{-1}s^{-1}	10^4	10^4
LM2	s^{-1}	0.07	0.02
	M^{-1}s^{-1}	10^2	10^2

Rate constants of DMA oxidation

	PL	MC
min^{-1}	6.7	1.2
M^{-1}min^{-1}	3·10^6	2·10^6

NADPH-dependent reactions:

Rate constants of cytochromes reduction

		PL	MC
b5	s^{-1}	4.2	1.1
	M^{-1}s^{-1}	10^3	10^3
	s^{-1}	0.1	0.03
	M^{-1}s^{-1}	10^2	10^2

Rate constants of DMA oxidation

	PL	MC
min^{-1}	15.1	6.1
M^{-1}min^{-1}	8·10^6	9·10^6

Increasing Emulgen 913 concentration up to 0.25 g/l and higher caused further dissociation of the investigated proteins aggregates. However, no complex formation was observed.

Therefore, the obtained results show that the phospholipids reactivating action observed during liver monooxygenase system reconstitution can be

Table 2. Aggregation State and Activity of NADPH-R and LM2 in the presence and absence of Emulgen 913

System	Aggregate State						Oxidation of DMA, min^{-1}
	NADPH-R		LM2		NADPH-R-LM2-complex		
	M. w. KD	Sub-units	M. w. KD	Sub-units	M. w. KD	Subunits	
Without Emulgen 913	1600	20	700	14	No		0
With Emulgen, 0.05 g/l	800	10	300	6	700	5 NADPH-R + 5 LM2	0.4

modelled with the use of detergents. It is shown that the stimulating effect of Emulgen upon the reaction of DMA oxidation is explained by its ability to dissociate cytochrome P-450 and NADPH-R oligomers and to provide the formation of the mixed complex of the enzymes with 1:1 stoichiometry possessing monooxygenase activity.

REFERENCES

1. Bachmanova G.I., Borodin E.A., Budennaya T.Yu., Karyakin A.V. and Archakov A.I. /1984/ Random distribution of NAD(P)H-specific electron carriers in liver microsomal and giant proteoliposomal reconstituted membranes. 6th Internat. Symp. on Microsomes and Drug oxidation. Abstracts.Taylor and Francis, London and Philadelphia, p. 31.
2. Borovyagin V.L., Tarachovsky Yu.S., Kanaeva I.P., Karyakin A.V., Bachmanova G.I. and Archakov A.I. /1984/ Study on the ultrastructure of rat liver microsomal reconstituted membranes. Biol. Membranes 1, 605-613.
3. Dignam J.D. and Strobel H.W. /1977/ NADPH-cytochrome P-450 reductase from rat liver: purification by affinity chromatography and characterization. Biochemistry 16, 1116-1123.
4. Karuzina I.I., Bachmanova G.I., Mengazetdinov D.E., Myasoedova K.N., Zhihareva V.O., Kuznetsova G.P. and Archakov A.I. /1979/ Isolation and characterization of rabbit liver microsomes. Biokhimia 44, 1049-1057.
5. Schater A. and Hulzquist D. /1980/ Purification of bovine liver microsomal NADH-cytochrome b5 reductase using affinity chromatography. Biochem. Biophys. Res. Commun. 95, 381-387.
6. Schwartz D., Gast K., Meyer H., Lachmann U., Coon M.J. and Ruckpaul K. /1984/ Incorporation of the cytochrome P-450 monooxygenase system into large unilamellar liposomes using octylglucoside especially for measurements of protein diffusion in membranes. Biochem. Biophys. Res. Commun. 121, 118-125.
7. Spatz L. and Strittmatter P. /1971/ A form of cytochrome b5 that contains an additional hydrophobic sequence 40 aminoacids residues. Proc. Natl. Acad. Sci. USA 68, 1042-1046.
8. Szoka F. and Papahadjopoulos D. /1978/ Procedure for preparation of liposomes with large internal aqueous space and high capture by reverse-phase evaporation. Proc. Natl. Acad. Sci. USA 75, 4194-4198.
9. Trudell T.R. and Bösterling B. /1983/ Interaction of cytochrome P-450 with phospholipids and proteins in endoplasmatic reticulum. In: Membrane Fluidity in Biology, v. 1, Concepts of Membrane Structure. /Aloia R.C.,ed./ Acad. Press, N.Y., 201-223.

EFFECTS OF KETOCONAZOLE ON THE TESTICULAR AND ADRENAL CHOLESTEROL SIDE-CHAIN CLEAVAGE

Gustaaf H.M. Willemsens and H. Vanden Bossche

Janssen Pharmaceutica Research Laboratories, B-2340 Beerse, Belgium

INTRODUCTION

The orally active antifungal imidazole derivative ketoconazole, binds to yeast and fungal cytochrome P-450 resulting in an inhibition of the microsomal 14α-demethylase, an enzyme which catalyzes the 14α-demethylation of lanosterol to form ergosterol (Vanden Bossche et al., 1984a). This inhibition of the cyt. P-450 dependent 14α-demethylase was obtained at low ketoconazole concentrations. In fact, 50 % inhibition of the ergosterol biosynthesis in exponentially growing Candida albicans cells was obtained after 1 h of contact with 4×10^{-9} M ketoconazole (Vanden Bossche et al., 1984b). At 2.3×10^{-8} M and 3.1×10^{-8} M, ketoconazole decreases for 50 % the peak height (448-490 nm) of the reduced CO-complex of microsomal cyt. P-450 from Saccharomyces cerevisiae and C. albicans respectively (Vanden Bossche et al., 1985a). At higher concentrations ketoconazole also affects cyt. P-450 species in pig testis and bovine adrenal cortex mitochondrial fractions. For example 1.2×10^{-6} M is needed to achieve 50 % decrease of the peak height of cyt. P-450 species in pig testis mitochondrial fractions (Vanden Bossche, et al., 1985b).

The present study was designed to examine in a reconstituted system, the effects of ketoconazole on the cholesterol side chain cleavage reaction from pig testis and bovine adrenal cortex mitochondria.

MATERIALS AND METHODS

Bovine adrenal glands and pig testis were collected at a local slaughterhouse and transported to the laboratory in ice cold 0.15 M KCl.

Adrenodoxin and adrenodoxin-reductase were isolated from the adrenal cortex according to Suhara et al. (1972a, b). Pig testis mitochondria were prepared as described by Luketich et al (1983). Pig testis and adrenal cortex mitochondrial cyt. P-450$_{scc}$ (Side-Chain-Cleavage) was isolated from in cholate solubilized, sonicated mitochondrial pellets as described by Katagiri et al. (1978) using aniline-Sepharose-4B column chromatography. The specific activities of cyt. P-450$_{scc}$ from pig testis and bovine adrenal cortex are shown in Table 1.

The effects of ketoconazole, dissolved in dimethylsulfoxide (DMSO), on the cholesterol side-chain cleavage was studied in a mixture containing in a final volume of 600 μl potassium phosphate buffer 0.1 M (pH 7.4): 240 nmol NADPH, 2.9 nmol adrenodoxin, 0.08 units of adrenodoxin-reductase, 6 μl ketoconazole and/or solvent and 60 pmol cyt. P-450$_{scc}$.

Table 1. Specific activity of cyt. P-450$_{scc}$ in pig testis and bovine adrenal cortex mitochondria before and after purification

	Content nmoles/mg protein	
	Adrenal cortex	Pig testis
Crude mitochondria	0.577	0.137
Aniline-Sepharose eluate	4.49	2.26

The reaction was initiated by the addition of 718 pmoles [4-^{14}C]-cholesterol (S.A. 55.7 mCi/mmol) dissolved in 20 μl potassium phosphate buffer 0.1 M (pH 7.4) containing 0.3 % Tween 20. Incubation was carried out at 30° C. The reaction was stopped by the addition 1.6 ml CH$_3$OH and steroids were extracted and separated on thin-layer-chromatography as described by Katagiri et al. (1978). Steroids were located and radio-activity determined as described by Vanden Bossche et al. (1984c).

RESULTS

Under the present experimental conditions, cyt. P-450$_{scc}$ isolated from pig testis and bovine adrenal cortex mitochondria, converts in the presence of solvent cholesterol into pregnenolone at a rate of respectively 17.5 and 10.7 pmoles/min/0.6 ml incubation mixture.

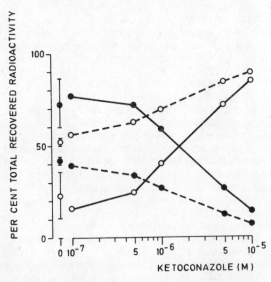

Fig. 1. Effects of ketoconazole on the conversion of cholesterol (o) into pregnenolone (●) in a reconstituted system from pig testis mitochondria (——) and bovine adrenal cortex (- - -) mitochondria.

As shown in Fig. 1, this pregnenolone synthesis is inhibited by ketoconazole. In fact, 50 % inhibition of the cholesterol side-chain cleavage reaction studied with cyt. P-450$_{scc}$ isolated from testis mitochondria, was obtained after 30 min of contact with 3×10^{-6} M of ketoconazole.

A similar inhibition was observed with 1.7×10^{-6} M ketoconazole when this reaction was studied with cyt. P-450$_{scc}$ isolated from adrenal cortex mitochondria.

At 10^{-5} M, the formation of prenenolone was inhibited for 81.2 and 83.1 % when studied with pig testis and bovine adrenal cortex cyt. P-450$_{scc}$.

CONCLUSION

The results presented in this study indicate that high concentrations of ketoconazole affects the cholesterol side-chain cleavage reaction in a reconstituted system from pig testis and bovine adrenal cortex mitochondria.

A complete inhibition of this reaction was not achieved even at concentrations higher than 10^{-5} M, indicating that this side-chain cleavage enzyme system is much less sensitive as the cyt. P-450 dependent 14α-demethylase system involved in the biosynthesis of ergosterol in yeast and fungi.

However, when ketoconazole is used at high doses in for example patients with prostate carcinoma, this effect on the cholesterol side-chain cleavage together with the previously described effects on the 17,20-lyase system (Vanden Bossche et al., 1985a) might be at the origin of the observed inhibition of androgen biosynthesis.

REFERENCES

Katagari, M., Takemori, S., Itagaki, E., Suhara, K. (1978). In Methods in Enzymology, Fleischer and Packer (Eds.). Academic Press, vol. LII part C., pp. 124-129.

Luketich, J.M., Melner, M.J., Guengerich, F.P., Puett, D. (1983). Effects of human choriogonadotropin on mitochondrial and microsomal Cytochrome P-450 levels in mouse testes. Biochem. Biophys. Res. Commun. 111, 424-429.

Suhara, K., Takemori, S., Katagiri, M. (1972a). Improved purification of bovine adrenal iron-sulfur protein. Biochem. Biophys. Acta 263, 272-278.

Suhara, K., Ikeda, Y., Takemori, S., Katagiri, M. (1972b). The purification and properties of NADPH-adrenodoxin reductase from bovine adrenocortical mitochondria. FEBS Lett. 28, 45-47.

Vanden Bossche, H., Lauwers, W., Willemsens, G., Marichal, P., Cornelissen, F., Cools, W. (1984a). Molecular basis for the antimycotic and antibacterial activity of N-substituted imidazoles and triazoles: the inhibition of isoprenoid biosynthesis. Pestic. Sci., 15, 188-138.

Vanden Bossche, H., Willemsens, G., Marichal, P., Cools, W. and Lauwers, W. (1984b). The molecular basis for the antifungal activities of N-substituted azole derivatives. Focus on R 51 211. In Mode of action of antifungal agents, A.P.J. Trinci, J.F. Ryley (Eds.). Cambridge University Press, Cambridge, pp. 321-341.

Vanden Bossche, H., Willemsens, G., Cools, W., Bellens, D. (1984c). Effects of etomidate on steroid biosynthesis in subcellular fractions of bovine adrenals. Biochem. Pharmacol. 33, 3861-3868.

Vanden Bossche, H., Lauwers, W., Willemsens, G., Cools, W. (1985a). The Cytochrome P-450 dependent C17,20-lyase in subcellular fractions of the rat testis. Differences in sensitivity to ketoconazole and itraconazole. In <u>Microsomes and drug oxidations</u>, A.R. Boobis, J. Caldwell, F. de Matteis, C.R. Elcombe (Eds.). Taylor & Francis Ltd., Basingstoke, U.K. (in press).

Vanden Bossche, H., De Coster, R. (1985b). Pharmacology and clinical uses of ketoconazole. In <u>Pharmacology and clinical uses of inhibitors of hormone secretion and action</u>, B.J.A. Furr, A. Wakeling (Eds.). Praeger Scientific, Eastbourne, U.K. (in press).

PROPERTIES OF RECONSTITUTED HYBRIDE CYTOCHROME P-450 SYSTEMS

SMETTAN, G., SHKUMATOV, V.M.*, POMMERENING, K. and K. RUCKPAUL

Central Institute of Molecular Biology, Academy of Sciences
of the GDR, 1115 Berlin-Buch, GDR
*Institute of Bioorganic Chemistry, Academy of Sciences of the
Byelorussian SSR, 220600 Minsk, USSR

INTRODUCTION

The analysis of interactions between the essential components
of cytochrome P-450 dependent monooxygenatic systems has proved
as useful approach to elucidate the catalytic function and its
regulation /Bernhardt et al. 1984, Blanck et al. 1984/. Most of
our knowledge originates from studies on naturally occuring or
reconstituted systems with different phospholipid composition
/Smettan et al. 1984/. The specificity of these interactions
however, including the exchangeability of the respective components are rather poorly understood. Such kind of investigations
requires the construction of hybride systems. That means the
combination of components from different species, different
cell compartments and different structure of the electron transfer systems (systems without or with ferredoxin as electron
shuttle).

In the present study the isolated and highly purified components of an isozymic system from rabbit liver microsomes (cytochrome P-450 LM2; NADPH-cytochrome P-450 reductase) were differently coupled with the isolated components of the mitochondrial system from the bovine adenocortical system (cytochrome
P-450$_{scc}$; cytochrome P-450$_{11\beta}$; adrenodoxin; adrenodoxin reductase) and the activity of the reconstituted hybride systems
determined.

RESULTS AND DISCUSSION

The enzymatic activity of the constructed hybride systems
is expected to depend on (i) the matrix used, (ii) the capability of the proteins to form catalytically active complexes
and (iii) the compatibility of the electrochemical potentials

as essential prerequisite for an exchangeability of the components of the electron transfer chain resulting in an enzymatically active system. The results of the influence of different
matrices on the conversion of cholesterol by several constructed hybride systems are summarized in Table 1.

Table 1   Formation of pregnenolone from cholesterol by cyto-
          chrome P-450$_{scc}$ with different electron transport
          chains

Experimental conditions: 0.05 M Na-phosphate buffer, pH 7.4,
t = 37°C; molar stoichiometry reductase/adrenodoxin/cytochrome
P-450: 0.5 µM/1.0 µM/0.5 µM

Matrix components			Electron transport proteins		
DLPC	Tween 20	chole-sterol	LM2 red.	LM2 red. + adrenodoxin	adrenodoxin red. + adrenodoxin
/mg/ml/	/%/	/µM/	nMol pregnenolone/nMol		P-450/min
-	0.03	25	0	0	1.4
-	0.3	150	0	0	3.5
0.2	-	25	0.14	0.1	1.4
0.2	-	150	0.06	0.11	1.7
0.2	0.03	150	0.31	0.33	1.6
0.2	0.3	150	0	0	4.7

The highest conversion rates of the reconstituted adrenocortical system were achieved in detergent micelles and in mixed
micelles from phospholipids and detergent whereas the activity
of the hepatic endoplasmic cytochrome P-450 reductase supported
reaction under these conditions completely disappeared. Incorporation of the protein components in dilauroylphosphatidylcholine leads to enzymatic activity in the pure and in the hybride
system as well. The highest product formation in the hybride
system is achieved in the presence of a mixture of phospholipid
and detergent in a molar stoichiometry of 12:1. This can be due
to an enhanced bioavailibility of the substrate for the enzyme
and/or an improved organization of an active system within mixed micelles compared to bilayer structures. Interestingly addition of adrenodoxin to the hybride system does not significantly enhance the activity.

   The diminished enzymatic activity of the hybride system
(~20 % of the mitochondrial activity within the same vesicles)
can be caused by an impaired electron transfer. Therefore the
NADPH dependent reduction reaction was analyzed by means of
stopped-flow-analysis (Table 2).

The reduction rate in both hybride systems as compared to the reconstituted one is lowered by about one order of magnitude despite the significant higher rate of the cytochrome P-450 LM2 system.

Table 2 Rate constants of the NADPH dependent reduction in hybride systems

Experimental conditions: 0.05 M Na-phosphat buffer, pH 7.4, t = 25°C, reconstitution into dilauroylphosphatidylcholine, the cytochrome P-450 LM2 system into microsomal lipid

System	k_{-1} / s^{-1} /
P-450$_{scc}$ + adrenodoxin + adrenodoxin-reductase	0.3
P-450$_{scc}$ + LM2 reductase	0.02
P-450 LM2 + LM2 reductase	3.0
P-450 LM2 + adrenodoxin + adrenodoxin-reductase	0.2

A comparison of the conversion (Table 1) with the reduction (Table 2) in the corresponding hybride system taking into account the temperature difference in both kinds of determination reveals that substrate conversion proceeds about 60-fold lower than reduction in the reconstituted and about 100-fold lower in the hybride system. This difference indicates that obviously the cytochrome P-450$_{scc}$ system is similarly regulated by the spin equilibrium controlled reduction as it has been shown for the P-450 LM2 system /Blanck et al. 1983/. The data further suggest that subsequent steps after the first reduction probable the insertion of the second electron are rate limiting. The results however do not allow to decide what finally is responsible for the decrease in activity: insufficient conformational adaptation of the electron transferring proteins due to steric hindrance and/or lacking compatibility of the redox potentials.

A complete reduction of cytochrome P-450$_{11\beta}$ by NADPH P-450 reductase is achieved analogously to cytochrome P-450$_{scc}$ but at an enhanced rate.

Certain properties of cytochrome P-450 isozymes have been shown to be affected by the complementary components e.g. the high spin amount resulting in changed substrate binding. As the lowered activity in hybride systems can emerge from lacking modulating effectuations of such interactions the substrate binding of cholesterol and ß-sitosterol in dependence on the presence of adrenodoxin was studied. The binding affinity of cytochrome P-450$_{scc}$ /Radyuk et al. 1982/ and ß-sitosterol is almost the same. However the ß-sitosterol induced high spin shift is enhanced by adrenodoxin up to about one order of magnitude (Table 3).

Table 3   Binding parameters of substrates to cytochrome P-450
Experimental conditions: 0.05 M Na-phosphat buffer,
pH 7.4

Substrate	in absence of adrenodox.		in presence of adrenodox.	
	K_s	ΔA_{max}	K_s	ΔA_{max}
cholesterol	2.3 µM	0.04	2.5 µM	0.04
ß-sitosterol	5.5 µM	0.004	4.3 µM	0.03

The parameters of the conversion reaction differ. Whereas the K_M values for both substrates range about 100 µM the V_{max} differs slightly: 2 nMol pregnenolone/nMol cytochrome P-450$_{scc}$/min for ß-sitosterol and 5 nMoles formed from cholesterol as substrate-again indicating a tightly correlation between ΔA_{max} and the conversion rate. These data suggest to assume a remarkably decreased activity in the constructed hybride system due to the very low ΔA_{max} value in the absence of adrenodoxin.

REFERENCES

Blanck, J., Smettan, G., Ristau, O., Ingelman-Sundberg, M. and Ruckpaul, K. (1984) Eur. J. Biochem. 144, 509-513

Bernhardt, R., Makower, A., Jänig, G.-R. and Ruckpaul, K. (1984) Biochim. Biophys. Acta 745, 140-148

Smettan, G., Kisselev, P.A., Kissel, M.A., Akhrem, A.A. and Ruckpaul, K. (1984) Biomed. Biochim. Acta 43, 1073-1082

Blanck, J., Rein, H., Sommer, M., Ristau, O., Smettan, G. and Ruckpaul, K. (1983) Biochem. Pharmacol. 32, 1683-1688

Lambeth, J.D. and Pember, S.O. (1983) J. Biol. chem. 258, 5596-5602

Radyuk, V.G., Shkumatov, V.M., Chashin, V.L. and Akhrem, A.A. (1982) Biochimija 47, 1792-1801

PURIFICATION AND CHARACTERIZATION OF A MICROSOMAL CYTOCHROME P-450 WITH ACTIVITY TOWARDS COUMARIN 7-HYDROXYLATION FROM RABBIT LIVER

Kaipainen P. K.

Department of Pharmacology and Toxicology, University of Kuopio, P. O. X. 6, Kuopio, Finland

INTRODUCTION

Recent studies have shown that inbred strains of mice such as D2 have a cytochrome P-450 isoenzyme with high activity towards coumarin 7-hydroxylation (Kaipainen et al. 1984). We have also reported that rabbit liver microsomes have coumarin 7-hydroxylase activity (Kaipainen et al. 1985). In this paper I describe a method for the purification of rabbit cytochrome P-450 isoenzyme with activity towards coumarin 7-hydroxylation. Furthermore I compare the immunological properties of the purified enzyme with those of the D2 mouse.

MATERIALS AND METHODS

Chemicals were purchased from following sources: coumarin and 7-ethoxycoumarin from Aldrich Chemical CO., Milwaukee, WI, USA., sodium cholate, dithiothreitol and NADPH from Sigma Chemical CO., St. Louis, MO, USA., cyanogen bromide, potassium phosphates and glycerol were from Merck, Darmstadt, F.R.G., the gel materials for purification from Pharmacia Company, Uppsala, Sweden., emulgen 911 was from Kao-Atlas CO., Tokyo, Japan and hydroxylapatite from Bio-Rad, Richmond, CA,USA.

Two adult male rabbits (2-3 kg) of the strain NZW (White) were used for purification of P-450. They were obtained from the Animal Center of University of Kuopio, Kuopio, Finland. Preparation of microsomes was carried out as described by Lang et al. (1981). Washed microsomes were stored at $-80^{\circ}C$ in small aliquots until used.

Purification procedure for P-450 is essentially described in Kaipainen et al. (1984). Minor modifications are described as follows. The
solubilized microsomes were centrifuged at 105,000 g for two hours.
The supernatant was applied to a 2.6 x 30 cm 8-aminooctyl-Sepharose 4B
column. Purification of P-450 was monitored by measuring the activity
of coumarin 7-hydroxylase and contents of P-450 and protein from each
fraction. Those fractions with coumarin 7-hydroxylase activity were
pooled and the potassium phosphate concentration of it was lowered by
dilution to 20 mM. The mixture was applied to a hydroxylapatite
column (2.6 x 5 cm). Elution was carried out with 200 mM potassium
phosphate buffer, pH 7.25, containing 20% glycerol, 0.5% sodium cholate and 0.5 mM dithiothreitol. Fractions with coumarin 7-hydroxylase
activity were pooled and the pool was applied to a second 8-aminooctyl
Sepharose 4B column (1.6 x 15 cm). Washing and elution were same as in
the first step. Fractions with coumarin 7-hydroxylase activity were
pooled and the pool was dialyzed against 200 mM potassium phosphate
buffer, pH 7.25, containing 20% glycerol, 0.5% sodium cholate and 0.5
mM dithiothreitol, for two days and stored in small aliquots at -80°C
until used.

Liver NADPH cytochrome c reductase was purified from untreated Wistar
rats according to Yasukochi and Masters (1976). Determination of P-450
and protein were carried out according to Omura and Sato (1964) and
Castell et al. (1979), respectively. Reconstitution experiment was
carried out as described by Kaipainen et al. (1985). Immunologic
techniques and inhibition studies were carried out according to Kaipainen et al. (1984).

RESULTS AND DISCUSSION

Each purification step and the elution profiles are described in
figure 1. The specific content of the purified P-450 was 14.5 nmoles
per mg of protein and exhibited a single band on SDS polyacrylamide
gel elecrophoresis (figure 2). The reconstituted activities of
coumarin 7-hydroxylase and 7-ethoxycoumarin O-deethylase of the
purified rabbit cytochrome P-450 were 50 and 389 pmoles per minute per
mg of protein.

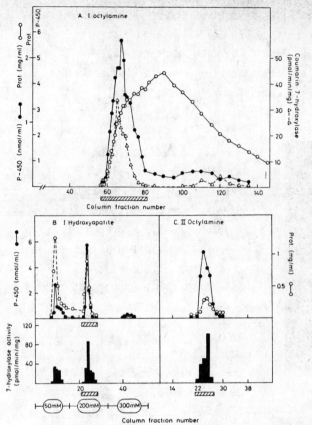

Figure 1. Chromatographic elutions and sequential purification of liver microsomal P-450 fractions from rabbit liver. Coumarin 7-hydroxylase activity is expressed in pmoles per mg of protein. Striped rectangles denote those fractions containing highest coumarin 7-hydroxylase activity. (A) First 8-aminooctyl-Sepharose 4B column; (B) The hydroxyapatite column eluted with 200 mM and 300 mM potassium phosphate buffers and; (C) Second 8-aminooctyl-Sepharose 4B column.

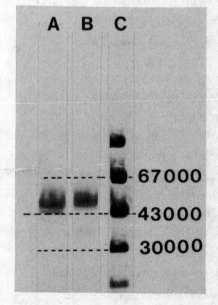

Figure 2. SDS-polyacrylamide gel electrophoresis of purified P-450s from untreated rabbit and phenobarbital treated D2 mouse liver. (A) 2-4 ug D2 P-450; (B) 2-4 ug rabbit P-450; (C) Standards: albumin (67,000)., ovalbumin (43,000) and carbonic anhydrase (30,000).

In Ouchterlony double immunodiffusion analysis there is no precipitin line against the purified rabbit cytochrome P-450 with activity of coumarin 7-hydroxylase by our previously developed anti-(D2 P-450$_{Coh}$) (figure 3). Our antibody did not block the reconstituted activities of coumarin 7-hydroxylase or 7-ethoxycoumarin O-deethylase in rabbit (figure 4).

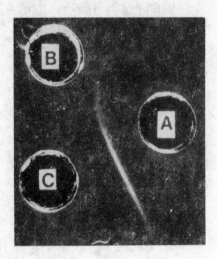

Figure 3. Ouchterlony double immunodiffusion analysis of purified P-450 from untreated rabbit and phenobarbital treated D2 mouse against our previously developed anti-(D2 P-450$_{Coh}$). (A) D2 P-450 (50 pmoles); (B) rabbit P-450 (50 pmoles); (C) antibody (0.1 mg).

Figure 4. Figure shows the inhibition of rabbit and pheno-barbital treated D2 mouse reconstituted coumarin 7-hydroxy-lase and 7-ethoxycoumarin O-de-ethylase activities. The starting coumarin 7-hydroxylase activities of rabbit and D2 mouse were 50 and 1056 pmoles per min per mg of protein, respectively and the corresponding values of 7-ethoxy-coumarin O-deethylase were 389 and 1640 pmoles per min per mg of protein, respectively. Experiment contained duplicate samples at each concentrations of immuno-globulin.

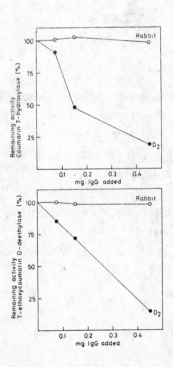

All these findings suggest that cytochrome P-450 responsible for coumarin metabolism in rabbit and D2 mouse is different.

Acknowledgements: This study was supported by the Finnish Academy of Natural Sciences, grant no: 1091-5. I would like to thank prof Matti Lang for his constructive criticism for preparing this article.

REFERENCES

Castell, J. V., Cervera, M and Marco, R. (1979) A convenient micro-method for the assay of primary amines and proteins with fluorescamine. A reexamination of the conditions of reaction. Anal. Biochem. 99:379-91.

Kaipainen, P. K., Koivusaari, U and Lang, M. A. (1985) Catalytic and immunological comparison of coumarin 7-hydroxylation in different species. Comp. Biochem. Physiol. in press.

Kaipainen, P. K., Nebert, D. W and Lang, M. A. (1984) Purification and characterization of a microsomal cytochrome P-450 with high activity of coumarin 7-hydroxylase from mouse liver. Eur. J. Biochem. 144:425-31.

Lang, M. A., Gielen, J. E and Nebert, D. W. (1981) Genetic evidence for many unique liver microsomal P-450-mediated monooxygenase activities in heterogeneic stock mice. J. Biol. Chem. 256:12068-12075.

Omura, T and Sato, R. (1964) The carbon-monoxide binding pigment of liver microsomes. J. Biol. Chem. 239:2379-2385.

Yasukochi, Y and Masters, B. S. (1976) Some properties of a detergent-solubilized NADPH cytochrome c (cytochrome P-450) reductase purified by biospecific affinity chromatography. J. Biol. Chem. 251:5337-5344.

Transcriptionally Controlled Mechanism of Induction of Cytochrome P-450

MECHANISM OF INDUCTION OF SEX SPECIFIC FORMS OF CYTOCHROME
P-450 IN LIVER MICROSOMES OF RATS

Ryuichi Kato and Tetsuya Kamataki

Department of Pharmacology, School of Medicine, Keio
University, 35 Shinanomachi, Shinjuku-ku, Tokyo 160, Japan

INTRODUCTION

Marked sex-related differences in the duration of actions
of drugs have been known with certain drugs. Sex-difference
in the oxidation of drugs and steroids by liver microsomes
have also been noted. Thus, the differences in the
properties of mixed-function oxidase system were regarded as
the cause of the drug actions (Kato et al. 1974). In the
past two decades, many efforts were paid to clarify the
enzymatic mechanisms responsible for the occurrence of sex
difference. Since cytochrome P-450 was proven to be a
terminal oxidase in the electron transport system in liver
microsomes, extensive works on cytochrome P-450 in liver
microsomes of male and female rats were conducted to account
for the sex-related differences (Kato et al. 1968; Kato and
Kamataki 1982a). Although only a slight differences in the
amounts of cytochrome P-450 were detected between male and
female rats, considerable differences were found to occur in
the affinity of microsomal cytochrome P-450 to bind to
certain drugs and in the reduction rate enhanced by the
addition of drugs. These differences can now be explained in
part by a concept that the population of forms of cytochrome
P-450 in respective liver microsomes from male and female
rats are different (Kato and Kamataki 1982a).

The difference between male and female rats in the
population of cytochrome P-450, particularly on the existence
of sex-specific forms of cytochrome P-450, was first
demonstrated by us(Kato and Kamataki 1982a; Kamataki et al.
1981; Kato and Kamataki 1982b). We purified one of each form
of cytochrome P-450, namely P-450-male and P-450-female, from
liver microsomes of male and female rats, and proved
immunochemically that these forms of cytochrome P-450 occurs
specifically in each sex, indicating that the syntheses of P-
450-male and P-450-female are regulated by androgens and
estrogens, respectively (Kamataki et al. 1983). The present
study was thus undertaken to strengthen this view.
Furthermore, the mechanism proposed so far through analysis
of drug and steroid hydroxylations, in which androgens and
estrogens induce sex specific hydroxylases through the
stimulation or depression of the function of pituitary gland,

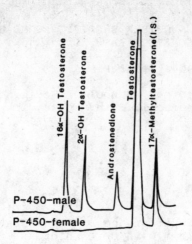

Fig. 1. HPLC elution profile of metabolites of testosterone
        incubated with P-450-male and P-450-female.

was examined by the immunochemical quantification of P-450-
male and P-450-female. We report here that induction of P-
450-femle by estradiol is mediated through pituitary gland.

RESULTS AND DISCUSSION

Characterization and Identification of P-450-male and P-450-
female

    After we demonstrated the existence of P-450-male and P-
450-female as sex specific forms of cytochrome P-450, the
purification of forms of cytochrome P-450 probably identical
to P-450-male and P-450-female have been reported. The
reported forms of cytochrome P-450 demonstrated as identical
to P-450-male are P-450UT-A by Guengerich et al. (1982) P-
450h by Ryan et al. (1984b) and P-450 2c by Waxman, (1984)
based on spectral and catalytic properties and the yield
through chromatographic procedures. In addition,
hydroxylation of testosterone at 16α position has been
proposed to be characteristic of these forms of cytochrome P-
450 (MacGeoch et al. 1984).

    As can be seen in Fig. 1, P-450-male but not P-450-female
showed high activities in the hydroxylation of testosterone
at 16α and 2α positions. P-450 15β and P-450i have been
reported to catalyze 15β-hydroxylation of 5α-androstane-
3α,17β-diol-3,17-disulfate (MacGeoch et al. 1984; Ryan et al.
1984a). Although we have not examined whether P-450-female
catalyzes the 15β-hydroxylation, our results showing that P-
450-female appeared in a certin period (around 25 days) after
birth in liver microsomes of male rats strongly support the
view that P-450i is identical to P-450-female (Maeda et al.
1984). Recent studies have confirmed the N-terminal amino

P-450-male:

Met-Asp-Pro-Val-Leu-Val-Leu-Val-Leu-Thr-Leu-Ser- X -Leu-Leu-Leu-Leu-

P-450-female:

Met-Asp-Pro-Phe-Val-Val-Leu-Val-Leu-Ser-Leu-Ser- X -Leu-Leu-

Fig. 2. N-Terminal amino acid sequence of P-450-male and P-
     450-female.

acid sequence of P-450 2c, P-450h and P-450i (Waxman 1984;
Haniu et al. 1984). The N-terminal amino acid sequence of P-
450-male and P-450-female is shown in Fig. 2.

The sequence of P-450-male corresponded to those of P-450h
and P-450 2c and the sequence of P-450-female to that of P-
450i. Based on these lines of evidence, we concluded that P-
450-male was identical with P-450h and P-450 2c, and that P-
450-female was identical with P-450i and P-450 15β.

Postnatal Development of P-450-male and P-450-female

As reported previously, P-450-male and P-450-female were
detectable in respective liver microsomes from male and
female rats. Castration of male rats decreased the amounts
of P-450-male, and treatment of castrated male rats with
testostrone restored the level (Table 1) (Kamataki et al.
1983).

Treatment of ovariectomized female rats with testosterone
resulted in the disappearance of P-450-female and the
appearance of P-450-male, thus indicating that P-450-male was
apparently inducible by testosterone. On the other hand,
treatment of male rats with estradiol caused the
disappearance of P-450-male and the appearance of P-450-
female, indicating that P-450-female was apparently induced
by estradiol (Kamataki et al. 1983). These sex-specific
forms of cytochrome P-450 were, thus assumed to appear with
sexual maturation. Postnatal development of P-450-male and
P-450-female is shown in Fig. 3 (Maeda et al. 1984).

P-450-male and P-450-female appeared in respective liver
microsomes from male and female rats with sexual maturation.
Interestingly, as mentioned above briefly, P-450-female
appeared in male rats at a period around 25 days after birth
prior to the occurrence of P-450-male. The mechanism by
which P-450-female is induced in male rats is not known as
yet.

Table 1. Effects of castration and treatment with
testosterone or estradiol on the content of
P-450-male and P-450-female in liver microsomes
of male and female rats

Treatment	P-450-male (nmol/mg)	%*	P-450-female (nmol/mg)	%*
Male rats				
Untreated	0.31	35	< 0.03	< 4
Testosterone	0.37	43	< 0.03	< 3
Estradiol	< 0.03	< 4	0.21	29
Sham operated	0.26	27	< 0.03	< 3
Castrated	0.15	18	< 0.03	< 3
Castrated plus testosterone	0.23	27	< 0.03	< 3
Castrated plus estradiol	< 0.03	< 4	0.17	26
Female rats				
Untreated	< 0.03	< 4	0.25	36
Testosterone	< 0.03	< 5	0.23	39
Estradiol	< 0.03	< 4	0.22	35
Sham operated	< 0.01	< 1	0.32	39
Ovariectomized	< 0.01	< 1	0.29	33
Ovariectomized plus testosterone	0.08	14	< 0.01	< 2
Ovariectomized plus estradiol	< 0.03	< 4	0.23	33

n=5, *percent of total cyt. P-450 obtained from CO-spectra

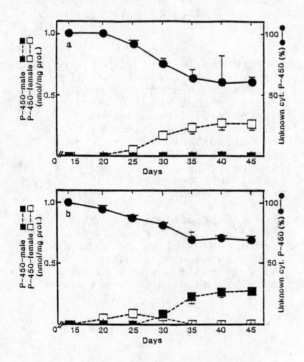

Fig. 3.
Postnatal development of P-450-male and P-450-female in liver microsomes from male and female rats.
a: female rats
b: male rats

Neonatal Testosterone Imprinting of P-450-male

Neonatal treatment of castrated male rats with
testosterone is known to 'imprint' male specific drug and
steroid hydroxylases. Thus, the possibility of whether
castration and/or testosteron treatment of neonatal male rats
affects the induction of P-450-male was studied together with
drug hydroxylases. Although the extents of changes varied
with the substrate used, castration of male rats 1 day after
birth depressed drug hydroxylase activities when measured at
8 weeks of age. The activities were partially restored in
rats castrated 7 days after birth. Treatment of castrated
rats with testosterone three time after castration also
restored the activities. The levels of the activities in the
testosterone/castrated rats were almost the same among rats
castrated 1,7 and 14 days after birth.

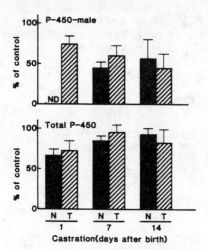

Fig. 4
Effects of neonatal
castration and treatment
with testosterone on the
content of P-450-male and
P-450-female.
Male rats were castrated
1,7 or 14 days after
birth and the halves of
rats were treated
subcutaneously with
testosterone propionate
(T, 20mg/kg) 2,4 and 6
days after castration.

Changes in the amounts of P-450-male in these rats are
shown in Fig. 4 (Kamataki et al. 1984). The amount of P-450-
male altered in a pattern similar to those seen in drug
hydroxylation activities, except that P-450-male could not be
detected in liver microsomes of rats castrated 1 day after
birth, in which a considerable amount of P-450-female lasts
for a life unless the synthesis is inhibited by androgens.
Although not shown here, we found that P-450-female appeared
and P-450-male disappeared in eldery male rats (Kamataki et
al. 1985). The change in the population from P-450-male to
P-450-female in the eldery male rats was associated with the
decrease of the ratio of testosterone to estradiol levels in
plasma.

Mechanism of Induction of P-450-male and P-450-female

Forms of cytochrome P-450 are induced by 3-
methylcholanthrene and other compounds. Among inducers of
cytochrome P-450, polychlorinated biphenyl (PCB) is known to
induce many forms of cytochrome P-450. The effects of

treatment with 3-methylcholanthrene, phenobarbital, or PCB on
the contents of P-450-male and P-450-female were examined
using male and female rats. These chemicals induced neither
P-450-male in female rats nor P-450-female in male rats.
Moreover, the amounts of P-450-male in male rats and of P-
450-female in female rats were decreased by these inducers.
Therefore, we concluded that sex steroids, but not so called
drug metabolism inducers, were specific inducers of P-450-
male and P-450-female.

Testosterone and estradiol apparently induce P-450-male
and P-450-female, respectively. The mechanisms by which
testosterone and estradiol induce sex-specific steroid
hydroxylases have been proposed to be regulated by pituitary
gland (Gustafsson et al. 1982). Therefore, we examined the
effects of hypophysectomy and treatment with testosterone or
estradiol on drug hydroxylation activities and the amounts of
P-450-male and P-450-female. Hypophysectomy decreased the
content of P-450-male, drug hydroxylation activities and
total content of cytochrome P-450. Hypophysectomy of female
rats resulted in a remarkable increase in the drug
hydroxylase activities.

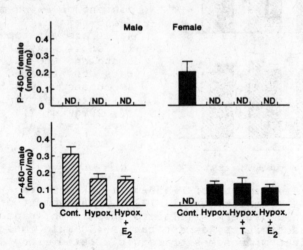

Fig. 5. Effects of hypophysectomy and treatment with
        testosterone or estradiol on the content of P-450-
        male and P-450-female.
        Hypophysectomy was carried out 17 days before the
        sacrifice. Testosteron propinate (T, 5mg/kg) or
        estradiol benzoate (E, 500μg/kg) was given
        substaneously for 10 days.

The increase in the activites was closely correlated with
the change in the population of cytochrome P-450; namely P-
450-female disappeared and P-450-male appeared in female rats
by hypophysectomy (Fig. 5). Again the mechanism involved in
the appearance of P-450-male in hypophysectomized female rats
is not elucidated.

ACKNOWLEDGEMENT

This work was supported in part by a Grant-in-Aid for Scientic Research from the Ministry of Education, Science and Culture of Japan.

REFERENCES

Guengerich, F. P., Dannan, G. A., Wright, S. T., Martin, M. V., Kaminsky, L. S. (1982) Purification and characterization of liver microsomal cytochromes P-450: electrophoretic, spectral, catalytic and immunochemical properties and inducibility of eight isozymes isolated from rats treated with phenobarbital or β-naphthoflavone. Biochemistry, 21, 6019-6030.

Gustafsson, J.-Å., Mode, A., Norstedt, G., Skett, P. (1983) Sex steroid induced changes in hepatic enzymes. Ann. Rev. Physiol., 45, 51-60.

Haniu, M., Ryan, D. E., Iida, S., Lieber, C. S., Levin, W., Shively, J. E. (1984) NH_2-Terminal sequence analyses of four rat hepatic microsomal cytochrome P-450. Arch. Biochem. Biophys., 235, 304-311.

Kamataki, T., Maeda, K., Kato, R. (1981) Partial purification and characterization of cytochrome P-450 responsible for the occurrence of sex difference in drug metabolism in the rat. Biochem. Biophys. Res. Commun., 103, 1-7.

Kamataki, T., Maeda, K., Yamazoe, Y., Nagai, T., Kato, R. (1983) Sex difference of cytochrome P-450 in the rat: Purification, characterization, and quantitation of constitutive forms of cytochrome P-450 from liver microsomes of male and female rats. Arch. Biochem. Biophys., 225, 758-770.

Kamataki, T., Maeda, K., Shimada, M., Nagai, T., Kato, R. (1984) Neonatal testosterone imprinting of hepatic microsomal drug metabolism and a male-specific form of cytochrome P-450 in the rat. J. Biochem., 96, 1939-1942.

Kamataki, T., Maeda, K., Shimada, M., Kitani, K., Nagai, T., Kato, R. (1985) Age-related alteration in the activities of drug metabolizing enzymes and contents of sex-specific forms of cytochrome P-450 in liver microsomes from male and female rats. J. Pharmacol. Exp. Ther., 233, 222-228.

Kato, R. (1974) Sex-related differences in drug metabolism. Drug Metab. Rev., 3, 1-32.

Kato, R., Kamataki, T. (1982a) Cytochrome P-450 as a determinant of sex difference of drug metabolism in the rat. Xenobiotica, 12, 787-800.

Kato, R., Kamataki, T. (1982b) Studies on sex-related difference of cytochrome P-450 in the rat: Purification of a constitutive form of cytochrome P-450 from liver microsomes of untreated rats. Cytochrome P-450, Biochemistry, Biophysics and Environmental Implications, Eds. Hietanen, E., Laitinen, M, Hänninen, O., Elsevier Biomedical Press B.V., Amsterdam, 421-428.

Kato, R., Takanaka, A., Takayanagi, M. (1968) Studies on mechanism of sex difference in drug-oxidizing activity of liver microsomes. Japan. J. Pharmacol., 18, 482-489.

MacGeoch, C., Morgan, E. T., Halpert, J., Gustafsson, J.-Å. (1984) Purification, characterization, and pituitary regulation of the sex-specific cytochrome P-450 15β-hydroxylase from liver microsomes of untreated female rats. J. Biol. Chem., 259, 15433-15439.

Maeda, K., Kamataki, T., Nagai, T., Kato, R. (1984) Postnatal development of constitutive forms of cytochrome P-450 in liver microsomes of male and female rats. Biochem. Pharmacol., 33, 509-512.

Ryan, D. E., Dixon, R., Evans, R. H., Ramanathan, L., Thomas, P. E., Wood, A. W., Levin, W. (1984a) Rat hepatic cytochrome P-450 isozyme specificity for the metabolism of the steroid sulfate, 5α-androstane-3α,17β-diol-3,17-disulfate. Arch. Biochem. Biophys., 233, 636-642.

Ryan, D. E., Iida, S., Wood, A. W., Thomas, P. E., Lieber, C. S., Levin, W. (1984b) Characterization of three highly purified cytochromes P-450 from hepatic microsomes of adult male rats. J. Biol. Chem., 259, 1239-1250.

Waxman, D. J. (1984) Rat hepatic cytochrome P-450 isoenzyme 2c. Identification as a male-specific, developmentally induced steroid 16α-hydroxylase and comparison to a female-specific cytochrome P-450 isoenzyme. J. Biol. Chem., 259, 15481-15490.

Ah RECEPTOR REGULATING INDUCTION OF CYTOCHROME P_1-450: SPECIES DIFFERENCES IN MOLECULAR PROPERTIES BETWEEN MOUSE AND RAT HEPATIC Ah RECEPTORS

Allan B. Okey, Lynn M. Vella and Michael S. Denison

Division of Clinical Pharmacology
The Hospital for Sick Children
Toronto, Ontario   CANADA   M5G 1X8

INTRODUCTION

The Ah (aromatic hydrocarbon receptor) mediates induction of aryl hydrocarbon hydroxylase (AHH; cytochrome P_1-450) and other structural gene products associated with the Ah gene complex (Eisen et al. 1983; Poland and Knutson 1982). Chemicals that are potent P_1-450 inducers, such as 3-methylcholanthrene (MC) and 2,3,7,8-tetrachlorodibenzo-p-dioxin (TCDD), initially bind to Ah receptor in the cytoplasmic compartment, after which the inducer:receptor complex translocates into the nuclear compartment (Okey et al. 1980). In the nucleus the inducer:receptor complex appears to associate with specific chromatin regions, thereby enhancing synthesis of mRNA species which code for cytochrome P_1-450 and related products of the Ah gene complex (Tukey et al. 1982; Jones et al. 1985).

It previously has been shown that the cytochrome P-450 protein induced in rats by MC treatment is very similar, perhaps identical to the P_1-450 protein induced by MC in C57BL/6N mice (Chen et al. 1982). We compared several molecular properties of Ah receptor from C57BL/6N mice with properties of Ah receptor from Sprague-Dawley rats to determine if the receptors which regulate P_1-450 are identical macromolecules in the two rodent species.

MATERIALS AND METHODS

Molecular properties of mouse and rat hepatic Ah receptor were determined by velocity sedimentation on sucrose gradients (Tsui and Okey, 1982) and by gel permeation chromatography on Sephacryl S-300 (Poellinger et al. 1983.)

RESULTS AND DISCUSSION

Molecular Size of Mouse and Rat Ah Receptors in Conditions of Low Ionic Strength

Molecular parameters of mouse and rat Ah receptors are compared in Table 1. In conditions of low ionic strength the Ah receptor from hepatic cytosol of C57BL/6N mice exhibited a significantly (p<0.01) higher relative molecular mass (M_r = 277,000) than the Ah receptor from rat hepatic cytosol (M_r = 257,000).

16*

TABLE 1

MOLECULAR PARAMETERS OF Ah RECEPTORS FROM MICE AND RATS

| | In 0.1 M KCl | | In 0.4 M KCl | | |
	C57BL/6N Mouse	Sprague-Dawley Rat	Mouse: Non-Dissociated Form	Mouse: Dissociated Form	Rat
Sedimentation Coefficient ($S_{20,w}$)	9.4	8.8	9.7	5.3	5.6
Stokes radius (nm)	7.1	7.0	6.8	5.2	5.2
Relative Molecular Mass (M_r)	277,000	257,000	274,000	105,000	121,000

Response to High Ionic Strength Environments

The major difference between mouse and rat Ah receptors appeared during their analysis in conditions of high ionic strength. In the presence of 0.4 M KCl the Ah receptor from rat hepatic cytosol readily dissociated from a form which had sedimented at 9.4 S in low ionic strength to a ligand-binding subunit which sedimented at 5.6 S (Table 1). The Ah receptor from mouse hepatic cytosol resisted dissociation into subunits when exposed to high ionic strength conditions. Exposure of mouse hepatic cytosol to 0.4 M KCl for 16 hours prior to analysis on sucrose gradients produced specific binding peaks both at 9.7 S (non-dissociated form) and at 5.3 S (dissociated form). This is in contrast to rat Ah receptor which was completely shifted from 9.4 S to the dissociated 5.6 S subunit by exposure to 0.4 M KCl for only one hour prior to sucrose gradient centrifugation. The M_r of the dissociated subunit of rat Ah receptor (121,000) was significantly (p<0.01) larger than the dissociated subunit from mouse hepatic cytosol (M_r = 105,000).

Although rat Ah receptor dissociated more easily into subunits than did mouse Ah receptor, the rat Ah receptor also was more susceptible to loss of ligand binding function when exposed to high ionic strength conditions. Exposure of rat hepatic cytosol to 0.4 M KCl prior to addition of ligand ($[^3H]$TCDD) destroyed = 50% of specific binding capacity in the rat cytosol, whereas specific binding in mouse hepatic cytosol was completely stable at KCl conentrations up to 1.0 M.

Ligand Binding Preferences

$[^3H]$TCDD is the most widely used radioligand for detecting and characterizing the Ah receptor, but the Ah receptor also can be detected with $[^3H]$MC (Hannah et al. 1982; Okey and Vella 1982; Poellinger et al. 1983) and with $[^3H]$dibenz(a,h)anthracene (Okey et al. 1984). $[^3H]$Benzo(a)pyrene (BP) successfully detects Ah receptor in Sprague-Dawley rats (Poellinger et al. 1983; Okey et al. 1984a) and cynomolgus monkeys (Okey et al. 1984b). Although Poellinger et al. (1983) have reported that $[^3H]$BP binds to Ah receptor in hepatic cytosol from C57BL/6 mice, we have not been able to detect significant specific binding of $[^3H]$BP in mouse hepatic cytosol (Okey et al. 1984a). We also tested seven nonradioactive Ah-receptor agonists for their potency as inhibitors of $[^3H]$TCDD binding to Ah receptors in mouse and rat hepatic cytosols. By these competition studies the ligand-binding preferences of rat Ah receptor were seen to be different from the ligand-binding preferences of mouse Ah receptor. Benzo(a)anthacene, in particular, was much more potent (30 times more) as a competitor for Ah receptor sites in the rat (IC_{50} = 2 x 10^{-8} M) than in the mouse (IC_{50} = 6 x 10^{-7} M).

CONCLUSIONS

By criteria of molecular size, response to high ionic strength conditions and ligand binding preferences, mouse and rat Ah receptors appear to be similar but not identical molecular species. These studies suggest diversity in structure and function of Ah receptor from different animal species. The degree of diversity in structure and the significance of the diversity for regulatory function requires considerable further study.

ACKNOWLEDGEMENTS

Supported by a grant from the Medical Research Council of Canada to ABO. MSD is the recipient of a National Research Service Award from the National Institute for Environmental Health Sciences, National Institutes of Health, U.S.A.

REFERENCES

CHEN, Y.T., M.A. LANG, N.M. JENSEN, R.H. TUKEY, E. SIDRANSKY, T.M. GUENTHNER & D.W. NEBERT. (1982). Similarities between mouse and rat-liver microsomal cytochromes P-450 induced by 3-methylcholanthrene. Eur. J. Biochem 122:361-38.

EISEN, H.J., R.R. HANNAH, C. LEGRAVEREND, A.B. OKEY, & D.W. NEBERT. (1983) The Ah receptor:controlling factor in the induction of drug-metabolizing enzymes by certain chemical carcinogens and other environmental pollutants. Biochem. Actions of Hormones (G. Litwack, ed.) vol. X, pp, 227-257. Academic Press, N.Y.

HANNAH, R.R., D.W. NEBERT, & H.J. EISEN (1982). Regulatory gene product of the Ah complex: comparison of 2,3,7,8-tetrachlorodibenzo-p-dioxin and 3-methylcholanthrene binding to several moieties in mouse liver cytosol. J. Biol. Chem. 256:4584-4590.

JONES, P.B.C., D.R. GALEAZZI, J.M. FISHER & J.P. WHITLOCK, Jr. (1985). Control of cytochrome P_1-450 gene expression by dioxin. Science 227:1499-1502.

OKEY, A.B., G.P. BONDY, M.E. MASON, D.W. NEBERT, C.J. FORSTER-GIBSON, J. MUNCAN & M.J. DUFRESNE (1980). Temperature-dependent cytosol-to-nucleus translocation of the Ah receptor for 2,3,7,8-tetrachlorodibenzo-p-dioxin in continuous cell culture lines. J. Biol. Chem. 255:11415-11422.

OKEY, A.B. & L.M. VELLA (1982). Binding of 3-methycholanthrene and 2,3,7, 8-tetrachlorodibenzo-p-dioxin to a common Ah receptor site in mouse and rat hepatic cytosols. Eur. J. Biochem. 127:39-47.

OKEY, A.B., A.W. DUBE & L.M. VELLA (1984). Binding of benzo(a)pyrene and dibenz(a,h)anthracene to the Ah receptor in mouse and rat hepatic cytosols. Cancer Res. 44:1426-1432.

OKEY, A.B., L.M. VELLA & F. IVERSON (1984). Ah receptor in primate liver: binding of 2,3,7,8-tetrachlorodibenzo-p-dioxin and carcinogenic aromatic hydrocarbons. Canadian J. Physiol. Pharmacol. 62:1292-1295.

POELLINGER, L., J. LUND, M. GILLNER, L.-A. HANSSON & J.-A. GUSTAFSSON (1983). Physicochemical characterization of specific and nonspecific polyaromatic hydrocarbon binders in rat and mouse liver cytosol. J. Biol. Chem. 258:13535-13542.

POLAND, A. & J.C. KNUTSON (1982). 2,3,7,8-Tetrachlorodibenzo-p-dioxin and related halogenated aromatic hydrocarbons: examination of the mechanism of toxicity. Annu. Rev. Pharmacol. 22:517-554.

TUKEY, R.H., R.R. HANNAH, M. NEGISHI, D.W. NEBERT & H.J. EISEN (1982). The Ah locus: correlation of intranuclear appearance of inducer-receptor complex with induction of cytochrome P_1-450 mRNA. Cell 31:275-284.

THE MECHANISM OF THE SENESCENCE ASSOCIATED LOSS OF SEX DIFFERENCE IN DRUG METABOLIZING ENZYME ACTIVITIES IN RATS

Tokuji Suzuki, Shoichi Fujita, and Kenichi Kitani*

Department of Biopharmaceutics, Faculty of Pharmaceutical Sciences, Chiba University, 1-33 Yayoicho, Chiba, Japan 260. and
*Tokyo Metropolitan Institute of Gerontology, Itabashiku, Tokyo, Japan

Investigations on alterations in the activity of drug metabolizing enzymes due to the senescence are of urgent importance for the establishment of rational bases for the prescription of drugs to geriatric patients. However, only a few basic researches have been carried out in this area using experimental animals . Male rats have been typically used for these researches, and it is generally accepted that the hepatic drug metabolizing enzyme activities decline with old age in rats. To date, following four hypotheses have been proposed as the cause of this decline in drug metabolism: (1) decrease in cytochrome P-450 levels (Kato et al. 1968), (2) decrease in cytochrome c reductase activity (Schmucker et al. 1982), (3) altered microsomal membrane quality (Stier 1982, Armbrecht et al. 1982, Birnbaum et al. 1978), (4) Altered relative abundance of cytochrome P-450 species. There are sufficient evidences to support these alterations do occure in old age in rats. However, there has not been a conclusive evidence to support which alteration is the major cause of the senescence-associated decrease in drug metabolism. Current reserch was undertaken to obtain the informations on the nature of the alterations and to reevaluate which of these proposed causes can explain the characteristic nature of the alterations.

We have investigated the alterations in hepatic drug metabolism using male and female rats. Surprisingly, in sharp contrust to the marked decrease in drug metabolism in senescent male rats, very little alteration occured in female rats, resulting in the complete loss of sex difference in old age. It is also of interest then to clarify if this senescence associated decrease in drug metabolizing enzyme activities observed in male rats is due to the functional feminization of the liver or to the general deterioration of liver functions in old male rats.

Male and female Fischer 344 rats of ages ranging from 3 to 30 month were raised under S.P.F condition. They were killed by decapitation, the blood was collected in heparinized test tubes, the livers were excised and microsomes were prepared according to the method of Omura and Sato (1964). Diazepam 3-hydroxylase and N-demethylase activities, and androstenedione 5a-reductase activities were assayed according to the previously described methods (Cotler et al. 1981, Lax et al. 1979) with slight modifications. Plasma sex hormone levels were determined by radioimmuno assay methods (Arai et al. 1979).

Table 1 summarizes the patterns of age associated alterations in activities of various cytochrome P-450 mediated reactions observed in our laboratory. The activities of reactions which showed large sex differences (male>female) when young showed marked decrease with old age in male rats (pattern I), and that those which did not show large sex differences when young did not show age associated alterations (pattern III). The activities of the reactions which showed small sex differences when young resulted in the pattern of alterations intermediate of the two (pattern II).

It is noteworthy that alterations in the metabolism of the different positions of a single substrate (in lidocaine, imipramine, and diazepam)

Table 1. Generalised patterns of age-associated alterations in hepatic micro-somal drug metabolizing enzyme activities (expressed in P-450 turn over numbers).

Patterns of age-associated alterations	Reactions	References
Pattern I 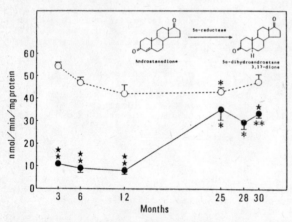	Aminopyrine N-demethylase Hexobarbital oxidase Lidocaine N-deethylase Imipramine N-demethylase Diazepam 3-hydroxylase	(1) (1) (2) (3) (3)
Pattern II	7-Ethoxycoumarine O-deethylase Diazepam N-demethylase	(1) (3)
Pattern III	Aniline hydroxylase Amaranth reducatse Lidocaine 3-hydroxylase Imipramine 2-hydroxylase	(1) (1) (2) (3)

———— male, and ---- female activities

(1) Fujita et al. (1982), (2) Fujita et al. (1985), (3) unpublished observations

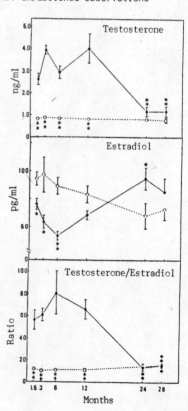

Fig.1 Age-associated alterations in androstenedione 5a-reductase activities in male and female rats.
*,** significantly different from 3 months old value ($P<0.05$, $P<0.01$, respectively. N=6/age group).
★,★★ significant sex difference ($P<0.05$, $P<0.01$).
Activities are expressed in nmol androstenedione reduced/min/mg liver microsomal protein (mean±SE).

Fig.2 (Right) Age associated alterations of plasma testo-sterone and estrogen levels and their ratio (mean ± SE).

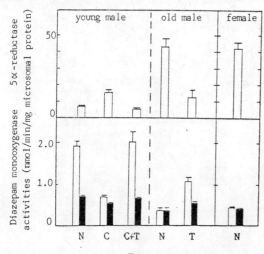

Fig.3 Effect of castration and administration of
testosterone on androstenedione 5a-reductase and
diazepam monooxygenase activities in young and
old rats.
Two groups of young (7 months old) rats were castrated and one group (N=4) received testosterone
(100mg/kg/week) suspended in corn oil (C+T) and
the other group received corn oil only (C). At
the age of 9 months, they were killed along with
non-castrated control (N) rats of 9 and 27 months
old and testosterone treated 27 months old rats
(T). Non-treated female rats of 12 months old was
also killed as the reference. Liver microsomes
were prepared and androstenedione 5a-reductase
activities (upper) and diazepam 3-hydroxylase
and N-demethylase activities (shaded bar) were
determined. Values are expressed in nmol/min/
mg microsomal protein (mean ± SE).

showed different patterns. Therefore, age-associated alterations in drug
metabolism are characterised by their sex, substrate, and position selecti-
veness. Inclusive of all of these characteristics are explained by the
hypothesis (4), the alterations in relative abundance of cytochrome P-450
species occure in the male rat liver during the process of aging. Other
hypotheses could not explain all of these characteristics of alterations.

It is also apparent from this table that the senescence-associated
disappearance of sex difference in drug metabolism is due to the decrease in
activities of male rat liver. Is it because liver function deteriorates
with old age only in male rats ?

Fig 1 shows the age associated alterations in androstenedione 5a-
reductase activity in male and female rat liver. The activity was higher in
females than in males when young, up to 12 months, but the male activity
increased to female levels by 24 months old. The female activity did not
show significant age-associated alterations. This loss of sex difference
due to the senescence associated increase in male activity observed in
androstenedione 5a-reductase supports that male rat liver is functionally
feminized rather than functionally deteriorated in old age.

In parallel with these loss of sex difference in enzyme activities, we
observed the decrease in male predominant p-450 isozymes and increase in
female predominant ones by directly applying solubilised rat liver microsomes to HPLC column (Fujita et al. 1985). The P-450 isozyme composition in
liver microsomes of old male rats became closer to that of young females
than to young males. Similarly, Kamataki et al. (1985) recently reported
the decrease in P-450 male and the appearance of P-450 female in senescent
male rats. These observations indicate that cytochrome P-450 composition of
liver microsomes from old male rats is feminized and this is the cause for
the senescence associated loss of sex difference in drug metabolism in rats.

The disappearance of sex difference in old age is not limited to cytochrome P-450 mediated reactions only, but it is observed in activities of
androstenedione 5a-reductase as mentioned above, and glutathione-S-transferase isozymes (unpublished observation).

To further investigate whether this feminization of male rat liver in
senescence is somehow related to plasma sex hormone levels, we measured
plasma testosterone and estradiol levels (Fig.2). The patterns of alterations in plasma testosterone levels as well as testosterone to estradiol
ratio were very similar to the pattern I of the alteration in table 1. This
suggests that the levels of testosterone regulate most strongly the magnitude of drug metabolizing enzyme activities, and that the decrease in the
testosterone levels with senescence causes the decrease in the drug metabolizing enzyme activities.

To clarify this possibility, testoterone was administered to the aged,
as well as to the castrated young rats and the drug metabolism in those rats

234

were investigated. Diazepam 3-hydroxylation, which showed a large sex difference (male>female) up to 12 months (Table 1), decreased in the aged (27 month old) as well as in the castrated adult (9 months old) male rats (Fig 3). Administration of testosterone reversed the effects of age and castration. Similar tendencies were observed in diazepam N-demethylation. Androstenedione 5a-reductase activity, on the other hand, increased in the aged as well as in the castrated adult male rats. Administration of testosterone reversed the effects of age and castration, i.e., caused a decrease in the activity, in this case. It is noteworthy that the activities in the aged male rats are closer to those in females than the castrated male rats.

It is concluded that the senescence associated loss of sex difference in hepatic drug metabolism is due to the functional feminization of male rat liver with old age, and that this is caused by the lack of androgenic influence on the liver function in the aged male rats due to the decrease in testosterone levels associated with old age. This causes the alterations in isozyme composition of drug metabolizing enzymes to the female pattern.

REFERENCES

Arai,Y., Demura,R., and Demura,H. (1979) Fundamental and clinical evaluation of ^{125}I-Testosterone Radioimmunoassay Kit "EIKEN". Horumon To Rinsho, 27, 527-531 (in Japanese).

Armbrecht,H.J., Birnbaum,L.S., Zenser,T.V., and Davis,B.B. (1982) Changes in hepatic microsomal membrane fluidity with age. Exp. Gelontol., 17, 41-48.

Birnbaum,L. and Baird,M.B. (1978) Induction of hepatic mixed function oxidases in senescent rodents. Exp. Gelontol., 13, 299-303.

Cotler,S., Puglishi,C.U. and Gustafson,J.H. (1981) Determination of diazepam and its major metabolites in man and in the cat by high-performance liquid chromatography. J. Chromatogr., 222, 95-106.

Fujita,S., Chiba,M, Suzuki,T., and Kitani, K. (1985) Age and sex associated differences of multiple species of cytochrome P-450 in rat liver microsomes. Biochem. Pharmacol., in press.

Kamataki,T., Maeda,K., Shimada,M., Kitani,K., Nagai,T., and Kato,K. (1985) Age-associated alteration in the activities of drug metabolizing enzymes and contents of sex-specific formes of cytochrome P-450 in liver microsomes from male and female rats. J. Pharmacol. Exp. Ther., in press.

Kato,R., and Takanaka,A. (1968) Metabolism of drugs in old rats (I): Activities of NADPH-linked electron transport and drug metabolizing enzyme system in liver microsomes of old rats. Jap. J. Pharmac., 18, 331-388.

Lax,E.R., Kreuzfelder,E. and Schriefers,H. (1979) Time-studies of the changes in the sex-dependent activities of enzymes of hepatic steroid metabolism in the rat following gonadectomy. Hoppe-Seyler's Z. Physiol. Chem., 360, 1799-1805.

Omura,T. and Sato,R. (1964) The carbon monoxide-binding pigment of liver microsomes. J. Biol. Chem., 237, 2370-2378.

Schmucker,D.L., Wang,R.K., and Kwong,P. (1982) Age-dependent alterations in rat liver microsomal NADPH cytochrome c (P-450) reductase. in Liver and Aging - 1982 (Kitani,K. ed.) pp 75-96, Elsevier Biomedical, Amsterdam.

Stier,A., Finch,S.A.E., Greinert,R., Hohne,M., and Muller,R. (1982) Membrane structure and function of the hepatic microsomal cytochrome P-450 system. ibid, pp 3-14.

HORMONAL REGULATION OF SEXUALLY DIFFERENTIATED CYTOCHROME P-450 ISOZYMES IN RAT LIVER

Catriona MacGeoch, Edward T. Morgan and Jan-Åke Gustafsson

Dept. of Medical Nutrition, Karolinska Institute, Huddinge University Hospital F69, S-141 86 Huddinge, Sweden

The hepatic metabolism of xenobiotics and steroids in the rat displays pronounced sex differences, hydroxylations being generally more efficient in liver in male rats. This sexual dimorphism of drug and steroid metabolism is not observed until the onset of puberty, and the sex differences in microsomal enzyme activities have been attributed to the action of androgens. However, steroid hormone control of steroid metabolism is mediated through the hypothalamo-pituitary axis. The male type of hepatic metabolism has been shown to be irreversibly "imprinted" at birth by testicular androgen, due to an action on the hypothalamus. The female metabolic type is also maintained by a pituitary factor.

The pituitary factor controlling the metabolic sexual differentiation of the liver is growth hormone, GH (Mode et al., 1981). Thus, the different effects of GH on hepatic metabolism in male and female rats have been shown to be related to the sexual dimorphism of the growth hormone secretory pattern (Edén, S. 1979).

We have purified a male-specific cytochrome P-450 isozyme, $P-450_{16\alpha}$, (androstenedione-16α-hydroxylase) as well as a female specific cytochrome P-450 isozyme, $P-450_{15\beta}$, (androstenediol disulphate-15β-hydroxylase) from rat liver. Since the aforementioned steroid hydroxylating activities have been shown to be sexually differentiated and regulated by the pituitary, we have studied the effects of gonadectomy, hypophysectomy (Hx), and treatment with pituitary and gonadal hormones on $P-450_{16\alpha}$ and $P-450_{15\beta}$ apoprotein levels in rat liver. In addition, the ontogenesis of the hepatic expression of $P-450_{16\alpha}$ and $P-450_{15\beta}$ in rats of both sexes was investigated.

METHODS

The purification of the P-450 isozymes, and preparation of specific, immunoabsorbed, IgG or antiserum against these proteins have been described previously (MacGeoch et al., 1984; Morgan et al., 1985). The specific levels of $P-450_{15\beta}$ and $P-450_{16\alpha}$ in microsomes were assayed by an immunoblot procedure (MacGeoch et al., 1984). Intact and Hx rats of both sexes were administered androgenic, oestrogenic and hypophyseal hormones by injection or by continuous infusion from osmotic minipumps.

RESULTS AND DISCUSSION

In all cases, it was found that the levels of the two isozymes were oppositely affected by the various treatments. Thus, oestrogen treatment of males lowered the levels of the male-specific P-450$_{16\alpha}$, and increased levels of the female-specific P-450$_{15\beta}$, almost to female levels. Androgen treatment of female rats had the reverse effects (Table 1). The effects of sex steroids in masculinizing or feminizing the hepatic P-450 isozyme distribution are similar to those observed for 16α-hydroxylase and 15β-hydroxylase activity (data not shown).

Table 1   <u>Effects of androgen and estrogen administration on P-450$_{16\alpha}$ and P-450$_{15\beta}$ levels in intact adult rats.</u>

Estradiol valerate was administered as a single im injection of 125 µg to male rats, which were killed 14 days later. Methyltrienolone (R1881) was given sc once daily for 14 days, 125 µg in 0.1 ml of propylene glycol, to female rats. Each group contained 3 rats.

Group of animals	P-450 content of microsomes	
	16α	15β
	ng/µg microsomal protein	
♂	49.8 ± 5.1[b]	1.0 ± 0.6[b]
♂ + Estradiol valerate	1.9 ± 1.2[a,b]	10.4 ± 4.4[a]
♀	0.5 ± 0.2[a]	11.6 ± 4.8[a]
♀ + R1881	30.5 ± 4.3[a,b]	N.D.[b]

[a] significantly different from untreated male value.
[b] significantly different from untreated female value.
N.D. = not detectable.

Castration of male rats in the neonatal period led to an increase in the levels of P-450$_{15\beta}$ and a decrease in levels of P-450$_{16\alpha}$ in the adult animals (Fig. 1). However, castration of male rats postpubertally had no effect on the specific microsomal content of either isozyme. This demonstrates that expression of hepatic P-450$_{15\beta}$ and P-450$_{16\alpha}$ in adult male rats is imprinted by androgens in the neonatal period. Next, we investigated the development of P-450$_{15\beta}$ and P-450$_{16\alpha}$ expression in liver microsomes of male and female rats from the ages of 6 days to 8 months. P-450$_{15\beta}$ and P-450$_{16\alpha}$ levels were low in both female and male animals until the age of puberty. At 35 days, just after onset of puberty and up to 8 months of age, the P-450$_{15\beta}$ level in the female and the P-450$_{16\alpha}$ level in the male became fully developed. The development of P-450$_{16\alpha}$ and P-450$_{15\beta}$ expression in rat liver coincided with the development of the corresponding microsomal steroid hydroxylase activities (data not shown) and of the sexually dimorphic pattern of GH secretion.

Administration of human GH (hGH) in a continuous fashion, mimicking the female pattern of blood GH levels, led to a feminization of the levels of both of the apoproteins in intact male animals and Hx rats of both sexes (Fig. 2). Thus, P-450$_{15\beta}$ levels increased while P-450$_{16\alpha}$ levels decreased. However, injecting hGH intermittently into males every 12 h, mimicking a male-type pattern of blood GH levels, had no effect on the apoprotein levels (not shown). These results emphasize the importance of continuous levels of GH in the blood for a feminization of the $P-450_{15\beta}$ and $P-450_{16}$ levels. Hypophysectomy abolished the sex difference, so that Hx rats of either sex had $P-450_{16\alpha}$ and $P-450_{15\beta}$ levels intermediate to those seen in intact rats of each sex (Fig. 2).

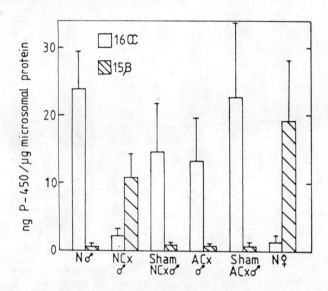

Fig. 1 <u>Effect of castration of neonatal and adult male rats on the specific microsomal content of $P-450_{15\beta}$ and $P-450_{16\alpha}$.</u>

Rats were castrated or sham operated in the neonatal period (NCx, ShNCx) or adult period (ACx, ShACx). All animals were killed at 63 days of age, and microsomes prepared. Each group contained 3 animals. Values are expressed as mean ± S.D.

Since hGH can bind to both prolactin (Prl) and GH receptors in rat liver we tested the specific somatogenic hormone, rat GH (rGH) for effects on $P-450_{15\beta}$ and $P-450_{16\alpha}$ levels. rGH infusion in male rats caused a feminization of both $P-450_{15\beta}$ and $P-450_{16\alpha}$ levels, although this hormone was less potent than hGH. Ovine Prl infusion had no effect (not shown). It is concluded that GH, the major regulatory factor for the expression of the two sexually differentiated enzymes, probably exerts its effects via somatogenic receptors.

Our results clearly indicate that the effects of sexual development, androgen and oestrogen treatment and GH on the specific microsomal content of the $P-450_{15\beta}$ and $P-450_{16\alpha}$ apoproteins in every case correlate with the known effects on the microsomal $P-450_{15\beta}$ and $P-450_{16\alpha}$ activities. Thus, the pituitary gland and the gonadal hormones control hepatic metabolism by modulation of levels of specific P-450 isozymes in rat liver.

238

The possible mechanisms whereby the pattern of blood GH levels
concomitantly regulates the upregulation of one isozyme, and the down-
regulation of another are intriguing. Development of cDNA probes for these
enzymes will be required to provide the tools with which to begin to
answer these questions.

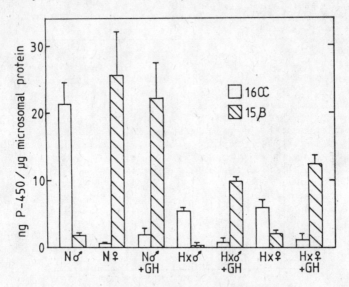

Fig. 2 **Effects of hGH infusion and hypophysectomy on hepatic microsomal
levels of P-450 16α and P-450 15β in the rat.**

hGH was infused at a rate of 5 µg/h for 7 days to intact or hypophysectomized (Hx) male and female rats. Hx rats were operated at 8 weeks, and
all animals were 10 weeks old at initiation of the treatment. After 7
days' treatment, microsomes were prepared from the three different animals
in each group, and $P-450_{16\alpha}$ and $P-450_{15\beta}$ quantified.

REFERENCES

Edén, S. 1979. Age and sex related differences in episodic growth hormone
secretion in the rat. Endocrinology 105:555.

MacGeoch, C., Morgan, E.T., Halpert, J., Gustafsson, J.-Å. 1984.
Purification, characterization and pituitary regulation of the sex
specific cytochrome P-450 15β-hydroxylase from liver microsomes of
untreated female rats. J. Biol. Chem. 259, 15433.

Mode, A., Norstedt, G., Simic, B., Eneroth, P., Gustafsson, J.-Å. 1981.
Continuous infusion of growth hormone feminized hepatic steroid metabolism
in the rat. Endocrinology 108:2103.

Morgan, E.T., MacGeoch, C., Gustafsson, J.-Å. 1985. Sexual differentiation
of cytochrome P-450 in rat liver. Evidence for a constitutive isozyme as
the male-sepcific 16α-hydroxylase. Mol. Pharmacol. 27;471.

INFLUENCING OF HEPATIC NADPH-CYTOCHROME C REDUCTASE (p-450) IN RATS BY DIFFERENT DOSES OF SEROTONIN

Medhat F. Mahmoud

Biochemistry Department, Faculty of Vet. Medicine, Zagazig University, Zagazig , Egypt.

Acute physiological stress leads to a mobilisation of a different anabolic and catabolic hormones (Schalch and Reichlim, 1968). Such hormonal changes in the body result in a coresponding effect on the flavine enzymes with subsequent changes in the redox-status of the glutathione (Asada and Kanematsu, 1976; Harisch and Mahmoud, 1980).

Thus, the scope of this investigation is intended to deduce the effect of different doses of serotonin on the hepatic NADPH-cytochrome c reductase in rats as one of the most important flavoproteins which might reflect the hormonal changes in the body.

MATERIAL AND METHODS

Male albino rats were used for the experiment. They were brought at a weight of 50g and underwent a period of adaptation to the conditions of the animal quarter with commercial pellet feed and water ad.libitum until reached the desired test weight of $100 \pm 5g$.

The test animals obtained intraperitonealy the serotonin (Roth, Karlsruhe, West Germany) in three different doses; a relatively low of 15 mg/kg, a dose of 45 mg/kg and a third one of 75 mg/kg body weight.

After the desired time (10, 30, 60 and 120 minutes post injection), the rats were anaesthetized with Nembutal (Pentobarbital-Sodium Abbot, 40mg/kg i.p.). Then, using the freeze-stop procedure (Faupel et al., 1972), the liver was taken and pulverised under liquid nitrogen in which it kept in small plastic bottles untill required for working up. This method ensured short sampling under constant condition without alteration of the enzyme activities (Harisch et al., 1980).

The activity of NADPH-cytochrome c reductase was verified in whole liver homogenate according to the method of Phillips and Langdon (1962) and the results given as nmol. cytochrome c (reduced) per mg protein of the whole homogenate per minute.

RESULTS

As given in table 1, the results can be summarised as follows:
1. Intraperitoneal administration of serotonin (15 mg/kg , B.W.): The activity of NADPH-cytochrome c reductase was decreased significantly after ten minutes with maximum decreasing after 30 minutes. The activity of the enzyme return again to the control range after 60 minutes of application.

2. Administration (i.p.) of serotonin in a dose of 45 mg/kg B.W. : The activity of NADPH-cytochrome c reductase increased significantly after 30 minutes followed by decrease activity of the enzyme significantly after 60 minutes.

3. Intraperitoneal administration of serotonin (75 mg/kg , B.W.): This relatively high dose of serotonin induced a significant increase in the activity of the enzyme and such increase was more pronounced after 30 minutes of application, followed by a significant decrease of the enzyme activity after 60 minutes, which persist,but to a lesser extent,untill 120 minutes post injection.

DISCUSSION

Table 1 revealed that the 5-hydroxy tryptamine (serotonin) in a relatively low dose (15 mg/kg) induced an anabolic run which signed with significant decrease in the activity of hepatic NADPH-cytochrome c reductase began as soon as 10 minutes and was more pronounced at 30 minutes after i.p. application of serotonin. Schole et al., (1978) showed that serotonin (10 mg/kg) resulted in a distinct anabolic course manifested by significant increase of glutathione quotient (GSH/GSSG) in the liver of rats which was detectable after 10 minutes and increased with the time untill 60 minutes post administration.

On the other hand, the large doses of serotonin (45 and 75 mg/kg) created two-phases reaction manifested by a sharp increase of the activity of the hepatic NADPH-cytochrome c reductase with the maximum increase of the activity at 30 minute post injection, followed by another anabolic phase with significant decrease of the enzyme activity at 60 minute and this decrease of the enzyme activity persisted significantly under the control level untill 120 minutes post injection of serotonin. The strength of this two - phases reaction was dosis-dependant (Table 1). Furthermore, Harisch et al. (1971) used serotonin (50 mg/kg) in chicken to produce two-phases alarm reaction and showed that the

241

Table (1): NADPH-cytochrome c reductase (p-450) in the liver of rats following i.p. administration of different doses of serotonin.

	Control	Minutes after application			
		10	30	60	120
Serotonin (15 mg/kg) (nmol cytochrome c-mg.$^{-1}$min^{-1}) n	10.02±0.91 12	++ 8.81±0.82 8	+++ 8.41±0.99 8	9.89±1.04 10	10.13±1.12 8
Serotonin (45 mg/kg) (nmol cytochrome c-mg.$^{-1}$ min^{-1}) n		10.94±1.11 8	++ 11.98±1.09 10	+++ 8.11±1.03 8	+ 9.12±0.92 12
Serotonin (75 mg/kg) (nmol cytochrome c-mg.$^{-1}$min^{-1}) n		+ 11.12±1.22 8	+++ 13.02±1.48 8	+++ 8.02±1.01 8	++ 8.98±0.98 10

male albino rats; 100 ± 5 g body weight; comparison with control group using students t-test; + P < 0.05; ++ P<0.01; +++ P<0.001; n=number of animals.

17 Vereczkey

top of the anabolic phase was after 60 minutes. This two -
phases alarm reaction can only clarified through the two -
component theory of endocrine regulation of metabolism
(Schole, 1966), which postulated that an anabolic component
(STH, insulin, or a tissue-specific anabolic component
dependent upon STH) and a catabolic component (corticosteroid and/or thyroxin) regulate cellular metabolism with
the help of the glutathione system. In other words,the
administration of serotonin in relatively large doses (45 and
75 mg/kg) may acted as acute stress which stimulate the
hypothalamus-hypophysis-adrenal axis with subsequent induction of a catabolic corticosteroid phase, directly followed
by the anabolic STH-phase (Selye, 1973).

Furthermore, the hypersenstivity of the hepatic cytochrome p-450 to the different hormonal changes in the body,
as manifested above in this work, makes this enzyme one of
the most important known yellow enzymes which play a key
regulatory role in the cellular metabolism and subsequently
the body resistance as described by Selye (1961).

SUMMARY

While relatively low dose of serotonin (15 mg/kg)
caused a significant decrease of the activity of the hepatic
NADPH-cytochrome c reductase (p-450) of rats, a high doses
of serotonin (45 and 75 mg/kg) induced a two-phases alarm
reaction and top of the catabolic phase was at 30 minute
(increase of the enzyme activity), whereas, that of the
anabolic phase was at 60 minute (decrease of the enzyme
activity).

These results are of great value when discussed in
relation to the regulation of the cellular metabolism.

REFERENCES

Asada, K. and S. Kanematsu,(1976): Reactivity of thiols
      with superoxide radicals. Agr. Biol. Chem., 40,
      1891 - 1892.

Faupel, R.P.; H.J. Seitz and W. Tarnowski,(1972): The
      problem of tissue sampling from experimental
      animals. Arch. Biochem. Biophys., 148, 509-522.

Harisch, G.; A. Dey-Hazra, K. Enigh and J. Schole, (1971):
      Einfluss von Strahlenschutzsubstanzen auf
      Überlebens-rate und Glutathionquotient von
      Hühnerküken nach einer Eimeria-necatrix Infektion.
      Zbl. Vet. Med. B, 18, 359 - 365.

Harisch, G. and M.F. Mahmoud, (1980): The glutathione
      status of the rat liver following different doses
      of cortisol. Zbl. Vet. Med. A, 27, 433-436.

Harisch, G.; M.F. Mahmoud and J. Schole, (1980):
    Flavoproteins, glutathione and NADH oxidation rate
    in the liver of rats of various ages. Zbl. Vet. Med.
    A, 27, 829 - 834.

Phillips, A.H., and R.G. Langdon,(1962): Hepatic triphospho-
    pyridine nucleotide-cytochrome c reductase.
    J. Biol. Chem., 237, 2652 - 2660.

Schalch, D.S. and Reichlin, S., (1968): Stress and Growth
    Hormone Release. In: Growth Hormone, Ed.: Pecile,
    A.; Müller, E.E.; Milan: Excerpta Med. Found. pp.
    211 - 225.

Schole, J.,(1966): Theorie der Stoffwechselregulation unter
    besonderer Berücksichtigung der Regulation des
    Wachstums. Verlag Paul Parey, Berlin und Hamburg.

Schole, J. ; G. Harisch; A. Dey-Hazra and K. Enigk, (1972):
    Resistenzphase nach unspezifischer Vorbelastung
    bzw. nach ACTH-Gaben. Zbl. Vet. Med. B, 19, 776-778.

Schole, J.; G. Harisch and H.-P. Sallmann, (1978):
    Belastung, Ernährung und Resistenz. Verlag Paul
    Parey, Berlin and Hamburg.

Selye, H., (1961): Non-specific resistence. Erg. All. Pathol.
    Pathol. Anat., 41, 208-241.

Selye, H., (1971): Hormones and resistence. Springer Verlag
    Berlin, Heidelberg, New York.

Selye, H., (1973): The evolution of the stress concept.
    Am. Scientist, 61, 692-699.

Toxicological Consequences of Induction
of Cytochrome P-450 and its Clinical Relevance

MULTIPLE FORMS OF HUMAN HEPATIC CYTOCHROME P-450

Olavi Pelkonen, Markku Pasanen, Kirsi Vähäkangas and Eero A. Sotaniemi

Departments of Pharmacology and Internal Medicine (Clinical Research Unit), University of Oulu, SF-90220 Oulu (Finland)

SUMMARY

Although a number of P-450 proteins have been purified from human liver and other tissues, it is not possible at present to correlate the properties of these purified proteins with the known features of human xenobiotic metabolism at the microsomal or tissue level. Instead, we try to evaluate, on the basis of indirect evidence, the presence or absence of some specific isozymes in human liver and compare the postulated properties of the tentative isozymes with those purified from livers of experimental animals.

INTRODUCTION

Almost ten different isozymes of cytochrome P-450 from rat and rabbit livers have been purified into homogeneity and characterized more or less extensively (for recent reviews see Lu and West 1980; Guengerich et al. 1982; Ryan et al. 1982). Antibodies prepared against these isozymes have been extensively used to delineate the contributions of different isozymes to total xenobiotic-metabolizing activities in different rodent tissues. It seems likely that in the very near future a combination of techniques will lead to a total characterization of the isoenzymic pools of cytochrome P-450 in the tissues of experimental animals.

ATTEMPTS TO PURIFY HUMAN P-450 ISOZYMES

Several attempts have been made to purify and characterize cytochromes P-450 from human liver (Beaune et al. 1979; Kitada and Kamataki 1979; Wang et al. 1980, 1983; Guengerich et al. 1981, 1982; Dayer et al. 1984; Gut et al. 1984). Although several electrophoretically homogeneous proteins have been obtained, at present it is not possible to correlate the properties of these purified P-450 proteins with the known features of human xenobiotic metabolism, except perhaps in the case of debrisoquine 4-hydroxylase. Still the study of multiplicity of P-450 and the characterization of different tentative isozymes requires indirect approaches, for example, studies on substrate selectivity and kinetics, preferential inhibitors, differential induction behaviour, gel electrophoretic pattern and use of heterologous antibodies in immunoinhibition and immunoprecipitation experiments (e.g. Lu and West 1980). Although naturally the purification is the final proof for the presence of the specific isozyme indirect approaches may give very useful information and, because of technical and ethical limitations inherent in human studies, may in some cases yield the only information we can obtain. In the next presentation we will discuss the evidence suggesting the presence (or absence) of some specific isozymes in human liver and also try find out whether any one among the already purified isozymes corresponds to the form tentatively identified in human liver. At present the data available are not convincing enough to make any definite conclusions, but the following discussion might give a useful starting point for better defined questions.

ON THE PRESENCE OF PHENOBARBITAL-INDUCED P-450 IN HUMAN LIVER

Ever since the pioneering studies of Remmer and coworkers, phenobarbital has been known to induce the elimination of a number of compounds in humans. Many of those compounds the metabolism of which have been shown to be induced by phenobarbital in man have been shown to be metabolized by the phenobarbital-induced form(s) of P-450 in experimental animals. Furthermore, Sotaniemi et al. (1978) showed that P-450 content is much higher in liver biopsy samples from epileptic patients receiving phenobarbital and phenytoin than in those from controls. Consequently, on the basis of indirect evidence it seems quite certain that human liver contains one or more phenobarbital-induced P-450 isozymes. Recent studies with polyclonal antibodies to phenobarbital-pretreated rat liver P-450 (Guengerich et al. 1981) have also demonstrated that human liver contains isozymes recognized by these antibodies. As to the possible differences in catalytic and other properties of this isozyme between humans and experimental animals, nothing certain can be said before the purification of the appropriate isozyme from the human liver. Wang et al. (1980, 1983) purified from human liver an isozyme that was very active in catalyzing benzphetamine N-demethylation (a typical phenobarbital-inducible activity), but its relationship with induction remains unclear.

ON THE PRESENCE OF 3-METHYLCHOLANTHRENE-INDUCED P-450 IN HUMAN LIVER

Another classical P-450 isozyme, which is induced by PAH or cigarette smoking (if they indeed induce the same form), has been shown to be present or induced by PAH or cigarette smoking in a number of human cells and tissues (Harris et al. 1984). The presence of this isozyme in the human liver has been uncertain for a long time. Although early studies tentatively suggested that AHH activity, presumably linked with this isozyme, was induced by cigarette smoking, later, more definite studies addressed to this question (Vähäkangas et al. 1983) failed to reveal the induction of AHH or P-450 by cigarette smoking in human liver, although a very clear effect on antipyrine elimination was evident. It seems that this discrepancy can be explained by the substrate and metabolic pathway specificity of cigarette smoking. A recent study demonstrated, that 7-ethoxyresorufin O-deethylase is clearly elevated in livers from smokers and that the enzyme is inhibited by the monoclonal antibody against the rat hepatic isozyme from 3-methylcholanthrene-pretreated animals (Pelkonen et al., unpublished results). The immunoinhibition was more marked in liver preparations from smokers. Whether the isozyme catalyzing the O-deethylation of 7-ethoxyresorufin is also responsible for the metabolism of antipyrine and other compounds shown to be inducible in vivo by cigarette smoke exposure is not known at the present moment. Among the purified human liver P-450 isozymes there is no good candidate for this form.

ON THE NATURE OF P-450 ASSOCIATED WITH HIGH COUMARIN 7-HYDROXYLASE

Studies during the recent decade have established, that mouse liver contains an isozyme very active in coumarin 7-hydroxylation, responsive to phenobarbital induction, and not found in rat liver microsomes (for ref. see Pelkonen et al. 1985). This isozyme was purified into homogeneity and the antibody prepared in goats (Kaipainen et al. 1984). Studies with human liver preparations indicated the presence of an analogous isozyme, because human liver seemed to be very active in hydroxylating coumarin (Pelkonen et al. 1985) and the antibody prepared to the murine isozyme recognized the human hepatic activity in immunoinhibition experiments (Pelkonen, Lang and others, unpublished). However, the human hepatic activity was not statistically significantly higher in patients receiving inducing compounds (Pelkonen et al. 1985). In a later study, a monoclonal antibody prepared against the rat hepatic phenobarbital-induced P-450 failed to inhibit coumarin 7-hydroxylation in human liver, a finding which is in accordance to the relative absence of this activity in rat liver (Pelkonen et al., unpublished results).

ON THE NATURE OF DEBRISOQUINE 4-HYDROXYLASE

English and German researches found in the late seventies, that some persons are deficient in metabolizing debrisoquine and sparteine (see Boobis and Davies 1984; Vesell 1984). This defect was ultimately traced to the absence (or inactivity) of a specific P-450 isozyme, which catalyzes the metabolism of a large number of other substances besides debrisoquine or sparteine, for example metoprolol, phenformin and perhexilene. This isoenzyme is recognized in immunoinhibition experiments with the antibody to rat hepatic debrisoquine-hydroxylating isozyme (Distlerath et al. 1984). As to the relationship of this isozyme to other forms in human liver, studies with "preferential" inhibitors suggest, that the enzyme is distinctly different from those forms associated with phenobarbital or cigarette smoke induction or from those forms participating in the metabolism of antipyrine or coumarin (Boobis et al. 1983, Pasanen and Pelkonen, unpublished).

ON THE NATURE OF ISOZYMES PARTICIPATING IN ANTIPYRINE METABOLISM

Antipyrine has been widely used as a model drug for pharmacokinetic studies (Vesell 1984). It is almost totally metabolized, presumably in the liver, into three main metabolites. Studies have been accumulating to indicate, that these three metabolic pathways are catalyzed by different P-450 isozymes (Breimer 1983). Recent family studies on the excretion of antipyrine metabolites give convincing support to the notion, that the three pathways are indeed separately regulated and they are catalyzed by distinct P-450 isozymes (Vesell 1984). Both phenobarbital and cigarette smoking affect antipyrine elimination, but studies on the specific isoenzymic effects of these inducers are lacking.

ON THE NATURE OF ALCOHOL-INDUCED P-450 FORM

Ethanol has been shown to induce P-450-associated metabolic reactions in the liver of experimental animals and alcohol drinking enhances the elimination of a number of substances in man, too (see Pelkonen and Sotaniemi 1982). The alcohol-induced P-450 isozyme has been purified from the livers of experimental animals and it seems to be a distinct isoenzyme. Human studies are difficult, because alcohol drinking also has a well-known

250

propensity to damage the liver with consequent decrease of total P-450 content and the activity of associated enzymes (Pelkonen and Sotaniemi 1982).

ON THE NATURE OF STEROID-METABOLIZING P-450 FORMS

Studies with experimental animals have shown that some P-450 isozymes metabolize also steroids, although it is generally held, that those isozymes in steroidogenic tissues that are responsible for specific steroid biosynthesis do not participate in xenobiotic metabolism. Pasanen and Pelkonen (unpublished results) have purified a homogeneous P-450 form from human placental mitochondria and made a polyclonal antibody against it. This antibody does not inhibit any xenobiotic-metabolizing activity. Thus this placental mitochondrial P-450 is probably a distinct isozyme not participating in xenobiotic metabolism.

PERSPECTIVES

Indirect evidence clearly demonstrate the presence of a number of distinct P-450 isozymes in human liver. Although ultimately it may be possible to purify all these forms from human liver (or other tissues, if some of them are present), we think that also in the future one of the most promising approaches is to use heterologous antibodies prepared against P-450 isozymes in experimental animals in the characterization and quantitation of human P-450 forms. Also DNA recombination technology gives powerful tools to study the presence or absence of specific P-450 genes in the human genome and to characterize the whole gene-action system in greater detail than is now possible.

ACKNOWLEDGEMENTS

Original studies from our laboratories were supported by the Finnish Cancer Research Foundation and the Academy of Finland (Medical Research Council).

REFERENCES

Beaune, P., Dansette, P., Flinois, J. P., Columelli, S., Mansuy, D., Leroux, J. P. (1979) Partial purification of human liver cytochrome P-450. Biochem. Biophys. Res. Commun. 88: 826-832.
Boobis, A. R., Davies, D. S. (1984) Human cytochromes P-450. Xenobiotica 14: 151-185.
Boobis, A. R., Murray, S., Kahn, G. C., Robertz, G., Davies, D. S. (1983) Substrate specificity of the form of cytochrome P-450 catalyzing the 4-hydroxylation of debrisoquine in man. Mol. Pharmacol. 23: 474-481.
Breimer, D. D. (1983) Interindividual variations in drug disposition. Clinical implications and methods of investigation. Clin. Pharmacokin. 8: 371-377.
Cresteil, T., Beaune, P., Kremers, P., Flinois, J.-P., Leroux, J.-P. (1982) Drug-metabolizing enzymes in human foetal liver: partial resolution of multiple cytochromes P-450. Pediat. Pharmacol. 2: 199-207.
Dayer, P., Gasser, R., Gut, J., Kronbach, T., Robertz, G., Eichelbaum, M., Meyer, U. A. (1984) Characterization of a common genetic defect of cytochrome P-450 function (debrisoquine-sparteine type polymorphism) - Increased Michaelis constant (Km) and loss of stereoselectivity of bufuralol 1'-hydroxylation in poor metabolizers. Biochem. Biophys. Res. Commun. 125: 372-380.

Distlerath, L. M., Guengerich, F. P. (1985) Characterization of a human liver cytochrome P-450 involved in the oxidation of debrisoquine and other drugs by using antibodies raised to the analogous rat enzyme. Biochemistry, in press.

Guengerich, F. P., Wang, P., Mason, P. S., Mitchell, M. B. (1981) Immunological comparison of rat, rabbit, and human microsomal cytochromes P-450. Biochemistry 20: 2370-2378.

Guengerich, F. P., Dannan, G. A., Wright, S. T., Martin, M. V. (1982) Purification and characterization of microsomal cytochrome P-450s. Xenobiotica 12: 701-716.

Guengerich, F. P., Wang, P., Davidson, N. K. (1982) Estimation of isozymes of microsomal cytochrome P-450 in rats, rabbits, and humans using immmunochemical staining coupled with sodium dodecyl sulfate-polyacrylamide gel electrophoresis. Biochemistry 21: 1698-1706.

Gut, J., Gasser, R., Dayer, P., Kronbach, T., Catin, T., Meyer, U. A. (1984) Debrisoquine-type polymorphism of drug-oxidation: purification from human liver of a cytochrome P-450 isozyme with high affinity for bufuralol hydroxylation. FEBS Lett. 173: 287-290.

Harris, C. C., Autrup, H., Vähäkangas, K., Trump, B. F. (1984) Interindividual variation in carcinogen activation and DNA repair. In: Banbury Report 16: Genetic Variability in Responses to Chemical Exposure. Cold Spring Harbor Laboratory, Cold Spring Harbor, pp. 145-154.

Kaipainen, P., Nebert, D. W., Lang, M. A. (1984) Purification and characterization of a microsomal cytochrome P-450 with high activity of coumarin 7-hydroxylase from mouse livers. Eur. J. Biochem. 144: 425-431.

Kamataki, T., Sugiura, M., Yamazoe, Y., Kato, R. (1979) Purification and properties of cytochrome P-450 and NADPH-cytochrome c (P-450) reductase from human liver microsomes. Biochem. Pharmacol. 28: 1993-2000.

Kitada, M., Kamataki, T. (1979) Partial purification and properties of cytochrome P-450 from homogenates of human fetal livers. Biochem. Pharmacol. 28: 793-797.

Lu, A. Y. H., West, S. B. (1980) Multiplicity of mammalian microsomal cytochromes P-450. Pharmac. Rev. 31: 277-295.

Pelkonen, O., Sotaniemi, E. (1982) Drug metabolism in alcoholics. Pharmac. Ther. 16: 261-268.

Pelkonen, O., Sotaniemi, E. A., Ahokas, J. T. (1985) Coumarin 7-hydroxylase activity in human liver microsomes. Properties of the enzyme and interspecies comparisons. Br. J. Clin. Pharmacol. 19: 59-66.

Ryan, D. E., Thomas, P. E., Reik, L. M., Levin, W. (1982) Purification, characterization and regulation of five rat hepatic microsomal cytochrome P-450 isozymes. Xenobiotica 12: 727-744.

Sotaniemi, E. A., Pelkonen, R. O., Ahokas, J., Pirttiaho, H. I., Ahlqvist, J. (1978) Drug metabolism in epileptics: in vivo and in vitro correlations. Br. J. Clin. Pharmacol. 5: 71-76.

Vähäkangas, K., Pelkonen, O., Sotaniemi, E. A. (1983) Cigarette smoking and drug metabolism. Clin. Pharmacol. Ther. 33: 375-380.

Vesell, E. S. (1984) Pharmacogenetic perspectives: Genes, drugs and diseases. Hepatology 4: 959-965.

Wang, P., Mason, P. S., Guengerich, F. P. (1980) Purification of human liver cytochrome P-450 and comparison to the enzyme isolated from rat liver. Arch. Biochem. Biophys. 199: 206-219.

Wang, P. P., Beaune, P., Kaminsky, L. S., Dannan, G. A., Kadlubar, F. F., Larrey, D., Guengerich, F. P. (1983) Purification and characterization of six cytochrome P-450 isozymes from human liver microsomes. Biochemistry 22: 5375-5383.

PHARMACOLOGY OF FLUMECINOL (ZIXORYN), A DRUG ACTING THROUGH THE INDUCTION OF LIVER CYTOCHROME P-450 SYSTEM

Sz.SZEBERÉNYI*

Chemical Works of Gedeon Richter Ltd.. Budapest, Hungary

Zixoryn (flumecinol, RGH-3332, 3-trifluoromethyl-α - -ethylbenzhydrol) was synthesized from about 120 other benzhydrol derivatives, and developed for therapeutic use as a selective inducer of microsomal polysubstrate monooxygenase enzyme system (PSMO).

Selective inducer meets the criteria of a "non drug" drug described by Mannering (5): it does not elicit sufficient pharmacodynamic effects of its own to permit its inclusion in other common classes of drugs, but interacts with living systems so as to alter the activities of other drugs.

This work is based on the following conceptions:
a) Although a relative newcomer in the group of the adaptive structures and the protective systems of "milieu intérieur" it has been accepted that PSMO plays a key role in the defense mechanisms against lipophilic xenobiotics by catalyzing the biotransformation of several drugs, toxicants and endogenous substances. In case of impaired biotransformation, the subnormal activity of the liver microsomal PSMO may decrease generally the nonspecific resistance of the organism. Under nonphysiologic and pathological conditions, when the drug metabolizing reactions are suppressed by various factors, the protective capacity of this enzyme system is insufficient (2) although the liver afflicted by parenchymal injury is inducible by PSMO inducing compounds. The aim of our work is to support or normalize the decreased function of PSMO and/or facilitate the regeneration of this enzyme system. There is no doubt that hepatic microsomal enzyme induction can be useful in the treatment of certain spontaneous diseases in man (3,7,9). However, in spite of therapeutic use of PSMO induction, it is not sufficiently widespread yet. The cause of the aversion is that every known inducing compound has other, in this respect undesirable pharmacological effects. These effects limit the application of these drugs as therapeutic enzyme inducers.
b) More than 300 drugs and various chemicals are known to induce PSMO. The pharmacological activities and chemical structures of these agents are extremely diverse and at

*Present address: National Institute of Occupational Health, Budapest, Hungary present there is no apparent relation between the structure and inducing activity (1). On this basis I have supposed that it is possible to find such compounds which can induce PSMO without any other pharmacological actions.

The purpose of this presentation is to summarize briefly our studies on the inducing effects of Zixoryn (10,11,6).

Analysis of the effect of Zixoryn on the duration and intensity of action of various drugs known to be metabolized in the liver.

The sleeping or muscle relaxation (paralysis) times are substantially shortened in rats pretreated with Zixoryn (Fig. 1).

Repeated administration of Zixoryn has no effect on the blood coagulation. However, this treatment can strongly interfere with the hypoprothrombinemic activity of aceno-cumarol by the more rapid inactivation of this anticoagulant

To estimate the effect of Zixoryn on the anticonvulsive activity of diphenylhydantoin and antinociceptive potency of ethylmorphine, the half lives of the activities were calculated.

Table 1. Effect of pretreatment with Zixoryn (150 μmol/kg p.o. for 3 days) on anticonvulsive and antinociceptive potency (10).

	t 1/2 of effect (min)		
	Placebo	Zixoryn	P
Diphenylhydantoin 12 mg/kg i.p.	120	43	<0.001
Ethylmorphine 40 mg/kg s.c.	104	32	<0.001

n = 20 in each group

Determination of the effect of Zixoryn on the metabolic rate of various drugs in vivo.

Data obtained following a single dose of several compounds known to be metabolized or removed by the liver are shown in Table 2. The differences between the control and Zixoryn influenced data pairs are significant (P < 0.05 by comparison of slopes). Zixoryn does not influence the apparent distribution volume of any drug.

Table 2. Effect of Zixoryn on the plasma disappearance of various compounds in rats.

		Placebo	Zixoryn pretreatment	$\Delta\%$
Hexobarbital	t 1/2	40 min	13 min	−68
	K *)	1.73	5.33	
Meprobamate	t 1/2	176 min	84 min	−52
	K	0.39	0.83	
Mephenesine	t 1/2	66 min	42 min	−36
	K	1.05	1.65	
Nikethamid	t 1/2	81 min	41 min	−50
	K	0.86	1.69	
Canrenone	t 1/2	57 min	34 min	−40
	K	1.22	2.04	
Metyrapone	t 1/2	17 min	9 min	−47
	K	4.08	7.70	

Antipyrine	t 1/2	137 min	82 min	-40
	K	0.51	0.85	
Bromsulphalein	t 1/2	7.4 min	5.4 min	-31
	K	9.36	13.59	
Indocyanin green	t 1/2	7.5 min	4.2 min	-44
	K	9.24	16.50	
Bilirubin	t 1/2	17 min	11 min	-35
	K	4.08	6.48	

Zixoryn 150 umol/kg p.o. once. K*): disappearance rate.

Determination of the effect of Zixoryn on the urinary excretion of certain normal metabolites.

Ascorbic acid excretion was increased by 371 $\pm$ 18 % under the influence of Zixoryn (75 μmol/kg p.o.daily) pretreatment for 3 days (P < 0.001).

Determination of the effect of Zixoryn on the amount/activity of components of microsomal electron transport chain.

Zixoryn combines with microsomal cytochrome P-450 to give a type I spectral change (9) max 387 nm, min 420 nm (Fig.2.)

Zixoryn administration resulted in a substantial increase of the amount/activity of the microsomal cytochromes P-450 and b5 as well as NADPH: cytochrome c reductase (EC 1.6.2.4) (Table 3. and 4.).

Table 3. Influence of Zixoryn on the cytochrome P-450 in rat liver.

	CO-reactive cytochrome P-450 nmol/g liver	Metyrapone (5) binding P-450 nmol/g liver
Untreated	19.8 $\pm$ 1.73	5.4 $\pm$ 0.31
	100 $\pm$ 9 %	
Placebo	102 $\pm$ 9 %	103 $\pm$ 5 %
Zixoryn 150 μmol/kg p.o. once	149 $\pm$ 12 %*	324 $\pm$ 15 %*
Zixoryn 150 μmol/kg p.o. for 3 days	253 $\pm$ 19 %*	456 $\pm$ 28 %*

* = P < 0.001   n = 10 in each group

Table 4. Influence of Zixoryn on the components of the microsomal transport chain in rat liver.

Treatment	Cytochrome b5 nmol/g liver	EC 1.6.2.4. nmol/g liver/min
Placebo	10.7 $\pm$ 1.01	3240 $\pm$ 205
	100 $\pm$ 9 %	100 $\pm$ 6 %
Zixoryn 150 μmol/kg p.o. once	157 $\pm$ 8 %*	165 $\pm$ 11 %*
Zixoryn 150 μmol/kg p.o. for 3 days	149 $\pm$ 11 %*	193 $\pm$ 14 %*

* = P < 0.01;   n = 10 in each group

Assessment of the drug metabolizing enzyme activities after Zixoryn administration.

Treatment with Zixoryn increased substantially the N-demethylase, O-demethylase and ring hydroxylase activities in rat liver microsomes (Table 5). Zixoryn induced the activity

256

of aminopyrine N-demethylase to the same extent as it increases the content of cytochrome P-450 in the liver microsomes. An induction mechanism similar to that of phenobarbital was observed after Zixoryn pretreatment; it was proved
furthermore by the data obtained with metyrapone-ferrocyto-
chrome P-450 complex determinations (Table 4.).

Table 5. Effect of Zixoryn on liver microsomal enzyme
activities in rat liver

	Placebo	Zixoryn
Aminopyrine	100 + 2 %	238 + 12 %*
N-demethylase	219.3 + 4.92	
nmol/g liver/min.		
Aniline p-hydroxylase	100 + %	164 + 9 %*
nmol/g liver/min.	16.4 + 0.61	
p-Nitroanisol	100 + 6 %	208 + 16 %*
O-demethylase	38.5 + 2.26	

Zixoryn 150 μmol/kg p.o. for days; n = 12 in each group;
* = P< 0.01

Investigation of the interaction of Zixoryn and protein
synthesis inhibitors.
Cycloheximide, puromycin, D,L-ethionin and chloramphenicol were themselves incapable of modifying the hexobarbital
sleeping time, meprobamate elimination or cytochrome P-450
concentration in the liver microsomes. However, these protein synthesis inhibitors suppressed the enzyme inducing
effect of Zixoryn (in the same doses). The findings support
the view that changes in drug metabolizing activity following administration of Zixoryn is the result of de novo enzyme synthesis (induction) (1) (Fig. 3.).

Other experiments not reported in detail here demonstrated that Zixoryn acts as a quick inducer of PSMO, its effect
can be detected 8-18 hours after a single oral dose. The
minimal effective dose is 1.25 mg/kg (4.5 umol/kg) p.o. in
rats. The inducing effect can be increased after repeated
administration and is lasting for several days dependent on
the dose, but it is reversible. Zixoryn is more (2-3 fold)
effective given p.o. than s.c. or i.p. Its inducing potency
was not blocked by hypophysectomy, adrenalectomy, castration
or ovariectomy. Zixoryn is excreted with milk and increases
cytochrome P-450 content in newborn rat liver.The biological
t 1/2 of Zixoryn in mature rats is about 180 min. Zixoryn
also enhances (induces) PSMO in various species (rats, mice,
guinea pigs, rabbits, cats, dogs, chicken, calves, frogs)
and also in humans. Zixoryn causes a marked proliferation of
smooth endoplasmic reticulum in hepatocytes, characteristic
for enzyme induction. No other subcellular structure was effected. Toxicity of Zixoryn is 10-fold less than that of
phenobarbital, its acute LD50 p.o. is 2332 mg/kg, its
therapeutic index is 58.4, while that of phenobarbital is
only 5.8 as inducing agent. Zixoryn had no effect on the
central nervous system in the 5-fold of the maximum inducing
dose as tested by open field test, ataxia and muscle coordi-

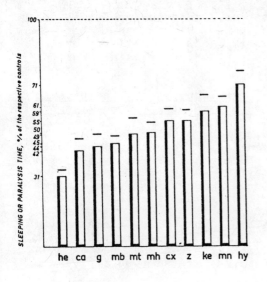

Fig.1. Effect of Zixoryn
(150 μmol/kg, p.o.) on
the sleeping or paralysis
time caused by drugs
he: Hexobarbital,
ca: Carisoprodol,
g: Glutethimide,
mb: Meprobamate
mt: Methaqualone,
mh: Methohexitone,
cx: Chlorzoxazon,
z: Zoxazolamine,
ke: Ketamine,
mn: Mephenesin
hy: Hydroxidione

n = 10-14 rats in each
    group

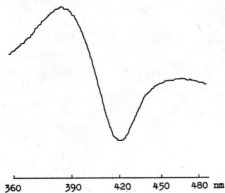

Fig.2. Type I difference
spectrum induced by the
addition of Zixoryn to
suspensions of rat liver
microsomes.

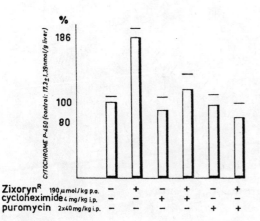

Fig.3. The prevention by
protein synthesis inhibi-
tors of Zixoryn induced
increase in the cyto-
chrome P-450 content.

nation tests, locomotor activity, thiosemicarbazide- and strychnine convulsions, electroshock (6). Zixoryn is markedly choleretic. Zixoryn induces the PSMO of the liver during regeneration after partial hepatectomy, endotoxin treatment and cholestasis.

<u>Pharmacological reversal of cholestasis associated decrease in hepatic microsomal polysubstrate monooxygenase (PSMO) system.</u>
In the rat cholestasis produced by bile duct ligation (BDL) is associated with a significant reduction in hepatic cytrochrome P-450, cytochrome b5 content NADPH: cytochrom c reductase EC 1.6.2.4, aminopyrine N-demethylase, aniline p-hydroxylase and p-nitroanisole O-demethylase activities. We have investigated the potency of Zixoryn for reversing the depression of the components of PSMO due to experimental cholestasis (bile duct ligation). Continued administration of Zixoryn (360 μmol/kg p.o. for 7 days) resulted in a significant inducing effect and could almost restore the content/activity of the enzymes investigated (Table 6.).

The plasma bilirubin level was highly increased after BDL. Zixoryn stimulated the glucuronidation, increased the amount of the conjugated bilirubin and reduced the unconjugated, and to a lesser extent, the total bilirubin.

<u>Table</u> 6. Effect of Zixoryn on liver microsomal PSMO in bile duct ligated (BDL) rats.

	Controls	% of sham-operated, untreated controls	
		BDL	BDL + Zixoryn
Cytochrome P-450 nmol/g liver	17.4 ± 1.05	47 ± 4.6 %	92 ± 8.3 %
Cytochrome b5 nmol/g liver	9.3 ± 0.49	65 ± 4.9 %	98 ± 7.5 %
EC 1.6.2.4 nmol/g liver/min.	3042 ± 119	58 ± 9.3 %	99 ± 5.6 %
Aminopyrine N-demethylase nmol/g liver/min.	171.9 ± 6.71	51 ± 4.1 %	92 ± 8.0 %
Aniline p-hydroxylase nmol/g liver/min.	20.4 ± 0.84	61 ± 3.9 %	74 ± 7.3 %
p-Nitroanisole O-demethylase nmol/g liver/min.	33.4 ± 0.94	46 ± 6.9 %	100 ± 7.5 %

Zixoryn 360 μmol/kg p.o. for 7 days.
n = 10-17 in each group

*

From the data collected during our study it is clear that Zixoryn may meet the requirements of an effective and well-tolerable inducing agent for human therapy.
*

References:

1. Conney,A.H.:
   Pharmacological implications of microsomal enzyme induction.
   Pharmacol.Revs. 19: 317-366 (1967)
2. Kato,R.:
   Drug metabolism under pathological and abnormal physiological states in animals and man.
   Xenobiotica 7: 25-92 (1977)
3. Luoma,P.V., Sotaniemi,E.A., Pelkonen, R.O.:
   Inverse relationship of serum LDL cholesterol and the LDL/HDL cholesterol ratio to liver microsomal enzyme induction in man.
   Res.Commun.Chem.Pathol.Pharmacol. 42: 173-176 (1983)
4. Luu-The,V., Cumps,J., Dumont,P.:
   Metyrapone-reduced cytochrome P-450 complex: A specific method for the determination of the phenobarbital inducible form of rat hepatic microsomal cytochrome P-450.
   Biochem.Biophys.Res.Commun. 93: 776-781 (1980)
5. Mannering,G.J.:
   Biotransformation. In: Antiepileptic drugs, Woodbury,D.M Penroy,J.K., Schmidt,R.P. (ed.), New York, Raven Press, 1972. pp. 23-43
6. Pálosi,É., Szeberényi,Sz., Szporny,L.:
   Effects of 3-trifluoromethyl-α-ethylbenzhydrol (RGH-3332), a new enzyme inducer on the central nervous system of rats.
   Arzneim.-Forsch./Drug Res. 28: 669-672 (1978)
7. Rautio,A., Sotaniemi,E.A., Pelkonen,R.O., Luoma,P.V.:
   Treatment of alcoholic cirrhosis with enzyme inducers.
   Clin.Pharmacol.Therap. 28: 629-637 (1980)
8. Schenkman,J.B., Remmer,H., Estabrook,R.W.:
   Spectral studies of drug interaction with hepatic microsomal cytochrome.
   Mol.Pharmacol. 3: 113-123 (1967)
9. Sotaniemi,E.A., Arranto,A.J., Sutinen,S., Stengard,J.H.:
   Treatment of noninsulin dependent diabetes mellitus with enzyme inducers.
   Clin.Pharmacol.Therap. 33: 826-835 (1983)
10. Szeberényi,Sz., Pálosi,É., Szporny,L.:
    Effects of 3-trifluoromethyl-α-ethylbenzhydrol (RGH-3332), a new enzyme inducer, on the microsomal drug metabolism.
    Arzneim.-Forsch./Drug Res. 28: 66323-668 (1978)
11. Szeberényi,Sz., Ungváry,Gy., Szporny,L.:
    3-Trifluoromethyl-α-ethylbenzhydrol (RGH-3332, Zixoryn) a new inducer of drug metabolizing enzymes.
    Abstr. no. 71. Eight International Congress of Pharmacology, Tokyo, 1981. p. 254

18*

EFFECT OF FLUMECINOL[R] ON THE INDUCTION OF CYTOCHROME P-450 IN RATS AND RABBITS

P. BENDZKO, L. VERECZKEY*, J. FRIEDRICH, W. SCHWARZE AND K. RUCKPAUL

Central Institute of Molecular Biology, Academy of Sciences of the GDR, 1115 Berlin-Buch, GDR
* Department of Pharmacokinetics and Drug Metabolism, Gedeon-Richter LTD., 1475 Budapest, Hungary

INTRODUCTION

Human newborns are incapable to excrete the toxic bilirubin due to lacking distinct post-oxidation enzymes. Bilirubin is mainly excreted after conjugation with glucuronic acid which is catalyzed by UDP-glucuronyl transferase. In order to avoid subsequent injury in health an increased bilirubin level in human newborns (icterus neonatorum) is usually treated with compounds which accelerate the excretion of bilirubin. These inducers are known to effect not only a selective induction of a single enzyme but to induce also additional enzymatic activities.

Besides the treatment with phenobarbital or some of its derivatives Flumecinol[R] has proved effective in treatment of icterus neonatorum. With regard to undesired side effects it is of clinical relevance to clearly define the inductive effect of Flumecinol[R] on hepatic endoplasmic cytochrome P-450 isozymes. Consequently in the present paper the induction of cytochrome P-450 isozymes in dependence on age, sex and species has been analyzed.

MATERIALS AND METHODS

Flumecinol[R] suspended in 0.9 %NaCl and 15 % Tween 80 was administered to adult and newborn Wistar rats (weight: 120-150 g and 10-15 g, respectively) intraperitoneally at a dose of 100 mg/kg body weight three times at intervalls of 24 h. Each group of adult rats comprised 10 animals. The newborn rat group comprised 150 animals. Flumecinol[R] was administe-

red to adult rabbits (2.5-3.0 kg body weight) with the same
dose as to adult rats, seven times at intervalls of 24 h.
Each group comprised 16 rabbits. The microsomes from uninduced and induced animals were obtained after homogenization
of the liver by differential centrifugation according to commonly used procedures. The 100.000 xg pellets were resuspended in 0.1 M potassium phosphate, 0.25 M saccharose buffer,
pH 7.4.

Benzphetamine N-demethylase activity was assayed by formaldehyde release according to NASH /1953/. O-deethylation activity of p-nitrophenetole was determined by the method described previously /SHIGEMATSU 1976/. SDS-polyacrylamide gel
electrophoresis was performed using a modified LAEMMLI /1973/
discontinuous buffer system. Protein was determined by the
method of LOWRY /1951/ with Serulat (VEB Sächsisches Serum-
werk Dresden) as standard. The cytochrome P-450 content was
calculated according to the method of OMURA and SATO /1964/
from the CO-reduced difference spectrum based on an extinction coefficient of 91 $mM^{-1}cm^{-1}$.

ß-naphthoflavone (ß-NF) in corn oil was administered intraperitoneally once 40 h before removal of the rat and rabbit
livers (80 mg/kg body weight), usually phenobarbital (PB) was
injected in 0.9 % sodium chloride (40 mg/kg body weight) at
three consecutive days analogously to the Flumecinol[R] induction of rats and was given for seven consecutive days to
rabbits (48 mg/kg body weight).

RESULTS AND DISCUSSION
Flumecinol[R] Induction of Cytochrome P-450 in Adult Male and
Female and Newborn Rats

Flumecinol[R] treatment of adult male and female rats results
in a significant increase of the total P-450 content. The
changes in the cytochrome P-450 level are sex dependent. Contrary to phenobarbital induction where the induction rate in
male rats (130 % over control) is significantly higher than
in females (80 % over control) Flumecinol[R] effects a 90 %
higher induction in female than in male rats (Table 1).

Table 1   Influence of different inducers on the total cyto-
           chrome P-450 content in adult and newborn rat liver
           microsomes

| | | Total cytochrome P-450 content (nmol/mg protein) | | |
		Male	Female	Newborn
Flumecinol[R]	c	0.99 ± 0.2	0.54 ± 0.2	0.48 ± 0.3
	i	1.45 ± 0.3	1.28 ± 0.3	0.64 ± 0.3
Phenobarbital	c	1.03 ± 0.3	0.8 ± 0.2	–
	i	2.36 ± 0.3	1.5 ± 0.3	–
ß-Naphthoflavone	c	0.92 ± 0.2	0.8 ± 0.2	–
	i	1.9 ± 0.3	1.7 ± 0.3	–

c = control; i = induced

The total P-450 content in newborn rats after Flumecinol[R]
treatments is only slightly increased (about 30 % in comparison with the untreated animals).

a) Electrophoretic Analysis: The SDS-gel electrophoresis of
solubilized microsomes from male and female rats revealed that
the increase in the total p-450 content in adult male and female rats after Flumecinol[R] induction is essentially due to an
increase of the b-form. The increased induction is reflected
in the staining intensities of bands from corresponding standards as derived from control and induced rats. Densitometric
tracings of band patterns indicate a 2-fold increase of the
b-form in adult female rats and an increase of about 50 % in
adult male rats, as compared to untreated adult animals (Fig.1).
Induction of the c-form in adult male and female rats can be
excluded. The electrophoretic band pattern of the microsomal
fraction of newborn rats shows additional protein bands indicating a qualitative difference as compared to the band distribution in adult male and female rats. One band with a molecular weight in the range of 50 kD represents an isozyme which
has not been described so far in the band pattern of adult
rats (see arrow in Fig. 1). The electrophoretic pattern of
control and Flumecinol[R] treated newborn rats does not show a
distinct protein band corresponding to cytochrome P-450c thus

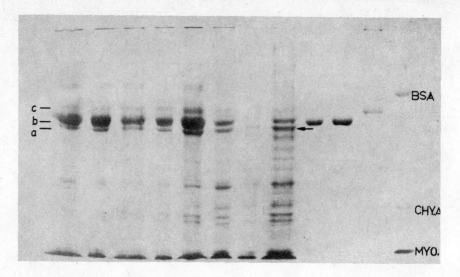

Fig.1: SDS-polyacrylamide slab gel electrophoresis. Liver
microsomal cytochrome P-450 from rats: adult male (lane 1-un-
treated, 4.5 µg; lane 2-Flumecinol[R] induced, 4.5 µg) adult
female (lane 3-untreated, 4.5 µg; lane 4,5-induced, 1.7 µg
and 4.5 µg); newborn (lane 6-control, 5.0 µg; lane 7,8-indu-
ced, 1,5 µg and 5.0 µg). As standards: lane 9,10 cytochrome
P-450 LM2 (4.7 µg, 7.3 µg); lane 11 cytochrome P-450 LM4
(7.0 µg); lane 12 standard proteins (bovine serum albumin,
chymotrypsinogen A and myoglobin from the top to the bottom
were used.

supporting the finding of a low content of form c in
newborn rats /THOMAS et al. 1981/.
b) Enzymatic Activity: The substrate specificity of the induced cytochrome P-450 forms was determined using benzphetamine
/BPh/ and p-nitrophenetole /p-NP/ as specific substrates for
cytochrome P-450b and P-450c, respectively. The BPh-turnover
rate in male rats is about 100 % higher than in untreated animals, whereas the p-NP-turnover rate considerably decreases
by 60 %. In accordance with the electrophoretic data the BPh-
turnover rate in female rats is by 60 % higher than in male
rats. The induction rate of induced female rats compared to
untreated animals is increased by 160 %. Just as in male rats
the p-NP-turnover rate in females is considerably decreased by

70 %. Therefore the observed increase in the BPh-turnover in both adult male and female rats refers to an induction of the b-form of the rat liver cytochrome P-450. The concomitant diminution of the conversion of ß-naphthoflavone /ß-NF/ evidences that the induction process is not only due to activity changes of the b-form.

Concomitant with a significant lower total cytochrome P-450 content in untreated newborn rats the respective BPh-turnover is by 200 % lower than that in adult male and by 100 % lower than that in adult female animals (Tab.2). By Flumecinol^R treatment, however, the BPh-turnover rate in newborn rats is about 6-fold increased. This remarkably enhanced turnover is assumed to reflect an age dependent induction. The increase of the total cytochrome P-450 content after Flumecinol^R induction in newborn rat liver microsomes most probably is explicable to originate from the induction of isozyme(s) with N-demethylase activity. The discrepancy between the increased BPh-turnover which is not reflected in the electrophoretic pattern cannot be explained till now.

The pNP-turnover rate in newborn rats decreases as in adult rats to about 70 %.

Flumecinol^R Induction of Cytochrome P-450 in Adult Male and Female Rabbits

Different from adult rats Flumecinol^R treatment of adult rabbits only slightly enhanced the total P-450 content (Tab.3). In the case of female trabbits the content increases to about 10 %, whereas the total cytochrome P-450 content in male rabbits increases by about 30 %. Taking into account the standard deviation of the individual values the changes of the total cytochrome P-450 content don't show any sex dependences as in the case of adult rats. Untreated adult rabbits show a much higher level of cytochrome P-450 than untreated adult rats which may be explained by a higher sensitivity of rabbits towards various environmental factors acting as inducers.

The SDS-gel electrophoretic band pattern after Flumecinol^R treatment of adult rabbits displays a significant smaller induction of cytochrome P-450 LM2 and LM4 than in adult rats but

Table 2 Comparison of the induction effect of different inducing agents on the phenobarbital and p-nitro-phenetole turnover rates in microsomes in rats and rabbits

		rats male	female	rabbits male	female	newborn rats
Flumecinol[R]						
BPh-turnover	c	2.1 ±0.5	1.05±0.4	2 ±0.8	2.1 ±0.9	0.53±0.2
(nmol/nmol min)	i	4.2 ±0.7	2.6 ±0.7	2.3±0.7	2.1 ±0.7	3.37±0.1
p-NP-turnover	c	1.76±0.2	1.26±0.4	2.07±0.3	1.86±0.4	3.56±0.4
(pmol/pmol min)	i	1.07±0.1	0.92±0.3	1.92±0.3	1.88±0.4	2.65±0.1
Phenobarbital						
BPh-turnover	c	2.9 ±0.5	1.5 ±0.3	2.1 ±0.4	2.1 ±0.4	-
(nmol/nmol min)	i	5.7 ±0.5	2.9 ±0.4	6.5 ±0.5	6.0 ±0.5	-
p-NP-turnover	c	1.4 ±0.3	2.2 ±0.2	2.9 ±0.3	3.5 ±0.4	-
(pmol/pmol min)	i	1.2 ±0.3	2.1 ±0.3	4.5 ±0.5	4.2 ±0.5	-
ß-naphthoflavone						
BPh-turnover	c	2.9 ±0.5	1.29±0.2	2.3 ±0.3	3.2 ±0.3	-
(nmol/nmol min)	i	1.2 ±0.2	0.4 ±0.02	1.9 ±0.3	1.5 ±0.2	-
p-NP-turnover	c	3.3 ±0.5	1.5 ±0.3	1.44±0.3	1.73±0.3	-
(pmol/pmol min)	i	13.7 ±0.6	11.9 ±0.6	2.7 ±0.3	2.95±0.4	-

Table 3 Influence of different inducers on the total cytochrome P-450 content in liver microsomes from adult rabbits

		Total cytochrome P-450 content (nMol/mg protein)	
		Male	Female
Flumecinol[R]	c	1.4 ± 0.2	1.4 ± 0.3
	i	1.9 ± 0.4	1.6 ± 0.2
Phenobarbital	c	1.5 ± 0.3	1.3 ± 0.3
	i	2.6 ± 0.4	3.0 ± 0.4
ß-Naphthoflavone	c	2.3 ± 0.3	1.8 ± 0.3
	i	3.7 ± 0.4	3.9 ± 0.4

c = control; i = induced

the appearance of new protein bands between the forms LM2 and LM4 after Flumecinol^R treatment clearly indicates the induction of other cytochrome P-450 isozymes.

A small increase (by 20 %) of the BPh-turnover rate in male rabbits is observed, which correlates with the increase (by 30 %) of the total cytochrome P-450 content (see Tab. 2). However the electrophoretic protein band pattern does not allow to derive any conclusions on the induction of cytochrome P-450 LM2 by Flumecinol^R in adult male rabbits. In female rabbits neither the BPh- nor the p-NP-turnover rates are increased by Flumecinol^R corresponding to an almost unchanged total cytochrome P-450 content which agree with the same microsomal protein band pattern of the SDS-gels before and after Flumecinol^R treatment.

The present results evidence Flumecinol^R as phenobarbital analogous inducer of cytochrome P-450 in rat liver microsomes. The Flumecinol^R induced increase of the total cytochrome P-450 content in microsomes shows sex and age dependent differences ($\female > \male$; adult $>$ newborn; induction rate: adult $<$ newborn). Liver microsomes from Flumecinol^R induced adult rats revealed a similar increase of N-demethylase activity towards benzphetamine ($\male < \female$) as in microsomes from phenobarbital induced animals ($\male > \female$) but with an opposite sex dependence.

In microsomes from newborn rat liver Flumecinol^R induction mainly enhances N-demethylase activity. Induction of cytochrome P-450 c by Flumecinol^R has proved ineffective in adult and in newborn rats as well.

Flumecinol^R treatment of rabbits only slightly increases the cytochrome P-450 content in liver microsomes whereas by phenobarbital and ß-naphthoflavone the content of the total cytochrome P-450 content in liver microsomes from male adult rabbits is increased 1.7-fold and 1.6-fold, respectively. The respective values in female rabbits yielded a 2.2-fold increase in response to both inducers. Correspondingly the enzymatic activites with benzphetamine and p-nitrophenetole as substrates are less enhanced than in rats.

The p-NP-turnover in male and female rabbits is not increased by Flumecinol[R] reflecting an unchanged cytochrome P-450 LM4 content.

The observed species dependence of Flumecinol[R] induction needs further investigation.

REFERENCES

Laemmli, U.K. and M. Favre (1973) J. Mol. Biol. 80,575-599

Lowry, O.H., Rosebrough, N.J., Farr, A.L. and R.J. Randall (1951) J. Biol. Chem. 193, 265-275

Nash, T. (1953) Biochem. J. 55, 416-421

Omura, T. and R. Sato (1964) J. Biol. Chem. 239, 2379-2385

Shigematsu, H., Yamano, S. and H. Yoshimura (1976) Arch. Biochem. Biophys. 137, 178-185

Thomas, P.E., Reik, L.M., Ryan, D.E. and W. Lewin (1981) J. Biol. Chem. 256, 1044-1057

CLINICAL EXPERIENCES WITH 3-TRIFLUOROMETHYL-ALPHA-
-ETHYLBENZHYDROL / ZIXORYN /,A HEPATIC MICROSOMAL
ENZYME INDUCING AGENT

Gachályi,B.

Postgraduate Medical School,I.Department of Medicine,
H-1389 Budapest,P.O.Box: 112.,Hungary

The intensity and duration of biological actions of various
xenobiotics are largely influenced by the activity of drug
metabolizing enzymes in the endoplasmic reticulum of the liver cells.These enzymes constitute the so called microsomal
mixed function monooxygenase system which metabolizes various
drugs and normal body constituents making them more water soluble in order to enhance their clearance from the body.About
200 lipid soluble compounds are known to stimulate the activity of the hepatic monooxygenase system.
The ideal hepatic enzyme inducer would be a potent but otherwise pharmacologically inactive substance.Looking for such a
substance 3-trifluoromethyl-alpha-ethylbenzhydrol was synthesized among a hundred of other benzhydrol derivatives by Tóth
et al. / RGH-3332,Zixoryn,Chemical Works of Gedeon Richter Ltd.
Hungary / .
According to pharmacokinetic studies peak plasma concentration
of Zixoryn occurs at 2.1 hours after ingestion in man.Zixoryn's
half life is 17.6 hours,corresponding to a clearance of 94.0
litre/hour in humans.
Animal studies proved the strong inducing effect of Zixoryn on
the hepatic microsomal monooxygenase system.
We have studied the effect of Zixoryn / 300 mg/day orally for
7 days / in man / Gachályi et al.,1978 /.Zixoryn significantly reduced the half life of antipyrine and that of tolbutami-
de,furthermore greatly enhanced the urinary excretion of D-
-glucaric acid.These results clearly indicate that Zixoryn is
a powerful inducer of phase I. and phase II. reactions of the
human liver.No side effects were observed during the administration of Zixoryn.
Vezekényi et al. / 1982 / carried out human phase I and II
clinico-pharmacological studies on Zixoryn.According to their
results the recommended dosage scheme of Zixoryn in order to
achieve the stimulation of phase I. reactions is a single
600 mg dose orally every 7th day.For rapid onset of induction
800 mg every 7th day is preferable.Phase II. reactions are
effectively induced by 400 mg or 600 mg Zixoryn in a single
dose every 7th - 10th day.

In a series of studies we administered Zixoryn / 300 mg/day for 7 days / and/or alpha-methyldopa / Dopegyt,EGIS Chemical Works, Hungary / in a dose of 1000 mg/day for 7 days / Gachályi et al., 1983 /.During combined treatment the two agents were given simultaneously for one week in the doses mentioned.Zixoryn alone significantly decreased the half life of antipyrine and increased urinary D-glucaric axcretion.Dopegyt alone had opposite effects.Given together the two agents counteracted eachother's effects resulting in no change in the hepatic microsomal monooxygenase activity.
These studies confirm Zixoryn's strong inducing potency on the activity of the human hepatic microsomal monooxygenase system.
Lewis and Friedman / 1979 / published the results of a randomized controlled trial of antipyrine in the prophylaxis of neonatal jaundice.Treatment of mothers from the 38th week of pregnancy reduced neonatal plasma bilirubin concentrations on the 4th day after birth by an average of 44%.Antipyrine is a weak inducer of the hepatic microsomal mixed function monooxygenase system besides being its substrate too.
Marosvári and Neubauer / 1984 / studied the effect of Zixoryn in the therapy of premature infant's non-haemolytic hyperbilirubinaemia.Zixoryn was administered in the form of syrup and in a dose of 30 mg/kg/day for 3 days to premature infants.Zixoryn proved to be appropiate for the conservative treatment of hyperbilirubinaemia.The appearance of extremely high bilirubin levels can be prevented by using Zixoryn alone or in combination with the traditional blue-light therapy.
Sas / RGD: 13733 / administered 3x100 mg Zixoryn to 50 parturient women 24 hours prior to delivery.After delivery,the serum bilirubin levels of the newborns were recorded until discharge from the hospital.Zixoryn pretreatment did not eliminate the elevation of serum bilirubin in newborns but had a modulating effect on it.This modulation was very effective,as serum bilirubin levels exceeding the dangerous 20 mg% value were not found in any single case.
Karmazsin / RGD: 13565 / reported good results on the prophylactic administration of Zixoryn to prevent hyperbilirubinaemia in newborns in a dose of 10 mg/kg/day.Zixoryn caused a moderate T-cell depletion without any kind of infection in the treated group.Zixoryn was of therapeutic value in the treatment of a boy suffering in Crigler-Najjar syndrome and had beneficial effect in the treatment of a cirrhotic patient whose disease was due to congenital occlusion of the biliary duct.
László / 1979 / reported on the clinical investigations with Zixoryn in patients with various hepatic diseases.Zixoryn had a favourable effect in Gilbert syndrome but was ineffective in acute B-viral hepatitis.Zixoryn given in combination with thioctanacid,cocarboxylase and vitamin B complex was useful in acute alcoholic hepatitis.The drug had no effect on contraceptive induced cholestatic icterus but markedly relieved pruritus.Only minor side effects like parageusia,gastric pain were observed. One of his patients developed urticaria.
Vezekényi et al. / 1983 / experienced the transient beneficial effect of Zixoryn in tumorous diseases with hepatic metastases.

Guba / RGD: 12931 / carried out electronmicroscopic studies on
liver biopsy samples obtained from patients treated with Zixo-
ryn.Liver cells were not destroyed by Zixoryn and what is more,
he observed a process of extremely high regeneration rate which
was shown by the increase of endoplasmic reticulum with rough
surface,as well as that of the Golgi apparatus.
Latest clinical experiences indicate that Zixoryn treatment may
improve the glucose balance of patients with non insulin dependent diabetes mellitus.
In conclusion we can state that Zixoryn is a potent inducer of
the hepatic microsomal monooxygenase system with practically no
side effects.The application of the drug in the treatment of
hyperbilirubinaemias,various liver diseases,non insulin dependent diabetes mellitus,Cushing's syndrome,thyrotoxicosis and
intoxications with various agents is very promising.

References

1. Gachályi,B.,Káldor,A.,Szeberényi,Sz. / 1978 / m-trifluoro-
   methyl-alpha-ethylbenzhydrol: a new enzyme inducer
   Europ.J.Clin.Pharmacol. 13: 299-302.
2. Gachályi,B.,Vas,Á.,Káldor,A. / 1983 / Combined effect of
   alpha-methyldopa and a new enzyme inducer,Zixoryn on the
   hepatic mixed-function monooxygenase system in humans
   Intern.J.Clin.Pharmacol.Therap.Toxicol. 21: 183-185.
3. Guba,F. Enzyme-inducing effect of Zixoryn /RGH-3332/
   Liver biopsy and electron microscopic studies RGD: 12931
4. Karmazsin,L. Investigation of the enzyme inducing effect of
   Zixoryn /RGH-3332/ on newborns RGD: 13565
5. László,B. / 1979 / Report on the clinical investigations
   with an enzyme inducer RGH 3332 /Zixoryn/ in patient with
   liver disease RGD: 20484
6. Lewis,P.J.,Friedman,L.A. / 1979 / Prophylaxis of neonatal
   jaundice with maternal antipyrine treatment
   The Lancet, February 10, 300-302.
7. Marosvári,I.,Neubauer,K. / 1984 / Zixoryn therapy of hyperbilirubinaemia of the newborns RGD: 30348
8. Sas,M. Effects of Zixoryn on serum bilirubin level in newborns RGD: 13733
9. Vezekényi,Zs.,Ferenci,J.,Radnai,B.,Horváth,T.,Past,T,,Nagy,
   L.,Jávor,T. / 1982 / Human phase I-II. clinico-pharmacological studies on Zixoryn RGD: 28043
10. Vezekényi,Zs.,Jávor,T.,Kovács,Á. / 1983 / Some possibilities
    of application the new selective enzyme inductor Zixoryn
    capsule in internal medicine RGD: 28058

BETA-ADRENOCEPTOR BLOCKING DRUGS AND CYTOCHROME P-450:
THE INFLUENCE OF PROPRANOLOL TREATMENT ON CAFFEINE-,
ANTIPYRINE-, D-GLUCARIC ACID AND INDOCYANINE GREEN
ELIMINATION IN PATIENTS WITH HEPATIC DISORDERS

Traeger, A., T. Horváth, C. Rechenbach, T. Jávor

1st Department of Medicine, University Medical School Pécs,
Hungary, and Department of Clinical Pharmacology, Institute
of Pharmacology and Toxicology, Friedrich Schiller University Jena, G.D.R.

INTRODUCTION

In the beginning of an alcoholic liver disease there is
often an increase in drug metabolizing activity caused by
the induction effect of alcohol. As the survival time of
patients with liver diseases is increasing the incidence of
other diseases is a high risk. Therefore it is of great interest to know the biotransformation capacity of the
patients.
Beta-adrenoceptor blocking drugs are used not only in cardio-
vascular diseases but also in the treatment of portal hyper-
tension.
We used caffeine as a marker drug for the 3-methylcholanth-
rene inducible subtypes of cytochrome P-450, and antipyrine,
which represents the phenobarbital inducible subtypes. D-
glucaric acid excretion as an endogenous compound which
served as an indicator of induced state of drug metabolizing
capacity of the liver was measured as well as indocyanine
green elimination as test material for liver blood flow
limited substances.

MATERIAL AND METHODS

10 alcoholic patients of both sexes, 35 to 60 years old,
were studied. In their history there was no indication of
any earlier drug treatment. 6 patients without liver disorders were taken as control group.
After initial investigation of the test parameters the patients got for 10 days 80 mg propranolol daily divided into
two applications. Thereafter the test parameters were measured again.

Caffeine was determined by HPLC according to Bonati (1982)
after 200 mg caffeine perorally and blood sampling 2 to 8
hours later.
Antipyrine was determined according to Brodie (1949) after
oral application of 20 mg/kg b.w. antipyrine and blood sampling 3 to 12 hours later.
D-glucaric acid was determined according to Marsh (1963) in
the 24 hours collected urine.

<u>Indocyanine green</u> disappearance from blood was measured 0.5 to 10 min. after 0.3 mg/kg b.w.

<u>Mathematical treatment</u>

The kinetic parameters were calculated according to a one-compartment model. There are given arithmetic means and standard error. Statistical differences were calculated by the Student's t-test or the paired t-test. The limit of significance was set 5 %.

RESULTS AND DISCUSSION

The results are given in table 1.

Table 1: Metabolic parameters ($\bar{x} \pm s_{\bar{x}}$) before (I) and after (II) 10 days treatment with propranolol 80 mg daily. ( ) = number of patients. $\triangle$= paired differences. + = significant, Student's t-test. $*$= significant, paired t-test.

	alcoholic liver disease			without hepatic disease		
	I (ml/min)	II (ml/min)	$\triangle$ (%)	I (ml/min)	II (ml/min)	$\triangle$ (%)
Antipyrine clearance (ml/min)	47.5 ± 8.5 (9)	34.3 ± 3.1 (9)	−14.7 ±13.1 (9)	35.4 ± 5.4 (6)	33.8 ± 6.5 (6)	−31.9 ±11.4 (6)
Caffeine clearance (ml/min)	55.9 ± 7.6 (6)	71.6 ± 7.7 (6)	+32.0$*$ ±10.3 (6)	109.7 (2)	118.1 (2)	+ 6.5 (2)
D-Glucaric acid excretion (µM/die)	101.4 ±16.1 (10)	122.9 ±16.7 (10)	+52.2 ±34.1 (10)	120.2 ±13.1 (6)	73.4$^+$ ±11.9 (6)	−40.0$*$ ± 7.7 (6)
Indocyanine-green clearance (ml/min)	4601 ± 526 (10)	3650 ± 372 (10)	−14.4 ±10.8 (10)	not determined		

Evidently in alcoholic liver patients propranolol has different effects on antipyrine- and caffeineelimination. While after 10 days propranolol treatment the clearance of antipyrine is slightly diminished, the clearance of caffeine is significantly enhanced. D-glucaric acid excretion is slightly enhanced in liver patients, significantly diminished in the control group without liver diseases. The clearance of ICG which only was determined in liver patients showed a small decrease as the consequence of 10 days propranolol treatment.

The effect of beta-adrenoceptor blocking drugs on drug metabolism was first reported in animal experiments and was correlated to the lipidsolubility of the drugs and their binding affinity to cytochrome P-450 by Facino and Lanzani 1979. Greenblatt et al. 1978, Conrad et Nyman 1980 and Bax et al. 1981 showed the effect of high therapeutic doses in men.

Doses less than 100 mg propranolol per day inhibited the
oxidation of diazepam but not the conjugation of lorazepam
(Ochs and Greenblatt 1984), did not influence the elimination of antipyrine after 3 weeks but after 3 months of treatment (Adamska et al. 1983). Markiewicz et al. (1984) observed in acute experiments after 80 mg propranolol in 3 patients an increase in the metabolic clearance rate of antipyrine.
In a dosage of 80 mg propranolol per day over 10 days in our
investigation the parameters antipyrine as a marker of the
phenobarbital inducible subtype of Cytochrome P-450 and
caffeine as a marker of the 3-MC inducible subtype were differently influenced. While the elimination of antipyrine was
slightly reduced, the elimination of caffeine was significantly increased, which could be caused by a different
effect of beta-adrenoceptor blocking drugs on both subtypes
of cytochrome P-450.
The excretion of glucaric acid was slightly increased in
correlation to the increase in caffeine clearance in patients with liver disease, but significantly decreased in
the control patients in whom the influence of propranolol on
caffeine clearance is still under investigation. Possibly
there exists a different influence of the drug on drug metabolizing activity depending on the activity stage of the
mixedfunctional oxidase system of the liver. In early stages
of alcoholic liver diseases often an induction of biotransformation reactions is to be seen (Lieber 1982, Horvath 1983).
Conrad and Nyman (1980) registered different effects of propranolol on theophylline kinetics in smokers and nonsmokers
what could be caused by the same fact.
As ICG clearance in our experiments by 80 mg propranolol is
reduced to a very small percentage only, a reduced liver
blood flow had not be taken into consideration.
In further investigations in men as well as in animal experiments we continue to investigate the influence of different beta-adrenoceptor blockers in different patient groups
and on different subspecies of cytochrome P-450.

LITERATURE

1. Adamska-Dyniewska, H.; Pruszczynski, J.; Wichan, P. (1983)
8th Congr. Pol. Pharmacol. Soc. Warsaw;
Antipyrineelimination in patients treated with ß-blockers

2. Bax, N.D.S.; Lennard, M.S.; Tucker, G.T. (1981) Br. J.
Clin. Pharmac. 12, 779 Inhibition of antipyrine metabolism by ß-adrenoceptor antagonists

3. Bonati, M.; Latini, R.; Galletti, F.; Young, J.F.; Togno-
ni, G.; Garattini, S. (1982) Clin. Pharmacol. Ther. 32,
98 Caffeine disposition after oral doses.

4. Brodie, B.B.; Axelrod, J.; Soberman, R.; Levy, B.B. (1949)
J. biol. Chem. 179, 25 The estimation of antipyrine in
biological materials.

5. Conrad, K.A.; Nyman, D.W. (1980) Clin. Pharmacol. Ther. 28, 463 Effects of metoprolol and propranolol on theo-phylline elimination.

6. Facino, R.M.; Lanzani, R. (1979) Pharmacol. Res. Comm. 11, 433 Interaction of a series of ß-adrenergic blocking drugs with rat hepatic microsomal monooxygenase.

7. Horváth,T.; Pár, A.; Berő, T.; Kádas, I.; Fábián, Cs.; Jávor, T. (1983) Acta Med. Hungar. 40, 203 Correlation between biochemical tests, parameters of drug elimina-tion and hepatic enzyme induction in chronic liver diseases.

8. Lieber, C.S. (1982) The liver annual, ed. Arias, J.M.; Frenkel, M.; Wison, J.H.P. Excerpta medica Amsterdam, p. 92.

9. Markiewicz, A.; Hartleb, M.; Rudzki, K.; Boldys, H. (1984) 9th Symp. Clin. Pharmac. Szeged, Time dependent effect of propranolol on hepatic extraction of 99mTc-HIDA and antipyrine clearance in healthy man.

10. Marsh, C.A. (1963) Biochem. J. 86, 77 Metabolism of D-glucuronolactone in mammalian systems.

11. Ochs, H.R.; Greenblatt, D.J. (1984) Clin. Pharmacol. Ther. 35, 263 Propranolol impairs diazepam oxidation but not lorazepam conjugation.

EFFECT OF TOBANUM ON HEPATIC MICROSOMAL ENZYME ACTIVITIES IN RATS

Vas,Á.,Gachályi,B.,Tihanyi,K.

Postgraduate Medical School,I.Department of
Medicine,Budapest P.O.B.112.,1389 Hungary

It is well known that beta-adrenoceptor blocking agents have
influence on the activity of the hepatic microsomal monooxygenase enzyme system.Animal studies proved that propranolol
/Hermansen,1969/ and alprenolol /Peters,1972/ inhibit hepatic microsomal monooxygenase activity.Propranolol and metoprolol /Bax et al.,1981/ as well as labetalol /Daneshmend
and Roberts,1982/ have similar effects in man.
We studied the effect of cloranolol /Tobanum,Gedeon Richter
Chemical Works,Ltd.,Budapest,Hungary/,a new non-selective
beta-blocking agent on the hepatic microsomal enzyme activities in rats.

Methods and material

Female Wistar rats weighing 180-220 g were used for the determination of hepatic microsomal monooxygenase activities.
Tobanum was administered as an aquous suspension in a daily dose of 40 mg/kg bw. for 5 days via gastric tube.Controls
received distilled water.Microsomes were prepared by the
ultracentrifugation method.Protein content was 25-30 mg/ml.
The cytochrome P-450 content was measured according to Omura and Sato /1964/.The cytochrome b_5 content was determined
from the oxydized-reduced difference spectrum /Omura and
Sato,1964/.The reaction mixture for the determination of
N-demethylase activity consisted of 0.5mM NADPH,10mM gluco-
se-phosphate,0.5U/ml glucose-6-phosphate dehydrogenase,10mM
Na-pirophosphate,5mM $MgCl_2$,8mM amidazophen in 0.1M pH 7.5
Tris-HCl buffer.Microsomal protein content was 0.5-0.8 mg/ml.
Formaldehyde originated from the reaction was measured according to Nash /1953/.The aniline hydroxylase activity was
determined under similar condition only the substance being
4mM aniline.The p-aminophenol originated was measured by the
method of Imai et al. /1966/.The NADPH-cytochrome c /P-450/
reductase activity was determined in 0.1M pH 7.8 phosphate
buffer containing 100 microM cytochrome c and 0.1mM KCN.
Microsomal protein content was 0.06-0.08 mg/ml.The reaction
was started with the addition of 0.1mM NADPH and followed

photometrically on wavelength 550 nm.The protein contents
were measured by the method of Lowry et al. /1951/.
In another set of experiments Wistar rats of both sexes,
weighing 150-230 g were used.The animals received 60mg/kg
bw. Tobanum in aquous suspension daily for 14 days via gastric tube.Controls received saline.24 hours after the last dose all animals were given 100 mg/kg bw. antipyrine ip. and
were killed.The plasma half-life of antipyrine was calculated from the determination of antipyrine concentrations according to Brodie et al. /1949/.For the determination of
hexobarbitone sleeping time Wistar rats of both sexes were
pretreated with 60 mg/kg bw. Tobanum suspension via gastric
tube for 14 days.Controls received saline.At the end of treatment all animals were given 40 mg/kg bw. hexobarbitone iv.
Sleeping time was measured.Some of these rats were treated
for another week's period with 60 mg/kg bw. Tobanum and then
bile secretion rate was measured in 1.8 g/kg bw. urethane
narcose.In an acute experiment Wistar rats of both sexes
weighing 220-310 g received a single dose of 60 mg/kg bw.
Tobanum and one hour later hexobarbitone sleeping time was
measured.

Results

Tobanum significantly decreased the hepatic microsomal cytochrome P-450 content but no other changes were observed
/Table 1./.

Table 1. Effect of Tobanum on the hepatic microsomal cytochrome contents and enzymatic activities in rats

	Control/n=10/ $\bar{x}$+SD	Treated/n=11/	P
cytochrome P-450 /nmol/mg protein/	0.384 ± 0.056	0.309 ± 0.071	p<0.05
cytochrome b$_5$ /nmol/mg protein/	0.411 ± 0.089	0.368 ± 0.099	N.S.
N-demethylase /nmol HCHO/min/mg protein/	4.21 ± 0.67	4.38 ± 0.65	N.S.
aniline hydroxylase /nmol p-aminophenol/ min/mg protein/	0.289 ± 0.081	0.286 ± 0.075	N.S.
NADPH-cytochrome c reductase /nmol red. cyto.c/min/mg protein/	79.2 ± 19.8	76.0 ± 19.0	N.S.

N.S.: not significant

There were no significant changes in the half-life of anti-
pyrine.hexobarbitone sleeping time and bile flow rate after
Tobanum treatment /Table 2./.

Table 2. Effect of Tobanum on the half-life of antipyrine, hexobarbitone sleeping time and bile flow rate in rats

	Control $\bar{x} \pm SD$ / n /	Treated	P
Antipyrine half-life /hour/	3.12 ± 0.6 /36/	3.21 ± 0.6 /44/	N.S.
Hexobarbitone sleeping time /min/ Female	21.2 ± 5.5 /11/	19.6 ± 6.1 /8/	N.S.
Male	11.9 ± 2.6 /8/	11.4 ± 3.7 /8/	N.S.
Hexobarbitone sleeping time, acute experiment /min/ Female	20.0 ± 4.5	26.3 ± 7.5	N.S.
Male	7.0 ± 2.0 /5/	8.3 ± 2.3 /5/	N.S.
Bile flow rate /ml/hour/	0.578 ± 0.14 /5/	0.578 ± 0.1 /5/	N.S.

N.S.: not significant

Discussion

In the present study cloranolol, a new non-selective beta-adrenoceptor blocking agent significantly decreased the hepatic microsomal cytochrome P-450 content in rats but failed to exert any effect on microsomal enzyme activities. Changes in cytochrome P-450 content do not directly reflect hepatic microsomal enzyme activities.

References

Bax,N.D.S.,Lennard,M.S.,Tucker,G.T./1981/ Inhibition of antipyrine metabolism by beta-adrenoceptor antagonists. Br.J.clin.Pharmacol. 12: 779-784.
Brodie,B.B.,Axelrod,J.,Soberman,R.,Levy,B.B./1949/ The estimation of antipyrine in biological materials. J.Biol.Chem. 179: 25-29.

Daneshmend,T.K.,Roberts,C.J.C./1982/ The short term effects
    of propranolol,atenolol and labetalol on antipyrine
    kinetics. Br.J.clin.Pharmacol. 13: 817-820.
Hermansen,K./1969/ Effect of different beta-adrenergic recep-
    tor blocking agents on hexobarbital induced narcosis in
    mice. Acta Pharmacol.Toxicol./Copenhagen/ 27: 453-460.
Imai,Y.,Ito,A.,Sato,R./1966/ Evidence for biochemically dif-
    ferent types of vesicles in the hepatic microsomal frac-
    tion. J.Biochem. 60: 417-428.
Lowry,O.H.,Rosebrough,N.J.,Farr,A.L.,Randall,R.J./1951/
    Protein measurement with the Folin phenol reagent.
    J.Biol.Chem. 193: 265-275.
Nash,T./1953/ The colorimetric estimation of formaldehyde by
    means of the Hantzsch reaction. Biochem.J. 55: 416-421.
Omura,T.,Sato,R./1964/ The carbon monoxide-binding pigment
    of liver microsomes.I.Evidence for its hemoprotein natu-
    re. J.Biol.Chem. 239: 2370-2378.
Peters,M.A./1972/ Possible mechanism/s/ of alprenolol /beta
    adrenergic receptor blocking agent/ prolongation of pen-
    tobarbital hypnosis in mice. J.Pharmacol.Exptl.Therap.
    181: 417-4°4.

INDUCTION OF CYTOCHROME P-450 LINKED MONOOXYGENASES BY THE SALTS OF Co, Cd AND Zn

M.Kadiiska, Ts.Stoytchev, L.Krustev[*]

Institute of Physiology,Institute of Morphology[*]
Bulgarian Academy of Sciences
1113 Sofia, Bulgaria

It is well known that different inhibitors of drug metabolism exert an enzyme-inducing effect when applied in
smaller doses or after chronic and subchronic treatment(Serro
ne,Fujamoto 1962.,Kato et al.1964).In our previous experiments we established that 24h after subcutaneous injection of
high toxic doses the salts of Co,Cu,Cd,Pb and Ni inhibited
the activity of rat liver monooxygenases(Kadiiska,Stoytchev
1980) while after subchronic treatment with subtoxic doses
the salts of Co,Cd and Zn increased the activity of ethylmor-
phine-N-demethylase and the cyt.P-450 content(Stoytchev,Ka-
diiska 1983).In further experiments we studied the effect of
different heavy metal salts on the activity of 5-aminolevuli-
nic acid(ALA)synthetase,heme oxygenase and NADPH-dependent
lipid peroxidation which are related to the synthesis,degra-
dation and stability of cyt.P-450(Kadiiska et al.1984,1985).
    The aim of the present work is to summerize the effects
of the salts of Co,Cd and Zn on cyt.P-450 related enzyme activities in the rat liver,to elucidate the mechanisms involved
in the induction processes and to asses whether or not the
ultrastructural morphological changes in the rat liver correlate with the induction of cyt.P-450 dependent monooxygenases.
    Materials and methods
All experiments were carried out on male Wistar albino rats.
At the beggining of the experiment their mean weight was
70 ± 20g and at the end-180 ± 15g.Heavy metal salts were given
with the drinking water for 30 days in daily doses as follows:
$Co(NO_3)_2$-20mg/kg, $CdSO_4$-10mg/kg, $ZnSO_4$-100mg/kg body wt.Twen-
ty four hours after the last intake of the metal salts was
followed the duration of hexobarbital (Hb)sleeping time/hexo-
barbital-100mg/kg body wt.i.p./and the following enzyme
assays were carried out:N-demethylation of ethylmorphine(EMD)
and benzphetamine(BND) as substrates - by the method of Nash
/1953/for estimation of formaldehyde production;the cyt.P-450
content-by Omura and Sato/1964/;microsomal heme content - by
the method of Paul et al./1953/;the activity of ALA-syntheta-
se-according to Marver et al./1966/;heme oxygenase activity-
by Tenhunen et al./1968/;NADPH-dependent lipid peroxidation-
by Porter et al/1976/.In all experiments the protein content
was determined with the Biuret method.All results were

statistically processed by the Student "t" test.The morphological investigations were carried on liver tissue through
light and electron microscopy.For light microscopy the classical histological techniques were performed.For electron microscopy the tissue was fixed in Glutaraldehyde,emmbeded in
Durcupan,contrasted with uranylacetate and lead citrate/Krus-
tev 1971/and studied on Opton 9 and Hitachi HS-7S.

Results

The effects of heavy metal salts after 30 day treatment with
the salts of Co,Cd and Zn on the duration of Hb sleeping time
and cyt.P-450 linked monooxygenases is shown in Table 1.

Table 1.

	Controls	$Co(NO_3)_2$	$CdSO_4$	$ZnSO_4$
Hb sleeping time/min/	$27,1\pm0,6$	$18,1\pm1,1$	$20,5\pm1,8$	$17,5\pm1,9$
EMD /nmoles/mg/min/	$3,59\pm0,05$	$4,11\pm0,07$	$4,03\pm0,07$	$3,99\pm0,03$
BND /nmoles/mg/min/	$1,45\pm0,24$	$3,10\pm0,38$	$2,18\pm0,17$	$2,39\pm0,23$
Cyt.P-450 /nmoles/mg/	$0,671\pm0,04$	$0,828\pm0,03$	$0,851\pm0,03$	$0,890\pm0,07$
Heme /nmoles/mg/	$0,895\pm0,07$	$1,180\pm0,05$	$0,897\pm0,03$	$1,448\pm0,01$
ALA-synthetase /μgALA/h/g.liver/	$23,6\pm0,67$	$27,4\pm0,90$	$34,7\pm0,99$	$29,2\pm0,50$
Heme oxygenase /nmoles/mg/min/	$1,74\pm0,11$	$1,12\pm0,18$	$1,20\pm0,13$	$1,18\pm0,13$

It was established that the salts of Co,Cd and Zn significantly shorten the Hb sleeping time,increased the activity of EMD
and BND,increased the cyt.P-450 and heme content/except for
$CdSO_4$ which did not change significantly the content of heme/,
increased the activity of ALA-synthetase and decreased that
of heme oxygenase.

In Fig.1 is shown the effect of heavy metal salts studied
on NADPH-dependent lipid peroxidation.We found that the salts
of Co,Cd and Zn decreased significantly lipid peroxidation.

Ultrastructurally,the changes of the liver cell organelles
were not always quite the same.However as far as the main
structure of drug biotransformation is concerned namely the
endoplasmic reticulum/ER/,several points are to be mentioned.
First of all is the diminution of the granular ER/GER/in the
perinuclear reagion of the hepatocyte.Although in variance,
but generally,in all other parts of the cytoplasm the GER is
well expressed.There are some portions of 5-10 cysterns with
preserved attachment of the ribosomes.At the same time in
other locations there are signs of degranulations ,vacuolisa-
tion etc./Fig.2/.As far as the smooth ER/SER/has to be taken
into account,the general picture is that of good preservation
even hyperthrophy.

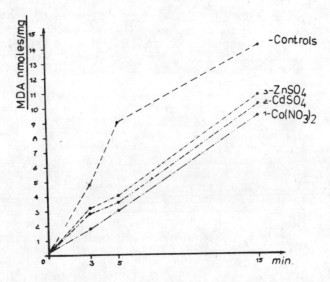

Fig.1.Effect of heavy metal salts on NADPH-
dependent lipid peroxidation.

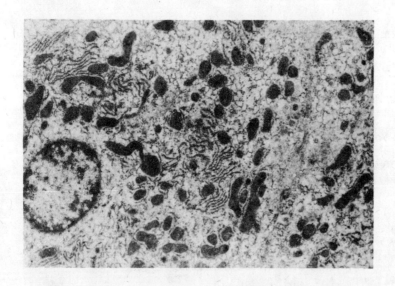

Fig.2.Ultrastructural changes in the rat
liver after treatment with $ZnSO_4$.

Discussion

Our results show that subtoxic doses of Co,Cd and Zn given
daily with the drinking water for 30 days produce neither
behavioral changes nor changes in the body weight of the
animals as compared with the control group.The shortening of

the Hb sleeping time,the increased activities of EMD and BND
as well as cyt.P-450 content and microsomal heme suggest an
enzyme-inducing action of these metal salts.We established
that the salts of Co,Cd and Zn increased the activity of ALA-
synthetase-the rate limiting enzyme in the heme biosynthesis
pathway,and decreased the activity of heme oxygenase-the
first enzyme in the breakdown of heme.These data suggest that
the increased cyt.P-450 content is due both to the increased
synthesis and to the decreased breakdown of this hemoprotein
/Kadiiska et al.,1985/.Our results show that these metal salts
decrease NADPH-dependent lipid peroxidation.Probably this
effect is one of the mechanisms for the decreased breakdown
of cyt.P-450 and for the increased activity of the mixed
function oxydising enzyme systems.When trying to put in correlation the enzyme/and biochemical in general/investigations
with the data of electron microscopy some problems shoorly
arise not only in this particular case.Anyway,the good preservation and even hyperthrophy of the SER once more correlates
with the functional performance of the liver cell.Not so well
is everything with the GER.Although abundant in the periphery
of the cell,in other places it is diminished and with signs
of degranulation and desorganisation.Even if we find the link
between the latter with the process of subcellular reorgani-
zation,there still are some minor structural facts which on
this stage of our knowledge are not perfectly claimed to be
understood.

References
Kadiiska M,Stoytchev Ts(1980)Effect of acute intoxication
with some heavy metals on drug metabolism.Arch.Toxicol.,
Suppl. 4, 363-365
Kadiiska M,Stoytchev Ts,Serbinova E(1984)Influence of mul-
tiple application of some heavy metals on NADPH-dependent
lipid peroxidation.Eksperim.med.i morf. 3,123-126
Kadiiska M,Stoytchev Ts,Serbinova E(1985)On the mechanism
of the enzyme-inducing action of some heavy metal salts
Arch.Toxicol 56, 167-169
Kato R,Chiesara E,Vassaneli P(1964)Further studies on the
inhibition and stimulation of microsomal drug metabolizing
enzymes of rat liver by various compounds Biochem.Pharmacol
13,69-83
Krustev L(1971)Lisosomi Raichev R.,ed,BAN,Sofia,pp 41-46
Marver HS,Tschudy D,Perloth M,Collins A(1966)Aminolevulinic
acid synthetase.Studies on liver homogenates J.Biol Chem
241,2803-2809.
Nash T(1953)The colorimetric estimation of formaldehyde by
means of Hantzsch reaction J.Biol.Chem 55,416-421.
Paul KG,Teorell H ,Akeson A(1953)The molar light absorption
of pyridine ferroprotoporphyrin Acta Chem Scand 7,1284-1287.
Serrone DM,Fujumoto JM(1962)The effect of certain inhibitors
in producing shortening of hexobarbital action Biochem.
Pharmacol 11,609-615.
Stoytchev Ts,Kadiiska MB(1983)Heavy metals and enzyme
induction Acta Physiol et Pharmacol Bulg 9,n.1,29-34.
Tenhunen R,Marver H,Schmid R(1968)The enzymatic conversion
of heme to bilirubin by microsomal heme oxygenase
Biochemistry,61,748-755.

INTERACTION BETWEEN IMIPRAMINE AND HALOPERIDOL INVOLVES ENZYMATIC COMPETITION

DANIEL, W., MELZACKA, M.

Institute of Pharmacology, Polish Academy of Sciences, 31-343 Kraków, Poland

A joint administration of neuroleptics and tricyclic anti-depressants (TAD) concomitantly has been recommended in the treatment of patients who exibit depressive and schizoaffec-tive symptoms (Davis et al. 1970; Del Zompo et al. 1984). It was found that such combination of drugs induced an increase in plasma and brain levels of TAD in both man and the rat (Gram et al. 1974; Fuller et al. 1974; Bock et al. 1983), moreover, acc. to investigators, those pharmacokinetic changes were due to the inhibition of hydroxylation and/or glucurona-tion of TAD by concurrent administration of neuroleptics (Gram et al. 1974).

The main step in TAD biotransformation of the tertiary amines structure in rat liver microsomes is not hydroxylation but N-demethylation (Bickel, 1971; Gigon and Bickel, 1971), which could also be affected by concurrent administration of neuroleptics. As far as we know, there are no literature data concerning the effect of neuroleptics on the N-demethylase activity, therefore, we have carried out a study of imipramine (IMI) demethylase activity in the liver microsomes of rats treated concurrently with IMI and haloperidol (HAL) for two weeks. Simultaneously with the enzymatic study, we investigat-ed the IMI pharmacokinetics in the same rats, in order to find whether concurrent administration of HAL affected the level of IMI and its metabolite desipramine (DMI) in the central ner-vous system (CNS) of rats.

Methods

The experiments were carried out on male Wistar rats (250-300 g), kept under standard laboratory conditions at a fixed 12:12 h light/dark cycle. The animals were fed on a standard granulated diet (Bacutil) until 18 h before the experiment, and had free access to tap water. They received orally (with drinking water) either IMI, in a dose of 20 mg/kg/day, or HAL (0.2 mg/kg/day), or a combination of both drugs i.e. IMI (20 mg/kg/day) and HAL (0.2 mg/kg/day). The drugs were administer-ed for two weeks. The control animals received water.

Livers were excised 24 h after the withdrawal of drugs. Microsomes were prepared acc. to the conventional methodology by a differential centrifugation in TRIS/KCl buffer (pH = 7.4).

The cytochrome P_{450} level was assayed acc. to Omura and Sato (1964). Protein was assessed acc. to Lowry et al. (1951). The in vitro study of IMI metabolism was carried out acc. to Daniel et al. (1984). The amount of IMI used as a substrate was 200 nM/ml. The mixture was incubated for 25 min at 37°C. Formaldehyde was measured acc. to Nash (1953). In order to prepare Lineweaver-Burk plots the following amounts of IMI were used for in vitro demethylation: 40, 50, 60, 100, 200 and 300 nM/ml. The incubation was performed either in the absence or in the presence of 100 or 300 nM/ml of HAL.

Pharmacokinetic studies were carried out 1 or 24 h after the withdrawal of IMI and HAL. IMI and its metabolite DMI were assayed spectrofluorometrically acc. to Dingell et al. (1964).

The results were eleborated statistically using the Duncan or Student t-tests.

Results

HAL alone, given to rats for two weeks in a dose of 0.2 mg/kg/day p.o., elevated the level of cytochrome P_{450} in rat liver microsomes, however, the effect was not statistically significant. IMI, in a dose of 20 mg/kg/day p.o., increased the level of cytochrome P_{450} in a statistically significant manner. When IMI and HAL were given jointly to rats for two weeks, the level of microsomal cytochrome P_{450} increased more markedly than after chronic IMI or HAL (Table 1).

Prolonged administration of HAL to rats increased the rate of IMI demethylation in rat liver microsomes, while IMI (14 x 20 mg/kg/day p.o.) showed a tendency to inhibit IMI demethylase activity; concurrent administration of both drugs resulted in a significant acceleration of IMI demethylation, in comparison with the group treated with IMI alone (Table 1). The relative liver weight was nearly the same in all investigated groups of animals.

The Lineweaver-Burk plots which illustrate the relationship between 1/v and 1/s when different amounts of IMI (40-300 nM/ml) were incubated with 100 or 300 nM/ml of HAL indicated competition between IMI and HAL (straight lines are characterized with the same value of 1/v for 1/s = 0) (Fig. 1).

The brain level of IMI in rats treated with IMI (14 x 20 mg/kg/day p.o.) and Hal (14 x 0.2 mg/kg/day p.o.) 1 h after withdrawal of drugs, was elevated in a statistically significant manner when compared with IMI treated animals, and was followed by decrease in the ratio of the brain concentration of DMI and IMI. When the assessement of IMI and DMI level in the rat brain was performed 24 h after the withdrawal of IMI and HAL, neither the brain level of IMI and DMI, nor the ratio of the DMI/IMI brain concentration differed significantly from the control (Table 2).

Discussion

The elevated brain level of IMI followed by the decreased ratio of the DMI/IMI brain concentration in rats treated with IMI and HAL for two weeks, suggested inhibition of the IMI demethylase activity by a concurrent administration of the neuroleptic. Our results of enzymatic studies indicate the

Table 1. The level of cytochrome P_{450} in rat liver microsomes and the rate of IMI in vitro demethylation in rats treated with IMI (14 x 20 mg/kg/day p.o.), HAL (14 x 0.2 mg/kg/day p.o.), and jointly with IMI and HAL. The results are the mean $\pm$ SEM of 6 rats. Statistical significance: *p 0.05; **p / 0.01 in comparison with water treated animals (cytochrome P_{450}) or IMI (14 x 20 mg/kg/day p.o.) treated animals (demethylase activity) (Duncan test)

Group of animals	cytochrome P_{450} (nM.mg protein^{-1})	formaldehyde formation from IMI (nM.mg protein^{-1}.30 min^{-1})
Control	0.4709 $\pm$ 0.0346	9.64 $\pm$ 0.84
IMI	0.6870 $\pm$ 0.0536*	7.88 $\pm$ 1.66
IMI + HAL	0.8072 $\pm$ 0.0247**	12.89 $\pm$ 1.99*

Fig. 1. Kinetics of the inhibition of IMI demethylation by HAL in vitro experiments. V = velocity of formaldehyde formation from IMI (M.mg protein^{-1}.30 min^{-1}.10^{-8}); s=concentration of IMI in the incubation mixture (M.ml^{-1}.10^{-7}). 1 - control; 2 - IMI + 100 nM of HAL; 3 - IMI + 300 nM of HAL.

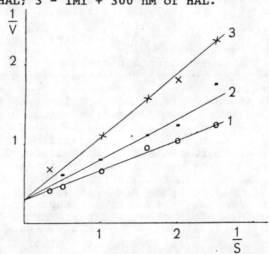

Table 2. The brain level of IMI and DMI and the ratio of DMI/ IMI brain level in rats treated with IMI (14 x 20 mg/kg/day p.o.) and HAL (14 x 0.2 mg/kg/day p.o.) 1 and 24 h after the withdrawal of IMI and HAL. The results are the mean $\pm$ SEM of 5 rats. Statistical significance *p 0.05, in comparison with IMI (14 x 20 mg/kg/day p.o.) treated animals (Student t-test).

	concentration (μg/g)					
	IMI		DMI		DMI/IMI	
	1 h	24 h	1 h	24 h	1 h	24 h
Control (IMI)	2.48$\pm$0.20	0.44$\pm$0.13	4.52$\pm$0.25	2.72$\pm$0.13	1.82	6.18
IMI + HAL	4.47$\pm$0.50*	0.49$\pm$0.19	3.54$\pm$0.61	2.38$\pm$0.44	0.79	4.85

inhibition did not proceed via hindrance of IMI demethylase activity in the rat liver microsomes, since prolonged admi-nistration of IMI and HAL to rats accelerated the rate of IMI demethylation in in vitro experiments. Those results were contrary to our pharmacokinetic data. However, the microsomes used for enzymatic studies were devoid of IMI and HAL, as the method of microsome preparation excluded the presence of pre-viously administered drugs, and the above mentioned pharmaco-kinetic studies were carried out on rats killed 1 h after the withdrawal of IMI and HAL, i.e. at the time when the level of both drugs in rats was very high (Lewi et al. 1970; Daniel et al. 1982). In order to simulate the in vivo situation in in vitro experiments, IMI and HAL were added jointly to the in-cubation mixture, the former as a substrate of the demethyla-tion reaction. That procedure resulted in inhibition of IMI demethylation, and the Lineweaver-Burk plots indicated that the inhibition proceeded via a competitive mechanism.

Interestingly, when pharmacokinetic studies of IMI and DMI were conducted 24 h after the withdrawal of IMI and HAL, i.e. at the time when the concentration of HAL in rats was low (Lewi et al. 1970), and the concentration of IMI still high (Daniel et al. 1982) (due to the differences in their $t_{0.5}$ values and different doses used in the experiments), the brain level of IMI and DMI was nearly the same as in control animals. This confirms our assumption that the inhibition of IMI demethylation by HAL requires the presence of a high con-centration of HAL in rats.

Another problem is the mechanism of acceleration of IMI demethylation in in vitro experiments by a concurrent ad-ministration of HAL to rats. It was found before (Daniel et al. 1984). and confirmed by the present results that IMI given to rats for two weeks increased the microsomal level of cyto-chrome P_{450} and showed a tendency to inhibit its own demethy-lation. HAL alone, when given to rats for two weeks, slightly increased the level of cytochrome P_{450} in rat liver microsomes and accelerated in some degree the IMI demethylation. When both drugs were administered jointly to rats for two weeks, the procedure resulted in a higher increase in the cytochrome P_{450} level than in the rats receiving only IMI or HAL, and in a significant acceleration of the IMI demethylation in compari-son with the IMI treated animals.

In our earlier experiments (Daniel and Melzacka 1985) it was found that IMI demethylation proceeded via two different oxygenase systems, one dependent and the other independent of cytochrome P_{450}. On the basis of the present results it is difficult to predict which system is activated by HAL. The higher elevation of cytochrome P_{450} level induced by concur-rent administration of IMI and HAL in comparison with IMI or HAL treated animals, followed by acceleration of the IMI de-methylation, suggested activation of the cytochrome P_{450}-de-pendent oxygenase system; however, an alteration of the acti-vity of another system, independent of cytochrome P_{450} cannot be excluded.

Apart from these assumptions, it is obvious that pharmaco-kinetic changes of IMI, induced by a concurrent administration of HAL were mainly due to an enzymatic competition between

these two drugs, and not to the changes of IMI demethylase
activity. However, it does not exclude the involvement of
other enzymatic changes in the discussed pharmacokinetic alteration such as inhibition of hydroxylation and/or glucuronation
of IMI by HAL, as it was proposed by Gram et al. (1974).

References

Bickel, M.H. (1971) Xenobiotica 1 : 313-319
Bock, J.L., Nelson, J.C., Gray, S., Jatlow, P.I. (1983) Clin.
    Pharmacol. Ther. 33: 322-328
Daniel, W., Adamus, A., Melzacka, M., Szymura, J. (1982) J.
    Pharm. Pharmacol. 34: 678-680
Daniel, W., Friebertshauser, J., Steffen, C. (1984) Naunyn-
    Schmiedeberg's Arch. Pharmacol. 328: 83-86
Daniel, W., Melzacka, M. (1985) J. Pharm. Pharmacol. in press
Davis, J.M., Klerman, G.L., Schildkraut, J.J. (1970) in:
    Psychopharmacology. A review of progress. Ed. D.H. Efron,
    Washinghton Public Health Service Publication. p. 1957-1967
Del Zompo, M., Ferini-Strambi, L., Martis, G., Corsini, G.U.
    (1984) Proceedings of 14th CINP Congress, Florence, S502
Dingell, J.V., Sulser, F., Gillette, J.R. (1964) J. Pharma-
    col. Exp. Ther. 143: 14-22
Fuller, R.W., Snoddy, H.D., Slater, I.H. (1974) Toxicol. Appl.
    Pharmacol. 29: 259-269
Gigon, P.L., Bickel, M.H. (1971) Biochem. Pharmacol. 20: 1921-
    1931
Gram, L.F., Christiansen, J., Fredricson Overo, K. (1974)
    Acta Pharmacol. Toxicol. 35: 223-232
Lewi, P.J., Heykants, J.J.P., Allewijn, F.T.N., Dony, J.G.H.,
    Janssen, P.A.J. (1970) Arzneimittel-Forsch. 20: 943-948
Lowry, O.H., Rosebrough, N.J., Farr, A.L., Randall, R.J. (1951)
    J. Biol. Chem. 193: 265-275
Nash, T. (1953) Biochem. J. 55: 416-421
Omura, T., Sato, R. (1964) J. Biol. Chem. 239: 2370-2378.

THE EFFECT OF INDUCERS ON α-AMANITIN POISONED MICE

KLEBOVICH,I., KIRÁLY,Á. AND VERECZKEY,L.

Chemical Works of Gedeon Richter Ltd., H-1475
Budapest, Hungary

Introduction

It is well known that the ingestion of toadstool (amanita phalloides) meal is accompanied by severe kidney and liver lesions leading to death in the majority of cases. Whereas kidney impairment can be effectively treated by the modern therapy there is no therapy for the liver and thus toadstool poisoning generally results in death 4-6 days after the ingestion due to hepatic coma. As the main poison material of the toadstool, ∠-amanitin (Wieland and Wieland, 1972) inhibits the RNA-synthesis of the cells (Stirpe and Fiume, 1967) and the inducers of the cytochrome P-450 system act through de novo RNA and protein synthesis it seemed to be reasonable for us to investigate the possibility of their protective effect against ∠-amanitin poisoning.

We have investigated the action of phenobarbitone, one of the most widely used inducers in both animal experiments and in human therapy (Trolle, 1968) and that of flumecinol (Zixoryn[R], 3-trifluoromethyl-∠-ethyl benzhydrol)[x] developed for the treatment of jaundice of neonates and proved to induce the cytochrome P-450 system (Szeberényi et al. 1978, Gachályi et al. 1978). Flumecinol has no sedative effect and

---

[x]Manufacturer: Chemical Works of Gedeon Richter Ltd., Budapest, Hungary

its half-life is shorter than that of phenobarbitone
(Klebovich et al. 1982), nevertheless it brings about the
phenobarbitone-type induction.

Materials

α-amanitin was the product of Serva (Heidelberg, GFR).
Phenobarbitone-Na and Tween 80 were the products of Reanal
(Budapest, Hungary).

Animal treatment

Male CFLP mice weighing 20 g were administered i.p.
0.4 mg/kg of α-amanitin in physiological saline solution.
After 24 hours the mice received phenobarbitone-Na (Mw: 254)
in distilled water or flumecinol (Mw: 280) in 5 % Tween 80
suspension in a single oral dose of either 143 or 286
μmol/kg, corresponding to 36.3 or 72.6 mg/kg doses of phenobarbitone-Na and 40 and 80 mg/kg doses of flumecinol. Every
group contained ten mice. The control group (I) received
α-amanitin only.

Results and discussion

Table 1. shows the mortality of mice treated i.p. with
0.4 mg/kg α-amanitin (I) as well as the mortality of mice
treated with α-amanitin and two doses of phenobarbitone-Na
(II,III) and of flumecinol (IV,V). It can be seen that all
mice have died in every group between 4-7 days.

Our results show that there is no difference in the
mortality between the phenobarbitone-treated, flumecinol-
-treated and control animals, suggesting that inducers cannot
be used in the treatment of amanitin poisoning. Our results
also show that the inhibition of RNA synthesis brought about
by α-amanitin cannot be counteracted by inducers like
phenobarbitone or flumecinol.

293

Table 1.

Mortality of mice treated with 0.4 mg/kg i.p. α-amanitin alone (control) and α-amanitin and two doses of phenobarbitone-Na or flumecinol p.o.

Mortality

Days	I. Control	II. Phenobarbitone-Na 143 μmol/kg	III. Phenobarbitone-Na 286 μmol/kg	IV. Flumecinol 143 μmol/kg	V. Flumecinol 286 μmol/kg
1	–	–	–	–	–
2.	–	–	–	–	–
3.	–	–	–	–	–
4.	4	4	4	3	4
5.	5	4	4	4	5
6.	1	1	1	1	–
7.	–	1	1	2	1

Groups

Acknowledgements

Thanks are due to Mrs.I.Kókai for her excellent technical assistance.

References

Gachályi,B., Káldor,A. and Szeberényi,Sz., (1978): "m-trifluoromethyl-α-ethyl benzhydrol: A new enzyme inducer". Europ.J.Clin.Pharmacol., 13, 299-302.

Klebovich,I., Vereczkey,L., Kerpel-Fronius,S., Ringwald,G. and Bodrogligeti,I., (1982): "Pharmacokinetic studies of flumecinol in man and dog". Eur.J.Drug Metab. Pharmacokin., 7, 99-104.

Stirpe,F. and Fiume,L. (1967): "Studies on the pathogenesis of liver necrosis by α-amanitin. Effect of α-amanitin on ribonucleic acid synthesis and on ribonucleic acid polymerase in mouse liver nuclei". Biochem.J., 105, 779-782.

Szeberényi,Sz., Pálosi,É. and Szporny,L. (1978): "Effects of 3-trifluoromethyl-α -ethyl benzhydrol (RGH-3332), a new enzyme inducer, on the microsomal drug metabolism". Arzneim.-Forsch. (Drug Res.), 28, 663-668.

Trolle,D. (1968): "Decrease of total serumbilirubin concentration in newborn infants after phenobarbitone treatment". The Lancet, 28, 705-708.

Wieland,T. and Wieland,O. (1972): "The toxic Peptides of amanita species". in Microbial Toxins, Vol 8. (Kadis,S., Cigler,A. and Alj,S., eds.), Academic Press, New York, pp 249-280.

INDUCTION OF CYTOCHROME P-450 BY DIETHYLAMIDE OF NICOTINIC ACID

Bushma,M.I., Lukienko,P.I., Nikitin,V.S.

Division of Metabolic Regulation Byelorussian SSR Academy of Sciences, 50 Lenin Komsomol Boulevard, Grodno 230009, USSR

ABSTRACT

The experiments on male albino rats have shown that administration of diethylamide of nicotinic acid (DANA) per os at a dose of 73 mg/kg for 45 days provokes hepatomegaly, proliferation of the smooth endoplasmic reticulum in hepatocytes an increase in the microsomal fraction release a rise in the cytochrome P-450 content and in the rates of ethylmorphine demethylation and aniline hydroxylation by 34.7, 109.8, 40.6, 43.2, 50.0, 61.7, respectively. The spectral magnitude of cytochrome P-450 binding with aniline and DANA does not change under the effect of the latter drug, while the interaction of the enzyme with ethylmorphine decreases by 63.0%. The rate of the recovery of the cytochrome P-450--ethylmorphine complex increases 3-fold.

It is assumed that the increase in the content of cytochrome and in the rate of xenobiotic hydroxylation in microsomes after the prolonged administration of DANA may be regarded as substrate induction in nature, whereas the hepatomegaly and proliferation of the endoplasmic reticulum are considered as an adaptation of the liver cell directed to an enhancement of the liver hydroxylating function.

Diethylamide of nicotinic acid (DANA) can considerably stimulate activity of liver UDP-glucuronosyl transferase (EC 2.4.1.17) that catalyzes conjugation of many exogenous and endogenous substances with glucuronic acid (Skakun, 1976). Since the products of the cytochrome P-450-catalyzed reactions aften serve as substrates for UDP-glucuronosyl transferase, it has been suggested that DANA causes induction of the hemoprotein.

Methods

Thirty-two male albino rats weighing about 80-130 g were used in these studies. DANA was administered intragastrically at a dose of 73 mg/kg daily for 45 days. Control animals received the same volume of water. The rats were killed by decapitation 24 hours after the last administration of the drug.

After the livers from one of the groups of animals had been perfused with a cold 1.15% KCL solution, the microsomal fraction was isolated (Karuzina, Archakov, 1972) in which we determined the content of cytochrome P-450 (Omura, Sato, 1964), the activities of N-demethylation of ethylmorphine (type I hydroxylation substrate) (Nash, 1973) and p-hydroxylation of aniline (type II substrate) (Kato, Gillette, 1965). The binding capacity of cytochrome P-450 with the substrates (2 mM) was found using the magnitudes of the spectral changes in the enzyme-substrate complex. The overall catalytic activity of the complex and the rate of the recovery of the cytochrome P-450-substrate complex (Kato et al., 1970) were estimated from the rations of the substrate oxidation rates to the cytochrome P-450 content and to the magnitudes of the spectral changes. The protein concentration was measured according to Lowry et al. (1951).

Livers of rats from the other group were not perfused. The relative weight of the liver (g/100 g body weight), the yield of the microsomal fraction were determined and the ultrastructure of hepatocytes studied.

Results and Discussion

The content of cytochrome P-450, the rates of ethylmorphine N-demethylation and aniline p-hydroxylation were increased by 43.2, 50.0 and 61.7%, respectively in the liver microsomal fraction of rats administered with DANA (Table).

In calculating the value of metabolic rate per 1 nmol of cytochrome P-450, we found that the overall catalytic activity was not essentially changed both with respect to ethylmorphine N-dealkylation and aniline p-hydroxylation in experimental animals.

However, the study on the interaction between the substrates hydroxylated and cytochrome P-450 enabled us to find a difference in the mechanisms of activation of their metabolism. The binding of ethylmorphine to cytochrome was found to decrease in the experimental animals both in calculating per 1 nmol of P-450 (by 62.8%) and per 1 nmol of microsomal protein (by 47.1%). At the same time the rate of the cytochrome P-450-ethylmorphine complex recovery raised approximately 3-fold the control values. The spectral magnitude of enzyme binding to aniline and DANA was increased insignificantly when calculated per 1 nmol of cytochrome P-450 and considerably (by 60.3 and 74.3%) in estimating per 1 mg of microsomal protein. The intensity of the rate of the recovery of the cytochrome P-450-aniline complex did not differ from the control values (Table).

Table. Effect of DANA administration (per os, for 45 days) on the cytochrome P-450 content, the spectral magnitude of enzyme binding to substrates and the rate of their metabolism in rat liver microsomes

Index	Control	DANA	%	P
Cytochrome P-450, nmol/mg	0.74 ± 0.07	1.06 ± 0.09	143.24	<0.02
Ethylmorphyne-cytochrome P-450, $\Delta O.D.\times10^3$/mg,	7.89 ± 1.39	4.17 ± 0.75	52.85	<0.05
$\Delta O.D.\times10^3$/nmol P-450	10.70 ± 1.52	3.98 ± 0.78	37.20	<0.002
Ethylmorphine N-demethylation, nmol/min/mg,	9.60 ± 1.20	14.40 ± 1.20	150.00	<0.05
nmol/min/nmol P-450,	12.97 ± 1.62	13.58 ± 1.13	104.70	>0.5
nmol/min/$\Delta O.D.\times10^3$	1.22 ± 0.15	3.45 ± 0.19	282.79	<0.002
Aniline-cytochrome P-450, $\Delta O.D.\times10^3$/mg,	17.31 ± 1.61	27.74 ± 2.93	161.85	<0.01
$\Delta O.D.\times10^3$/nmol P-450	22.85 ± 1.91	26.57 ± 1.68	116.28	>0.1
Aniline p-hydroxylation, nmol/min/mg,	0.69 ± 0.08	1.11 ± 0.10	161.73	<0.01
nmol/min/nmol P-450, nmol/min/$\Delta O.D.\times10^3$	0.93 ± 0.14	1.05 ± 0.09	112.90	>0.25
DANA-cytochrome P-450, $\Delta O.D.\times10^3$/mg,	17.34 ± 1.19	30.42 ± 2.46	175.43	<0.001
$\Delta O.D.\times10^3$/nmol P-450	23.77 ± 2.13	28.13 ± 2.60	118.34	>0.25

The deficiency of the ethylmorphine binding sites in the cytochrome P-450 molecule makes it difficult to unequivocally interpret the increase in its N-demethylation. We may only suggest that the rise in ethylmorphine metabolism under the effect of DANA occurred due to the increase in turnover of the reactions of its interaction with cytochrome P-450. The acceleration of aniline p-hydroxylation is likely to be related to the elevation of the cytochrome P-450 microsomal level.

Enhancing the activity of the microsomal hydroxylating function, the DANA administration provoked hepatomegaly (34.7% increase) and an increase in the microsomal fraction release (by 40.6% from 1 g of liver tissue). However, the specific content of microsomal protein remained unchanged.

The electron microscopic studies of hepatocytes from the control animals showed considerable proliferation of membranes of the smooth endoplasmic reticulum. Their fraction was increased 2-fold in comparison with the control values

(0.172 as opposed to the control value of 0.082, p∠0.001). the high degree of polymorphism was characteristic of the smooth endoplasmic reticulum. The increase in its volume was related to the more tight localization of the elements and the distribution in the cytoplasm.

The hepatomegaly, proliferation of the smooth endoplasmic reticulum in hepatocytes, the increase in the activity of the hydroxylating system, which is localized in it, after the long-term administration of DANA were probably due to the substrate induction of the systems of xenobiotic metabolism. To a greater extent this concerns the systems which interact with type II substrates. This suggestion has been confirmed by our data on DANA showing the capacity to bind to cytochrome P-450 according to type II (K_s = 0.59 mM) and the evidence of its metabolism (N-oxidation) in the liver hydroxylase system (Cowan et al., 1978). Since DANA is type II substrate, one can suggest that its prolonged administration causes induction of cytochrome P-450 which mainly interacts with type II substrates.

The data obtained enable one to assume that prolonged administration of DANA in clinic and out-patient clinic may increase the hydroxylating function of the liver and, consequently, change the metabolism, therapeutic effect and toxicity of not only DANA itself but also other substances which undergo biotransformations in the monooxygenase system.

REFERENCES

Cowan D.A., Damani L.A., Corrod J.W.: Metabolic N-Oxidation 3-Substituted Pyridines: Identification of Products by Mass Spectrometry. Biomed. Mass Spectrometr. 5, 551-556.

Karuzina I.I., Archakov A.I.: Isolation of liver microsomal fraction and characterization of its oxidative systems. In: Modern Methods in Biochemistry (V.N. Orekhovich, ed.), pp. 49-63. Moscow 1977

Kato R., Gillette J.R.: Effect of starvation on NADPH-dependent enzymes in liver microsomes of male and female rats. Sex differences in the effects of abnormal physiological states on the metabolism of drugs by rat liver microsomes. J. Pharmacol. exp. Therap. 150, 279-291 (1965)

Kato R., Takanaka A., Takahashi A.: Effect of Thyroid Hormone on the Substrate Interaction with P-450 in the Oxidation of Drugs by Liver Microsomes. J. Biochem. 68, 613-623 (1970)

Lowry O.H., Rosebrough N.J., Farr A. et al.: Protein measurement with the folin phenol reagent. J. Biol. Chem. 193, 265-275 (1951)

Nash T.: Colorimetric estimation of formaldehyde by means of the Hantzsch reaction. Biochem. J. 55, 416-421 (1953)

Omura T., Sato R.: The carbon Monoxide-binding Pigment of Liver microsomes. II Solubilisation, Purification and Properties. J. Biol. Chem. 230, 2378-2385 (1964)

Skakun N.P. The Fundamentals of Pharmacogenetics, pp. 107-111. Kiev 1976

MONOOXYGENASE ACTIVITY OF MICE HEPATOCYTES AND IMMUNOCYTES
IN DYNAMICS OF DEVELOPMENT OF THE IMMUNE RESPONSE AND UNDER
THE EFFECT OF IMMUNOMODULATORS

N.Ya.Golovenko, B.N.Galkin

Physico-Chemical Institute after A.V.Bogatsky Academy of
Sciences of the Ukrainian SSR Odessa, USSR

Immune reactions are functionally connected with the processes of
microsomal oxidation involving liver cytochrome P-450 (Kovalev, 1977).
This hemoprotein takes part both in xenobiotics oxidation and in
synthesis and metabolism of different endogenous regulators (steroids,
prostaglandins, leucotrienes, etc.) which in their turn regulate
immunologic processes. Monooxygenases of immunocompetent organs and cells
are also able to take part in realization and maintenance of chemical
homeostasis of organism.

The present work is aimed at the study of the change of the
activity of monooxygenase systems of isolated immunocytes and hepatocytes
in the dynamics of development of immune response and under the effect of
immunomodulators which may prove the participation of these enzymes in
immunity regulation.

Isolated cells of thymus, spleen and liver were used as well as
peritoneal macrophages. The activities of monooxygenase reactions were
estimated in them (dimethylaniline N-demethylation and benz(a) pyrene
hydroxylation) and the content of cell protein and nucleic acids was
determined. To determine enzymatic activities NADPH-generating system was
used. Corpuscle antigen (sheep erythrocytes) was administered at
immunogenic dose, intraperitoneally. The rates of monooxygenase reactions
were determined daily following the immunization up to the appearance of
the maximum of antigens in synthesis (4th day) and then on the days of
altering one immunoglobulin class by another one (7th-14th days) as well
as on the day of the end of primary immune response (21st day) and in
one and in four days after the secondary immunization (22nd and 25th
days, respectively). Fluorene-9-one and 9,10-anthraquinone derivatives
were administered at the dose 50 mg/kg of the body weight intraperitoneally, three times. Experiments were carried out the day after the
last injection. Common bean lectine (PHA) was applied once at the dose of
15 mg/kg of the body weight, intraperitoneally. Enzyme activity was
determined in 1, 3, 4 and 6 days after mitogen administration.

Experimental data obtained show that monooxygenase activity in macrophages decreases almost twice as large during the first day of immune
response development. At the same time it increases by 3 fold in thymus
cells to decrease to the norm the next day. The days followed antigen
administration the reduction of the activity of cytochrome P-450
dependent enzymes was observed in all immunocompetent cells its value
being 20-60 % of that of intact mice. Secondary immunization results in
the decrease of the activity in microphages and thymocytes as compared to

the level of the 21-st day (Golovenko et al.,1984). In isolated cells of
liver the activity of N-demethylase reaction was 1.5-2 times less than the
control during all the studied periods of primary response. Maximal
inhibition is observed on the day after the administration of sheep
erythrocytes. Meanwhile, the rate of benz(a)pyrene hydroxylation is much
higher than that of the intact animals. Maximal increase (by 75 %) of the
activity is registered on the first day. During the period of the
productive phase of antibody synthesis (4th-7th days) the activity of
N-demethylase somewhat increases approaching the control value. The rate
of benz(a)pyrene oxidation during this period is 1.6 times greater than
that of the control. It should be noted that by the 14th day, i.e., by
the moment of the considerable decrease of the level of the main class
antibodies - the IgG activity of the studied reactions increases
considerably. N-demethylase is activated by 33 %, while benz(a)hydroxylase
- 4 times. By the end of the primary response (21st day) the rates of
the studied reactions come back gradually to the norm. Secondary immune
response characterized by longer duration and more rapid increase of antibody concentration is accompanied by the decrease of N-demethylase
activity by 60 %. The character of the change on the 4th day after the
repeated immunization corresponds to those of the 14th day of the primary
immune response. As compared to the previous period of investigation the
rate of N-demethylase reaction increases 5.3 times and benz(a)pyrene
hydroxylase - 1.7 times (Galkin et al.,1983).

Fluorene-9-one and 9,10-anthraquinone derivatives exhibit different
effect on the activities of monooxygenase systems in the same cell
depending on the type of their effect on the immunity. Compounds
possessing immunostimulating properties (tilorone, IS-33, IS-22, IS-24)
intensify the activities of enzymes in thymus cells considerably - 5.6
times. In the rest immunocompetent cells of these preparations do not
practically change monooxygenase activity though in macrophages the
intensity of cytochrome P-450 dependent reactions is somewhat lowered.
Under the effect of IS-16 and IS-30 which inhibit the effect on the immune
system, the rate of benz(a)pyrene oxidation in splenocytes increases
almost 3 times. Somewhat less increase is observed at macrophages. In
thymocytes tendency towards the decrease of enzyme activity has been
observed (Bogatsky et al.,1983; Golovenko et al.,1984). In hepatocytes
under the effect of immunomodulators, the decrease in N-demethylase
activity is observed and the activation of benz(a)pyrene hydroxylase
takes place, providing that the compounds showed immunostimulating effect,
while under the effect of immunodepressants the rate of dimethylaniline
oxidation increases and the rate of benz(a)pyrene hydroxylation
decreases (Bqgatsky et al.,1981; Bogatsky et al.,1983; Filippova et al.,
1983).

The activity of monooxygenase enzymes after PHA administration
decreases in thymus cells on the first day reaching the least value
(45 % of the control level) on the 4th day. In splenocytes there was no
change in the rate of benz(a)pyrene and dimethylaniline oxidation under
the effect of lectine in vivo. In macrophages the activity of enzymes
increases twice in 24 hours, while on the fourth day it reduces
considerably as it was also observed in case of thymocytes. The sixth day
is characterized by the return of all the studied parameters to the level
of the intact values. In isolated hepatocytes during the first day after
mitogenic lectine administration the intensity of the studied reactions
is rather reduced. On the third day a 2.5-4 times activation of monooxygenase systems was observed. On the fourth day the rate of dimethylaniline N-demethylation is 1.6 times greater than the control while the

activity of benz(a)pyrene hydroxylase is on the level of the control. On the sixth day the activity of N-demethylase comes back to the control values, and the rate of benz(a)pyrene oxidation is reduced by 46 % (Golovenko et al., 1983).

By comparing the results of the effect of phytohemagglutinin on the different types of cells it should be noted that the studied parameters in hepatocytes and immunocompetent cells change, as a rule, in a contrary direction.

The study of the effect of immunomodulators of different nature on monooxygenase systems of immunocytes and hepatocytes allowed to find a number of regularities for revealing the mechanism of the effect of these compounds. Low molecular immunostimulators, like corpuscle antigen show similar effects on these systems in immunocompetent cells and in hepatocytes. We suggest that activation of cytochrome P-450 (P-448) dependent enzymes in thymus cells and the reducing of these reaction intensity in splenocytes and macrophages are necessary for stimulation and development of immune response. In hepatocytes monooxygenase reactions depending only on hemoprotein P-450 are also blocked.

R E F E R E N C E S

Bogatsky A.V., Galkin B.N., Golovenko N.Ya., Filippova T.O., Litvinova L.A.,(1981) Changes in the enzymic activity of monooxygenase components of rat hepatocytes with administration of tilorone. Ukr.Biokhim.J.,53, p.108-110.

Bogatsky A.V., Filippova T.O., Kovalev I.E., Andronati S.A., Golovenko N.Ya., Galkin B.N., Litvinova L.A. (1983) Benz(a)pyrene hydroxylase activity of immunocompetent cells. Bull.Exp.Biol.Med., 46, p.23-24.

Filippova T.O., Andronati S.A., Galkin B.N., Golovenko N.Ya., Shurduk Zh.N. (1983). Influence of low molecular immunomodulators on changes in monooxygenase activity of mice hepatocytes. Dokl.Akad.Nauk UkrSSR, N 1, p.71-73.

Galkin B.N., Filippova T.O., Golovenko N.Ya. (1983) Effect of high molecular immunomodulators on the activity of monooxygenases in mice liver tissue. Voprosy Med.Khimii, 29, p.60-63.

Golovenko N.Ya. (1981) Mechanism of the reactions of xenobiotics metabolism in biological membranes. Naukova Dumka, Kiev, 220 p.

Golovenko N.Ya., Galkin B.N., Filippova T.O. (1984) Characterization of monooxygenases system in the immunocompetent cells. Usp.Sovrem.Biol., 97, p.268-278.

Golovenko N.Ya., Galkin B.N., Filippova T.O., Vasilenko L.S. (1983) Interaction of lectins with the receptors of immunocyte and hepatocyte membranes as a factor regulating the activity of cytochrome P-450 dependent enzymes. In: Biological membranes: structure and function, Tashkent, p.148.

Golovenko N.Ya., Karaseva T.L. (1983) Comparative biochemistry of foreign compounds. Naukova Dumka, Kiev, 200 p.

Kovalev I.E. (1977) Immunity as a function of the body system inactivating heterologous chemical compounds. Khim.-Farm.Zh., 11, N 12, p.3-14.

EFFECTS OF ETHANOL ON CYT. P-450 MICROSOMAL MONOOXYGENASE AND VINYL CHLORIDE METABOLISM IN RATS

Justyna M. Wiśniewska-Knypl and Jan Kołakowski

Institute of Occupational Medicine, Lodz, Poland

INTRODUCTION

Hepatotoxic and carcinogenic properties of vinyl chloride (VC) depend on its biotransformation in the liver endoplasmic reticulum to an electrophilic epoxide metabolite mediated by cyt. P-450 monooxygenase with following coupling of the bio-derivative to non-protein sulfhydryls (Jaeger et al. 1974; Green and Hathway, 1977; Tarkowski et al. 1980). Stimulation of cyt. P-450 monooxygenase may lead to increased hepatotoxicity of VC due to stimulation of VC bioactivation (Jaeger et al. 1974; Tarkowski et al. 1980). Ethanol intake, due to inductive effects on cyt. P-450 monooxygenase (Lieber and DeCarli, 1974) may potentiate hepatotoxic (Miller et al. 1982) and carcinogenic (Radike et al. 1977) activity of VC.

This work was aiming of investigating chronic, combined effects of ethanol-VC interaction in relation to cyt. P-450 monooxygenase, VC metabolism and ultrastructural alteration in the liver.

MATERIALS AND METHODS

Animals and treatment. Experiments were carried out on male Wistar rats with initial body weight of 230-250 g, fed a laboratory chow Murigran (Bioveter. Indust., Gorzów Wlkp., Poland) and either tap water or 10 % (w/v) ethanol (EtOH) ad lib. The water- and EtOH-fed rats were concurrently exposed in 1 m^3 dynamic chambers either to air flow or 500 mg/m^3 VC vapour, 5 h daily, 5 days a week for 5 months according to schedule: air + water; air + EtOH; VC; EtOH + VC.

Microsomal monooxygenases. Aniline p-hydroxylase (E.C. 1.14.1.1) was measured in 9000 g supernatant of the liver (Wiśniewska-Knypl et al. 1980). Microsomal ethanol oxidizing system (MEOS) was determined with microsomes (Lieber and DeCarli, 1970). Microsomes were prepared by $CaCl_2$-aggregation method, protein was determined with Folin phenol reagent and cytochrome P-450 from CO-difference spectra of dithionite-reduced microsomes (cf. Wiśniewska-Knypl et al. 1980).

VC metabolism. Non-protein sulfhydryl content in the liver homogenates was determined with Ellman reagent and thiodiglycollic acid (TDGA), a metabolite of VC was determined in 24 h urine of rat by gas-chromatographic method (cf. Tarkowski

304

et al. 1980).
Ultrastructure of the livers, after standard procedure,
was examined in a JOEL JEM-100C electron microscope.
Student's t-test was used for statistics of the data.

RESULTS AND DISCUSSION

Chronic, 5-month treatment of rats with 10 % ethanol (daily
ethanol intake ~7 g/kg body wt.; average ethanol blood level
~40 mg%) resulted in the stimulation of microsomal monooxygenase activity - aniline p-hydroxylase and MEOS - with an
increase of cytochrome P-450 level in the liver (Table).
The stimulatory effects of ethanol were accompanied by proliferation of smooth- (SER) and degranulation of rough endoplasmic reticulum (RER) appearing as atypical, condensed mitochondria (Fig.1). The stimulation of aniline p-hydroxylase
and MEOS was maintained after 24 h since withdrawal of
ethanol but cytochrome P-450 level in microsomes decreased.
Chronic inhalatory exposure of rats to VC results in stimulation of activity of aniline p-hydroxylase without any
effects on MEOS and cytochrome P-450 (cf.Table and Wiśniewska-
Knypl et al. 1980). Metabolism of VC, as determined by TDGA
excretion in 24-h collected urine did not deplete the level
of non-protein sulfhydryls in the liver, most probably due to
the "rebound" mechanism and stimulated recovery of -SH group
by repeated exposure to VC (Tarkowski et al. 1980). Ultrastructural examination of the liver revealed moderate proliferation of SER similarly as in the group exposed to ethanol
(cf. Fig. 1).
Combined, chronic exposure of rats to ethanol and VC resulted in increased by 50 % excretion of TDGA with urine (Table);
the level of non-protein sulfhydryl in the liver - adopted as
an index of coupling of epoxide metabolite of VC - was unaffected although activity of microsomal monooxygenase in the
liver was still stimulated. Withdrawal of ethanol for 24 h
abolished the stimulatory effect of ethanol on TDGA excretion
with urine, but activity of microsomal monooxygenase was
still increased. Combined effect of ethanol and VC on liver
ultrastructure resulted in greater variation of the shape of
the condensed mitochondria and greater proliferation of SER,
in comparison with the group exposed to ethanol or VC exclu-
sively (Fig. 2). Proliferation of SER may be regarded as a
functional response to metabolic demands of VC and ethanol,
both of compounds metabolized in monooxygenase system in the
liver. Pathological alteration in mitochondria may be regarded
as a symptom of toxic response (cf. Miller et al. 1982) to
NAD/NADH and NADP/NADPH unbalance due to ethanol oxidation
followed by a decrease in the tricarboxylic acids cycle activity and generation of ATP.
Stimulatory effect of chronic ethanol intake, when coexists
with foreign chemical, VC (cf. Table), on microsomal monooxygenase and VC metabolism may potentiate the risk of carcinogenesis from the combined exposure (cf. Radike et al. 1977).

Acknowledgements. This study was done under the program
"Health and Social Problems of Alcohol Abuse" MR-II-22.

TABLE

EFFECT OF PROLONGED EXPOSURE OF RATS TO ETHANOL AND VC (SINGLY OR COMBINED) ON MICROSOMAL CYT. P-450 MONOOXYGENASE AND VC METABOLISM IN THE LIVER

Exposure		Microsomal monooxygenase nmol/mg microsomal protein			VC metabolism	
Vapour	Drink	Aniline p-hydroxylase	MEOS	Cytochrome P-450	Non-protein sulfhydryls /umol/g liver	TDGA /ug/24 h urine per 100 g rat
Air	Water	8.3 ± 0.4	4.4 ± 0.3	0.80 ± 0.07	7.5 ± 0.4	0
Air	EtOH (+)[1]	12.5 ± 0.3[c]	5.9 ± 0.3[a]	1.05 ± 0.08[a]	7.4 ± 0.1	0
Air	EtOH (-)[2]	11.9 ± 0.8[b]	5.5 ± 0.3[a]	0.95 ± 0.08	7.2 ± 0.4	0
VC	Water	11.5 ± 0.5[c]	4.8 ± 0.6	0.94 ± 0.10	7.1 ± 0.2	110.4 ± 8.1
VC	EtOH (+)	12.3 ± 0.3[c]	6.1 ± 0.7[a]	0.96 ± 0.08	7.4 ± 0.1	169.7 ± 12.8[b]
VC	EtOH (-)	9.9 ± 0.4[a]	6.6 ± 0.7[a]	0.90 ± 0.12	7.8 ± 0.2	132.2 ± 11.1

Regimen of exposure: 10 % ethanol in drinking water for 5 months; 500 mg/m^3 VC by inhalation, 5 h daily, 5 days a week for 5 months.
Microsomal monooxygenase and -SH groups level were determined immediately after termination of VC-exposure. TDGA was determined in 24-h urine collected thereafter.
1 - Ethanol-treated rats drank ethanol up to the day before death.
2 - Ethanol-treated rats drank water in place of ethanol only on the day before death.
Each group consisted of 8 animals. Results are the mean ± S.E.M.
a, b, c - Significantly different from air- and water-exposed group at $p = 0.05$, 0.01 and 0.001, respectively.

306

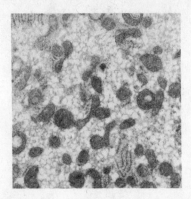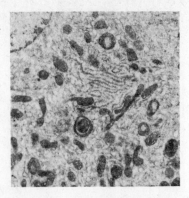

Fig. 1 (left). Hepatocyte of rat exposed to ethanol for
5 months. Variation in the shape of mitochondria (not seen in
the group exposed to VC only) and moderate proliferation of
SER (similarly as in the group exposed to VC only). X 6.600
Fig. 2 (right). Hepatocyte of rat subject to combined exposure
to ethanol and VC for 5 months. Variation in the shape of
condensed mitochondria and greater proliferation of SER
(in comparison with the group exposed to ethanol or VC exclu-
sively). X 5.000

REFERENCES

Green, T., and Hathway, D.E. (1977). The chemistry and bioge-
nesis of the S-containing metabolites of vinyl chloride in
rats. Chem.-Biol. Interact. 17, 137-150.
Jaeger, R.J., Reynolds, E.R., Conolly, R.B., Moslen, M.T.,
Szabo, S. and Murphy, S.D. (1974). Acute hepatic injury by
vinyl chloride in rats pretreated with phenobarbital.
Nature, 252, 724-726.
Lieber, C.S. and DeCarli, L.M. (1970). Hepatic microsomal
ethanol oxidizing system. In vitro characteristic and
adaptive properties in vivo. J. Biol. Chem. 245, 2505-2512.
Miller, M.L., Radike, M.J. Andringa, A. and Bingham, E.
(1982). Mitochondrial changes in hepatocytes of rats
chronically exposed to vinyl chloride and ethanol.
Environ. Res. 29, 272-279.
Radike, M.J., Stemmer, K.L., Brown, P.G., Larson, E. and
Bingham, E. (1977). Effect of ethanol and vinyl chloride on
the induction of liver tumors: Preliminary report.
Environ. Health Persp. 21, 153-155.
Tarkowski, S., Wiśniewska-Knypl, J.M., Klimczak, J.,
Dramiński, W. and Wróblewska, K. (1980). Assessment of
urinary excretion of thiodiglycollic acid and hepatic free
thiols content in rats at different levels of exposure to
vinyl chloride. J. Hyg. Epidemiol. Microbiol. Immunol. 24,
253-261.
Wiśniewska-Knypl, J.M., Klimczak, J. and Kołakowski, J.
(1980). Monooxygenase activity and ultrastructure changes
of liver in the course of chronic exposure of rats to
vinyl chloride. Int. Arch. Occup. Environ. Health 46,
241-249.

INTERACTION OF ETHANOL WITH CARBON DISULPHIDE IN MICROSOMAL MONOOXYGENASE SYSTEM AND ITS BIOLOGICAL CONSEQUENCES

Teresa Wrońska-Nofer, Justyna Wiśniewska-Knypl,
Jan Stetkiewicz
Institute of Occupational Medicine, Lodz, Poland

INTRODUCTION

Toxic effects of the chemicals often appear as the result of their metabolic activation in the microsomal cyt. P-450 system. Ethanol has been found to stimulate the activity of the microsomal monooxygenases (Rubin and Lieber, 1968; Villeneuve et al. 1976) and in consequences potentiates the toxicity of the chemicals activated in this system (as reviewed by Strubelt, 1980). CS_2 undergoing biotransformation in cyt. P-450 system might be one of the case.

The present study was undertaken to investigate whether chronic ethanol intake enhances the depressive effect of CS_2 on the cyt. P-450 enzymes and whether, via this mechanism, ethanol may alter the susceptibility of animals to toxic action of CS_2.

MATERIAL AND METHODS

Male Wistar rats (230-250 g body weight) fed on a laboratory chow "Murigran" and ethanol (10 % aqueous solution) or water ad lib. inhaled in toxicological chamber air containing CS_2 vapour at conc. 0,4 and 1,5 g/m^3 for 6 months, 5 days a week, 5 h daily. Reference groups of rats drinking water or ethanol concurrently were kept in toxicological chamber inhaling clean air. At the end of 6th month of exposure, directly after termination of exposure to CS_2 and after ethanol withdrawing, biochemical and morphological examinations were carried out. Microsomal cyt. P-450 enzymes, duration of hexobarbital sleeping time and lipid peroxidation were measured as described previously (Klimczak et al. 1984). Microsomal ethanol oxidizing system (MEOS) was tested acc. to Lieber and DeCarli (1970). Total- and lipoprotein-cholesterol was estimated with Boehringer Mannheim tests.

Ultrastructure of the liver after standard preparation was examined under JEOL JEM-100C electron microscope.

Statistics: Student's t-test was used and p-values smaller than 0.05 were considered at significant.

RESULTS AND DISCUSSION

The results of the present study are consistent with the earlier findings which indicated that CS_2 inhibited the liver cyt. P-450 monooxygenases and induced loss of P-450 hemoproteins (Bond and DeMatteis, 1969; Torres et al. 1981).

21*

The decrease in microsomal monooxygenase activity and
depression of cyt. P-450 level found in CS_2 exposure, both at
concentration of 0,4 and 1,5 g/m³, were paralleled with the
minor lesion of hepatocytes - local degranulation of rough
andoplasmic reticulum (RER) observed in electron microscope
(Table 1). Prolonged ethanol consumption potentiated the
depressive effect of CS_2 on the activity of monooxygenases
in the liver with the decrease in cyt. P-450 level and its
conversion into catalytically inactive cyt. P-420 (Table 1).
This alteration was accompanied by ultrastructural lesion of
hepatocytes: moderate increase in the amount of SER induced
by ethanol, degranulation of RER and the occurence of giant
mitochondria (Fig. 2). An underlying mechanism for the ethanol potentiation may be related to its inducing or activating
effect on microsomal enzymes. Ethanol has generally been
known to stimulate the activity of liver microsomal enzymes
(Rubin and Lieber, 1968; Villeneuve et al. 1976). In this
study the induction of liver microsomal cyt. P-450 enzymes
accompanied by proliferation of SER, segmentary degranulation
of RER and distortion in shape of some mitochondria has been
shown after prolonged consumption of 10 % solution of ethanol
(Table 1, Fig. 1).

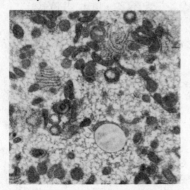 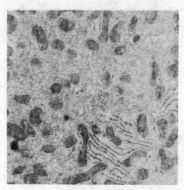

Fig. 1 (left). Hepatocyte of rat exposed to ethanol for
6 months. Marked polymorphism of mitochondria and decreased
RER amount and increased SER. X 5.000
Fig. 2 (right). Hepatocyte of rat exposed jointly to ethanol
and CS_2 for 6 months. Moderate proliferation of SER comparing
to ethanol fed animals. X 6.600

The present study revealed that ethanol, analogically to
phenobarbital treatment increased the biotransformation of
CS_2. CS_2 undergoes metabolic activation via cyt. P-450 to
toxic derivatives which leads to the destruction of the apoprotein moiety of cyt. P-450. Thus cyt. P-450 "suicidally"
destroys itself by converting CS_2 to derivative(s) which
combine covalently with apocytochrome P-450 and cause its
destruction. An additional consequence of chronic ethanol -
CS_2 interaction leading to biochemical disturbances in hepatocytes may be assumed to be the process of lipid peroxidation, stimulated by CS_2 exposure and aggravated in combined
exposure (Table 1). It was also found, that the rise of

TABLE 1
MICROSOMAL MONOOXYGENASES, CYTOCHROME P-450 AND LIPID PEROXIDATION IN THE LIVER MICROSOMES OF RATS JOINTLY EXPOSED TO CS_2 AND ETHANOL

Groups	CS_2 conc. g/m^3	Aniline p-hydroxylase nmol/mg protein/15min	MEOS nmol/mg protein/min	Hexobarbital[1] sleeping time min	Cyt. P-450 nmol/mg protein	Lipid peroxidation /umol MDA/mg protein
Control	-	13.4 ± 0.5	6.3 ± 0.4	35.8 ± 3.5	0.75 ± 0.035	2.0 ± 0.1
CS_2	0.4	5.5 ± 0.3^c	-	70.8 ± 5.5^c	0.55 ± 0.03	-
	1.5	3.9 ± 0.3^c	4.2 ± 0.3^b	120.0 ± 10.0^c	0.51 ± 0.02	3.2 ± 0.3^a
EtOH	-	27.1 ± 1.6^c	10.7 ± 0.5^c	44.8 ± 1.8	0.95 ± 0.04	2.8 ± 0.2
CS_2 + EtOH	0.4	3.9 ± 0.3^{cA}	-	89.0 ± 9.3^c	0.50 ± 0.02	-
	1.5	2.8 ± 0.2^{cA}	2.9 ± 0.2^c	151.0 ± 13.0^c	0.43 ± 0.02	4.2 ± 0.7^a

The rats were exposed concurrently to CS_2 and ethanol (10 % solution in drinking water) for 6 months. Tests were performed directly after termination of CS_2 exposure.
Figures are the mean $\pm$ S.E. for 8-10 rats.
a, b, c - significantly different from the control at $p = 0.05$, 0.01 and 0.001, respectively.
A - significantly different from the CS_2-exposed group at $p = 0.05$

[1] Challenging dose: 125 mg/kg intraperitoneally.
MDA - malondialdehyde

HDL-cholesterol produced by ethanol consumption underwent regression in case of combined CS_2 and ethanol exposure (Table 2). This might be associated with the inhibitory effect of CS_2 on the microsomal enzyme system, a site of HDL-cholesterol production.

TABLE 2

CHOLESTEROL LEVEL IN PLASMA AND LIPOPROTEINS IN RATS JOINTLY EXPOSED TO CS_2 (0.4 mg/m^3) AND ETHANOL

Groups	Total cholesterol mg/dl	HDL-cholest. mg/dl	HDL-cholest. VLDL+LDL cholest.
Control	63.0 ± 14	38.1 ± 5.9	1.6
EtOH	77.0 ± 7.6[a]	54.2 ± 6.2[c]	1.1
CS_2	78.2 ± 10[a]	33.0 ± 5.4	2.3
EtOH + CS_2	69.3 ± 11	42.1 ± 8.5	1.8

Figures are the mean ± S.D. for eight rats.

Acknowledgements. This study was done under the program "Health and Social Problems of Alcohol Abuse" MR-II-22.

REFERENCES

Bond, E.J. and DeMatteis, F. (1969). Biochemical changes in rat liver after administration of carbon disulphide, with particular reference to microsomal changes. Biochem. Pharmacol. 18, 2531-2549.

Klimczak, J., Wiśniewska-Knypl, J.M. and Kołakowski, J. (1984). Stimulation of lipid peroxidation and heme oxygenase activity with inhibition of cytochrome P-450 monooxygenase in the liver of rats repeatedly exposed to cadmium. Toxicology 32, 267-276.

Lieber, C.S., DeCarli, L.M. (1970). Hepatic microsomal ethanol-oxidizing system. In vitro characteristic and adaptive properties in vivo. J.Biol. Chem. 245, 2505-2512.

Rubin, E., Lieber, C.S. (1968). Hepatic microsomal enzymes in man and rat: Induction and inhibition by ethanol. Science 162, 690-691.

Strubelt, O. (1980). Interactions between ethanol and other hepatotoxic agents. Biochem. Pharmacol. 29, 1445-1449.

Torres, M., Jarvisalo, J., Hakim, J. (1981). Effect of in vivo carbon disulfide administration on the hepatic monooxygenase systems of microsomes from rats. Exp. Molec. Pathol. 35, 36-41.

EFFECT OF CYANIDANOL (CATERGEN®) ON HEPATIC DRUG BIOTRANSFORMATION AND ELIMINATION IN PATIENTS WITH CHRONIC ALCOHOLIC HEPATITIS AND CIRRHOSIS

Pár,A., T.Horváth, T.Beró, I.Kádas, F.Pakodi, I.Wittmann and T.Jávor
First Department of Medicine, University Medical School Pécs, County Hospital Baranya, Pécs, Hungary

There are some controversies regarding the effect of the flavonid substance, cyanidanol (3,4,5,7-tetrahydroxy-flavan-3-ol) on the hepatic drug metabolism. While Steffen et al (1981) have found cyanidanol to inhibit the microsomal mixed function mono-oxidase activity, Kovách et al (1982) reported on enzyme induction by the drug in patients with liver disease. The aim of the present study was to determine how cyanidanol influences the hepatic microsomal and mitochondrial enzyme functions (hydroxylation, glucuronide production, D-glucaric acid excretion and acetylation) as well as indocyanine green elimination in patients with chronic alcoholic hepatitis and cirrhosis.

PATIENTS AND METHODS

Fourteen patients with biopsy-proven chronic alcoholic liver disease were studied. Eight had alcoholic hepatitis and six had cirrhosis. Their age ranged between 3o and 53 (mean 4o). Previously they had not been treated with any drug. The diagnosis was compatible with their previous alcohol consumption and was estabilished according to the results of biochemical liver function tests and percutaneous liver biopsy.

To evaluate the hepatic drug metabolism capacity the following methods were used:
 1. antipyrine kinetics: metabolic clearance rate (MCR) was determined after oral antipyrine, 2o mg/kg (Brodie et al, 1949) for assessment of hydroxylation activity;

 2. menthol-loading test was carried out after 2 g menthol orally (Bitter and Muir,1962) for determination of glucuronide production;

 3. urinary D-glucaric acid excretion (Marsh,1963) was measured as an index of "enzyme induction state" and of glucuronide metabolism;

 4. sulphadimidine kinetics: the ratio of SD/acetylated SD and total body clearance were determined after oral and i.v. sulphadimidine (Evans,1965) to study the mitochondrial

acetylation function

5. <u>indocyanine green elimination</u> (ICG clearance) was measured according to Gilmore et al (1982) using o,3 mg/kg
Ujoviridin (Germed,DDR).

<u>Treatment shedule</u> was the following: after the initial exa-
minations patients were treated with cyanidanol (CATERGEN,
Zyma licence, BIOGAL,Debrecen) orally at a daily dose of
3,o g. The first treatment period lasted for lo days, then
liver function tests and drug metabolism studies were repeated. After it the treatment continued for a further 3o
days with the same dose of cyanidanol. Following the second
period, patients were studied again. Cirrhotic patients
were treated only for lo days.
<u>Statistical analysis</u> was done by using Student's one sample
t-test

RESULTS

<u>Biochemical liver function</u> tests improved significantly
after the first lo-day treatment period, later a slight
worsening occured in some cases (Fig.l.).

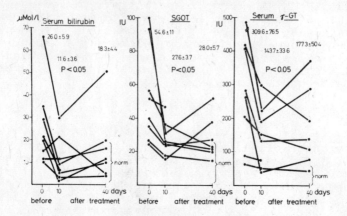

Fig.l. Bilirubin, GOT, GGT changes in chronic alcoholic
       hepatitis during CATERGEN treatment

<u>Antipyrine</u> metabolic clearance rate (MCR) decreased after
lo days, later it remained unchanged on this level (Fig.2.)

<u>Menthol-glucuronide</u> production and urinary <u>D-glucaric acid</u>
excretion markedly reduced after the two courses of cyanidanol teratment (Fig.3.).

<u>Sulphadimidine</u> kinetics studies (Fig.4.) revealed only
slight changes after cyanidanol. Although in the oral test
an increase in the SD/acetylated SD ratio was found, this
change could not be regarded as of biological significance.

<u>Indocyanine green clearance</u> increased in the half of the
cirrhotic patients and decreased on the other half(Tabl.1).

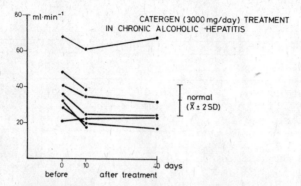

Fig.2. Antipyrine metabolic clearance rate changes in chronic alcoholic hepatitis patients

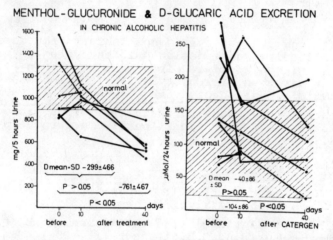

Fig.3. Menthol-glucuronide and urinary D-glucaric acid excretion during cyanidanol treatment in alcoholic hepatitis

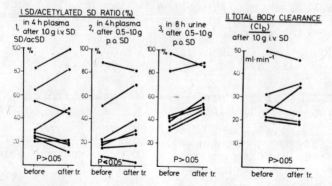

Fig.4. Sulphadimidine kinetics studies in chronic alcoholic hepatitis

Tabl.1.

Antipyrine MCR, D-glucaric acid excretion and ICG clearance
in alcoholic cirrhosis during cyanidanol treatment

Pat.s	Antipyrin MCR ml/min		D-glucaric acid excr. umol/24 h.		ICG clearance ml/min	
	before	after	before	after	before	after
No.1.	28,49	23,52	159,3	281,6	5o5,1	8o1,2
No.2.	15,81	27,46	264,6	154,5	4o59,o	1775,5
No.3.	22,34	22,83	98,8	136,o	880,4	814,9
No.4.	21,79	12,57	363,6	127,2	1599,2	32o8,3
No.5.	37,57	14,25	152,7	9o,3	1o39,5	541,4
No.6.	1o,oo	15,71	2o1,9	121,4	1673,6	1714,o

CONCLUSION

Our results suggest that cyanidanol inhibits the activity
of the microsomal mixed function oxidase system in patients
with chronic alcoholic liver disease. At the same time,
however, the mitochondrial acetylation function is not influenced by this hepatoprotective flavonoid.

ICG clearance studies suggest that liver blood flow of
cirrhotic patients is not significantly influenced by
cyanidanol.

REFERENCES

1. Bitter,T., Muir,H.M. (1962): A modified uronic acid
   carbazole reaction. Analyt.Biochem. 4, 33o

2. Brodie,B.B. et al (1949): The estimation of antipyrine in
   biological materials. J.Biol.Chem. 179, 25

3. Evans,D.A.P. (1965): An improved and simplified method of
   detecting the acetylation phenotype. J.med.Genet. 6, 4o5

4. Gilmore,I.T. et al (1982): Half-time or clearance of
   indocyanine green in patients with liver disease.
   Hepato-gastroenterol. 29, 55

5. Kovách,G. et al (1982): Effect of Catergen on the mixed
   type oxygenase system. Catergen Symposium. Biogal-Zyma.
   Debrecen.

6. Marsh,C.A. (1963): Metabolism of D-glucurinilacton in
   mammalian systems. Biochem.J. 86, 77

7. Steffen,C. et al (1981): Interaction of cyanidanol with
   the hepatic drug metabolizing system. In:Conn.H.O.(Ed):
   Cyanidanol in diseases of the liver. Academic Press.Inc.
   (London) and Roy.Soc.Med.

INFLUENCE OF DRUGS ON THE BILIRUBIN UDP-GLUCURONYLTRANSFERASE
ACTIVITY AND THE CONCENTRATION OF Y AND Z ACCEPTOR PROTEINS
IN RAT LIVER

Grimmer, Ingrid; R. Moller, J. Gross

Institute of Hygiene of Children and Adolescence, Berlin, GDR
and Department of Clinical and Pathological Biochemistry,
Humboldt University, Berlin, GDR

Jaundice is a frequent and important sign in the neonatal
period and could be due to an accumulation of either unconjugated or conjugated bilirubin. Neonatal unconjugated hyperbilirubinemia appears to result from an increase in bilirubin
load (Gartner et al. 1977), a decrease in hepatic bilirubin
uptake, intracellular transport to the microsomes (Levi et al.
1969), a decreased conjugation (Lathe and Walker 1958) and a
decreased excretion of bilirubin (Schröter 1970).
In vitro studies of bilirubin conjugation demonstrated deficiency of the microsomal enzyme bilirubin UDP-glucuronyl
transferase (bili UDP-GT) in the liver of human fetus and
neonates (Grimmer et al. 1978). According to Gartner et al.
two phases of nonconjugated hyperbilirubinemia can be distinguished in monkeys: phase I appears to result from a high
bilirubin load after birth connected with marked deficiency
in bili UDP-GT activity; phase II seems to result from the
deficiency in acceptor protein Y and a decreased excretion of
conjugated bilirubin. In the human newborn infant serum bilirubin concentration appears to subdivide physiological jaundice in to phases, as in the rhesus monkey.
In order to prevent the bilirubin encephalopathia, phenobarbital (PB) was introduced in the treatment of the nonconjugated hyperbilirubinemia in human newborn infants (Trolle
1968). As PB has a sedative effect, the treatment was modified by the additional administration of nicethamide (N), an
inductive and analeptic drug (Kintzel et al. 1971). This combination has proved to be useful in the prophylaxis of neonatal hyperbilirubinemia because of the distinct effect this
combination has.
Methylphenobarbital (MPB) was introduced into the therapy of
epilepsy 1932. Büch et al. (1968) showed that the pharmacological action of both isomers differs. The (+)isomer has no
sedative effect in contrast to the (-)isomer. In 1971 Gordis
showed an inductive action of MPB.

The aim of this paper is to report on the influence of the
combination of PB/N and the (+)Isomer of MPB on the activity
of bili UDP-GT and the concentration of intrahepatic Y and Z
proteins in rat liver

Animals and Methods

3-month-old male Sprague Dawley rats were divided into five
groups and injected intraperitoneally durin 4 consecutive
days with:
gruop I (n=20): 1,5 ml isotonic NaCl solution (control)
group II (n=19): 80 mg/kg body weight PB
group III(n=22): 80 mg/kg body weight PB and
                 200 mg/kg body weight N
group IV (n=10): 80 mg/kg body weight MPB racemate
group V (n=10): 80 mg/kg body weight (+)isomer of MPB
On the 5th day the rats were anestetized with ether and the
liver was removed.
The estimation of bili UDP-GT was performed within 1 h after
the death of the rat, according to the method of Black et al.
1970. The acceptor proteins Y and Z were separated by
Sephadex G 75 filtration of the 105,000 g supernatant with
rose bengal as acceptor, a modification of the method of Levi
et al. 1969. The protein content in the liver homogenate and
supernatant was estimated according to the method of Popov
et al. 1975.

Results

Fig. 1 shows the influence of the drugs on the bili UDP-GT
activity in the liver homogenate. The enzyme activity increases after the treatment in all groups. Pretreatment with
PB/N results in the highest enzyme activity using wet weight
or body weight as reference system. MPB (+)isomer as well as
racemate shows similar effect as PB. The use of protein as
reference showed no differences among the treatetd animals.
The concentration of hepatic Y acceptor protein using the
volume of the supernatant as reference (Fig. 2) increased in
all pretreated groups by about 30%. When protein was used as
reference a significant increase was found only in the (+)
MPB group.
The drugs did not increase the concentration of Z protein.
The concentration of Z protein even decreased in the PB/N
group in relation to the control group and the MPB groups.
When using the protein concentration in the supernatante as
reference a significant difference between PB/N and MPB was
found.

Conclusion

All drugs tested shows an increasing effect on bili UDP-GT
activity and the acceptor protein Y concentration. When using
liver wet weight as reference is is seem that the combination
of PB/N increases the activity of bili UDP-GT by about 60%
more than PB only; when using protein as reference no difference can seen. The increase of enzyme activity using liver
wet weight as reference is due rather to an unspecific induction of liver proteins than to a specific influence on the
bili UDP-GT. Our results concerning the influence of the (+)
isomer or the racemate of MPB on the activity of bili UDP-GT
and Y acceptor protein concentration suggest that both

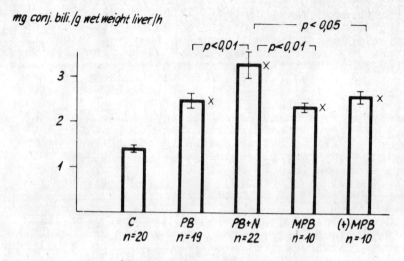

x: p < 0,01 in relation to the control

Fig. 1 : Influence of drugs on the bili UDP-GT

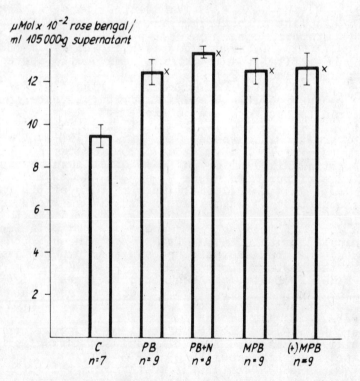

x: p < 0,01 in relation to the control

Fig. 2 : Influence of drugs on the hepatic Y protein

318

isomeric forms act as inducers. This finding is in accordance
with the results of Gordis (1971) who was able to show an
inductive effect of both isomers an the -aminolevolinic acid
sythetase. On the ather hand, it is clear from these studies
that the sedative action is connected with the (-)isomer.
With respect to the (+)MPB isomer it should be stressed that
this drug increases the activity of bili UDP-GT and the concentration of Y protein to the same degree as PB, without a
decreasing effect on the dye-binding Z protein fraction and
without sedative effect. It is, therefore, likely that the
(+)isomer of MPB is better suited as inductive drug in the
prophylaxis of hyperbilirubinemia and in the treatment of
Gilbert syndrome than PB or PB/N.

References

Büch, H.; W. Buzello, O. Neurohr, W. Rummel: Vergleich von
    Verteilung, narkotischer Wirksamkeit und metabolischer
    Elimination der optischen Antipoden von Methylphenobar-
    bital. Biochem. Pharmac. 1968 17: 2391 - 2398.
Black, M.; B.H. Billing, K.P.M. Heirwegh(1970): Determination
    of bilirubin UDP-glucuronyltransferase activity in
    needle biopsy specemens of human liver. Clinica chim.
    Acta 29: 27 - 35.
Gartner, L.M.; K.-S. Lee, S. Vaisman, D. Lane, I. Zarafu(1977)
    Development of bilirubin transport and metabolism in
    the newborn rhesus monkey. J. Pediat. 90: 513 - 531.
Gordis, E.(1971): Tolerance to the hypnotic effect of l-methyl
    phenobarbital induced by its nonhypnotic stereoisomer.
    Biochem. Pharm. 20: 246 - 248.
Grimmer, I.; R. Moller, D. Gmyrek, J. Gross,(1978): Bilirubin
    UDP-Glukuronyltransferase-Aktivität im Leberhomogenat
    menschlicher Feten. Acta biol. med. germ. 37: 131 -
    135
Kintzel, H.W.; R. Schwarze, G.K. Hinkel (1971): Die additive
    Wirkung einer Kombination von Enzyminduktoren (Pheno-
    barbital/Nikethamid) auf den Serumbilurubinspiegel
    Neugeborener. Dt. GesundhWes. 26: 2424 - 2428.
Lathe, G.H.; M. Walker(1958): The synthesis of bilirubin
    glucuronide in animal and human liver. Biolchem. J.
    70: 705 - 712.
Levi, A.J.; Z. Gatmaitan, I.M. Arias (1969): Deficiency of
    hepatic organic anion-binding protein as a possible
    cause of nonhaemolytic unconjugated hyperbilirubinemia
    in the newborn. Lancet ii, 139 - 140.
Popov, N.; M. Schmitt, S. Schulzeck, H. Matthies (1975):
    Eine störungsfreie Mikromethode zur Bestimmung des
    Proteingehaltes in Leberhomogenaten. Acta biol. med.
    germ. 34; 1441 - 1446.
Schröter, W.(1970): Die transitorische Neugeborenenhyper-
    bilirubinämie und ihre biochemischen Grundlagen. Ergeb.
    inn. Med. Kinderheilk. 29: 220 - 277.
Trolle, D. (1968): Decrease of total serum-bilirubin conju-
    gation in newborn infants after phenobarbitone
    treatment. Lancet ii: 705 - 708.

CHARACTERIZATION OF BENZO/a/PYRENE HYDROXYLASE AND GLUTATHIONE S-TRANSFERASE ACTIVITIES IN GREEN MONKEY KIDNEY CELLS

Mikstacka R., Pawlak A.L.

Institute of Human Genetics, Poznań, Poland

INTRODUCTION

The intact cells are the most suitable experimental model for studies of the "balance" between toxifying and detoxifying reactions which may in fact be decisive for the toxic effect produced. Most toxic, mutagenic and carcinogenic compounds undergo biotransformation either to a biologically more active /mixed function oxygenase/ or to inactive form /conjugation reactions/. Reactive intermediates of benzo/a/-pyrene /BP/ - epoxides are converted by epoxide hydratase to dihydrodiols, or are conjugated with glutathione /GSH/ /Hesse and Jernström, 1984/. Kidney cells are especially involved in the metabolism of circulating GSH.
In this report studies of the metabolic activation and inactivation of BP in Green Monkey Kidney /GMK/ cells cultured in vitro are presented.

MATERIALS AND METHODS

GMK cells were received from the Serum and Vaccine Factory in Lublin. Cells were cultured in Eagle's solution supplemented with 10% calf serum and antibiotics /penicillin 100 units/ml and streptomycin 100 µg/ml/, and incubated at $37^{o}C$ as stationary monolayers. Inducers of enzyme activities were added to the cultures 24 h before the enzyme assay. The products of BP metabolism were examined by the spectro-

320

fluorimetric methods described earlier /Mikstacka and Pawlak
1981/. The GSH S-transferase activity was measured by the
procedure of Habig et al. /1974/, slightly modified.
1-Chloro-2,4-dinitrobenzene /CDNB/ was used as substrate.
Concentration of protein was measured by the method of Lowry
et al. /1951/.

RESULTS AND DISCUSSION

The activity of BP hydroxylase in uninduced cells was
37.7 ∓ 8.8 pmole equivalents of 3-OH-BP/min/mg of protein
/n=5/. The inducibility of BP hydroxylase by benzanthracene
/BA/, phenobarbital /PB/ and aminophylline is shown in Table 1. The highest level of BP nydroxylase induction did not
change in the range of BA concentration from 0.2/uM to 2/uM.
The influence of PB was studied in the range of concentration from 1 to 200/uM. The increase of inducibility was observed in the range up to 50/uM. In higher concentrations
a decline of inducibility was noted. Aminophylline /0.01%/
induced activity of BP hydroxylase in GMK cells to the same
extent as 2/uM BA. The effect of these two inducers was addi
tive.

Table 1. Inducibility of BP hydroxylase by BA, PB and
aminophylline in GMK cells. The results are expressed as the ratio of induced to uninduced activity.

Inducer	Concentration	Inducibility ratio ∓ S.D.
BA	2/uM	1.73 ∓ 0.27
PB	50/uM	1.65 ∓ 0.47
Aminophylline	0.01%	1.61 ∓ 0.28
Aminophylline + BA	as above	2.15 ∓ 0.76

In comparison to the results of Wiebel et al. /1981/ and
Abe et al. /1983/ in our experiments the constitutive activity of BP hydroxylase in GMK cells was very high but weakly
inducible by polycyclic hydrocarbon type inducers.

7,8-Benzoflavone /7,8-BF/ and indomethacin added to the test
suspension with BP, stimulated BP hydroxylase in GMK cells
/Table 2/. Our results suggest that in GMK cells cytochrome
P-450 - dependent form of monooxygenase /hepatic type/ is
expressed.

Table 2. The influence of 7,8-BF and indomethacin on BP
hydroxylation in GMK cells.

	Concentration / μM/	The rate of BP hydroxylation /% of control/
7,8-BF	0.3	184
	0.6	230
	1.5	411
	3.0	745
Indomethacin	0.5	116
	2.0	130
	10.0	129
	40.0	142

The constitutive activity of GSH S-transferases in GMK
cells measured in intact cells was 7.28 ∓ 0.49 nmoles of
GS-conjugate/min/mln of cells. The concentrations of substrates
were 0.5 mM for GSH and 1 mM for CDNB. When GSH was absent in the reaction mixture the GSH S-transferase activity
was much lower /1.04 nmoles of GS-conjugate/min/mln of cells/
and did not increase with time. The results have shown that
GMK cells can utilize extracellular GSH. Utilization of
extracellular GSH in lung cells was observed earlier by
Dawson et al. /1984/.
The GSH S-transferase activity towards CDNB was inhibited
by indomethacin. The induction of GSH S-transferase by PB
in the range of concentrations between 50 μM and 200 μM
was minimal.

Acknowledgments
This work was supported by the Polish Academy of Sciences
under project 09.7.3.1.2.2.

REFERENCES

1. Abe S., Nemoto N., Sasaki M. /1983/. Comparison of aryl hydrocarbon hydroxylase activity and inducibility of sister-chromatid exchange by polycyclic aromatic hydrocarbons in mammalian cell lines. Mutation Res. 122, 47-51.

2. Dawson J.R., Vähäkangas K., Jernström B., Moldeus P. /1984/. Glutathione conjugation by isolated lung cells and the isolated, perfused lung. Effect of extracellular glutathione. Eur. J. Biochem. 138, 439-443.

3. Habig W.H., Pabst M.J., Jakoby W.B. /1974/. Glutathione S-transferases. The first step in mercapturic acid formation. J. Biol. Chem. 249, 7130-7139.

4. Hesse S., Jernström B. /1984/. Role of glutathione S-transferases: Detoxification of reactive metabolites of benzo/a/pyrene-7,8-dihydrodiol by conjugation with glutathione. In: Biochemical Basis of Chemical Carcinogenesis, edited by H. Greim, R. Jung, M. Kramer, H. Marquardt and F. Oesch. Raven Press, New York, 1984, 5-12.

5. Lowry O.H., Rosebrough N.J., Farr A.L., Randall R.J. /1951/. Protein measurement with the Folin phenol reagent. J. Biol. Chem. 193, 265-275.

6. Mikstacka R., Pawlak A.L. /1981/. Influence of α-naphthoflavone, β-naphthoflavone and metyrapone on the activity of aryl hydrocarbon /benzo/a/pyrene/ hydroxylase in the mitogen-stimulated human lymphocytes. Bull. l'Acad. Sci. Pol. /biol./ 29, 201-208.

7. Wiebel F.J., Loquet C., Pawlak A.L., Singh J. /1981/. Enzymes invoved in activation and inactivation of chemicals in continous mammalian cell cultures. Postępy Mikrobiol. 20, 67-81.

HALOMETHANE HEPATOTOXICITY IN THE MONGOLIAN GERBIL

Richard E. Ebel and Elizabeth A. McGrath

Department of Biochemistry and Nutrition
Virginia Polytechnic Institute and State University
Blacksburg, Virginia 24061 USA

INTRODUCTION

The hepatotoxicity of the halomethanes, CCl_4 and $CHCl_3$, results from the cytochrome P-450 dependent production of reactive metabolites, trichloromethy radical and phosgene, respectively (Macdonald, 1983). In the rat, induction of the hepatic monooxygenase system by agents including phenobarbital (PB) and chlordecone (CD) potentiates the toxicity of these halomethanes while induction by other compounds such as 3-methylcholanthrene or mirex (MX) does not (e.g. Pohl, 1979; Noguchi et al.,1982; Mehendale, 1984). This selective response is presumably the result of selective induction of a cytochrome(s) P-450 capable of metabolism of CCl_4 and $CHCl_3$.

The situation is quite different with the Mongolian gerbil. We have previously shown that the control gerbil is very sensitive to CCl_4-hepatotoxicity while gerbils induced with PB, CD or MX are less sensitive (Ebel and McGrath, 1984). On the basis of elevations in SGOT and SGPT and decreased content of hepatic cytochrome P-450, the control gerbil is as sensitive to CCl_4 as the CD-induced rat. This report compares the gerbil and the rat with respect to $CHCl_3$-hepatotoxicity.

MATERIALS AND METHODS

Male Mongolian gerbils (50-70g) or male Sprague-Dawley strain rats (200-250g) were used. Induction was achieved by i.p. injection of PB (80mg/kg/day for 4 days), CD in corn oil (15mg/kg/day for 5 days) or MX in corn oil (10mg/kg/day for 5 days). Controls were either untreated or received vehicle alone. The animals were given $CHCl_3$ in corn oil or corn oil alone by i.p. injection 24 hrs after the last injection of inducer. In most cases, food was withdrawn and the animals were killed 24 hrs later.

SGOT and SGPT levels were determined using commercial kits (Sigma Chemical Co.). Hepatic microsomes were prepared as previously described (Ebel et al., 1977), except that the livers were not perfused. Cytochrome P-450 was determined by the method of Omura and Sato (1964). The caudal lobe of the liver was used for non-protein sulfhydryl group

(NPSH) analysis according to the method of Sedlak and Lindsay (1968). Protein was determined by the method of Lowry et al. (1951) using bovine serum albumin as the standard.

RESULTS

Serum transaminases

SGOT and SGPT activities were increased by $CHCl_3$ exposure in a dose dependent manner in the control gerbil and CD-treated rat (Table I). However, the increase was more dramatic in the CD-treated rat than in the control gerbil. For example, 200ul $CHCl_3$/kg increased these activities by about 90-fold in the CD-treated rat but only 3- to 6-fold in the control gerbil. Little if any change in activities was noted in control rats or PB-treated gerbils in response to $CHCl_3$ up to 500ul/kg.

TABLE I. SERUM TRANSAMINASE LEVELS 24 HRS AFTER EXPOSURE TO $CHCl_3$

$CHCl_3$ (ul/kg)	CONTROL RAT		CD RAT		CONTROL GERBIL		PB GERBIL	
	SGOT[a]	SGPT[a]	SGOT	SGPT	SGOT	SGPT	SGOT	SGPT
0	64 (50-69)[b]	18 (13-21)	88 (69-128)	28 (16-37)	170 (135-215)	22 (16-34)	154 (97-270)	21 (16-25)
50	62 (56-69)	22 (13-32)	274 (78-779)	92 (14-249)				
100					154 (121-183)	24 (18-31)	141 (96-170)	17 (12-21)
200	69 (57-80)	19 (14-28)	7660 (2640-12400)	2500 (858-4010)	476 (328-740)	140 (54-250)	167 (141-193)	36 (21-50)
500	80 (56-112)	16 (9-24)	9380 (7440-11300)	2820 (1940-3450)	1350 (700-2440)	960 (540-1580)	207 (116-317)	93 (30-199)

[a]Activity expressed as umole/min/l. [b]Range.

Microsomal enzymes

Treatment of gerbils with PB, CD or MX resulted in significant increases in microsomal levels of cytchrome P-450, cytochrome b_5 and NADPH-cytochrome P-450(c) reductase activity. Exposure of control or induced gerbils to $CHCl_3$ up to 500ul/kg caused no significant reduction in cytochrome P-450 (Fig.1), cytochrome b_5 or reductase activity.

In the CD-treated rat exposed to 50ul $CHCl_3$/kg, cytochrome P-450 was reduced by about 35% and at the highest dose (500ul/kg) by 65% (Fig.1). At the higher doses of $CHCl_3$ (200-500ul/kg), cytochrome b_5 and NADPH-cytochrome P-450 reductase were also depressed. With the control rat, cytochrome P-450 was decreased by about 30% at the highest dose of $CHCl_3$ while cytochrome b_5 and reductase were unaffected.

NPSH levels

Regardless of pretreatment, hepatic NPSH levels of rats and gerbils were within normal limits 24 hrs after receiving 200ul $CHCl_3$/kg. At shorter times (2 and 4 hrs), NPSH levels were depressed by 10- to 20% in control or PB-induced gerbils and control rats. In the CD-induced rat, NPSH levels were 39% and 57% of control at 2 and 4 hrs after CHCl$_3$, respectively (Fig.2). There were no dramatic changes in serum transaminase or microsomal enzyme levels in either species at 2 and 4 hrs after receiving 200ul CHCl$_3$/kg regardless of pretreatment.

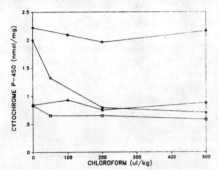

Fig.1 Hepatic microsomal cyto-chrome P-450 content 24 hr after chloroform exposure. (□ control rat, Δ -control gerbil X-CD rat, ＃-PB gerbil)

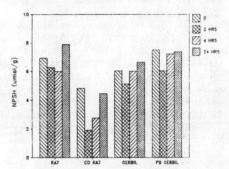

Fig.2 NPSH content of livers from fed animals after ex-posure to chloroform (200 μl/kg)

DISCUSSION

In the rat, CCl$_4$ and CHCl$_3$ hepatotoxicity are parallel phenomena in that agents which potentiate the toxicity of one of these halomethanes also potentiate the toxicity of the other. This observation is consistent with the induction of a particular cytochrome P-450 capable of activation of either CCl$_4$ or CHCl$_3$.

However, induction appears to protect the gerbil from the hepatotoxicity of CCl$_4$ (Ebel and McGrath, 1984). Given the association of CCl$_4$ and CHCl$_3$ hepatotoxicity in the rat, one might expect that the control gerbil would be more susceptible to CHCl$_3$ than the induced animal. On the basis of increases in serum transaminases this appears to be true. On the other hand, cytochrome P-450 and NPSH levels are either unaffected or only marginally altered by CHCl$_3$ exposure.

Based on these observations with the gerbil, the assumed association between induction and potentiation of halomethane hepatotoxicity as well as the association between CCl$_4$ and CHCl$_3$ toxicity must be reconsidered.

ACKNOWLEDGEMENT

This study was supported by Hatch project VA-0612304. The expert technical assistance of Rebecca Barlow was greatly appreciated.

REFERENCES

Ebel, R.E. and McGrath, E.A. (1984). CCl$_4$-hepatotoxicity in the Mongolian gerbil: Influence of monooxygenase system induction. Toxicol. Lett. 22, 205-210.

326

Ebel, R.E., O'Keeffe, D.H., and Peterson, J.A. (1977). Studies of the unique nature of the cytochrome P-450 active site, I. Electron spin resonance spectroscopy as a probe of the hemin iron. Arch. Biochem. Biophys. 183, 317-327.

Lowry, O.H., Rosebrough, N.J., Farr, A.L., and Randall, R.J. (1951). Protein measurement with the Folin phenol reagent. J. Biol. Chem. 193, 265-275.

Macdonald, T.L. (1983). Chemical mechanisms of halocarbon metabolism. CRC Crit. Rev. Toxicol. 11, 85-120.

Mehendale, H.M. (1984). Potentiation of halomethane hepatotoxicity: Chlordecone and carbon tetrachloride. Fund. Appl. Toxicol. 4, 295-308.

Noguchi, T., Fong, K.L., Lai, E.K., Alexander, S.S., King, M.M., Olson, L., Poyer, J.L., and McCay, P.B. (1982). Specificity of a phenobarbital-induced cytochrome P-450 for metabolism of carbon tetrachloride to the trichloromethyl radical. Biochem. Pharmacol. 31, 615-624.

Omura, T. and Sato, R. (1964). The carbon monoxide-binding pigment of liver microsomes, I. Evidence of its hemoprotein nature. J. Biol. Chem. 239, 2370-2378.

Pohl, L.R. (1979). Biochemical toxicology of $CHCl_3$. In Reviews of Biochemical Toxicology (J. Bend, R.M. Philpot, and E. Hodgson, eds.), Vol. 1, pp.79-107, Elsevier, New York.

Sedlak, J. and Lindsay, R.H. (1968). Estimation of total, protein-bound, and nonprotein sulfhydryl groups in tissue with Ellman's reagent. Anal. Biochem. 25, 192-205.

INTERACTION OF AZOLE-DERIVATIVES WITH LIVER MICRO-
SOMES IN RATS AND DOGS: STUDIES IN VIVO AND IN
VITRO

SCHMIDT, U. and SCHLÜTER, G.

Institute of Toxicology, BAYER AG, 5600 Wuppertal, FRG

INTRODUCTION

The main toxicological target for azole derivatives is the
liver. Hepatoxicity is found in all species which are used in
toxicological studies. Liver cell alterations have also been
reported (Mc Nair et al. 1981) to occur rarely in man after
oral treatment with Ketoconazole.

According to our experience the dog is the most sensitive
species with respect to azole induced liver cell damage. So
far, the relation between the liver pathology and inhibition
or induction of microsomal enzymes is still unclear. Therefore
we were interested to investigate a possible interference of
azole compounds with the mixed function oxidases in vivo. Antipyrine clearance is an appropriate method for the detection of
enzyme induction and inhibition as it measures indirectly but
in vivo the hepatic microsomal enzyme activity. Thus, this
method was used to demonstrate in dogs to what an extent the
primary inhibation is followed by induction using three antimycotic substances belonging to the class of azoles namely
Bifonazole (BAY h 4502, Mycospor®), Vibunazole (BAY n 7133)
and Ketoconazole (Nizoral®).

METHODS

In three differente studies four beagle dogs (two male and two
female) each received orally in capsules 10 mg/kg Bifonazole
for six days, 40 mg/kg Ketoconazole for six days and 40 mg/kg
Vibunazole for five days.

Antipyrine kinetics were determined following a 10 mg/kg dose
by intravenous injection before, the day after the 6 (5)-day
course of azole treatment and after a 6 (5)-day recovery period.

Blood samples were collected 5, 15, 30, 60, 90 and 120 minutes after injection and plasma antipyrine was measured by HPLC according to a method given by Danhof et al. (1979). The calculation of the half life was made by a computer programme.

For the determination of the influence on microsomal enzymes, indicated as inhibator concentrations necessary for 50 % inhibition (I 50) of ethoxycoumarindeethylase (EOD) and spectral interaction with cytochrom P-450 (Δ Absorbance) liver microsomes from phenobarbital pretreated rats and untreated dogs were used. EOD was measured by a modification of the procedure given by Ullrich and Weber (1972), the binding spectra by the method of Schenkman et al. (1967).

Table 1: Antipyrine kinetics in dogs (n = 4), after oral treatment with 10 mg/kg Bifonazole (mean ± S.D.)

	half life (min) [1]	factor (f)
before treatment	44.0 ± 5.8	1.00 ± 0
6 days, 10 mg/kg	75.5 ± 11.7	1.74 ± 0.43
after 6 day recovery period	43.6 ± 8.8	0.99 ± 0.13

1) f indicates the increase or decrease of half life as compared to the individual pretreatment values

Table 2: Antipyrine kinetics in dogs (n = 4), after oral treatment with 40 mg/kg Ketoconazole. (mean ± S.D.)

	half life (min)	factor
before treatment	41.0 ± 10.1	1.00 ± 0
6 days, 40 mg/kg	55.3 ± 15.5	1.38 ± 0.22
after 6 day recovery period	39.5 ± 7.0	0.99 ± 0.10

Table 3: Antipyrine kinetics in dogs, after oral treatment with 40 mg/kg Vibunazole. (mean ± S.D.)

	half life (min)	factor
before treatment	31.0 ± 4.0	1.00 ± 0
6 days, 40 mg/kg	23.6 ± 5.6	0.78 ± 0.23
after 6 day recovery period	33.3 ± 2.7	1.08 ± 0.15

Table 4: Interaction of azole derivatives with liver microsomes from rat and dog

Substance	Rat liver microsomes (phenobarbital pretreatment)		Dog liver microsomes (untreated)	
	1) I_{50} (M)	2) ΔA	1) I_{50} (M)	2) ΔA
Bifonazole	1.3×10^{-7}	0.023	1×10^{-5}	0.023
Ketoconazole	6×10^{-8}	0.020	1×10^{-5}	0.021
Vibunazole	2×10^{-6}	0.016	3×10^{-4}	0.007

1) I_{50} values for azole derivatives in microsomal monooxygenase reaction with 7-ethoxycoumarin

2) Λ Absorbance values in optical difference spectra of rat and dog liver microsomes after the addition of azole derivatives at concentrations of 1.6×10^{-6} M (rat) or 1.6×10^{-5} M (dog).

RESULTS AND DISCUSSION

As shown in Tab. 1 antipyrine half life is considerably increased (f=1.74) after six oral applications of 10 mg/kg Bifonazole which indicates an inhibition of liver monooxygenases. This strong affinity to the cytochrom P-450 system is supported also in vitro by the low I 50 concentrations and the high absorbance values (see tab. 4). All three azoles show a typ II binding spectrum. The values of phenobarbital pretreated rat liver microsomes differed considerably from those of untreated dog liver microsomes.

A comparable strong affinity to the cytochrom P-450 system of rat and dog liver microsomes can be demonstrated with Ketoconazole. In rats this inhibitory property was already shown in vivo (Niemegeers et al. 1981). As shown in tab. 2 the antipyrine half life after six oral applications of 40 mg/kg Ketoconazole in dogs is also clearly increased but to a lesser extend (f=1.38). Daneshmend et al. (1983) found in volunteers after five oral applications of about 8 mg/kg Ketoconazole an increased half life of the drug itself but no effect on the antipyrine half life.

Vibunazole had in vitro a much lower affinity to the cytochrom P-450 system of rat and dog liver microsomes then the other two azoles (see tab. 4). As in tab. 3 shown the antipyrine half life after five oral applications of 40 mg/kg was not increased but somewhat decreased (f=0.78) which indicates an enzyme induction.

With all three azoles, the antipyrine half life reached the pretreatment values after the 5 (6) day drug free period those indicating recovery.

The findings described before obviously correlate with histopathological observations. These show, that the strong enzyme inhibitors Bifonazole and Ketoconazole, but not Vibunazole cause liver cell damage in three month toxicity studies on dogs even with low oral doses.

REFERENCE

DANESHMEND, T.K., WARNOCK, D.W., ENE, M.D., JOHNSON, E.M., PARKER, G., RICHARDSON, M.D.: Multiple dose pharmacokinetics of Ketoconazole and their effects on antipyrine kinetics in man. 13. Int. Kong. of Chemotherapy, Wien (1983), Proceedings Part 40

DANHOF, M., DE GROOT-VAN DER VIS, E. and BREIMER, D.D.: Assay of antipyrine and its primary metabolites in plasma, saliva and urine by high-performance liquid chromatography and some preliminary results in man. Pharmacol. 18, 210-223 (1979)

MC NAIR, A.L., GASCOIGNE, E., HEAP, J., SCHUERMANS, V. and SYMOENS, J.: Hepatitis and Ketoconazole Therapy Brit. Med. J. 283, 1058 (1981)

NIEMEGEERS, C.J.E., LEVRON, J.C.I., AWOUTERS, F. and JANSSEN, P.A.J.: Inhibition and induction of microsomal enzymes in the rat: a comparative study of four antimycotics: miconazole, econazole, clotrimazole and ketoconazole. Arch. Int. Pharmacodyn, 251, 26-38 (1981)

SCHENKMAN, J.B., REMMER, H., ESTABROOK, R.W.: Spectral studies of drug interaction with hepatic microsomal cytochrome. Mol. Pharmacol. 3, 113-123 (1967)

ULLRICH, V., WEBER, P.: The O-dealkylation of 7-ethoxycoumarin by liver microsomes. A direct fluorimetric test. Hoppe-Seyler's Z. Physiol. Chem. 353, 1171-1177 (1972).

SELECTIVE INCREASE OF COUMARIN-7-HYDROXYLASE ACTIVITY IN DBA/2N MICE BY PYRAZOLE

1)JUVONEN, R., 2)KAIPAINEN, P. AND 2)LANG, M.

1)Department of Physiology, 2)Department of Pharmacology and Toxicology, University of Kuopio, P.O.B. 6. SF-70211 Kuopio, Finland

INTRODUCTION

Some inbred strains of mice such as D2 exhibit a high coumarin-7-hydroxylase activity and strains such as AKR low activity. It has been suggested that the Coh-locus controls the induction of liver microsomal coumarin-7-hydroxylase and that this induction takes place primarly by phenobarbital (1-3).

In this report we describe the induction of coumarin-7-hydroxylase by pyrazole which seems to have some features, different from those of the sofar well known inducers of the mono-oxygenase complex, phenobarbital (Pb) and 3-methylcholantrene (3-Mc). Induction is described in D2 and AKR mice and the coumarin-7-hydroxylase is characterized both in microsomes and purified cytochrome P-450 fractions.

MATERIALS AND METHODS

Male DBA/2N (D2) and AKR/N (AKR) mice, 7-10 weeks old, were from the animal center, University of Kuopio, Finland. Pyrazole, 100 mg/kg, once a day during five days was given as i.p. injection in 0.9 % NaCl. The microsomal fraction was isolated from mouse livers and stored as previously described (4,5).

Protein concentration was determined by the fluorometric method of Castell et al (6). Cytochrome P-450 content was determined as described by Omura and Sato (7). NADPH-cytochrome C-reductase according to Dallner et al (8), 7-ethoxycoumarin O-deethylase and coumarin-7-hydroxylase according to Aitio (9), benzo(a)pyrene hydroxylase according to Wattenberg et al (10) as modified by Nebert and Gelboin (11) and ethylmorphine N-demethylase (2 mM substrate) and dimethylnitrosamine demethylase (0.2 mM substrate) according to (12). In inhibition studies we used an antibody as described by Kaipainen et al (4). Antibody was preincubated 15 minutes at room temperature with antigen before starting reaction. Cytochrome P-450 was partially purified by a hydrophobic octylamino Sepharose CL 4B column and reconstituted as described by Kaipainen et al (4).

RESULTS AND DISCUSSION

Table 1 shows the effects of pyrazole on liver microsomal mixed function oxidase in D2 and AKR mice. As we can see following parameters of the mixed function oxidase are decreased in both strains of mice after pyrazole treatment: the microsomal cytochrome P-450 content, ethylmorphine N-demethylase and benzo(a)pyrene hydroxylase and slightly the activity of NADPH-cytochrome C-reductase. Dimethylnitrosamine demethylase activity was not affected. On the other hand an increase in the 7-ethoxycoumarin O-deethylase took place after pyrazole treatment both in D2 and AKR mice. Coumarin-7-hydroxylase activity behaved differently from all the other parameters determined: a very strong increase of this enzyme activity took place, but only in the D2 mice. In AKR mice pyrazole had hardly any effect on microsomal coumarin-7-hydroxylase activity.

Table 1. The amount of liver microsomal cytochrome P-450 (nmol/mg protein) and the activity of monooxygenase complex (nmol/min mg protein) as determined by five substrates. The data are mean $\pm$ sd of five mice.
a) Signifigantly different from control, P 0.05.

	AKR Control	Pyrazole	D2 Control	Pyrazole
Cytochrome P-450	1.00+0.10	0.70+0.10a)	0.90+0.11	0.66+0.10a)
cyt. C-reductace	190 +25	160 +15	175 +30	160 +10
ethylmorphine	8.0 +1.3	5.4 +1.0a)	12.3+1.8	5.4 +0.9a)
benzo(a)pyrene	0.24+0.05	0.12+0.04a)	0.26+0.06	0.12+0.08a)
coumarin	0.06+0.02	0.15+0.01a)	0.18+0.03	1.32+0.12a)
7-ethylcoumarin	3.0 +0.5	5.5 +1.1 a)	2.1 +0.3	5.8 +1.0a)
dimethylnitrosamine	3.2 +0.5	4.1 +0.7	3.8 +0.4	3.3 +0.6

According to the results it seems that pyrazole, as an inducer of microsomal drug metabolism has different properties from both Pb and 3-Mc, two well known inducers of drug metabolism. At first, because pyrazole decreases both microsomal benzo(a)pyrene hydroxylase and ethylmonrphine N-demethylase which are known to be increased by 3-Mc and PB -respectively (13). At second pyrazole also decreases the the total amount of microsomal cytochrome P-450 which is known to be increased by both 3-Mc and PB. Furthermore pyrazole increases coumarin-7-hydroxylase very strongly althought only in the D2 like Pb which causes some increase in the enzyme activity in both D2 and AKR.

In order to find out whether pyrazole affects the microsomal mixed fuction oxidase indirectly e.g. trough some modification of the microsomal membrane or directly by changing the microsomal cytochrome P-450 composition, we decided to purify the cytochrome P-450 after pyrazole treatment. Fig. 1 shows the purification from control and pyrazole treated D2 mice liver and also the reconstitution of the coumarin-7-hydroxylase from the P-450 containing fractions. As we can see a very strong increase in the enzyme activity expressed per nmol P-450 took place after pyrazole treatment. This indicates that pyrazole, althought decreasing the total cytochrome P-450 content, increases the content of a spesific P-450 isoenzyme with a high activity for coumarin-7-hydroxylation.

Figure 1. Partial purification of cytochrome P-450 from control and pyrazole treated D2 mouse liver microsomes and the coumarin-7-hydroxylase activity (pmol/min nmol P-450) after the reconstitution of fractions from hydrophobic octylamino Sepharose Cl 4B chromatography.

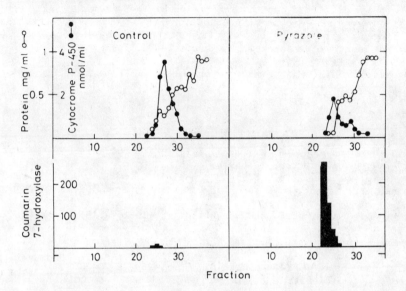

Figure 2. Inhibition of coumarin-7-hydroxylase activity of reconstituted monooxygenase by antibody. Cytochrome P-450 fractions from the figure 1 with the highest activity of coumarin-7-hydroxylase was used. The square is control and the triangel the pyrazole treated sample.

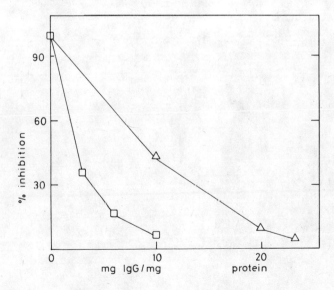

334

Finally in figure 2 we show the antibody cause inhibition of reconstituted coumarin-7-hydroxylase both in case of control and pyrazole treated D2 mice. The antibody used in the experiments was previously developed against a purified cytochrome P-450 fraction from D2 mouse liver with high activity for coumarin-7-hydroxylase as described by Kaipainen et al (4). As we can see more antibody is needed for the inhibition of the enzyme activity after pyrazole treatment than in control preparation. This further suggests that pyrazole indeed increases the activity of coumarin-7-hydroxylase by increasing the amount of a special cytochrome P-450 isoenzyme.

Acknowledgements: This study was supported by the Finnish Academy of Sciences grant no. 07/091.

REFERENCES

1. Wood, A. W. and Conney, A. H. (1974) Science 185, 612-614
2. Wood, A. W. and Taylor, B. A. (1979) J. Biol. Chem. 254, 5647-5651
3. Wood, A. W. (1979) J. Biol. Chem 254, 5641-5646
4. Kaipainen, P., Nebert, D. W. and Lang, M. (1984) Eur. J. Biochem. 144, 425-431
5. Lang, M. A., Gielen, J. E. and Nebert, D. W. (1981) J. Biol. Chem. 256, 12068-12075
6. Castell, J. V., Cervera, M. and Mario, R. (1979) Anal. Biochem. 99, 379-391
7. Omura, T. and Sato, R. (1964) J. Biol. Chem. 239, 2370-2378
8. Dallner, G., Siekewitz and Palade, G. E. (1966) J. Cell Biol. 30, 97-117
9. Aitio, A. (1978) Anal. Biochem. 85, 488-491
10. Wattenberg, W. C., Leong, J.C. and Strand, P. S. (1962) Cancer Res. 22, 1120
11. Nebert, D. W. and Gelboin, H. V. (1968) J. Biol. Chem. 244, 6242-6249
12 Jannetti, R. A. and Andersson, M. (1981) JNCI 67, 461-466
13. Paine , A. J. (1981) Essays in Biochemistry 17, 85-126

IMMUNOCHEMICAL STUDY ON IDENTITY OF MULTIPLE FORMS OF CYTOCHROME P-450 IN THE LIVERS OF RATS AND MICE

D.V. MISHINA, L.F. Guljaeva, N.I. Gutkina, V.M.Mishin

Institute of Clinical and Experimental Medicine, Academy of Medical Sciences of USSR, Siberian Department, 630091 Novosibirsk, U.S.S.R.

INTRODUCTION

In recent years different immunochemical methods have been used to study the regulation of induction of microsomal monooxygenases. The purification of some forms of cytochrome P-450 and the subsequent preparation of monospecific antibodies to these proteins made it possible to quantify the specific forms of cytochrome P-450 in liver microsomes after treatment of rats and mice by monooxygenase inducers(Harada, Omura,1981).One of the interesting problems in xenobiochemistry is an extent of immunological identity of different forms of cytochrome P-450, which like isoenzymes, have fairly common physio-chemical and functional properties. The purpose of the work presented in this communication was to gain data regarding the immunological identity of cytochrome P-448 induced with 3-methylcholantrene in the liver microsomes of rats and mice.

METHODS

Male Wistar rats and male C57BL mice were used in the experiments. Animals were induced as recommended by Ryan et al (1979). Microsomes were isolated from livers by the conventional methods. Cytochromes P-448rats and P-448mice were isolated as described by Guengrich and Martin(1980).Contents of the cytochromes were determined by the method of Omura and Sato(1964). SDS-polyacrylamide gel electrophoresis was carried out on 7,5 - 15 % acrylamide slab gels(Laemmli,1970). Antibodies against purified cytochromes were raised in rabbit as recommended in (Kamataki et al, 1976),Ouchterlony double immunodiffusion was performed accoding to Thomas et al(1981). Metabolic activities in microsomes and in reconstituted monooxygenase systems were estimated as recommended in (Lu, Levin, 1974). The fluorescent product of 3,4-BP was mearured fluorometrically (Robie et al , 1976),using 3-OH-BP as standard. The peptide mapping of the different hemoproteins was performed as described by Clivlend et al (1977).Metabolism of 7-ethoxyresorufin was assayed by direct fluorometric method(Prough et al,1978).

336

RESULTS AND DISCUSSION

All experiments have been carried out with antibodies
against the main forms of 3-MC-cytochrome P-448r and
P-448m, purified to homogenous state. The CO-reduced difference spectral peak of this cytochrome P-448 is at 448nm.
Both cytochromes catalyze the benzo(a)pyrene-hydroxylation
and 7-ethoxyresorufin-O-deethylation more efficiently than
demethylation of aminopyrine and benzphetamine. Each form
shows a single protein-staining band on SDS-poliacrylamide
gel. As shown by means of Ouchterlony double immunodiffusion
technique after induction of 3-MC the antibodies against
cytochrome P-448 from rat liver developed a clear precipitation line with homologous antigen,they also interacted with
cytochrome P-448m although in this case the precipitation
was less distinct. Antibodies against cytochrome P-448m
reacted with the rat cytochrome P-448r, but less distinctly
if compared to the mice cytochrome.Apparently,at least some
common antigenic determinants exist in cytochrome P-448 of
different species.(Fig.1).The monospecific antibodies expli-
city and rather specifically inhibit metabolism of substrate
catalyzed by these forms of cytochrome P-450 to which the
antobodies have been obtained. Such a specificity gives opportunity to evaluate the participation ofagiven form of cytochrome P-450 in metabolism of different substrates.Fig.2
shows that the anti-P-448r very effectively and to an equal
extent inhibits metabolism of 7-EthR and BP catalysed both
by cytochrome P-448r and P-448m. The same situation is observed with anti-P-448m.It may be noted that metabolism of BP
7-EthR when catalized by P-450 is insensitive to anti-P-448r
and P-448m.Thus, P-448 obtained from MC-treated mice and rats
have some common spectral, catalytic and immunological pro-
perties.Anti-P-448r and anti-P-448m gave a very weak precipitation line with heterologous antigens. At the same time
these antibodies inhibited similarly the metabolism,catalyz
ed by both P-448r and P-448m.The rather weak immunoprecipitation reaction may suggest a insignificant quantity of
the latter. By judging   the same effect of inhibition of
enzymatic reaction catalyzed by both homologous and heterologous antigens these common antigenic determinants have
direct relation to that part of a protein molecule, which
participates in enzymatic catalysis. Peptide maps obtained
by restricted chymotrypsin proteolysis of P-448r and P-448m
demonstrated some slight differences.It can be suggested
that common antigenic determinants revealed by the inhibition
of enzyme reaction is caused by identity of structures forming the active center of cytochrome P-450. Antigenic determinants common for P-448 can probably be arranged in protein
molecule region responsible for catalytical activity against
benzo(a)pyrene.

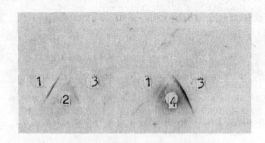

Fig.1 Ouchterlony double immunodiffusion test.
1 - cytochrome P-448m, 2 - anti-P-448m, 3 - cytochrome P-448r, 4 - anti-P-448r

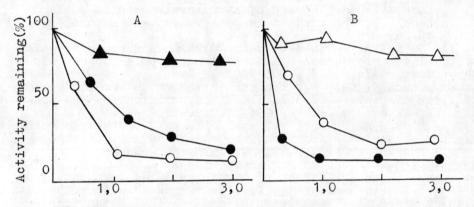

Fig. 2 Effect of anti-P-448r (-○-), anti-P-448m(-●-),
anti-P-450r(-▲-) and anti-P-450m(-△-) immunoglobulin on 7-ethoxyresorufin-O-deethylase activity,
catalyzed by P-448r(A) and P-448m(B).

338

REFERENCES

Clevland D.W.,Fischer S.G.,Kirscher M.W. et al - Peptide mapping by limited proteolysis in sodium dodesylsulfate and analysis by gel electrophoresis.-J.Biol.Chem., 1977,v.252,p.1102-1106.

Guengrich F.P.,Martin M.V. Purification of cytochrome P-450, NADPH-cytochrome P-450 reductase, and epoxide hydratase from a single preparation of rat liver microsomes.- Archives of Biochemistry and Biophysics,1980,v.205,p. 365-378

Harada H.,Omura T. Selective induction of two different molecular species of cytochrome P-450 by PB and 3-MC.- Journal of Biochemistry,1981,v.59,p237-248

Kamataki T.,Belcher D.H., Neal R.A.,Studies of the metabolism of parathion and benzphetamine using an apparently homogeneous preparation of rat liver cytochrome P-450:effect of a cytochrome P-450 antibody preparation.-Molecul.Pharmacol 1976,v.12,p.921-932

Laemmli U.R.Cleavage of structural proteins during the assembly by the head of bacteriophage T-4.-Nature,1970, p.265-275

Lu A.Y.M.,Levin W., The resolution and reconstitution of the liver microsomal hydroxylation system.-Biochem.,Biophys. Acta, 1974,v,344,p.205-240

Omura T.,Sato.R., The carbon monooxyde binding pigment of liver microsomes.II Solubilisation,purification and properties,-J. of Biol. Chem.,1964,v.239.p.2379-2385

Prough R.,Burke R.T.,Mayer, Direct fluorometric methods for measuring mixed-function oxidase activity.-Methods Enzymology,v.52,p.372-377,1978,ed.Fleischer.

Ryan D.E.,Thomas P.E.,Korseniowski D. Levin W.Separation and characterization of highly purified forms of liver microsomal cytochrome P-450 from rats treated with polychlorinated biphenyls, phenobarbital and 3-methylcholantrene.- J. of Biol. Chem.,1979,v.254,p.1365-1374.

Robie K.M.,Cha Y.N., Talcott R.E.,Schenkman J.B. Kinetic studies of benzpyrene and hydroxybenzpyrene metabolism. -Chemical and Biological Interaction,1976,v.12,p.285-297

Thomas P.E.,Korsenowski D., Ryan D.,Levin W.Preparation of monospecific antibodies against two forms of rat liver cytochrome P-450 and quantitation of these antigens in microsomes.-Archives of Biochemistry and Biophysics,1979, v.192,p.524-532.

INDUCTION OF THE LIVER MICROSOMAL CYTOCHROME P-450
FORMS IN NEWBORN RATS TREATED BY PHENOBARBITAL
AND 3-METHYLCHOLANTHRENE

Lyudmila F. Gulyaeva, Vladimir M. Mishin

Institute of Experimental Medicine, Academy of Medical
Sciences of the U.S.S.R., Siberian Division,
630091 Novosibirsk (U.S.S.R)

INTRODUCTION

Regulation of the activity of cytochrome P-450 multiple
forms in the microsomal monooxygenase system of animal liver
is poorly investigated, especially during ontogenesis. Some
studies demonstrated both quantitative and qualitative differences between newborns and adults (Short et al.1976, Guenthner & Mannering,1977). Direct evidence has been obtained
for the temporal control of two P-450 forms induced by TCDD
during the perinatal period in rats, mice (Guenthner & Nebert
1978) and rabbits (Atlas et al.1977). Immunochemical methods
using antibodies against some forms of P-450 allowed investigation of their activity in animals of different age. This
technique was used to determine the amount of P-450a, P-450b
and P-450c in liver microsomes of immature rats (Thomas et
al.1981), the main forms of P-450 in liver of rabbits (LM2
and LM4) during their ageing (Chiang et al.1982). However,
for the early neonatal period, which is critical for the development of the monooxygenase system, the activity of P-450
forms remains uninvestigated. In this work we report on the
attempt to characterize some induction parameters and activity of P-450 forms in newborn rats. To this end immunochemical methods are applied using antibodies against PB-cytochrome P-450 (induction by phenobarbital) and MC-cytochrome
P-448 (induction by 3-methylcholanthrene) of the liver microsomes of adult rats.

METHODS

3-, 6- and 12-day-old neonatal Wistar rats have been used in
the experiments. Microsomal enzyme induction was caused by
phenobarbital (PB) (80 mg/kg body weight) and 3-methylcholan-
threne (3-MC) (25 mg/kg). Microsomes were isolated using the
differential centrifuging method. Microsomes of unsexed neonates from a single female were pooled together for futher
study. Adult males (150-200 g weight) were similarly prepa-
rated. The content of the cytochrome P-450 was measured by
the method of Omura and Sato (1964). Cytochromes P-450 from
microsomes of the liver of adult rats treated by PB and 3-MC

were isolated by the method of Guengerich and Martin (1980).
Proteins of the liver preparation were determined according
to Lowry et al. (1951). The quantities of PB-cytochrome P-
450 and MC-cytochrome P-448 were determined by rocket immunoelectrophoresis as described by Pickett et al. (1981). The
hydroxylation rate of benzo(a)pyrene was measured by fluorescence method according to the output of 3-OH-BP (Robie et al.
1976). Benzphetamine N-demethylation was assayed by following
the formaldehyde formation (Nash, 1953). Metabolism of 7-
ethoxyresorufin was assaed spectrofluorimetrically with reso-
rufin as a standard (Burke & Mayer, 1974), while that of aldrin - by the method of Wolff et al. (1980).

RESULTS AND DISCUSSION

Characteristics of induction of the microsomal monooxygenase system in newborn and adult animals are presented in
table 1.

Table 1

Age (days)	Induction	Activity of NADPH-cyto-chrome c reductase (μmol min^{-1}mg^{-1})	P-450 hemopro-tein le-vels (nmol/mg)	Quantity of forms (nmol/mg) PB-P-450	MC-P-448
3	Untreated	0,10	0,15	n.d.	n.d.
	PB	0,09	0,35	0,014	n.d.
	3-MC	0,10	0,40	n.d.	0,020
6	Untreated	0,11	0,42	n.d.	n.d.
	PB	0,14	0,65	0,065	n.d.
	3-MC	0,13	0,73	n.d.	0,102
16	Untreated	0,19	0,54	n.d.	n.d.
	PB	0,24	1,30	0,221	n.d.
	3-MC	0,21	0,84	n.d.	0,420
adults	Untreated	0,18	0,64	n.d.	n.d.
	PB	0,22	2,20	0,924	n.d.
	3-MC	0,17	1,70	n.d.	1,190

It can be seen that the amount of P-450 determined by
its CO-reduced form increases after PB and 3-MC induction
both for newborns and adults. Rather high NADPH-cytochrome c
reductase activity is observed in liver microsomes for all
ages studied. The amount of specific forms of P-450 is considerably lower in newborns than in adults, indicating the
low ability of newborns for response to the external factors.
At the same time Ouchterlony double diffusion experiments did
not reveal qualitative differences between newborns and adults.
One of the parameters characterising separate forms of
P-450 in adults is their substrate specifity. The following

Table 2

Age (days)	Induction	ACTIVITY (nmol products / min / mg)			
		BPh	Aldrin	BP	7-ER
3	Untreated	0,5	0,9	0,17	0,05
	PB	1,5	1,1	0,35	0,05
	3-MC	0,5	0,7	2,50	3,70
6	Untreated	0,6	1,3	0,25	0,05
	PB	1,7	1,5	0,70	0,05
	3-MC	1,0	0,8	2,60	4,70
16	Unterated	0,9	1,0	0,25	0,05
	PB	1,8	2,1	1,10	0,05
	3-MC	1,0	0,7	3,90	6,00
adults	Unterated	1,7	2,0	0,50	0,20
	PB	17,0	14,0	0,80	0,20
	3-MC	1,3	1,5	4,50	7,00

substrates have been used in our study:benzphetamine (BPh) and aldrin - for the PB-cytochrome P-450; benzo(a)pyrene (BP) and 7-ethoxyresorufin (7-ER) - for the MC-cytochrome P-448. As seen from table 2, administration of 3-MC results in a notable increase of 3-OH-BP hydroxylase and 7-ER-O-deethylase activity despite the insignificant amount of MC-P-448 in newborns. In adults the same increase of activities is observed, but the amount of MC-P-448 is rather high. This can suggest a high activity of MC-cytochrome P-448 during early neonatal period.

To study how P-450 forms above participated in metabolism, inhibition by antibodies was used. 3-OH-hydroxylation of BP and 7-ER were inhibited by anti-P-448 by almost 100% similar to adults. Unusual was the case of BP-hydroxylation, metabolism of which was enhanced after administration of PB to newborns: neither anti-P-450, nor anti-P-448 inhibited it. In our opinion, this can give evidence for involment of some cytochrome P-450 forms in newborns, which absent in adults. PB, in contrast to 3-MC, did not cause such increase of the BPh-N-demethylase activity as was observed in adults. Analysis showed that anti-P-450 rather slightly inhibited this reaction, while anti-P-448 did not produce such effect at all. In 3- and 6-day-old rats anti-P-450 inhibited BPh metabolism by 10-20%, whereas in adults the effect was 60-70%. This can also indicate to participation of another forms of P-450 besides PB-P-450 in BPh metabolism. Aldrin epoxydase activity was also poorly induced by PB in contrast to adults. This probably suggests the absence of induced P-450 forms responsible for this metabolism during first weeks of life.

Thus, analysis of the majority of parameters shows that the microsomal monooxygenase system of newborn rats differs not only quantitatively from that in adults. It is important

342

to note that we have not observed BP inhibition in PB-microsomes, while no induction of 7-ER-O-deethylase activity was
found in this case. Another qualitative difference between
newborns and adults is poor inhibition of BPh metabolism in
PB-microsomes. This results can suggest the induction of special "neonatal" forms of Cytochrome P-450.

REFERENCES

Atlas S.A.,Boobis A.R.,Felton J.S.et al. Ontogenetic expression of polycyclic aromatic compound-inducible monooxygenase
activities and forms of cytochrome P-450 in rabbit.-J.Biol.
Chem.,1977, v.252, p.4712-4721.
Burke M.D., Mayer R.T. Ethoxyresorufin:direct assay of a microsomal O-dealkylation which is preferentially inducible by
3-methylcholanthrene.-Drug.Metab.Rev.,1974, v.2, p.583-586.
Chiang J.V.L., DiLella A.G.,Steggles A.V. Effect of inducers
and aging on rabbit liver microsomal drug-metabolizing systems.-Mol.Pharmacol.,1982, v.23, p.244-251.
Guengerich F.P., Martin M.V. Purification of cytochrome P-450
NADPH-cytochrome P-450 reductase and epoxide hydratase from
single preparation of rat liver microsomes.-Arch.Biochem.Bio-
phys.,1980, v.205, p.356-379.
Guenthner T.M., Mannering G.J. Induction of hepatic monooxyge
nase systems in fetal and neonatal rats with phenobarbital,
polycyclic hydrocarbons and other xenobiotics.-Biochem.Phar-
macol.,1977, v.26, p.567-575.
Guenthner T.M., Nebert D.W. Evidence in rat and mouse for tem
poral control of two forms of cytochrome P-450 inducible by
TCCD.-Eur.J.Biochem.,1978, v.91, p.449-456.
Lowry O.H.,Rosenbrough N.J.,Farr A.L.,Randal R.J. Protein measurement with the Folin phenol reagent.-J.Biol.Chem.,1951,
v.193, p.256-275.
Nash J., The colorimetric estimation of formaldehyde by means
of the Hantzch reaction.-Biochem.J.,1953, v.55, p.416-421.
Omura T.,Sato R. The carbon monooxide-binding pigment of liver microsomes.-J.Biol.Chem.,1964, v.23, p.2379-2385.
Pickett C.B., Jeter R.Z., Morin J.,Lu A.Y.H. Electroimmunochemical quantitation of cytochrome P-450 and P-448 and epoxide hydrolase in rat liver microsomes.-J.Biol.Chem.,1981,
v.256, p.8815-8820.
Robie K.M., Cha Y.N., Talkott R.E., Schenkmann J.B. Kinetic
studies of benzo(a)pyrene and hydroxybenzo(a)pyrene metabolism.-Chem.-Biol.Interact.,1978, v.12, p.285-297.
Short C.R.,Kinden D.A., Stith R. Fetal and neonatal development of the microsomal monooxygenase system.-Drug Metab.Rev.,
1976, v.5, p.1-42.
Thomas P.E., Reik L.M.,Ryan D.E.,Levin W. Regulation of three
forms of cytochrome P-450 and epoxyde hydrolase in rat liver
microsomes.-J.Biol.Chem.,1981, v.256, p.1044-1052.
Wolff T.,Greim H.,Huang M.-T. et al. Aldrin epoxidation catalyzed by purified rat liver cytochromes P-450 and P-448. High
selectivity for cytochrome P-450.-Eur.J.Bipchem.,1980, v.111,
p.545-551.

Cytochrome P-450 and its Implication
in Environmental Problems

INDUCTION OF RAT HEPATIC MICROSOMAL MIXED FUNCTION OXIDASES BY CLOSELY RELATED SUBSTITUTED UREA HERBICIDES

Bernadette Schoket and István Vincze

National Institute of Hygiene, Budapest, Hungary

INTRODUCTION

Substituted urea herbicides form an important class of pesticides, all the same there is not much information available concerning their inducing effects on hepatic drug metabolizing enzymes. In this work we have investigated the effects of eight structurally closely related substituted urea herbicides on rat hepatic microsomal cytochrome P-450 linked mixed function oxidases.

MATERIALS AND METHODS

Chemicals

Diuron, linuron, chlorbromuron, chlortoluron, metoxuron, monolinuron, metobromuron and isoproturon of 93-98% purity were obtained from Nitrokémiai Ipartelepek, Füzfőgyártelep, Hungary and Budapesti Vegyimüvek, Hungary.

Treatment of animals

Male Wistar rats /160-200 g/ were given 1/6 LD50 of herbicides by gavage on three consecutive days in the following doses: diuron 57, linuron 67, chlorbromuron 83, chlortoluron 166, metoxuron 53, monolinuron 38, metobromuron 33 and isoproturon 77 mg/ml/100 g body weight/day. The herbicides were freshly suspended in 4% gelatine solution. Control animals were given gelatine solution. The rats were sacrificed 24 hours after the last treatment. Each group included 6-8 animals. The results of two separate series of treatments were combined in the comparative study of the eight herbicides. Dose dependence of induction was studied with diuron and chlortoluron, and 1/20, 1/9, 1/6 and 1/4 LD50 were given according to the same protocol in separate series of experiments.

Assays

Cytochrome P-450 content (P-450) was determined according to Omura and Sato /1964/. Benzo/a/pyrene monooxygenase activity (BP-MOO) was measured by the method of DePierre et al. /1975/, 7-ethoxycoumarin O-deethylase activity (ECOD) by Matsubara et al. /1983/, 7-ethoxyresorufin O-deethylase activity (EROD) by the method of Pohl and Fouts /1980/, aminopyrine N-demethylase activity (AM-PY) according to Cochin and Axelrod /1959/, and aldrin epoxidase activity (A-EPOX) essentially by the method of Wolff et al. /1980/ with some modifications /Schoket and Vincze, 1985/. All the measurements were made from the 105000 x g microsomal fractions prepared in 0.15 M KCl.

RESULTS AND DISCUSSION

P-450 content, BP-MOO, ECOD, EROD, AM-PY and A-EPOX activities measured after the treatment with 1/6 LD50 of the eight herbicides can be seen in table 1.

Table 1.
Effect of eight sustituted urea herbicides on rat hepatic microsomal P-450 content and mixed function oxidase activities

Compound	P-450 %	BP-MOO %	ECOD %	EROD %	AM-PY %	A-EPOX %
Control	100	100	100	100	100	100
Diuron	155[a]	492[a]	1041[a]	4603[a]	132[c]	100[c]
Linuron	142[a]	641[a]	1296[a]	4375[a]	123	166[c]
Chlorbromuron	158[a]	799[a]	1800[a]	10819[a]	122	125
Chlortoluron	151[a]	348[a]	652[a]	1752[a]	130[c]	166[c]
Metoxuron	106	92	215[b]	653[a]	112	84
Monolinuron	125[c]	190[b]	385[a]	655[a]	123	116
Metobromuron	119[c]	230[a]	356[a]	1219[a]	113	75
Isoproturon	132[b]	118	337[a]	491[a]	122	309[a]

[a] $P \leqslant 0.001$; [b] $0.001 < P \leqslant 0.01$; [c] $0.01 < P < 0.05$ /With Student's t-test significantly different from the control values/. Control values: 0.53 ± 0.045 nmol P-450/mg protein; BP-MOO 0.73 ± 0.014, ECOD 0.27 ± 0.018, AM-PY 1.09 ± 0.085 and A-EPOX 0.32 ± 0.067 nmol/min/mg protein, and EROD 12 ± 1.7 pmol/min/mg protein.

Substituted phenylurea herbicides investigated here generally acted as inducers of mixed function oxidases. Diuron, linuron and chlorbromuron were the most effective inducers of BP-MOO, ECOD and EROD. AM-PY activity remained close around the control level. Among A-EPOX activities the greatest induction was obtained with isoproturon.

Dose dependence of induction was studied with diuron and
chlortoluron, a di- and a monosubstituted urea, respectively.
With chlortoluron P-450 content, BP-MOO, ECOD and EROD activities showed 'saturation' curves, in which saturation levels
were between 1/20-1/9 LD50. With diuron these parameters have
not reached saturation up to 1/4 LD50 /data not shown/.

Relationship between enzyme induction and chemical structure
of the eight substituted urea compounds was also examined
/table 2/.

Table 2.
Increment of P-450 content, BP-MOO, ECOD and EROD activities
relative to molar quantity of 1/6 LD50 of the herbicides.

Halogen	Compound	Index of induction /Mean $\pm$ S.E./			
		P-450	BP-MOO	ECOD	EROD
2	Diuron Linuron Chlorbromuron	0.20 $\pm$0.03	2.03 $\pm$0.35	4.73 $\pm$0.95	24.10 $\pm$9.75
1	Chlortoluron Metoxuron Monolinuron Metobromuron	0.10 $\pm$0.05	0.46 $\pm$0.38	1.20 $\pm$0.62	4.09 $\pm$2.69
$\emptyset$	Isoproturon	0.09	0.05	0.63	1.04

The index of induction was calculated according to the
following formula:

$$\left[\frac{I}{100} - 1 \right] \cdot \frac{6 \cdot Mw}{LD50} \text{, where}$$

I: percent enzyme content or activity relative to the control
Mw: molecular weight of the herbicide; LD50: acute oral LD50
of the herbicide for rats.

The compounds were arranged into three groups depending
on the number of halogen substituents on their phenyl ring.
Diuron, linuron and chlorbromuron which carry two halogens
caused significantly higher 'molar induction' of BP-MOO, ECOD
and EROD than those which have one halogen substituent or no
one. No significant difference could be established between
the influence of one and zero halogen substituent because
there was only one compound in the latter category.

Smaller structural differences, which do not concern the
phenylurea structure primarily, also influenced enzyme activities. The exchange of one chlorine substituent for bromine
either in the disubstituted linuron - chlorbromuron pair or
in the monosubstituted monolinuron - metobromuron pair increased induction of BP-MOO, ECOD and EROD activities.

AM-PY and A-EPOX activities cannot be analyzed in such
details because of the smaller range of induction. All the
same the highest A-EPOX activity caused by isoproturon might

348

be characteristic of the lack of halogen and/or the presence
of isopropyl substituent.

The substituted urea herbicides involved in this study –
with the exception of isoproturon - produced qualitatively
similar induction patterns of the different mixed function
oxidases. The activity patterns are more or less similar to
those obtainable by polycyclic aromatic hydrocarbon inducers.
A P-450 isoenzyme pattern study would give further information
on the type of induction by the substituted urea inducers.

ACKNOWLEDGEMENTS

We thank Mrs. Lévay for her excellent technical assistance
and Mrs. L. Gönczi for GLC measurements of aldrin epoxidase
activities.

REFERENCES

Cochin, J. and Axelrod, J. /1959/: Biochemical and pharmacological changes in the rat following chronic administration of
morphine, nalorphin and normorphine, J. Pharm. Exptl. Ther.
125, 105-110.

DePierre, J.W., Moron, M.S., Johannesen, K.A.M. and Ernster, L.
/1975/: A reliable, sensitive and convenient radioactive assay
for benzpyrene monooxygenase, Anal. Biochem. 63, 470-484.

Matsubara, T., Otsubo, S., Yoshihara, E. and Touchi, A. /1983/:
Biotransformation of coumarin derivatives /2/. Oxidative metabolism of 7-alkoxycoumarin by microsomal enzymes and a simple
assay procedure for 7-alkoxycoumarin O-dealkylase, Japan J.
Pharmacol. 33, 41-56.

Omura, T. and Sato, R. /1964/: The carbon monoxide-binding
pigment of liver microsomes, I. Evidence for its hemoprotein
nature, J. Biol. Chem. 239, 2370-2378.

Pohl, R.J. and Fouts, J.R. /1980/: A rapid method for assaying
the metabolism of 7-ethoxyresorufin by microsomal subcellular
fractions, Anal. Biochem. 107, 150-155.

Schoket, B. and Vincze, I. /1985/: Induction of rat hepatic
drug metabolizing enzymes by substituted urea herbicides,
Acta pharmacol. et toxicol, in press.

Wolff, T., Greim, H., Huang, M.T., Miwa, G.T. and Lu, A.Y.H.
/1980/: Aldrin epoxidation catalyzed by purified rat liver
cytochromes P-450 and P-448. High selectivity for cytochrome
P-450, Eur. J. Biochem. 111, 545-551.

THE EFFECT OF CIGARETTE SMOKING ON 7-ETHOXY-RESORUFIN
O-DEETHYLASE AND OTHER MONOOXYGENASE ACTIVITIES IN HUMAN
PLACENTA

Gachályi,B.,Pelkonen,O.,Pasanen,M.,Kuha,H.,Kairaluoma,M.,
Sotaniemi,E.A.,Park,S.S.,Friedman,F.K.,Gelboin,H.V.
Postgraduate Med.Sch.I.Dept.Med.,Budapest,Hungary,Univer-
sity of Oulu,Oulu,Finland,National Cancer Institute and
National Institutes of Health,Bethesda,Maryland,U.S.A.

Cigarette smoking increases the elimination of several drugs
/Jusko,1978/ and induces the metabolism of polycyclic aromatic
hydrocarbons and some other model substances in human tissues
and cultured cells.It is thought that the increase in the plasma elimination of a number of drugs is due to the induction of
hepatic drug metabolizing enzymes,but unequivocal proof in man
has not been provided.We studied various monooxygenase activities in human placentas.The assay using 7-ethoxyresorufin as a
substrate is said to reflect very specifically the cytochrome
P-450 form which is induced by polycyclic aromatic hydrocarbons
and cigarette smoking and in the present study we set out to
investigate the levels of 7-ethoxyresorufin O-deethylase /ERDE/
as well as other monooxygenase activities,like aryl hydrocarbon
hydroxylase /AHH/ and 7-ethoxycoumarin O-deethylase /ECDE/.Fur-
thermore,monoclonal antibodies /MAb/ against rat liver cytochromes P-450 were used to further characterize monooxygenases.
We used monoclonal antibodies prepared to methylcholanthrene
induced cytochrome P-450 /MAb-1-7-1/ and to phenobarbital induced cytochrome P-450 /MAb-2-66-3/ to determine the amount of
enzyme activity contributing to the total activity by MAb sensitive cytochrome P-450.

Materials and methods

The monoclonal antibodies were prepared by the hybridoma technique of Köhler and Milstein /1974/.The MAb towards methylcholanthrene induced form of rat liver cytochrome P-450 was from
the clone 1-7-1 and the MAb towards phenobarbital induced form
of cytochrome P-450 from the clone 2-66-3 /Song et al.,1984/.
As a control,an ascites fluid /NBS-1-48-5/ containing non-specific IgG was used /Park et al.,1982/.
Placentas were obtained from smoking mothers after normal delivery at term /smoking was verified by plasma cotinine assays/.
Microsomes were prepared by the standard ultracentrifugation
technique and stored at -70 C until assays.
In all enzyme assays the MAb concentration was related to the
microsomal protein content.The enzyme preparations were incubated with the MAb for 3 to 5 minutes at room temperature prior

350

to the beginning of the enzyme assay.
The ERDE activity was measured according to the end-point fluorometric method.Incubation mixture /1 ml/ contained the standard cofactors /MgCl$_2$ 2.5 mM,KCl 50 mM,glucose-6-phosphate 1.5
mM,NADP 62.5 microM and glucose-6-phosphate dehydrogenase 1 U/
and the substrate 100 microM 7-ethoxyresorufin.Reaction was
started by the addition of the substrate and after a 5 minutes
incubation at +37 C^o reaction was stopped by the addition of
2.5 ml methanol.Precipitated protein was separated by centrifuging and the fluorescence in the supernatant was measured by an
Aminco-Bowman spectrophotofluorometer with excitation and emission wavelengths of 530 and 585 nm,respectively.Resorufin /10
microM/ was used as an external standard in every series of
measurements.
The AHH and ECDE activities were measured with fluorometry too.
The method for measuring plasma cotinine was a modification of
an assay for urinary nicotine and cotinine.A 1 ml plasma sample
was diluted and made alkaline with 1 N NaOH and extracted with
chloroform.Organic phases were evaporated to dryness,dissolved
in a small volume of ethanol and cotinine was measured gas
chromatographically using the OV-17 column and nitrogen detec-
tor.Column temperature was 180 C^o and detector and injector
temperatures 250 C^o.External standardization was used.The limit
of the sensitivity of the method was about 1-2 ng/ml plasma.

Results

With human placental microsomes,the MAb-1-7-1 inhibited AHH and
ERDE activities /Fig.1./ as well as ECDE activity,whereas the
MAb-2-66-3 or NBS-1-48-5 had no effect on any activity.The inhibition of AHH and ERDE activities by the MAb-1-7-1 was uniformly strong,in average about 80 per cent inhibition,whereas
the percentage inhibition of the ECDE activity varied between
16 and 85 per cent.Plasma cotinine levels did not correlate
with any of the activities.

Discussion

The results of the present study have some interesting connec-
tions with our previous results obtained from human liver spe-
cimens /Pelkonen et al.,1985/.The ERDE activity is induced in
the placenta by cigarette smoking and it is inhibited by the
MAb-1-7-1.The characteristics of the ERDE activity in the liver are very similar,i.e. the induction by cigarette smoking
and the inhibition by MAb-1-7-1.An interesting contrast was provided by the experiments with AHH activity.The placental enzyme was inhibited by the MAb-1-7-1,whereas the hepatic activity
did not respond to the antibody.We could confirm erlier findings concerning the inhibition by MAbs of the placental and hepatic ECDE activities.The placental ECDE activity was partially
inhibited by the MAb-1-7-1,whereas the hepatic enzyme was not
inhibited by the antibody.These findings allow the following
conclusion: since hepatic AHH and ERDE activities are not parallel with respect to cigarette smoking and MAb-1-7-1 inhibition,these activities are catalyzed by unique forms of P-450 or

the MAb-1-7-1 epitope binding differentiates between the two substrates.

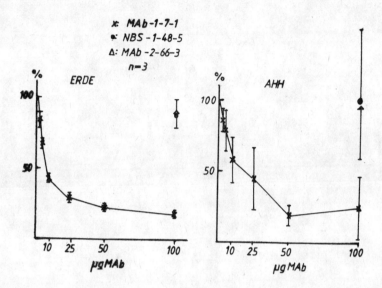

Fig.1. The inhibition of human placental ERDE and AHH activities by the monoclonal antibodies.Experiments were conducted with 3 placental samples from smokers with activities ranging between 15 and 26 /ERDE/ and 10 and 180 /AHH/ pmol/mg microsomal protein/min.

References

Jusko,W.J./1978/ Role of tobacco smoking in pharmacokinetics. J.Pharmacokin.Biopharm. 6: 7-39.
Köhler,G.,Milstein,C./1975/ Continuous cultures of fused cells secreting antibody of predefined specificity. Nature /London/ 256: 495-497.
Park,S.S.,Fujino,T.,West,D.,Guengerich,F.P.,Gelboin,H.V./1982/ Monoclonal antibodies that inhibit enzyme activity of 3-methylcholanthrene-induced cytochrome P-450. Cancer Res. 42: 1798-1808.
Pelkonen,O.,Pasanen,M.,Kuha,H.,Gachályi,B.,Kairaluoma,M.,
Sotaniemi,E.A.,Park,S.S.,Friedman,F.K.,Gelboin,H.V./1985/ Cigarette smoking and drug oxidations in human liver. Analyses with monoclonal antibodies. 26th Congress of the European Society of Toxicology,Kuopio,Finland,June 16-19.
Song,B.J.,Fujino,T.,Park,S.S.,Friedman,F.K.,Gelboin,H.V./1984/ Monoclonal antibody-directed radioimmunoassay of specific cytochromes P-450. J.Biol.Chem. 259: 1394-1397.

COMPARISON OF DEVELOPMENT OF LIVER MONOOXYGENASES IN TOLUENE
AND PHENOBARBITAL INDUCED RATS

Kaija Pyykkö

Department of Clinical Sciences, University of Tampere,
Tampere, Finland

INTRODUCTION

The methylbenzenes, toluene and xylene, are widely used
industrial organic solvents. They are metabolized in liver by
a cytochrome P-450 dependent enzyme system. They also induce
microsomal enzymes. The induction of various monooxygenases
by toluene depends on age and sex (Pyykkö 1983). The nature
of methylbenzenes as inducers is supposed to be similar to
that of phenobarbital (Toftgård 1983). In this study, the
development of responsiveness to toluene was compared to that
of phenobarbital in male rats from 20 to 180 days of age in
order to detect the similarities and differences in induction.

MATERIALS AND METHODS

Male Sprague-Dawley rats about 20, 40, 60, 90, 120 and 180
days old were used. Toluene 20 mmol/kg in corn oil was given
orally by gastric tube, and phenobarbital 80 mg/kg in 0,9%
NaCl by i.p. injection daily for three consecutive mornings.
Controls received oil or saline, respectively. On the fourth
morning the rats were killed and their livers removed. Microsomes were isolated and determinations performed by the
methods described previously (Pyykkö 1983). The results are
presented as means $\pm$ S.D. of 4 - 6 rats.

RESULTS AND DISCUSSION

The concentration of total cytochrome P-450 and the
activities of four cytochrome P-450 dependent monooxygenases
in liver microsomes of untreated and phenobarbital- or
toluene-treated male rats as a function of age are shown in
Fig. 1 - 5. In the control rats, both the cytochrome P-450
content and the aminopyrine N-demethylase and aryl hydrocarbon hydroxylase activities increased slightly or kept at
almost constant levels between 20 and 60 days of age; neither
did they change significantly thereafter. The activity of
aniline p-hydroxylase and 7-ethoxycoumarin O-deethylase in
the untreated rats decreased during the age period from 20 to
60 days, and then they remained at the low adult level. The
differences of developmental curves in the monooxygenase

activities reflect the multiplicity of cytochrome P-450
isozymes, the amounts of which vary with the age of animal
(Maeda et al. 1984). Various isozymes have different and more
or less specific affinity towards various substrates
(Guengerich et al. 1982).

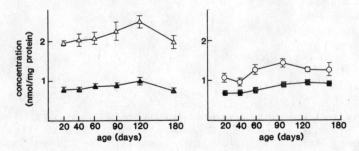

Fig.1. The effect of age on the concentration of cytochrome
P-450 in liver microsomes of phenobarbital-treated (△),
toluene-treated (o) and control (▲ , ●) male rats.

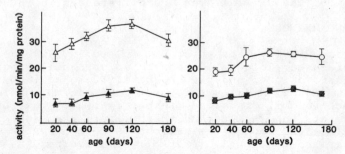

Fig.2. The effect of age on the activity of aminopyrine N-demethylase in the liver microsomes of phenobarbital-treated
(△), toluene-treated (o) and control (▲ , ●) male rats.

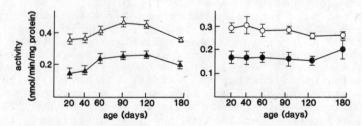

Fig.3. The effect of age on the activity of aryl hydrocarbon
hydroxylase in the liver microsomes of phenobarbital-treated
(△), toluene-treated (o) and control (▲ , ●) male rats.

355

The treatment with inducers emphasized the differences.
Both phenobarbital and toluene induced cytochrome P-450 and
all monooxygenases in all the age groups studied. The highest
induced levels were found at the age of 90 or 120 days. But
the response to induction varied depending on the enzyme and
inducer. With the doses of inducers used in this study,
phenobarbital increased total cytochrome P-450 concentration
over twice as much as toluene did. This difference was higher
than what occurred in monooxygenase activities. At the age of
90 days, phenobarbital and toluene increased the aminopyrine
N-demethylase 3.2- and 2.3-fold, aryl hydrocarbon hydroxylase
1.7- and 1.8-fold, aniline p-hydroxylase 2.5- and 2.4-fold
and 7-ethoxycoumarin O-deethylase 4.6- and 2.9-fold over the
control, respectively.

The most marked differences between toluene- and phenobar-
bital-induction were in the shapes of age-activity curves of
aniline p-hydroxylase and 7-ethoxycoumarin O-deethylase. The
toluene-induced levels, but not phenobarbital-induced ones,
of these enzymes decreased sharply from 20 to 40 days of age

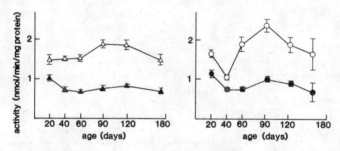

Fig.4. The effect of age on the activity of aniline p-hydroxylase in the liver microsomes of phenobarbital-treated (△),
toluene-treated (o) and control (▲ , ●) male rats.

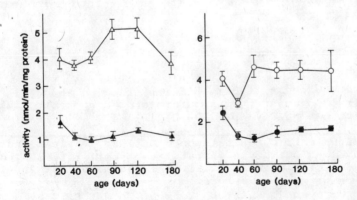

Fig.5. The effect of age on the activity of 7-ethoxycoumarin
O-deethylase in the liver microsomes of phenobarbital-treated
(△), toluene-treated (o) and control (▲ , ●) male rats.

24*

and increased then rapidly to the adult level. In accordance with the earlier findings (Schmucker 1981), the response to phenobarbital was weakened in the oldest age group, whereas the toluene-induced enzymes, except aniline p-hydroxylase, kept up the high levels until the age of 180 days.

Treatment by phenobarbital or 3-methylcholantrene has been found to induce several forms of cytochrome P-450, the proportion of which in induced hepatic microsomes is not the same as in untreated liver (Vlasuk et al. 1982). The quantitative differences between the inducibility of various monooxygenases and the total cytochrome P-450 in this study might mean that the several isozymes were synthetized and they were regulated individually. The occurrence of minima in toluene induced aniline p-hydroxylase and 7-ethoxycoumarin O-deethylase activities seems to suggest that the induced cytochrome P-450 forms presented in adult male rats were, at least partly, different from those in young rats. The minimum in the activities was at the junction of the decreasing immature and increasing mature forms of enzymes.

Though the methylbenzenes induce a cytochrome P-450 isozyme similar to the major isozyme induced by phenobarbital in the liver of adult male rat (Toftgård 1983), these inducers also seem to increase, especially in young rats, some other forms, which are not common to both of them.

ACKNOWLEDGEMENTS

This work was supported by a grant from the Finnish Academy.

REFERENCES

Guengerich FP, Dannan GA, Wright ST, Martin MV, Kaminsky LS (1982). Purification and characterization of microsomal cytochrome P-450s. Xenobiotica 12: 701-716.

Maeda K, Kamataki T, Nagai T, Kato R (1984). Postnatal development of constitutive forms of cytochrome P-450 in liver microsomes of male and female rats. Biochem Pharmacol 33: 509-512.

Pyykkö K (1983). Age- and sex-related differences in rat liver microsomal enzymes and their inducibility by toluene. Acta Pharmac Toxicol 53: 401-409.

Schmucker DL, Wang RK (1981). Effects of aging and phenobarbital on the rat liver microsomal drug-metabolizing system. Mech Ageing Dev 15: 189-202.

Toftgård R, Halpert J, Gustafsson J-Å (1983). Xylene induces a cytochrome P-450 isozyme in rat liver similar to the major isozyme induced by phenobarbital. Mol Pharmacol 23: 265-271.

Vlasuk GP, Ghrayeb J, Ryan DE, Reik L, Thomas PE, Levin W, Walz GW (1982). Multiplicity, Strain differences, and topology of phenobarbital-induced cytochromes P-450 in rat liver microsomes. Biochemistry 21: 789-798.

BIOMONITORING OF THE PETROLEUM SPILL IN VAASA ARCHIPELAGO, FINLAND BY BIOTRANSFORMATION ACTIVITIES IN FISH

LINDSTRÖM-SEPPÄ, P., HÄNNINEN, O. AND HUDD*, R.

Department of Physiology, University of Kuopio, P.O.B. 6, 70211 Kuopio, Finland and *Finnish Game and Fisheries Research Institute, Division in Vaasa, Rantakatu 28, 65120 Vaasa, Finland

INTRODUCTION

Cytochrome P-450 cored polysubstrate monooxygenase system in aquatic organisms is known to be induced by various environmental pollutants, e.g. petroleum (Payne and Penrose 1975, Stegeman 1981, Payne 1984). These early changes in enzyme activities have been expected to precede the onset of more serius damages at cellular and tissue level (Payne and Penrose 1975, Payne 1984). The induction can arise in less than a day as studied with rainbow trout (Salmo gairdneri) treated with beta-naphthoflavone (Andersson and Koivusaari 1984). According to Stegeman (1978) it can also exist for several years after one oil spill.

In the Vaasa Archipelago in the west coast of Finland there was an oil accident in autumn 1984. A ship (m/s Eira) run aground and oil was released to an area of 1500 km^2. The aim of this study was to clarify the metabolic effects of this oil spill on biotransformation ability in local wild fish (perch, Perca fluviatilis, white fish, Coregonus lavaretus, pike, Esox lucius, herring, Clupea harengus) and to biomonitor the duration of the spill effects, too. Simultaneously also laboratory studies were made by exposing rainbow trout (Salmo gairdneri) to the same oil. Here only the results from the wild perch and laboratory exposed rainbow trout are shortly presented.

MATERIALS AND METHODS

The fish were mainly perch (Perca fluviatilis) and of both sexes; length 23-31 cm, weight 180.1-416.9 g. Their livers were 2.15-8.57 g in weight. The perch studied were obtained from local fishermen in the Vaasa Archipelago in the west coast of Finland. The fish were frozen to -20 °C as whole and transported to Kuopio, where the liver samples were taken when the fish were still in frozen state. This unusual procedure of handling the fish was due to the circumstances. There were no possibilities in the remote sea to prepare the fish. The effects of the procedure on the biotransformation enzyme activities were controlled by studying rainbow trout at different storage conditions. The livers were let to melt in ice cold 0.25 M sucrose. The preparation of microsomes and cytosol fraction was carried out as reported previously (Lindström-Seppä et al. 1983).

The rainbow trout used in the laboratory experiments were immature (one year old) and about 250 g in weight. The livers weighted 2.43–6.15 g. They were bought from the Nilakkalohi Fish Farm Ltd in Tervo near Kuopio in Finland. The petroleum was given to the rainbow trout in olive oil by i.p. injection (200µl/kg). The control fish got olive oil only. The fish were sagrified four days after the injection. The livers were removed intact right after the slaughtering of the rainbow trout. The tissues were put to ice cold 0.25 M sucrose and the handling was carried out according to the same procedure as in the case of perch. The biotransformation activities were studied by measuring hepatic cytochrome P-450 content, monooxygenase enzyme activities (benzo(a)pyrene hydroxylase, 7–ethoxycoumarin O–de-ethylase, 7–ethoxyresorufin O–deethylase) and conjugation activities (UDPglucuronosyltransferase and glutathione S–transferase). The amount of protein was determined, too. The procedures were the same as reported previously (Lindström-Seppä et al. 1983).

RESULTS AND DISCUSSION

The handling of the fish by transporting them frozen as whole affects the enzyme activities measured. Monooxygenase activities decreased and glucuronidation activities increased. Glutathione S–transferase showed, however, no change.

Table 1. Hepatic biotransformation enzyme activities of perch (Perca fluviatilis) cought from the clean control and oil spill areas in Vaasa Archipelago, Finland.

	A		B	
	Control (n=4)	Oil spill (n=22)	Control (n=10)	Oil spill (n=13)
1.Benzo(a)pyrene hydroxylase	8.8± 7.1	22.9±12.1	11.9±11.0	14.1±10.9
2.7-Ethoxycoumarin O-deethylase	10.2± 2.1	13.3± 4.8	6.5±4.2	6.7± 3.7
3.UDPGlucuronosyl-transferase	ND	ND	61.2±42.9	66.9±36.7
4.Glutathione S-transferase	32.5± 2.7	79.7±15.3	69.6±24.5	54.3±19.6

A=four months after the oil spill
B=eight months after the oil spill
n=number of fish studied
ND=not detected
(activities measured at 18°C and expressed as pmol/min x mg prot (1,2 and 3) and nmol/min x mg prot (4), $\bar{x}\pm SD$)

The results showed only a slight monooxygenase (benzo(a)pyrene hydroxylase) induction in the case of wild perch when the samples were collected four months after the accident (Table 1). When eight months had passed no difference was seen between the fish from spill and control areas (Table 1).

This probably indicates the disappearance of induction, but a wide scatter of values was obvious. At the latter time the fish were beginning to rise from the deeper waters and to migrate to the spawning areas. There could be some mixing of the fish from the spill and control areas. It has been shown that spawning and the preparing to spawn affects the enzyme activities (Koivusaari 1984, Lindström-Seppä 1985). This can overshadow the influence of the possible induction and to increase the scatter.

The hepatic glutathione S-transferase was significantly induced in the oil exposed perch when the samples were collected four months after the accident. Eight months later no difference was seen.

The laboratory experiments with the spilled oil showed a clear induction in rainbow trout liver. Cytochrome P-450 content increased about 30 percent. 7-Ethoxycoumarin O-deethylase activities increased about ten and 7-ethoxyresorufin about fifteen fold. Benzo(a)pyrene hydroxylase showed only slight increase which was, however, not statistically significant (Table 2). Laboratory experiments showed no glutathione S-transferase induction in rainbow trout within the exposure time.

Table 2. Influence of spilled petroleum on the biotransformation enzyme activities in rainbow trout (Salmo gairdneri) liver in four days in laboratory conditions.

	Control (n=3)	Oil (n=5)
1.Cytochrome P-450 content	220.1+ 10.7 **	308.9+ 33.3
2.Benzo(a)pyrene hydroxylase	29.6+ 13.5	42.6+ 7.9
3.7-Ethoxycoumarin O-deethylase	61.5+ 17.7 ***	593.3+100.2
4.7-Ethoxyresorufin O-deethylase	260.5+227.8 ***	4018.8+537.6
5.UDPGlucuronosyltransferase	211.3+ 33.0	232.4+ 65.6
6.Glutathione S-transferase	430.5+ 90.7	339.6+ 80.6

(activities measured at 18°C and expressed as pmol/mg prot (1), pmol/min x mg prot (2,3,4 and 5), nmol/min x mg prot (6), $\bar{x}$+SD)

The petroleum from the ship spill had induction properties as indicated in the laboratory tests in rainbow trout. In the field conditions the effect had, however, disappeared either permanently or temporarily during the followery winter from the wild perch.

ACKNOWLEDGEMENTS

This study has been supported by the Finnish Cultural Foundation.

360

REFERENCES

Andersson,T. and Koivusaari,U. (1984) Influence of environmental temperature on the induction of xenobiotic metabolism by ß-naphthoflavone in Rainbow trout (Salmo gairdneri). Submitted for publication in Toxicol. Appl. Pharmacol.

Koivusaari,U., (1984) Xenobiotic biotransformation in Rainbow trout, Salmo gairdneri, with special reference to seasons and reproduction. Ph.D. Thesis, Publications of the University of Kuopio, Kuopio, Finland.

Lindström-Seppä,P. (1985) Seasonal variation of the xenobiotic metabolizing enzyme activities in the liver of male and female vendace (Coregonus albula L.). Aquatic Toxicology (in press).

Lindström-Seppä,P., Koivusaari,U. and Hänninen,O. (1983) Metabolism of foreign compounds in freshwater crayfish (Astacus astacus L.) tissues. Aquatic Toxicology, 3, 35–46.

Payne,J.F. (1984) Mixed-function oxygenases in biological monitoring programs: Review of potential usage in different phyla of aquatic animals. In: Proceedings of International Symposium on Ecotoxicological Testing for the Marine Environment (eds. Persoone,G., Jaspers,E. and Claus,C.), Ghent, Belgium.

Payne,J.F. and Penrose,W.R. (1975) Induction of arylhydrocarbon (benzo(a)pyrene) hydroxylase in fish by petroleum. Bull. Environ. Contam. Toxicol., 14, 112–116.

Stegeman,J.J. (1978) Influence of environmental contamination on cytochrome P-450 mixed function oxygenases in fish: implications for recovery in the wild Harbor Marsh. J. Fish. Res. Board. Can. 35, 668–674.

CHARACTERIZATION OF SUBCELLULAR FRACTIONS FROM RAINBOW TROUT KIDNEY WITH PARTICULAR EMPHASIS ON XENOBIOTIC METABOLIZING ENZYMES

PESONEN,M., ANDERSSON*,T AND HÄNNINEN,O.

Department of Physiology, University of Kuopio, 70211 Kuopio, Finland and
*Department of Zoophysiology, Box 25059, S-40031 Göteborg, Sweden

INTRODUCTION

Cytochrome P-450 monooxygenases catalyze the biotransformation of a great variety of foreign, as well as endogenous, lipid-soluble compounds to more water-soluble products. As in mammals, highest concentration of cytochrome P-450 in fish is found in the liver. However, previous studies have indicated that fish kidney contain relatively high cytochrome P-450--mediated activities (1). We have therefore prepared and characterized subcellular fractions from the kidney of rainbow trout by differential centrifugation in order to study cytochrome P-450-mediated reactions. The fractions were characterized with both electron microscopy and enzyme markers.

MATERIALS AND METHODS

Cultured immature rainbow trout, (Salmo gairdneri), ranging in weight between 200-400 g, were used in experiments. The fish were kept in laboratory basins with aerated recirculating fresh water without food at a temperature of 10 °C before sampling.
The fish was stunned with a blow on the head, and the whole kidneys were homogenized in 4 volumes of 0.15 M KCl or of 0.15 M phosphate buffer (pH 7.4) containing 0.15 M KCl using 10 up-and-down strokes of Potter-Elvehjem glass-teflon homogenizer. The homogenate was then subfractionated by differential centrifugation. The scheme is illustrated in Fig 1. All steps were performed at 0-4 °C and trypsin inhibitor (1.5 mg/ml homogenate) was used to prevent protease activity.

Chemical and enzymatic assays were carried out using freshly prepared subfractions and performed in dublicate. DNA, marker for the nuclei, was extracted according to Schmidt and Thannhauser (2) and determined by a diphenylamine reaction (3,4), cytochrome oxidase, marker for mito-chondria(5), catalase, marker for peroxisomes (6), p-nitrophenyl-α-mannosidase, marker for Golgi apparatus (7), ß-glyserophosphatase, marker for lysosome (8), AMP-ase and alkaline phosphatase, marker for plasma membrane (9), lactate dehydrogenase, marker for cytosol (10), glucose-6-phosphatase and NADPH-cytochrome-c-reductase, marker for endoplasmic reticulum (11,12), 7-ethoxycoumarin-O-deethylase (13), epoxide hydrolase (14), UDP-glucuronosyltransferase (15) glutathione transferase (16) were all determined according to published procedures. Protein contents were measured by method of Lowry et al. (17). Enzyme incubations were carried out at 20 °C and assays were linear with time and protein concentration under the conditions used.

The content of DNA and the enzymatic activities in subcellular fractions were calculated as relative specific activities (RSA) as DeDuve Plots.

RESULTS AND DISCUSSION

The kidney microsomal fraction contained mainly endoplasmic reticulum as revealed by the examination with electron microscopy.

The subcellular enrichment of various marker and xenobiotic enzymes among fractions are seen in Fig 2.

The kidney microsomal fraction was contaminated by 14% of lysosomes and 7% of mitochondria, calculated from activities of ß-glycerophosphatase and cytochrome oxidase, respectively. Furthermore 20% and 40% of total activities of both AMP-ase and alkaline phosphatase indicated the contamination of microsomes with plasma membrane fragments. The contamination by other cytoplasmic organelles was, however, relatively low and it was approximately similar to the contamination of endoplasmic reticulum fraction obtained as microsomes of rainbow trout liver (18) and the trunk kidney of the Northern pike (19).

With exception of epoxide hydrolase, the subcellular distribution of cytochrome P-450-dependent monooxygenases, UDP-glucuronosyltransferase and glutathione transferase closely resembled those reported for trout liver. Epoxide hydrolase was recovered in the mitochondrial and microsomal fractions almost to the same extent.

The specific activities of cytochrome P-450 dependent monooxygenases in the kidney microsomes were of the same order of magnitude as in trout liver.

In addition to the liver the kidney contributes significantly to the xenobiotic metabolism in rainbow trout.

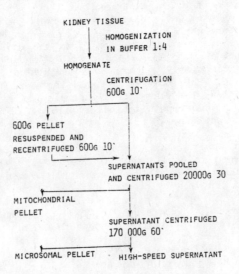

ISOLATION OF KIDNEY MICROSOMAL FRACTION

Fig 1. The scheme for the subcellular fractionation of the kidney homogenate for the isolation of microsomes from rainbow trout.

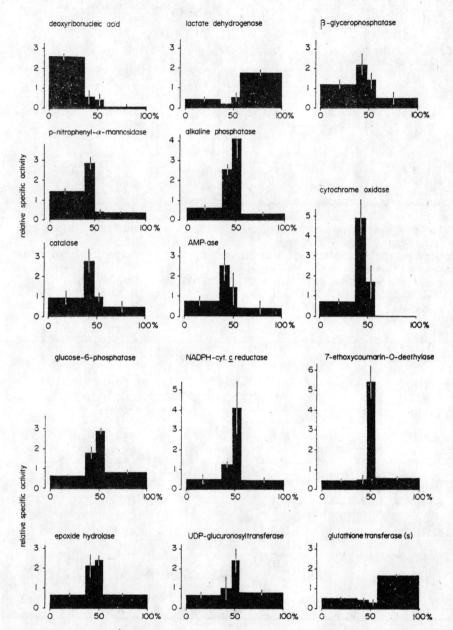

Fig 2. The distribution of DNA, marker enzymes and various xenobiotic biotransformation enzymes after subcellular fractionation of trout kidney by differential centrifugation. Fractions (left to right) are 600 g, 20,000 g, 170,000 g pellets and cytosol. The values are means of relative specific activities (RSA)±SE from 3 or 4 experiments each with pooled kidneys from 2–4 fishes.

REFERENCES

1. Lindström-Seppä, P., Koivusaari, U. and Hänninen O. (1981) Comp. Biochem. Physiol. 69C, 259-263.

2. Scmidt, G. and Thannhauser, J. (1945) J. Biol. Chem. 84, 161.

3. Burton, K. (1956) Biochem J. 62, 315.

4. Giles, K. W. and Mayers, A. (1965) Nature 93, 206.

5. Sottocasa, G., Kuylenstierna, G., Ernster, L. and Bergstrand, A. (1967) J. Cell Biol. 32, 415.

6. Bergmeyer, H. U. (1955) Biochem. J. 327, 255.

7. Dewald, B. and Touster, O. (1973) J. Biol. Chem. 248, 7223.

8. Appelmans, F., Wattiaux, R. and DeDuve, C. (1955) Biochem. J. 59, 438.

9. Song, C. S. and Bodansky, O. (1967) J. Biol. Chem. 242., 694.

10. Bergemayer, H. U. and Bernt, E. (1974) in: Methods in Enzymatic Analysis, Bergemayer H. U. (eds), Academic Press, New York 574.

11. Eriksson, L. C. (1973) Acta Pathologica et Microbiologica Scandinavica, supplement, No 239, section A.

12. Dallner, G., Siekewitz, P. and Plade, G. E. (1966) J. Cell Biol. 30.97.

13. Förlin, L. and Hansson, T. (1982) J. Endocrinol. 95, 245.

14. Jerina, D. M., Dansette, P. M., Lu, Ayh and Lewin, W (1977) Mol. Pharmac. 13, 342.

15. Hänninen, O. (1965) Rec. Prog. Horm. Res. Commun. 12, 134.

16. Habig, W. H., Pabst, M. J. and Jakoby, W. B. (1974) J. Biol. Chem. 249,7130.

17. Lowry, O. H., Rosebrough, N. J., Farr, A. L. and Randall, R. J. (1951) J. Biol. Chem. 193, 265.

18. Statham, C. N., Szyjka, S. P. and Lench, J. J. (1977) Biochem. Pharmac.26.1395.

19. Balk, L., Måner, S., Bergstrand, A. and DePierre, J. W. (1984) Biochem. Pharmac. 33, 2447.

INFLUENCE OF OCTYL GALLATE ON RAT MICROSOMAL DRUG METABOLIZING ENZYMES AND LIPID PEROXIDATION

J.Gnojkowski and W.Baer-Dubowska

Department of Biochemistry, Academy of Medicine
Grunwaldzka 6, 60-780 Poznań, Poland

INTRODUCTION

The administration of butylated hydroxyanisole/BHA/, butylated hydroxytoluene/BHT/ and other phenolic antioxidants to experimental animals results in protection against a number of chemical carcinogens/Wattenberg 1980/. Previous investigations on
the mechanism of the protective action of BHA have indicated
that this antioxidant affects the metabolism of carcinogens and
the activities of enzymes involved in the biotransformation of
foreign compounds. The inclusion of BHA in the diet of mice and
rats reduced the formation of benzo/a/pyrene/BP/ metabolites
that can bind to DNA and enhanced the activities of glutathione
S-transferase/GST/, epoxide hydrolase/EH/ and UDP-glucuronyltransferase that play an important role in the inactivation of
reactive forms of chemical carcinogens/De Long et al.1983, Kahl
and Kahl 1983/. Although in some countries gallic acid esters
are widely used as antioxidants in food their effects on drug-
metabolizing enzymes have been less defined. Inhibition of liver
microsomal monooxygenase by propyl, octyl and dodecyl gallate
has been observed in vitro. In vivo the effect of gallates on
hepatic drug metabolism was very moderated, only dodecyl gallate
decreased monooxygenase activity/Depner et al.1982/.
In the present study the effect of octyl gallate on the activities of several enzymes involved in the metabolism of xenobiotics was studied in the rat liver, kidney and lung. We also investigated the effect this antioxidant on the microsomal lipid
peroxidation.

MATERIALS AND METHODS

Male Wistar rats with an initial weight of 150g received octyl
gallate dissolved in sunflower oil in 2 ip. injections per week
for 2 weeks in a dose of 50 mg/kg. Preparation of microsomes was
performed as described previously/Gnojkowski et al.1984/. The
activity of aryl hydrocarbon hydroxylase/AHH/ was measured according to Nebert and Gelboin/1968/, NADPH-cytochrome c reductase
activity according to Phillips and Langdon/1962/, EH activity by
the fluorimetric procedure described by Dansette et al./1979/,
GST activity according to Habig et al./1974/. Protein was determined by the method of Lowry et al./1951/. The content of cytoch-

rome P-450 was measured according to Omura and Sato/1964/. Microsomal lipid peroxidation was measured by the thiobarbituric acid method/Slater and Sawyer 1971/.

RESULTS AND DISCUSSION

The effect of intraperitoneal administration of octyl gallate on the activities of microsomal AHH, NADPH-cytochrome c reductase, EH and cytosolic GST and on the content of cytochrome P-450 in the rat liver, kidney and lung is presented in table I.

Table I
Effect of octyl gallate on the drug-metabolizing enzymes in the rat liver, kidney and lung

	Liver		Kidney		Lung	
	Control	Octyl gallate	Control	Octyl gallate	Control	Octyl gallate
AHH /nmol/min/mg/	0,865 ±0,155	1,093 ±0,154	0,019 ±0,002	0,029[a] ±0,003	0,041 ±0,009	0,038 ±0,017
NADPH-cytochrome c reductase /nmol/min/mg/	180 ±12	206 ±18	27,7 -+4,8	35,4 ±10	33,4 ±2,0	28,4[a] -+1,9
Cytochrome P-450 /nmol/min/mg/	0,688 ±0,121	0,638 ±0,040	0,101 ±0,034	0,165[a] ±0,032	ND	ND
EH /nmol/min/mg/	2,40 ±0,91	3,09 ±0,81	0,338 ±0,068	0,456[a] ±0,065	0,046 ±0,009	0,049 -+0,010
Glutathione S-transferase /nmol/min/mg/	2497 ±106	3270 ±754	585 ±71	854[a] ±95	276 ±72	287 ±32

Values are means ± S.D. of five animals
[a]Significantly different from the control at $P < 0,05$
ND, not determined

The data indicate that pretreatment of animals with octyl gallate caused a significant increase in the activities of AHH, EH, GST and in the level of cytochrome P-450 and an increase in the activity of NADPH-cytochrome c reductase in the kidney. Slight but not statistically significant increase in the activities of these enzymes in the liver was also observed. In the lung only NADPH-cytochrome c reductase activity was decreased. Our results are in agreement with previous report/Depner et al. 1982/ that dietary of octyl gallate has no effect on hepatic drug-metabolizing enzymes. In vivo administration of octyl gallate results in significant increase of microsomal lipid peroxidation in the kidney /table II/. This is consistent with

Table II
In vivo and in vitro effect of octyl gallate on microsomal lipid peroxidation

	Lipid peroxidation /nmol/min/mg/	
	In vivo	In vitro[a]
Liver		
Control	0,178 ±0,032	0,178 ±0,017
Octyl gallate	0,154 ±0,032	0,135 ±0,023[b]
Kidney		
Control	0,181 ±0,028	0,241 ±0,008
Octyl gallate	0,352 ±0,109[b]	0,217 ±0,012[b]
Lung		
Control	0,171 ±0,035	0,262 ±0,008
Octyl gallate	0,195 ±0,024	0,239 ±0,004[b]

Values are means ±S.D. of five animals

[a]Microsome suspensions with 20 umol of octyl gallate were incubated for 15 min at $37^{o}C$

[b]Significantly different from the control at $P < 0,05$

Burton et al./1983/ suggestion that microsomal lipid peroxidation may depend on the content of cytochrome P-450 which can act as a propagating site for lipid peroxidation as well as an initiating site by producing free radicals intermediates. In vitro octyl gallate significantly inhibited microsomal lipid peroxidation in all tested organs. The data obtained in this paper suggest that octyl gallate may modify the activities of enzymes involved in the biotransformation of xenobiotics but this effect seems to be tissue depending.

REFERENCES

Burton G.W.,Cheesman K.H.,Ingold K.U. and Slater T.F./1983/ Lipid antioxidants and products of lipid peroxidation as potential tumour protective agents, Biochem. Soc. Trans. 11,261-262
Dansette P.M.,Dubois S. and Jerina D.M./1979/ Continuous fluorimetric assay of epoxide hydrase activity, Anal. Biochem. 97, 340-345
De Long M.J.,Prohaska H.J. and Talalay P./1983/ Substituted phenols as inducers of enzymes that inactivate electrophilic

368

compounds , in Protective Agents in Cancer /D.C.H. Mc Brien
and T.F. Slater eds./Academic Press, London, pp 175-196
Depner M.,Kahl G.F. and Kahl R./1982/ Influence of gallic acid
esters on drug-metabolizing enzymes of rat liver, Fd. Chem.
Toxic. 20, 507-511
Gnojkowski J.,Baer-Dubowska W., Klimek D. and Chmiel J./1984/
Effect of toluidines on drug metabolizing enzymes in rat liver,
kidney and lung, Toxicology 32, 335-342
Habig W.H.,Pabst M.J. and Jacoby W.B./1974/ Glutathione S-trans-
ferases.The first enzymatic step in mercapturic acid formation,
J. Biol. Chem. 249, 7130-7139
Kahl R. and Kahl G.F./1983/ Effect of dietary antioxidants on
benzo/a/pyrene metabolism in rat liver microsomes, Toxicology
28, 229-233
Lowry O.H., Rosebrough N.J., Farr A.L. and Randall R.J./1951/
Protein measurement with the Folin reagent, J. Biol. Chem.
193, 265-275
Nebert D.W. and Gelboin H.V./1968/ Substrate-inducible microsomal aryl hydroxylase in mammalian cell culture, J. Biol. Chem.
243, 6242-6249
Omura T. and Sato R./1964/ The carbon monoxide-binding pigment
of liver microsomes, J. Biol. Chem. 239, 2370-2378
Phillips A.H. and Langdon R.G./1962/ Hepatic triphosphopyridine
nucleotide-cytochrome c reductase: Isolation, characterization
and kinetic studies, J. Biol. Chem. 237, 2652-2660
Slater T.F. and Sawyer B.C./1971/ The stimulatory effects of
carbon tetrachloride and other halogenoalkanes on peroxidative
reactions in rat liver fractions in vitro, Biochem. J. 123,
805-814
Wattenberg L.W./1980/ Inhibitors of chemical carcinogens, J.En-
viron. Pathol. Toxicol. 3, 35-52

CYTOCHROME P-450 AND CARCINOGENESIS

MODULATION OF METABOLIC ACTIVATION OF BENZO(a)PYRENE BY AZO DYES

Shoichi Fujita, Takashi Matsunaga, and Tokuji Suzuki

Department of Biopharmaceutics, Faculty of Pharmaceutical sciences,
Chiba University, 1-33 Yayoicho, Chiba, Japan 260

A marked reduction in the incidence of 7,12-dimethylbenzanthracene (DMBA)-induced leukaemia in rats after pretreatment with an azo dye, Sudan III, was reported by Huggins et. al.(1978). Although the mechanism for the prevention of chemical carcinogenesis by sudan III has not been established to date, they implied that this may be due to increased detoxification of DMBA by quinone reductase induced by sudan III.

Conney and Levin (1966) observed increased metabolism of DMBA by sudan III pretreatment. Considering the existence of many pathways of DMBA metabolism, including detoxification and activation pathways, the induction of metabolism alone can not directly explain the prevention of metabolic activation.

With the use of microbial test system, we investigated the possible mechanism for the prevention of chemical carcinogenesis by sudan III. We also investigated whether other azo compounds possess similar ability to sudan III, and if they do, whether there is any structure-activity relationship.

Liver 9000g supernatant (S-9) fractions were prepared from rats treated with sudan III (40 mg/kg/day) for 1,2,3,5,and 10 days. Fig. 1 shows the menadione reductase activities of these S-9 fractions. As has been reported by Huggins et al.(1978), the activity increased with the increasing period of pretreatment. Benzo(a)pyrene (BP) was preincubated with these S-9 fractions prior to the microbial mutagenesis test (Ames test). On the contrary to the expectation, the numbers of the revertant colonies increased after preincubation of BP with S-9 fractions from sudan III treated rats with increasing days of pretreatment (Fig. 2). Apparently, sudan III does not show the effect observed in the animal study in the microbial assay system. Furthermore, the revertant colonies increased as menadione reductase activity increased. This does not necessarily mean that the activity of menadione reductase is responsible for the increase in the numbers of revertant colonies, but at least indicate that menadione reductase does not play a major role in prevention of metabolic activation of BP.

Before considering the mechanism for the prevention of metabolic activation of the carcinogen, we must consider what in liver S-9 fractions from sudan III-treated rats caused the increase of mutagenic potency of BP here. Fig. 3 shows the BP hydroxylase (AHH) activity of the fractions used for the test. AHH activity increased several to twenty fold by sudan III treatment. The number of mutant colonies and the levels of AHH activities of the S-9 fractions used showed a positive linear correlation (correlation coefficient r>0.9), suggesting that sudan III induced (an) enzyme(s) which activate(s)

25*

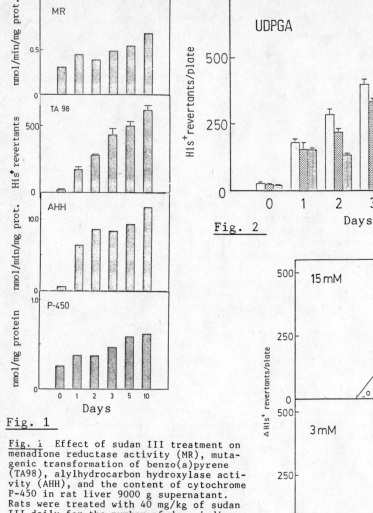

<u>Fig. 1</u>

<u>Fig. 1</u> Effect of sudan III treatment on
menadione reductase activity (MR), mutagenic transformation of benzo(a)pyrene
(TA98), alylhydrocarbon hydroxylase activity (AHH), and the content of cytochrome
P-450 in rat liver 9000 g supernatant.
Rats were treated with 40 mg/kg of sudan
III daily for the number of days indicated. Activities are expressed in nmol/
min/9000g supernatant protein.

<u>Fig. 2</u> Effect of UDP-glucuronic acid on
mutagenic transformation of benzo(a)pyrene
by liver 9000 g supernatant fractions from
sudan III treated rats.
UDP-Glucuronic acid (3 mM, 15 mM) was
added to the preincubation mixture containing liver 9000 g supernatant, NADPH-generating system, and benzo(a)pyrene. Ames
test was carried out using Salmonella TA
98 strain.

<u>Fig. 2</u>

<u>Fig. 3</u>

<u>Fig. 3</u> Relationship between UDP-glucuronyltransferase activities
and the decrease in mutagenicity of benzo(a)pyrene in the presence
of UDP-glucuronic acid in the preincubation mixture of Ames test.
Difference between the numbers of revertant colonies in the system
with and without added UDP-glucuronic acid in thepreincubation
mixture were plotted against the activities of UDP-glucuronyltransferase determined using p-nitrophenol as a substrate.

Table 1
Structures of azo dyes used

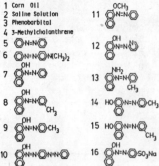

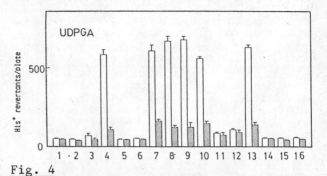

UDPGA

His⁺ revertants/plate

Fig. 4

Effect of various azo dye pretreatments on the
ability of liver S-9 fractions to mutagenically transform
benzo(a)pyrene in presence and absence of UDP-glucuronic
acid. The numbers correspond to the compound numbers
listed in Table 1. Open bars represent numbers of revertant
colonies without and shaded bars, with the presence of
added UDP-glucuronic acid in the preincubation mixture.

BP as mutagen in parallel with the induction of AHH. The CO difference
spectra of the reduced microsomes from livers of sudan III treated rats
showed absorptipon peak at 448 nm. The purification study of the induced
protein indicated that it is electrophoretically, immunologically, and
catalitically identical to the low spin cytochrome P-448 induced by 3-
methylcholanthrene.

It has been shown that cytochrome P-448 metabolically activates BP to
ultimate carcinogen /mutagen. The fact that sudan III induces P-448
explains the increase of mutagenesis, but does not explain the decrease in
chemically induced carcinogenic incidence by sudan III pretreatment observed
by Huggins et al.(1978). The difference between microbial mutagenicity test
and carcinogenesis test using animals is that in microbial test system, only
enzymes which are present in liver S-9 fraction and which utilize NADPH or
NADH are operative, while in animals, a whole set of enzymes including
detoxifying enzymes such as UDP-glucuronyltransferase (UDPGT) and glutathione-S-transferase (GST) are operative.

Sudan III induced prevention of carcinogenesis may be due to the
induction of these detoxifying enzymes. In fact, UDPGT and GST activities
tested using p-nitrophenol or 1-chloro-2,4,-dinitrobenzene as substrates
increased in S-9 fractions of liver from rats treated with sudan III for
more than two days. Therefore, we tested the effects of UDP-glucuronic acid
(UDPGA) or reduced glutathione (GSH), cofactors of UDPGT and GST, respectively, on the BP mutagenesis by adding them to the preincubation S-9 mixture
of the microbial test system. Fig. 2 indicates the effect of UDPGA on BP
mutagenesis. Marked reduction of mutagenic potency was observed when these
cofactores were added to the preincubation mixture as compared to that when
absence. Furthermore, the reduction in the numbers of revertant colonies
are highly correlated with the activities of UDPGT (Fig. 3) or GST of the S-
9 fractions used (r>0.9). Therefore, the reduction is due to metabolic
detoxification of active metabolic intermediates of BP by UDPGT or GST. It
should be noted that although the numbers of revertant colonies did not
decrease more than control levels, the half lives of BP as well as its
active metabolic intermediates in the animals treated with sudan III should
be greatly reduced. It is most likely that the effect of sudan III on
chemical carcinogenesis is a concerted effect of the activities of the
activating and the detoxifying enzymes, which are both induced by sudan III.

Is it only sudan III that has this effect or other azo compound too ?
We have screened about 40 azo compounds for the ability to induce cytochrome

P-448 and UDPGT activity. Several inducing and non-inducing azo compounds as well as archityrical inducers of cytochrome P-450, 3-methylcholanthrene (3-MC) and phenobarbital (PB) listed in table 1 were tested for their ability to modify the metabolic activation of BP.

As indicated in Fig. 4, only the inducing azo compounds and 3-MC caused a marked increase in mutagenic transformation of BP by the liver S-9 fractions from the rats treated with them. This effect was reversed by the addition of UDPGA (Fig. 4) or GSH (not shown). Non-inducing azo compounds did not alter mutagenic potency of BP with or without the addition of UDPGA or GSH. It is worth noting here, that 3-MC also showed similar effect to the inducing azo compounds, ie, prevention of mutagenesis by induced conjugating enzyme activities. This can explain the mechanism of prevention of chemical carcinogenesis by 3-MC pretreatment reported by Farber et al. (1968).

From our screening study it became apparent that only lipophilic azo compounds which possess 1-phenylazo-2-naphthol or 1-phenylazo-2-naphthylamine moieties were able to induce. An isomer of 1-phenylazo-2-naphthol, 1-phenylazo-4-naphthol does not induce, indicating that the position of hydroxyl group on naphthalene moiety is very important for the induction. Those 2-naphthol but not 4-naphthol azo compounds can form intramolecular hydrogen bonding to assume phenanthrene-like structure which is common to polycyclic hydrocarbons that induce cytochrome P-448. It may be this structure which is required for the induction of cytochrome P-448, UDPGT and GST activities.

Current study demonstrated the relationships between the structure of azo compounds - an ability to induce drug metabolizing enzymes - an ability to enhance BP elimination without causing marked build-up of active metabolic intermediates. Also this work directly demonstrated the importance of the balance between activating and detoxifying enzyme activities in metabolic activation of carcinogens. The latter could give an explanation to some of the unresolved discrepancies between the results of microbial mutagenesis test and carcinogenesis test using animals.

REFERENCES

Conney,A.H. and Levin,W. (1966) Induction of hepatic 7,12-dimethylbenz(a)anthracene metabolism by polycyclic aromatic hydrocarbons and aromatic azo derivatives. Life Sci., 5, 465-471.

Huggins,B.C., Ueda,N., and Russo,A. (1978) Azo dyes prevent hydrocarbon-induced leukemia in the rat. Proc. Natl. Acad. Sci. USA, 75, 4524-4527.

Marugami,M., Ito,N., Konishi,Y., Hiasa,Y., and Farber,E. (1968) Influence of 3-methylcholanthrene on liver carcinogenesis in rats ingesting DL-ethionine, 3'-methyl-4-dimethylaminoazobenzene, and N-2-fluorenylacetamide. Cancer Res., 27, 2011-2019.

CHARACTERIZATION OF THE RAT LIVER RECEPTOR FOR 2,3,7,8-TETRACHLORODIBENZO-p-DIOXIN

Lorenz Poellinger, Johan Lund and Jan-Åke Gustafsson

Dept. of Medical Nutrition, Karolinska Institute, Huddinge University Hospital F69, S-141 86 Huddinge, Sweden

INTRODUCTION.

A substantial amount of evidence indicates that several biological responses produced by 2,3,7,8-tetrachlorodibenzo-p-dioxin (TCDD) and its congeners are mediated by their binding to an intracellular, soluble receptor protein (Poland and Knutson, 1982). One interesting property of these compounds is their potency to induce specific forms of cytochrome P-450 (e.g. cytochrome P-450c in the rat) associated with aryl hydrocarbon hydroxylase activity (Poland and Knutson, 1982). A "two-step" model for the control of the expression of the cytochrome P-450c gene by the TCDD receptor has been postulated: subsequent to ligand-binding, the receptor undergoes a poorly understood process which results in the accumulation of the inducer-receptor complex within the cell nucleus (Poland and Knutson, 1982), followed by an increase in the rate of transcription of the cytochrome P-450c gene and in the content of cytochrome P-450c mRNA (Tukey et al., 1982; Israel and Whitlock, 1983, 1984).

Both this model for the mechanism of induction as well as the physicochemical properties of the TCDD receptor are strikingly similar to those reported for mammalian steroid hormone receptors (cf. Poellinger et al., 1985a). The salt-dissociated TCDD receptor has a sedimentation coefficient of 4-5 S, a Stokes radius of ~ 60 Å, a calculated relative molecular mass (M_r) of ~ 100,000 and a highly assymetrical form with a frictional ratio due to shape of ~ 1.7 (Poellinger et al., 1983). In further analogy to control of gene expression by steroid hormones, it seems as if at least two regulatory factors control the induction of the cytochrome P-450c gene expression. In addition to the TCDD-receptor complex, the presence of a cis-acting genomic control element flanking the 5' end of the cytochrome P-450c gene has recently been indicated (Jones et al., 1985).

MATERIALS AND METHODS.

The sources for all materials and details of all the used methods can be found in the following references: Poellinger et al., 1983, 1985b; Poellinger and Gullberg, 1985. Heparin-Sepharose chromatography was carried out essentially as described earlier (Poellinger et al., 1983) but in the presence or absence of 10-20 mM sodium molybdate. Non-labeled rat liver cytosol was applied to heparin-Sepharose columns, which were washed with 2-3 column volumes of ETG buffer (1 mM disodium EDTA, 20 mM Tris-HCl, pH 7.8, 10% (w/v) glycerol and 10 mM DTT), whereafter retained material was eluted with one column volume of buffer containing 0.5 M NaCl at a flow rate of approximately 4 ml/cm^2/h.

RESULTS AND DISCUSSION.

It has previously been shown that the TCDD receptor readily interacts with polyanions such as heparin-Sepharose (Poellinger et al., 1983) and DNA-cellulose (Carlstedt-Duke et al., 1981; Hannah, R.R., Lund, J., Poellinger, L., Gillner, M. and Gustafsson, J.-Å., submitted) indicating that the TCDD receptor, in further analogy to steroid hormone receptors, may in fact be a DNA-binding protein.

In the absence of ligand, the glucocorticoid receptor does not bind to crude DNA (cf. Radojcic et al., 1985). The unoccupied TCDD receptor elutes from heparin-Sepharose at an ionic strength of approximately 0.1 M KCl, whereas the occupied TCDD receptor elutes around 0.3 M KCl (Poellinger et al., 1983), suggesting that ligand binding increases the affinity of both steroid receptors and the TCDD receptor for polyanionic matrices. These properties might be of value for the development of a purification scheme for the TCDD receptor based on differential chromatography on either heparin-Sepharose or DNA-cellulose.

Sodium molybdate has remarkable effects on both functional and structural characteristics of steroid hormone receptors. For instance, molybdate has been shown to inhibit the process whereby steroid receptors acquire the ability to bind to cell nuclei or DNA and also to inhibit binding of steroid receptors to both specific and nonspecific DNA sequences (cf. Scheidereit et al., 1983). In the presence of 20 mM sodium molybdate, the liganded TCDD receptor elutesfrom heparin-Sepharose at an ionic strength of approximately 0.1 M KCl which is similar to the concentration of KCl required to elute the unoccupied TCDD receptor. Thus, sodium molybdate appears to detrimentally affect the polyanion-binding properties of both steroid hormone receptors and the TCDD receptor.

The presence of hydrophobic regions on the rat liver TCDD receptor can be demonstrated by its interaction with a series of n-alkyl agaroses. At 1 M NaCl, the TCDD receptor is adsorbed to uncharged pentyl Sepharose but not to butyl Sepharose. A chain length of C $\geq$ 5 thus appears to be sufficiently hydrophobic to retain the receptor. The TCDD receptor seems to be more hydrophobic than the androgen and the progesterone receptors since alkyl agaroses of longer chain-lengths (octyl-decyl agarose) at similar ionic strengths (Bruchowsky et al., 1981; Lamb and Bullock, 1983). The TCDD receptor is also adsorbed to phenyl Sepharose or Cibacron blue Sepharose at lower ionic strengths (0-0.15 M NaCl). Elution of adsorbed receptor cannot be achieved under mild conditions, i.e. decreasing salt concentration, increasing glycerol concentration. A concentration of 0.2% (w/v) of the zwitterionic detergent CHAPS is required to desorb the TCDD receptor from pentyl Sepharose. The recovery of receptor in the CHAPS elution step ranged between 11 and 14% of the applied receptor concentration, and the purification was 9- to 12-fold over that in crude cytosol, as estimated by velocity sedimentation analysis. The CHAPS-treated TCDD receptor exhibited the same hydrodynamic characteristics as the salt-dissociated receptor in crude liver cytosol

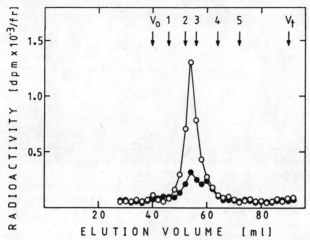

FIG. 1. Gel permeation chromatography of the TCDD receptor eluted from pentyl-Sepharose by CHAPS. Material eluted from pentyl-Sepharose by 0.2 % (w/v) CHAPS was labeled with 3 nM [^{3}H]TCDD in the presence (o) or absence (o) of 600 nM tetrachlorodibenzofuran and analyzed for specific binding on Sephacryl S-300 as described earlier (Poellinger et al., 1983). 1, thyroglobulin, (86 Å); 2, ferritin (62 Å); 3, catalase (51 Å); 4, albumin (36 Å) and 5, cytochrome c (18 Å).

(i.e. 4–5 S; Stokes radius ≅ 60 Å; M_r ~ 100,000; Fig. 1). The low yield in purification of receptor might not only reflect the efficiency of CHAPS in desorbing the receptor from pentyl-Sepharose, but also diffi- culties in labeling the partially purified receptor with ligand. Such difficulties have been observed during labeling of heparin-Sepharose eluates with [^{3}H]TCDD. Thus, hydrophobic interaction chromatography of the TCDD receptor may result only in a moderate purification of the receptor. However, this technique might be valuable as a step prior to affinity chromatography of the receptor with the aim to remove proteins interacting nonspecifically with the affinity ligand.

ACKNOWLEDGEMENTS.

This work was supported by grants from the Swedish Cancer So- ciety, the Swedish Board for Planning and Coordination of Research and the Swedish Society of Medical Sciences.

REFERENCES.

Bruchowsky, N., Rennie, P.S. and Comeau, T. (1981). Partial purification of nuclear androgen receptor by micrococcal nuclease digestion of chromatin and hydrophobic chromatography. Eur. J. Biochem. 120: 399-405.

Carlstedt-Duke, J., Harnemo, U.-B., Högberg, B. and Gustafsson, J.-Å. (1981). Interaction of the hepatic receptor protein for 2,3,7,8-tetrachlorodibenzo-p-dioxin with DNA. Biochim. Biophys. Acta 672: 131-141.

Israel, D.I. and Whitlock, J.P., Jr. (1983). Induction of mRNA specific for cytochrome P_1-450 in wild-type and variant mouse hepatoma cells. J. Biol. Chem. 258: 10390-10394.

Israel, D.I. and Whitlock, J.P., Jr. (1984). Regulation of cytochrome P_1-450 gene transcription by 2,3,7,8-tetrachlorodibenzo-p-dioxin in wild-type and variant mouse hepatoma cells. J. Biol. Chem. 259: 5400-5402.

Jones, P.B., Galeazzi, D.R., Fischer, J.M. and Whitlock, J.P., Jr. (1985). Control of cytochrome P_1-450 gene expression by dioxin. Science 227: 1499-1502.

Lamb, D.J. and Bullock, D.W. (1983). Hydrophobic interaction chromatography of the rabbit uterine progesterone receptor. J. Steroid Biochem. 19: 1039-1045.

Poellinger, L. and Gullberg, D. (1985). Characterization of the hydrophobic properties of the receptor for 2,3,7,8-tetrachlorodibenzo-p-dioxin. Mol. Pharmacol. 27: 271-276.

Poellinger, L., Lund, J., Gillner, M., Hansson, L.-A. and Gustafsson, J.-Å. (1983). Physico-chemical characterization of specific and nonspecific polyaromatic hydrocarbon binders in rat and mouse liver cytosol. J. Biol. Chem. 258: 13535-13542.

Poellinger, L., Lund, J., Gillner, M. and Gustafsson, J.-Å. (1985a). The receptor for 2,3,7,8-tetrachlorodibenzo-p-dioxin: similarities and dissimilarities with steroid hormone receptors. In: Molecular Mechanism of Steroid Hormone Action (Moudgil, V.K., ed.), Walter de Gruyter, New York, pp. 755-790.

Poellinger, L., Lund, J., Dahlberg, E. and Gustafsson, J.-Å. (1985b). A hydroxylapatite microassay for receptor binding of 2,3,7,8-tetrachlorodibenzo-p-dioxin and 3-methylcholanthrene in various target tissues. Anal. Biochem. 144: 371-384.

Poland, A. and Knutson, J.C. (1982). 2,3,7,8-tetrachlorodibenzo-p-dioxin and related halogenated aromatic hydrocarbons: examination of the mechanism of toxicity. Ann. Rev. Pharmacol. Toxicol. 22: 517-554.

Radojcic, M., Okret, S., Wrange, Ö and Gustafsson, J.-Å. (1985). Characterization of non-liganded glucocorticoid receptor from rat liver cytosol using an enzyme-linked immunosorbent assay. J. Steroid. Biochem., in press.

Scheidereit, C., Geisse, S., Westphal, H.M. and Beato, M. (1983). The glucocorticoid receptor binds to defined sequences near the promoter of mouse mammary tumour virus. Nature 304: 749-752.

Tukey, R.H., Hannah, R.R., Negishi, M., Nebert, D.W. and Eisen, H.J. (1982). The Ah locus: correlation of intranuclear appearance of inducer-receptor complex with induction of cytochrome P_1-450 mRNA. Cell 31: 275-284.

COMPARISON IN THE CHARACTER OF PURIFIED PULMONARY CYTOCHROME P-450 BETWEEN
3-METHYLCHOLANTHRENE-TREATED RAT AND HAMSTER

Watanabe, M., Sagami, I., Ohmachi, T. and Fujii, H.

The Research Institute for Tuberculosis and Cancer, Tohoku University,
Sendai 980, JAPAN

INTRODUCTION

Chemical carcinogens, such as polycyclic aromatic hydrocarbons and nitrosamines, are ubiquitous in the human environment, and enter the human body
through the surface epithelium. The surface area of respiratory tract is
directly exposed to some of these compounds during inhalation. Furthermore,
it is known that some chemical carcinogens, such as benzo[a]pyrene (BP),
3-methylcholanthrene (MC) and diethylnitrosamine, produce lung cancer in
mice, rats and hamsters (8).

Multiple forms of cytochrome P-450 (P-450) have been highly purified from
liver microsomes of several animals. On the other hand, only a few papers
have appeared concerning the characterization of purified pulmonary P-450
from MC-treated rats, and this form of P-450 was expressed high catalytic
activity toward BP hydroxylation (11,12). It is well known that hamster is
more susceptible than rat to grow tumor formation in the lung (6). Recently, Chiang and Steggles have described that   a P-450 was induced
in hamster lung microsomes by MC (1). In the present study, we have identified and characterized a major form of pulmonary P-450 in MC-treated
Syrian golden hamster, named as P-450$_{MC}$, and a comparison in the character
between rat and hamster pulmonary P-450$_{MC}$ was performed.

MATERIALS AND METHODS

Purification. Pulmonary microsomal P-450$_{MC}$ from MC-treated Syrian golden
hamsters and Buffalo rats was purified, respectively (10,11). NADPH-cytochrome c reductase (fp$_T$) from the hamster and rat liver microsomes was
purified by the modified method of Yasukochi and Masters (13). Epoxide
hydrolase was purified from the rat liver microsome by the modified method
of Guengerich et al. (3).

Analytical procedure. The content of pulmonary microsomal P-450 was
determined by the method of Johanessen and DePierre (4) to avoid the
interference of contaminating hemoglobin. Catalytic activity was assayed
in the reconstituted system containing P-450, fp$_T$, dilauroylphosphatidylcholine, NADPH and MgCl$_2$. Assay of BP hydroxylation activity was performed
by measuring the formation of 3-hydroxy BP according to Nebert and Gelboin
(7). The activities of 7-ethoxycoumarin O-deethylation and benzphetamine
N-demethylation were assayed by the modified method of Guengerich (2) and
Philpot et al. (9), respectively. Sodium dodecyl sulfate-polyacrylamide
gel electrophoresis (SDS-PAGE) was performed by the method of Laemmli (5).

Analysis of ^{14}C-BP metabolites. The reaction mixture contained P-450$_{MC}$,
fp$_T$, epoxide hydrolase, phospholipid, NADPH and MgCl$_2$. HPLC was performed
to analyze BP metabolites using a Japan Spectroscopic Liquid Chromatograph

Table 1. Comparison of pulmonary cytochrome P-450$_{MC}$ in 3-methylcholanthrene-treated rat and hamster

	Rat	Hamster
Specific content of purified forms (nmol/mg protein)	12.5	14.2
Purification fold	313	165
Molecular weight on SDS-PAGE (Mr. × 10³)	54	56
Maximum of reduced CO-complex (nm)	447.5	446.5
Spin state of heme iron	low	low
Catalytic activity in reconstituted system (mol/min/mol P-450)		
Benzo[a]pyrene 3-hydroxylation	11.9	11.4
7-Ethoxycoumarin O-deethylation	23.5	0.93
Benzphetamine N-demethylation	6.3	12.6
Ouchterlony double diffusion with antibody against rat hepatic P-450$_{MC}$	single fused line	faint precipitation line with spur formation

Trirotar and a Dupont Zorbax octadecyltrimethoxysilane column.

RESULTS AND DISCUSSION

The content of P-450 in the hamster lung was low, compared with that in the rat lung. At the present time no paper was available in the characterization of purified hamster lung P-450. As summarized in Table 1, a major form of hamster lung P-450$_{MC}$ was purified approximately 165 fold from the lung microsomes of MC-treated hamster, and was detected no activity of epoxide hydrolase. The minimum molecular weight of hamster lung P-450$_{MC}$ was estimated by SDS-PAGE to be 56,000 and was clearly different from that of rat lung P-450$_{MC}$. The BP hydroxylation activity in the reconstituted system was similar in hamster and rat lung P-450$_{MC}$, but the activities of 7-ethoxycoumarin O-deethylation and benzphetamine N-demethylation were apparently different. In the Ouchterlony double diffusion analysis the antibody prepared against the purified rat hepatic P-450$_{MC}$ reacted well with rat pulmonary P-450$_{MC}$ and formed a single fused precipitation line, whereas the purified hamster pulmonary P-450$_{MC}$ reacted with the antibody to form a faint precipitation line with spur formation, suggesting the presence of partial immunological similarity between the two P-450$_{MC}$. We tried to compare the HPLC patterns of BP metabolites formed by pulmonary microsome between rat and hamster. As shown in Table 2, the specific activity of total BP metabolites in hamster was apparently lower than that in rats, but higher relative formations of 9,10- and 7,8- diols, 1,6- and 3,6- diones and 6,12-dione were observed. In the reconstituted enzyme system containing epoxide hydrolase there were subtle different patterns of BP metabolites between the hamster and rat, as shown in Table 3, indicating that the increased specific activity for the formation of 7,8- and 9,10- diols and the decreased activity for the formation of 4,5-diol, 1,6- and 3,6- diones were detected in hamster P-450$_{MC}$, compared with the patterns of rat P-450$_{MC}$. These studies have identified a obvious relationship between BP metabolism by P-450 and the susceptibility in hamster lung to BP.

Table 2. Benzo[a]pyrene metabolites formed by lung microsomes of 3-methylcholanthrene-treated rat and hamster

Metabolites	Specific activity (pmol/min/nmol P-450)	
	Rat	Hamster
9,10-Diol	234 (6.5)[a]	129 (9.4)
4,5 -Diol	264 (7.3)	38 (2.8)
7,8 -Diol	146 (4.1)	135 (9.8)
1,6- and 3,6- Diones	463 (12.9)	278 (20.2)
6,12-Dione	126 (3.5)	102 (7.4)
9-Phenol	479 (13.3)	92 (6.7)
3-Phenol	1,886 (52.4)	502 (36.5)
Unknown	N.D.[b]	98 (7.1)
Total	3,598 (100)	1,374 (100)

[a] The percentage of each metabolite to total is shown in parentheses.
[b] N.D., Not detectable

Table 3. Benzo[a]pyrene metabolites formed by purified rat and hamster pulmonary P-450$_{MC}$ in the reconstituted system[a]

Metabolites	Specific activity (pmol/min/nmol P-450)	
	Rat	Hamster
Pre-9,10	261 (4.1)[b]	106 (1.5)
9,10-Diol	514 (8.1)	950 (13.0)
Unknown I	274 (4.3)	241 (3.3)
Unknown II	85 (1.3)	N.D.[c]
4,5-Diol	727 (11.5)	363 (5.0)
7,8-Diol	226 (3.6)	1,177 (16.1)
Unknown III	105 (1.7)	N.D.
Unknown IV	220 (3.5)	N.D.
1,6- and 3,6- Diones	1,554 (24.5)	955 (13.1)
6,12-Dione	144 (2.3)	257 (3.5)
9-Phenol	177 (2.8)	209 (2.9)
3-Phenol	1,981 (31.3)	2,801 (38.4)
Unknown V	69 (1.1)	239 (3.3)
Total	6,337 (100)	7,298 (100)

[a] The reconstituted system contained purified NADPH-cytochrome c reductase from rat or hamster liver, and purified epoxide hydrolase from rat liver.
[b] The percentage of each metabolite to total is shown in parentheses.
[c] N.D., Not detectable.

ACKNOWLEDGEMENTS
This research was supported, in part, by a Grant-in Aid for Cancer Research from the Ministry of Education and Culture, Japan, and by a grant from the Japan Tobacco and Salt Public Corporation.

REFERENCES
1) Chiang, J.Y.L. and Steggles, A.W. (1983) Identification and partial purification of hamster microsomal cytochrome P-450 isozymes. Biochem. Pharmacol., 32, 1389-1397.
2) Guengerich, F.P. (1978) Separation and purification of multiple forms of microsomal cytochrome P-450. J. Biol. Chem., 253, 7931-7939.
3) Guengerich, F.P., Wang, P., Mitchell, M.B. and Mason, P.S. (1979) Rat and human liver microsomal epoxide hydratase: Purification and evidence for the existence of multiple forms. J. Biol. Chem., 254, 12248-12254.
4) Johanessen, K.A. and DePierre, J.W. (1978) Measurement of cytochrome P-450 in the presence of large amounts of contaminating hemoglobin and methemoglobin. Anal. Biochem., 86, 725-732.
5) Laemmli, U.K. (1970) Cleavage of structural proteins during the assembly of the head of bacteriophage T₄. Nature (Lond.), 227, 680-685.
6) Mass, M.J. and Kaufman, D.G. (1983) A comparison between the activation of benzo[a]pyrene in organ cultures and microsomes from the tracheal epithelium of rats and hamsters.Carcinogenesis, 4, 297-303.
7) Nebert, D.W. and Gelboin, H.V. (1968) Substrate-inducible microsomal aryl hydroxylase in mammalian cell culture: I. Assay and properties of induced enzyme. J. Biol. Chem., 243, 6242-6249.
8) Nettesheim, P. and Griesemer, R.A. (1978) Experimental models for studies of respiratory tract carcinogenesis, In "Pathogenesis and Therapy of Lung Cancer", ed. C.C. Harris, Marcel Dekker, New York, pp. 75-188.
9) Philpot, R.M., Arinc, E. and Fouts, J.R. (1975) Reconstitution of the rabbit pulmonary microsomal mixed-function oxidase system from solubilized components. Drug Metab. Dispos., 3, 118-126.
10) Sagami, I., Ohmachi, T., Fujii, H. and Watanabe, M. Pulmonary microsomal cytochrome P-450 form 3-methylcholanthrene-treated hamsters; Purification, characterization and metabolism of benzo[a]pyrene. to be submitted.
11) Sagami, I. and Watanabe, M. (1983) Purification of pulmonary cytochrome P-450 from 3-methylcholanthrene-treated rats. J. Biochem., 93, 1499-1508.
12) Watanabe, M., Sagami, I., Ohmachi, T. and Fujii, H. (1985) Characteristics of purified cytochrome P-450s in microsomes of rat lung and Morris hepatoma 5123D. Gann Monograph, 30, 19-36.
13) Yasukochi, Y. and Masters, B.S.S. (1976) Some properties of a detergent-solubilized NADPH-cytochrome c reductase purified by biospecific affinity chromatography. J. Biol. Chem., 251, 5337-5344.

MUTAGENICITY OF POLYCYCLIC AROMATIC HYDROCARBONS:THE SIGNIFICANCE OF
A BAY-REGION AND GENETIC DIFFERENCES IN THE INDUCTION OF ARYL HYDROCARBON
HYDROXYLASE ACTIVITY

Raija Pahlman, Olavi Pelkonen and Veijo Raunio
Official address National Public Health Institute, SF-00280 Helsinki 28 and
Dep.Pharmacology, University of Oulu, Finland

Polycyclic aromatc hydrocarbons are widespread in environment and they
possess potential to cause cancer. All PAH are not, however, carcinogenic
and in fact this property is limited to a fraction of the many hundreds
of known PAH derivatives. Considerable variation of biological activity
within a series of PAH, and since the examination for oncogenic potential
of every compound is not possible, has led to attempts to predict carcinogenicity from chemical structure, differential metabolism, DNA reactivity
and/ or mutagenicity of active PAH metabolites in several short-term tests.
The bay-region theory is most widely accepted among structural hypotheses.
The metabolic activation of PAH by aryl hydrocarbon hydroxylase (AHH)
enzyme has also received considerable attention, because the induction of
AHH by PAH is under the control of the so called Ah locus.

We tested the mutagenicity of a number of PAH with or without the bay-
region with S. typhimurium tester strain TA 100 using as an activating
system S9 fraction from control and MC-induced rats or control, MC- and
TCDD-treated C57BL/ 6 (AHH responsive to PAH inducers) or DBA/ 2 (AHH non-
responsive to PAH inducers, but responsive to TCDD)mice.

There were significant differences between 1) bay-region and non-bay-
region PAHs, 2) between the activating effects of S9 fractions from
control and MC-pretreated B6 and D2 mice. These findings suggest,that
the Ah-locus-controlled inducible AHH enzyme metabolizes preferentially
bay-region PAH into metabolites mutagenic to TA 100 tester strain and
support the notion about the close linkage between the bay-region theory
and Ah locus theory of carcinogen activation.

HEPATIC MICROSOMAL HYDROXYLATION AND DNA BINDING OF A NON-AMINO AZO DYE, 1-PHENYLAZO-2-HYDROXYNAPHTALENE (SUDAN I) IN THE RAT

Z. Šípal, G. Befekadu, P. Hodek, E. Kovaříková and K. Stajner

Department of Biochemistry, Charles University, Albertov 2030
128 40 Prague 2, Czechoslovakia

INTRODUCTION

Covalent binding of an activated derivative of a carcinogenic substance with DNA is generally accepted as the initiating event in the process of chemical carcinogenesis (Miller
and Miller 1982). Polycyclic aromatic hydrocarbons (Levin et
al. 1982), nitrosamines (Magee et al. 1975), and aromatic a-
mines (Schut and Castonguay 1984) are activated to highly re
active electrophilic ultimate carcinogens through oxidative
biotransformation by microsomal monooxygenase systems. Molecular mechanisms of the activation and interaction with DNA
are well characterized for each of these three classes of c-
arcinogens; on the other hand there is no general mechanism
of activation of azo compounds known until now, in spite of
the fact that dimethylaminoazobenzene (DMAB) was among the f-
irst carcinogens shown to bind covalently with DNA   (Miller
and Miller 1947).

The biotransformation reactions of azo compounds may be
classified as follows: (a) reductive splitting of the N=N-
bond, (b) oxidative modification of functional groups,   (c)
hydroxylation of aromatic rings, and (d) oxidative splitting on the azo grouping. The first of the reactions results
in the formation of aromatic amines which may be the substrates for the second one; it is not, however, a general reaction of all azo compounds. Oxidative dealkylation followed by
N-oxidation and sulfatation is considered as a general mechanism of activation of amino azo compounds. It is not, however
related with the azo group and can not be the general mechanism for activation of all carcinogenic azo compounds.

In the present study we report the incorporation of a carcinogenic non-amino azo dye, 1-phenylazo-2-hydroxynaphtalene
(Sudan I, C.I. Solvent Yellow 14) into the calf thymus DNA
in vitro in connection with its biotransformation by the microsomal monooxygenase system from rat liver.

MATERIALS AND METHODS

^{14}C-4-dimethylaminoazobenzene (DMAB, 27.7 MBq/mmol) and
^{14}C-1-phenylazo-2-hydroxynaphtalene (PAHN, 23.4 MBq/mmol)
were prepared from ^{14}C-aniline (The Radiochemical Centre, A-

mersham) and purified by TLC on Silicagel G (Woelm). Male Wistar albino rats (Velaz n.p. Prague, Czechoslovakia) were
treated by 5,6-benzoflavone (5,6-BF, 20 mg/kg in corn oil
once daily i.p. for 2 days) or sodium phenobarbital (PB, 0.1
per cent solution in drinking water ad lib., 5-6 days). The
microsomal fraction of liver homogenate was prepared after
Remmer et al. (1967), the S9-fraction as the supernatant from 10-min centrifugation of the homogenate at 9000xg. Calf
thymus DNA was prepared after Kay et al. (1952). Incubations
were carried out in open tubes at 37°C. The incubation mixture contained 150 μmol phosphate buffer pH 7.5, 2 μmol NADPH
and 300 to 400 nmol of substrate in total volume 3.0 ml. In
binding studies 2.015 mg DNA was added. In the hydroxylation
studies the reaction was stopped by addition of 3.0 ml ethyl-
acetate, the dyes were extracted and analyzed by TLC in ben-
zene:ethylacetate (9:1) mixture as solvent. Separated substances were extracted by methanol and their amount was estimated by absorption spectroscopy at wavelengths corresponding
to absorption maxima. In the binding studies the reaction was
stopped by addition of 3 ml 80% phenol, DNA was isolated and
washed after King et al. (1975) and dissolved in 1.0 ml of
sodium citrate solution containing 0.25% of tetramethylammo-
nium hydroxide. 9.0 ml of PSC-scintillation solution (Amersh-
am/Searle) was added and the activity was measured by the I-
socape scintillation counter using standard program No. 2.
    The reconstituted system (Coon 1978) contained in 3.0 ml
of total volume 0.5 nmol cytochrome P-448, prepared from the
microsomal fractions of 5,6-BF-treated rats after Guengerich
et al. (1982) using Renex 690 instead of Lubrol PX as non-ionic detergent, 2.22 EU NADPH-cytochrome c reductase, 50 μg
L-dilauroylphosphatidylcholine and 150 μg sodium deoxycholate.

RESULTS AND DISCUSSION

    Using the S9-fraction of homogenate as the enzyme prepara-
tion comparable amounts of DMAB and PAHN were found to bind
with DNA (Table I.). Addition of ATP and sulfate raised the
incorporation of DMAB by about 40 per cent, but reduced the
incorporation of PAHN by about 20 per cent. While the later
result may be explained as a competition between DNA and ade-
nosine for the activated dye, the former one suggest the additive effect of the N-oxidation-sulfatation way of activation (Beland et al. 1980). This interpretation presumes an amino group-independent way of activation of azo compounds. The
binding of PAHN with DNA is completly suppressed in boiled
enzyme preparations and proceeds only in the presence of the
S9- or microsomal fraction of rat liver homogenate and NADPH.
    The activation reaction thus may be related with the mic-
rosomal monooxygenase system and quantified as the molecular
activity (v_{mol}) of the cytochrome P-450 present. This value
drops considerably with raising the concentration of the S9-
fraction in the incubation mixture; the same effect may be
observed when bovine serum albumin (BSA) is added to a microsomal preparation (Table II.), i.e. when only the total protein concentration is raised, without any change of cytochrome P-450 concentration. The effect may be interpreted in
both cases as the result of binding of the activated species
with proteins.

Table I. Binding of ^{14}C-azo dyes with DNA in the S9-fraction of liver homogenate from 5,6-benzoflavone-treated rats.

Substrate	Addition	Bound in 30 min (nmol g^{-1}DNA)	v_{mol} (min^{-1} g^{-1}DNA)
DMAB	none	68.7	2.01
	+ATP 1 mM +K$_2$SO$_4$ 1 mM	99.1	2.90
PAHN	none	116.8	3.42
	+ATP 1 mM +K$_2$SO$_4$ 1 mM	85.6	2.50

c_{prot} = 1.35 g l^{-1}, c_{P-450} = 3.8 x 10^{-7} mol l^{-1}
c_{NADPH} = 6.4 x 10^{-4} mol l^{-1}

Table II. Effect of protein concentration on the rates of binding of PAHN with DNA in S9- and microsomal fraction of liver homogenate from 5,6-benzoflavone-treated rats.

Enzyme preparation	c_{P-450} (µmol l^{-1})	c_{prot} (g l^{-1})	v_{mol} (min^{-1} g^{-1}DNA)
S9-Fraction	0.38	1.35	2,74
	1.23	4.38	0.85
Microsomes	1.61	1.49	2.76
	1.61	4.80	0.83

t = 30 min

The extent of binding is influenced by the pretreatment of the animals. Treatment with 5,6-BF affords microsomal preparation with rather constant molecular activity 2.16 ± 0.21 min^{-1}g^{-1}DNA (from 4 preparations at protein concentration about 4.5 g l^{-1}), while the activity of microsomal preparations from PB-treated rats is considerably lower (about 0.6 min^{-1}g^{-1}DNA, mean of 2 preparations). The incorporation into DNA is strongly inhibited by 7,8-benzoflavone and proadifen in 5,6-BF-treated rats, in PB-treated animals it is only slightly inhibited by proadifen, 7,8-benzoflavone being ineffective as inhibitor (Table III.). Mannitol was not an inhibitor in either case, thus precluding the participation of free hydroxyl radicals as the activating reagent.

It should be mentioned that the extent of binding with DNA is only about 0.03 per cent of the dose in 30 min; thus the activation is by no means the main biotransformation pathway

Table III. Effect of inhibitors of microsomal MFO-system on the rate of binding of PAHN with DNA in microsomes from rat liver.

Inducing agent	Inhibitor added	v_{mol} $(min^{-1}$ g^{-1}DNA)	Inhibition per cent
5,6-BF	none	2.52	
	7,8-benzoflavone, 10^{-4}M	0.23	91
	proadifen, 10^{-4}_2M	0.76	70
	mannitol, 10^{-2}M	2.88	--
PB	none	0.82	
	7,8-benzoflavone, 10^{-4}M	0.99	--
	proadifen, 10^{-4}_2M	0.61	26
	mannitol, 10^{-2}M	0.84	--

Table IV. Identification of hydroxylated metabolites of PAHN formed in hepatic microsomes from 5,6-benzoflavone-treated rats.

TLC in Bz:EtOAc=9:1 R_f	λ_{max}, nm	v_{mol} (min^{-1})	Corresponding standard
0.92	456		PAHN
0.52	392	1.71 ± 0.12	4-OH-PAHN
0.44	482	1.38 ± 0.06	6-OH-PAHN
0.08		< 0.2	unidentified

Table V. Rates of hydroxylation of PAHN in a reconstituted microsomal MFO-system containing cytochrome P-448.

Product	4-OH-PAHN	6-OH-PAHN	Total
v_{mol} (min^{-1})	3.75	6.31	10.06

of PAHN. After the incubation of this substrate with liver microsomes from control and PB-treated rats no hydroxylated product can be detected by TLC. With preparations from 5,6-BF -treated rats two principal products were isolated, identifi- ed as 1-(4-hydroxyphenylazo)-2-hydroxynaphtalene (4-OH-PAHN) and 1-phenylazo-2,6-dihydroxynaphtalene (6-OH-PAHN) by UV-VIS spectoscopy and TLC with comparison with synthetic standards

(Table IV.). Molecular activities (min^{-1}) were rather constant as well as the ratio of both trihydroxyderivatives. A minor colored product formed was not identified. Using labeled
substrate it was proved that the sum of these 4 substances e-
quals the amount of substrate used ($\pm$ about 1 per cent). Thus
the formation of any other substantial metabolite of PAHN in
microsomal preparation is precluded.

A reconstituted MFO-system containing electrophoretically
homogeneous hepatic microsomal cytochrome P-448 from 5,6-BF-
treated rats (corresponding most probably to isozyme βNF-B
of Guengerich et al. 1982) showed a high molecular activity
in hydroxylation of PAHN and a reversed ratio of the two isomeric products (Table V.). On the other hand the system was
absolutely inactive in activation of the dye for binding with
DNA.

In conclusion, PAHN may be converted to a derivative which
can bind with DNA in hepatic microsomes from 5,6-BF-treated
rats. The activity does not paralel thw hydroxylation of the
substrate, which is its main biotransformation reaction in
the preparation. As the mechanism of the activation may be
considered an oxidative splitting of the molecule on the azo
grouping, which was proved for DMAB in a model system containing cerium(IV) ions (Matrka and Pípalová 1982), producing
phenyldiazonium cation which can bind with DNA. The differen-
ceof the extent of hydroxylation of PAHN in microsomes and in
the reconstituted system suggests the presence of more than
one active isozyme of cytochrome P-450; the purified isozyme
used for reconstitution is ineffective in the activation reaction.

REFERENCES

Beland F.A., Tullis D.L., Kadlubar F.F., Straub K.M., and
    Evans P.E. (1980) Characterization of DNA adductcs of the
    carcinogen N-methyl-4-aminoazobenzene in vitro and in vivo.
    Chem.-Biol. Interact. 31, 1-17
Coon M.J. (1978) Reconstitution of the cytochrome P-450-con-
    taining mixed-function oxidase system of liver microsomes.
    Methods Enzymol. 52(Biomembranes Part C), 200-206
Guengerich F.P., Dannan G.A., Wright S.T., Martin M.V., and
    Kaminsky L.S. (1982) Purification and characterization of
    liver microsomal cytochromes P-450: electrophoretic, spec-
    tral, catalytic, and immunochemical properties and induci-
    bility of eight isoenzymes isolated from rats treated with
    phenobarbital or beta-naphtoflavone. Biochemistry 21,
    6019-6030
Kay G.P.M., Simmons N.S., and Doume H.L. (1952) An improved
    preparation of sodium desoxyribonucleate. J.Am.Chem.Soc.
    74, 1724-1726
King H.W.S., Thompson H.H., and Brookes P. (1975) Benzo(a)py-
    rene desoxyribonucleoside products isolated from DNA after
    metabolism of benzo(a)pyrene by rat liver microsomes in
    the presence of DNA. Cancer Res. 35, 1263-1269
Levin W., Wood A., Chang R., Ryan D., Thomas P., Yagi H.,
    Thakker D., Vyas K., Boyd C. and Chu S.-Y. (1982) Oxidati-
    ve metabolism of polycyclic aromatic hydrocarbons to ulti-

mate carcinogens. Drug Metabol. Rev. <u>13</u>, 550-580

Magee P.N., Pegg A.E., and Swann P.F. (1975) Molecular mechanisms of chemical carcinogenesis. In: Handbuch der Algemei-
ner Pathologie (Ed. E. Grundman) Springer Berlin Heidelberg
New York, pp. 329-420

Matrka M., and Pípalová J. (1982) Formation of arenediazonium
ion in oxidation of N,N-dimethyl-4-aminoazobenzene and some meta-substituted derivatives with cerium(IV) ion in acid
medium. Collection Czechoslovak Chem. Commun. <u>47</u>, 2711-2715

Miller E.C., and Miller J.A. (1947) The presence and significance of bound aminoazo dyes in the livers of rats fed p-
dimethylaminoazobenzene. Cancer Res. <u>7</u>, 468-480

Miller E.C., and Miller J.A. (1981) Searches for ultimate chemical carcinogens and their reactions with cellular macromolecules. Cancer (Philadelphia) <u>47</u>, 2327-2345

Remmer H., Greim H., Schenkman J.B., and Estabrook R.W. (1967)
Methods for the elevation of hepatic microsomal mixed function oxidase levels and cytochrome P-450. Methods Enzymol.
<u>10</u>(Oxidation and Phosphorylation), 703-708

Schut H.A., and Castonguay A. (1984) Metabolism of carcinogenic amino derivatives in various species and DNA alkylation
by their metabolites. Drug Metabol. Rev. <u>15</u>, 753-838

CARCINOGEN BINDING PROTEINS IN THE RAT PROSTATE

Peter Söderkvist, Lorenz Poellinger and Jan-Åke Gustafsson

Dept. of Medical Nutrition, Karolinska Institute, Huddinge University
Hospital F 69, S-141 86 Huddinge, Sweden

INTRODUCTION

As an environmental etiology of prostatic cancer has been suggested
(Goldsmith et al. 1980) several studies have addressed the issue whether
the prostatic gland possesses enzymes necessary for the metabolic
activation of chemical carcinogens. Following treatment with β-naphthoflavone (BNF) or 2,3,7,8-tetrachlorodibenzo-p-dioxin (TCDD) an approximately 500-fold induction of cytochrome P-450c is detectable by both
immunochemical techniques and aryl hydrocarbon hydroxylase (AHH) determinations in the rat ventral prostate (Söderkvist et al. 1982, Haaparanta
et al. 1983). Moreover, this induction was accompanied by an increased
capability to form mutagenic metabolites from the promutagens 2-amino-
fluorene and benzo(a)pyrene (BP) as determined in the Ames' Salmo-
nella/microsome assay (Söderkvist et al. 1983). The induction of AHH
activity is generally conceived to be mediated by a soluble receptor
protein, the TCDD-receptor, since TCDD is one of the most potent agonists
known today for both the enzyme-induction response (Poland and Knutson
1982) and for receptor binding (Poland et al. 1976). Potent agonists for
receptor binding include TCDD and such well-known carcinogens as 3-methyl-
cholanthrene (3-MC) and BP.

MATERIALS AND METHODS

Rats were killed and the ventral prostates were excised, minced and
homogenized in 2 volumes of buffer (20 mM potassiumphosphate, 1 mM
ethylenediaminetetraacetic acid, 2 mM 2-hydroxyethylmercaptan, 10% (w/v)
glycerol, pH 7.2 in a teflon/glass Potter-Elvehjem homogenizer. The
homogenate was centrifuged at 140,000 x g for 45 min. The supernatant,
hereafter termed cytosol, was removed and care was taken to avoid the
floating lipid layer. Purification of PSP was accomplished by anion-
exchange and gel permeation chromatography of prostatic cytosol as
described by Heyns et al. (1982). The sources for all materials and
details of all the used methods can be found in the following references:
(Forsgren et al. 1974; Lund et al. 1982; Poellinger et al. 1983).

RESULTS AND DISCUSSION

By means of gel permeation chromatography of ^{3}H-TCDD-labeled rat ventral prostatic cytosol on Sephacryl S-300 it was possible to distinguish between two discrete ^{3}H-TCDD binding species which both were included into the gel (Fig. 1). The first peak of radioactivity included into the gel, with an R_s of approxiamtely 60 Å, was displaceable by presence of a 200-fold molar excess of TCDF. High levels of nondisplaceable (non-specific) ^{3}H-TCDD binding eluted in the 25-28 Å region of the column, and disturbed further characterization of the 60 Å binding entity. Sedimen-tation in sucrose gradients showed that the specific ^{3}H-TCDD binding (60 Å) peak sedimented as an 8-10 entity in low ionic strength whereas in high ionic strength (0.4 M KCl) it sedimented as a 4-5 S entity. This salt-dependent sedimentation shift is characteristic of the TCDD receptor in several tissues (Lund et al. 1982; Poellinger et al. 1983). A Stokes' radius of 60 Å and a sedimentation coefficient of 4-5 S indicate a highly assymmetric molecule of approximately M_r 100,000, a frictional ratio due to shape, f/f_0, of 1.7-1.8 and an axial ratio of a prolate ellipsoid, a/b, of 12-15. All these hydrodynamic properties are indistinguishable from the physico-chemical characteristics of the TCDD receptor present in other tissues of the rat (Poellinger et al. 1983; Lund et al. 1982). These date indicate that this particular binding species represents the TCDD receptor in the rat prostate. The concentrations of the TCDD receptor ranged between 5-20 fmol/mg cytosolic protein as estimated by gel permeation chromatography on Sephacryl S-300. The hydrodynamic properties of the major nonspecific ^{3}H-TCDD binding entity in the rat ventral prostate (R_s of 25-28 Å, a sedimentation coefficient of 3.6-3.8 S) indicate an apparent M_r of 40,000-45,000, a frictional ration f/f_0 of 1.0-1.1 and an axial ratio, a/b, of 2-5 which is characteristic of a nearly symmetric molecule.

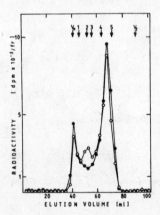

Fig. 1 Gel permeation chromatography of [^{3}H]TCDD labeled rat prostatic cytosol. Cytosol was labeled with 3 nM [^{3}H]TCDD in the presence (o) or absence (o) of 600 nM 2,3,7,8-tetrachlorodibenzofuran and analyzed for specific binding on Sephacryl S-300. 1, thyro-globulin, (86 Å); 2, ferritin (62 Å); 3, catalase (51 Å); 4, albumin (36 Å) and 5, cytochrome c (18 Å).

These data closely resemble the physico-chemical characteristics of a major secretory protein in the rat ventral prostate, referred to as PSP (prostatic secretory protein) or prostatic steroid binding protein (Heyns et al. 1977, 1978). Interestingly, PSP has been shown to exhibit a marked affinity (K_d ~ 30 nM) for estramustin, a nitrogen mustard derivative of estradiol used in the treatment of prostatic cancer (Forsgren et al. 1979).

Sedimentation analysis of purified PSP labeled with ^{3}H-TCDD revealed a single symmetrical peak of radioactivity in the same region of the gradients (3.5-4 S) as the non-saturable binding entity in crude cytosol. It was also possible to label purified PSP with ^{3}H 3-MC or ^{3}H BP as determined by velocity sedimentation. Scatchard analysis of the binding of ^{3}H-TCDD and ^{3}H-estramustine to diluted prostatic cytosol yielded linear plots with the same maximal number of binding sites. The K_d were calculated to 4.1 nM for estramustine and to 1.9 nM for ^{3}H-TCDD.

In the present study, ^{3}H-TCDD bound to a saturable binding species in crude cytosol with high affinity (K_d 2 nM). Several lines of evidence support the contention that this binding species, in fact, represents PSP: i) both ^{3}H-estramustine and ^{3}H-TCDD interacted with the same binding sites in crude cytosol as determined by competitive binding analysis; ii) the maximal number of binding sites were the same for both ligands; iii) the physico-chemical properties of the major ^{3}H-TCDD binding entity in crude cytosol closely resemble those reported for PSP; and iv) ^{3}H-TCDD (as well as ^{3}H 3-MC and ^{3}H-BP) readily interacted with purified PSP as assayed by velocity sedimentation analysis. It may be speculated that PSP is of importance in tissue specific accumulation of chemical carcinogens or in the induction process of cytochrome P-450 and in the metabolic activation of such compounds in the prostate.

REFERENCES

Forsgren, B., Björk, P, Carlström, K., Gustafsson, J.-Å., Pousette, Å., and Högberg, B. (1979) Purification and distribution of a major protein in the rat prostate that binds estramustine, a nitrogen mustard derivative of estradiol-17β. Proc. Natl. Acad. Sci. U.S.A. 76; 3149-3153.

Goldsmith, D.F., Smith, A.H., and McMichael, A.J. (1980) A case-control study of prostate cancer within a cohort of rubber and tire workers. J. Occup. Med. 22; 533-541.

Haaparanta, T., Halpert, J., Glaumann, H., and Gustafsson, J.-Å. (1983) Immunochemical detection and quantification of microsomal cytochrome P-450 and reduced nicotinamide adenine dinucleotide phosphate:cytochrome P-450 reductase in the rat ventral prostate. Cancer Res. 43; 5131-5137.

Heyns, W., and DeMoor, P. (1977) Prostatic binding protein; a steroid-binding protein secreted by rat prostate. Eur. J. Biochem. 78; 221-230.

Heyns, W., Peeters, B., Mous, J., Rombouts, W., and DeMoor, P. (1978) Purification and characterization of prostatic binding protein and its subunits. Eur. J. Biochem. 89; 181-186.

394

Heyns, W., Bossyns, D., Peeters, B., and Rombauts, W. (1982) Study of a proline-rich polypeptide bound to the prostatic binding protein of the rat ventral prostate. J. Biol.Chem. 257; 7407-7413.

Lund, J., Kurl, R.N., Poellinger, L., and Gustafsson, J.-Å. (1982) Cytosolic and nuclear binding for 2,3,7,8-tetrachlorodibenzo-p-dioxin in the rat thymus. Biochem. Biophys. Acta 716; 16-23.

Poellinger, L., Lund, J., Gillner, M., Hansson, L.-A., and Gustafsson, J.-Å. (1983) Physicochemical characterization of the specific and non-specific polycyclic hydrocarbon binders in the rat and mouse liver cytosol. J. Biol. Chem. 258; 13535-13542.

Poland, A., and Glover, E. (1974) Comparison of 2,3,7,8-tetrachlorodi-benzo-p-dioxin, a potent inducer of aryl hydrocarbon hydroxylase, with 3-methylcholanthrene. Mol. Pharmacol. 10; 349-359.

Poland, A., Glover, E., and Kende, A.S. (1976) Stereospecific, high-affinity binding of 2,3,7,8-tetrachlorodibenzo-p-dioxin by hepatic cytosol. Evidence that the binding species is a receptor for induction or aryl hydrocarbon hydroxylase. J. Biol. Chem. 251; 4936-4946.

Söderkvist, P., Toftgård, R., and Gustafsson, J.-Å. (1982) Induction of cytochrome P-450-related metabolic activities on the rat ventral prostate. Toxicol. Lett. 10; 61-69.

Söderkvist, P., Busk, L., Toftgård, R., and Gustafsson, J.-Å. (1983) Metabolic activation of promutagens, detectable in Ames' Salmonella ssay, by 5000 x g supernatant of rat ventral prostate. Chem. Biol. Interact. 46; 151-163.

DIFFERENCES BETWEEN THE Ah RECEPTOR AND STEROID HORMONE RECEPTORS IN
RESPONSE TO SODIUM MOLYBDATE

MICHAEL S. DENISON, LYNN M. VELLA and ALLAN B. OKEY

Division of Clinical Pharmacology, Department of Paediatrics, The Hospital
for Sick Children, 555 University Ave, Toronto, Ontario, CANADA, M5G 1X8

INTRODUCTION

The Ah receptor regulates induction of aryl hydrocarbon hydroxylase
(AHH, cytochrome P_1-450) by polycyclic aromatic hydrocarbons such as
3-methylcholanthrene and 2,3,7,8-tetrachlorodibenzo-p-dioxin (TCDD). After
binding of the inducing chemical, the association of the inducer:receptor
complex with DNA appears to stimulate the production of cytochrome P_1-450
mRNA (Jones et al., 1985; Poland et al., 1976; Tukey et al., 1982).

The overall mechanism of AHH induction by TCDD and the physiochemical
properties of the Ah receptor are in many ways similar to characteristics
of steroid hormones and their receptors (note however that competition
studies demonstrated that the Ah receptor is not simply one of the previously identified receptors for steroid hormone receptors (Poland et al.,
1976; Okey et al., 1979; Poellinger et al., 1983)). In general steroid
receptors sediment between 7-9S in low salt sucrose gradients while in
high salt conditions (0.4M KCl) the receptors dissociate to forms which
sediment between 4-5S (Sherman and Stevens, 1984). In addition, the presence of heparin has been observed to shift sedimentation of steroid receptors to the 4-5S form, a conversion which can be blocked by the presence of
molybdate (Thorsen, 1981; Yang et al., 1982). The rat hepatic Ah receptor
sediments between 8-10S in low salt conditions and dissociates to a form
which sediments between 4-6S in the presence of heparin or high salt (Denison et al., 1984). The work reported in this paper was performed to investigate the effect of molybdate on sedimentation of the rat hepatic Ah
receptor in order to determine if the Ah receptor responds to molybdate in
the same fashion as do receptors for steroid hormones.

MATERIALS AND METHODS

Hepatic cytosol from Sprague-Dawley rats was prepared as previously described (Okey and Vella, 1982) using homogenizing buffers with or without
20 mM molybdate. Cytosol was incubated with 10 nM [^{3}H]TCDD and assayed by
sucrose density centrifugation in 10-30% (v/v) sucrose gradients prepared
with or without 20 mM molybdate. Sedimentation coefficients (S) for different forms of the receptor were estimated from the sedimentation position of [^{14}C]BSA and [^{14}C]catalase included in each gradient as internal
sedimentation markers.

RESULTS AND DISCUSSION

Incubation of rat hepatic cytosol with [^{3}H]TCDD and subsequent sucrose
gradient analysis resulted in the formation of a single specific binding

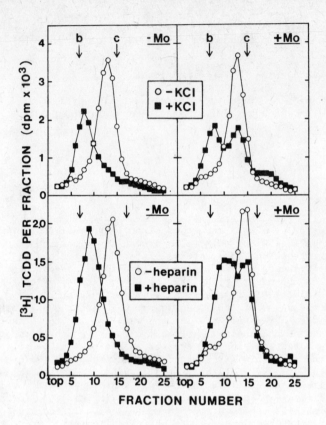

Figure 1. EFFECT OF MOLYBDATE ON THE SEDIMENTATION OF RAT HEPATIC Ah
RECEPTOR IN HIGH SALT CONDITIONS OR IN THE PRESENCE OF HEPARIN. Cytosol,
prepared in homogenizing buffer with or without 20 mM molybdate (Mo), was:
A) incubated in the presence or absence of 0.4M KCl for 15 min at 4°C
followed by the addition of 10 nM [3H] TCDD and further incubation for 1h
at 4°C; or B) incubated with 10 nM [3H] TCDD for 1 h at 4°C followed by
the addition of heparin (1mg/ml) and further incubation for 30 min at 4°C.
Aliquots (300ul) were analyzed by sucrose density centrifugation. Arrows
indicate the peak sedimentation of [14C]BSA (b, 4.4S) and catalase (c,11.3
S) included in each gradient as internal sedimentation markers.

peak which sedimented between 9-10S (near fraction 13) in low salt conditions in the presence or absence of molybdate (Fig. 1; Table 1). Treatment of cytosol with heparin or high salt (Fig. 1) shifted sedimentation of the specific binding peak to between 5-6S (near fraction 8) in the absence of molybdate (Table 1). In the presence of molybdate, treatment of cytosol with heparin or high salt produced biphasic peaks (Fig. 1) one a ("dissociated") peak at 5-6S and the other the original ("stabilized") peak at 9-10S (Table 1). Molybdate appears to stabilize a portion of the total Ah receptor pool in the higher sedimenting form while apparently not affecting another portion of receptor which shifts sedimentation to the 5-6S position.

TABLE 1

EFFECT OF MOLYBDATE ON THE SEDIMENTATION COEFFICIENT OF THE RAT HEPATIC Ah RECEPTOR IN THE PRESENCE OF HEPARIN OR HIGH SALT.[a]

Conditions	Sedimentation Coefficient ($S_{20,w}$)		
	low KCl	0.4M KCl	Heparin (1 mg/ml)
without molybdate	9.5 ± 0.7	5.5 ± 0.6	5.5 ± 0.8
with molybdate	9.6 ± 0.6	5.7 ± 0.3[b]	5.5 ± 0.6[b]
		9.7 ± 0.5[c]	9.6 ± 0.6[c]

a. Values are expressed as the mean ± SD of at least 4 individual determinations.
b. Sedimentation coefficient of "dissociated" form of the Ah receptor.
c. Sedimentation coefficient of "stabilized" form of the Ah receptor.

In contrast to these results with the Ah receptor, molybdate treatment of steroid receptors has been reported to stabilize the all receptors in the higher sedimenting (7-9S) form (Thorsen, 1981; Dougherty and Toft, 1982; Yang et al., 1982). In addition to these differences in response to molybdate, recent evidence from Poellinger and Gullberg (1985) indicates that the Ah receptor has more pronounced hydrophobic properties than those reported for steroid hormone receptors. In conclusion, although the Ah receptor and the receptors for steroid hormones may belong to a common family, the Ah receptor's response to molybdate indicates that the Ah receptor has certain properties distinct from the properties of receptors for steroid hormones.

ACKNOWLEDGEMENTS
Supported by a grant from the Medical Research Council of Canada to ABO. MSD is the reciepient of a National Research Service Award from the National Institute of Environmental Health Sciences, National Institutes of Health, U.S.A.

REFERENCES
M.S.Denison, L.M.Vella, A.B.Okey, Ah receptor for 2,3,7,8-tetrachlorodibenzo-p-dioxin: Species differences in reponse to high ionic strength conditions, presented at the Ninth Internat. Congr. Pharmacol., IUPHAR, London, England, 1984.
J.J.Dougherty and D.O.Toft, Characterization of two 8S forms of chick oviduct progesterone receptor, J. Biol. Chem. 257, 3113-3119 (1982).

P.B.C.Jones, D.R.Galeazzi, J.M.Fisher and J.P.Whitlock jr., Control of cytochrome P_1-450 gene expression by dioxin, Science 227, 1499–1502 (1985).

A.B.Okey and L.M.Vella, Binding of 3-methylcholanthrene and 2,3,7,8-tetrachlorodibenzo-p-dioxin to a common Ah receptor site in mouse and rat hepatic cytosols, Eur. J. Biochem. 127, 39–47 (1982).

A.B.Okey, G.P.Bondy, M.E.Mason, G.F.Kahl, H.J.Eisen, T.M.Guenthner and D.W.Nebert, Regulatory gene product of the Ah locus. Characterization of the cytosolic inducer-receptor complex and evidence for its nuclear translocation, J. Biol. Chem. 254, 11636–11648 (1979).

L.Poellinger and D.Gullberg, Characterization of the hydrophobic properties of the receptor for 2,3,7,8-tetrachlorodibenzo-p-dioxin, Molec. Pharmacol. 27, 271–276 (1985).

L.Poellinger, J.Lund, M.Gillner, L.-A.Hansson and J.-A.Gustafsson, Physiochemical characterization fo specific and nonspecific polyaromatic hydrocarbon binders in rat and mouse liver cytosols, J. Biol. Chem. 258, 13535–13542 (1983).

A.Poland, E.Glover and A.S.Kende, Stereospecific high affinity binding of 2,3,7,8-tetrachlorodibenzo-p-dioxin by hepatic cytosol. Evidence that the binding species is the receptor for induction of aryl hydrocarbon hydroxylase, J. Biol. Chem. 251, 4936–4946 (1976).

M.R.Sherman and J.Stevens, Structure of mammalian steroid receptors: evolving concepts and methodological developments, Ann. Rev. Physiol. 46, 83–105 (1984).

T.Thorsen, Interaction of heparin with the cytosolic progesterone receptor from human mammary tumours, J. Ster. Biochem. 14, 221–227 (1981).

R.H.Tukey, R.R.Hannah, M.Negishi, D.W.Nebert and H.J.Eisen, The Ah locus: correlation of intranuclear appearance of inducer-receptor complex with induction of cytochrome P_1-450 mRNA, Cell 3, 271–284 (1982).

C.R.Yang, J.Meister, A.Wolfson, J.-M.Renoir and E.-E.Balieu, Activation of the chick oviduct progesterone receptor by heparin in the presence or absence of hormone, Biochem. J. 208, 399–406 (1982).

ENHANCED ACTIVATION OF 4-AMINOBIPHENYL AND 2-AMINOFLUORENE BY CHRONIC ALCOHOL ADMINISTRATION TO HAMSTERS

Costas Ioannides and Christine M. Steele

Biochemistry Department, University of Surrey,
Guildford, Surrey, GU2 5XH, U.K.

INTRODUCTION

A positive association between alcohol consumption and tumour incidence at various sites has been established in a number of epidemiological studies (Lowenfels, 1974). Much of human cancer results from exposure to carcinogenic environmental chemicals, the majority of which require metabolic activation to intermediates that express their carcinogenicity. One of the major enzyme systems involved in the bioactivation of chemical carcinogens is the hepatic microsomal cytochromes P-450-dependent mixed-function oxidases. Although the association of alcohol intake with cancer incidence is likely to be multifactorial, a possible mechanism of the action of alcohol is enhanced bioactivation of precarcinogens. Such a mechanism of action has been demonstrated using the carcinogens dimethylnitrosamine (Garro et al., 1981), benzo(a)pyrene (Seitz et al., 1978) and nitrosopyrrolidine (McCoy et al., 1979). In the present paper we report the effect of chronic oral administration of alcohol on the bioactivation of two aromatic amines, namely 2-aminofluorene and 4-aminobiphenyl, to mutagens in the Ames test.

ANIMALS AND METHODS

Male golden Syrian hamsters were maintained on liquid diets containing 36% of total calories as alcohol for four weeks, while control animals were pair-fed isocaloric diets.

Post-mitochondrial supernatant (9,000 g, S_9) and microsomal fractions were prepared and the following determinations were carried out on the S_9 fraction, the N-demethylation of benzphetamine (Lu et al., 1972) and p-hydroxylation of aniline (Guarino et al., 1969); on the microsomal fraction, the O-deethylation of ethoxyresorufin (Burke and Mayer, 1974), cytochromes P-450 (Omuro and Sato, 1964) and protein (Lowry et al., 1951). Metabolic activation of the carcinogens to their mutagenic intermediates was determined using the Ames Test and employing Salmonella typhimurium strain TA 1538 and activation systems containing S_9 fractions at a final concentration of 10% (Ames et al., 1975).

RESULTS

The p-hydroxylation of aniline was markedly enhanced following chronic ingestion of alcohol (Table 1). The N-demethylation of benzphetamine and O-deethylation of ethoxyresorufin were also increased but to a much lesser extent while the microsomal levels of total cytochromes P-450 were unaffected by this treatment.

TABLE 1   HAMSTER HEPATIC MICROSOMAL MIXED-FUNCTION
OXIDASES FOLLOWING CHRONIC ALCOHOL ADMINISTRATION

Results are presented as mean $\pm$ SEM for five animals, * $p < 0.05$.

PARAMETER	CONTROL	ALCOHOL
Benzphetamine N-demethylase (nmol/min per mg protein)	11 ± 2	13 ± 1
Aniline p-hydroxylase (nmol/min per mg protein)	2.2 ± 0.4	4.6 ± 0.5*
Ethoxyresorufin O-deethylase (pmol/min per mg protein)	34 ± 7	49 ± 3
Cytochromes P-450 (nmol/mg protein)	1.0 ± 0.1	1.0 ± 0.1
Microsomal protein (mg/g liver)	16 ± 0.4	19 ± 0.1*

Hepatic S_9 preparations from the alcohol-treated animals were markedly more efficient in activating 4-aminobiphenyl and 2-aminofluorene to mutagens in the Ames test (Figure 1).   However, no increase was seen in the activation of the polycyclic aromatic hydrocarbons benzo(a)pyrene and 3-methylcholanthrene (results not shown).   Dimethylsulphoxide (DMSO) at a final concentration of 156-470 mM did not inhibit the alcohol-induced increase in the mutagenicity of the two aromatic amines (results not shown).

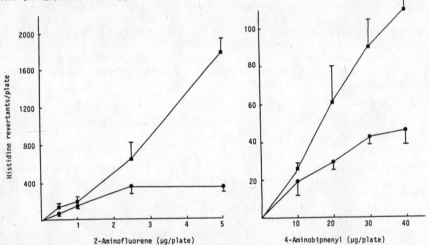

Figure 1:   Metabolic activation of 4-aminobiphenyl and 2-aminofluorene into mutagens by hepatic preparations from alcohol-treated (■———■) and pair-fed (●———●) hamsters in Salmonella typhimurium strain TA 1538.   The spontaneous reversion rate of 7 ± 2 has been substracted.   Results are presented as Mean $\pm$ SD for three plates.   Experiment was repeated twice with similar data.

DISCUSSION

Among a number of laboratory animals, the hamster was the most effective species in activating aromatic amines to mutagens (Phillipson and Ioannides, 1983) and was therefore used in the present study.   The initial step in the activation of this group of chemical carcinogens is believed to be an N-hydroxylation which may be catalysed by the microsomal mixed-function oxidases (Masson et al., 1983), the FAD-monooxygenase system (Pelroy and Gandolfi, 1980) and active forms of oxygen (Kadlubar et al., 1982). The activation systems employed in the present study were devoid of acetylase and phase II activities (McCann and Ames, 1977) and since the ring-hydroxylated products of aromatic amines are non-mutagenic (Masson et al., 1983), it can be inferred that ethanol enhances mutagenicity by facilitating the generation of N-hydroxylated metabolites.

Chronic alcohol administration has been shown to stimulate the oxidation of alcohols and other substrates by two pathways, one cytochrome P-450-dependent and one that involves interaction of the substrate with hydroxyl radicals (OH·) produced during the electron transfer (Krikun and Cederbaum, 1984). The OH· radical scavenger dimethylsulphoxide had no effect on the activation of either aromatic amine by control or alcohol-treated microsomal preparations indicating that the latter pathway is not responsible for the increased mutagenicity observed following ethanol consumption.

As previously reported in the rabbit (Morgan et al., 1982), ethanol administration had no effect on total cytochromes P-450 but caused changes in mixed-function oxidase activities indicating that this treatment has caused redistribution of the various cytochrome P-450 isozymes, inducing preferentially those favouring the N-hydroxylation of the aromatic amines. The minimal increases in ethoxyresorufin O-deethylase activity catalysed by cytochrome P-448 (Phillipson et al., 1984), and in benzphetamine N-demethylase, catalysed by the phenobarbital-induced cytochromes P-450 (Lu et al., 1972), cannot account for the marked increases in the activation of 4-aminobiphenyl and 2-aminofluorene. Furthermore, the activation of the two polycyclic aromatic hydrocarbons, which is specifically catalysed by cytochrome P-448 (Levin et al., 1977; Phillipson et al., 1985) were not stimulated by alcohol pretreatment. A unique form of cytochrome P-450 exhibiting high affinity towards aniline has been isolated from rabbits treated with alcohol (Morgan et al., 1982). Aniline hydroxylase was markedly induced in the present study, providing indirect evidence that such a form may also be induced by alcohol in the hamster. The possible contribution of the alcohol-induced form of cytochrome P-450 in the activation of aromatic amines merits investigation.

Finally, the activation of aromatic amines in the liver can be catalysed by the FAD-monooxygenase system, and the effect of alcohol on this enzyme system has not yet been investigated.

REFERENCES

Ames, B.N., McCann, J. and Yamasaki, E. (1975) Methods for detecting carcinogens and mutagens with the Salmonella/mammalian microsome mutagenicity test. Mutation Res., 31, 347-364.

Burke, M.D and Mayer, R.T. (1974) Ethoxyresorufin: Direct fluorimetric assay of a microsomal O-dealkylation which is preferentially inducible by 3-methylcholanthrene. Drug Metab. Disp., 2, 583-588.

Garro, A.J., Seitz, H.K. and Lieber, C.S. (1981) Enhancement of dimethylnitrosamine metabolism and activation to a mutagen following chronic ethanol consumption. Cancer Res., 41, 120-124.

Guarino, A.M., Gram, T.E., Gigon, P.L., Greene, F.E. and Gillette, J.R. (1969) Changes in Michaelis constants for aniline in hepatic microsomes from phenobarbitone-treated rats. Mol. Pharmacol., 5, 131-136.

Kadlubar, F.F., Frederick, C.B., Weis, C.C. and Zenser, T.V. (1982) Prostaglandin endoperoxide synthetase-medicated metabolism of carcinogenic aromatic amines and their binding to DNA and protein. Biochem. Biophys. Res. Commun., 108, 253-258.

Krikun, G. and Cederbaum, A.I. (1984) Stereochemical studies on the cytochrome P-450 and hydroxyl radical dependent pathways of 2-butanol oxidation by microsomes from Chow-fed, phenobarbital-treated and ethanol-treated rats. Biochemistry, 23, 5489-5494.

Levin, W., Wood, A.W., Lu, A.Y.H., Ryan, D., West, S., Conney, A.H., Thakker, D.R., Yagi, H. and Jerina, D.M. (1977) Role of purified cytochrome P-448 and epoxide hydrase in the activation and detoxication of benzo(a)pyrene, In Drug Metabolism Concepts, D.M. Jerina, ed.; ACS Symposium numbers 44, pp.99-126.

Lowenfels, A.B. (1974) Alcohol and cancer. N.Y. State J. Med., 74, 56-59.

Lowry, O.H., Rosebrough, N.J., Farr, A.L. and Randall, A.J. (1951) Protein measurement with the Folin-phenol reagent. J. Biol. Chem., 193, 265-275.

Lu, A.Y.H., Kuntzman, R., West, S., Jacobson, M. and Conney, A.H. (1972) Reconstituted liver microsomal enzyme system that hydroxylates drugs, other foreign compounds and endogenous substrates. II. Role of the cytochrome P-450 and P-448 fractions in drug and steroid hydroxylations. J. Biol. Chem., 247, 1727-1734.

Masson, H.A., Ioannides, C., Gorrod, J.W. and Gibson, G.G. (1983) The role of highly purified cytochrome P-450 isozymes in the activation of 4-aminobiphenyl to mutagenic products in the Ames test. Carcinogenesis, 4, 1583-1586.

McCann, J. and Ames, B.M. (1977) The Salmonella/microsomes mutagenicity test: Predictive value for animal carcinogenicity, In Origins of Human Cancer, H.H. Hiatt, J.D. Watson and J.A. Winsten, eds.; Cold Spring Harbor, New York, pp.1431-1450.

McCoy, G.D., Chen, C.B., Hecht, S.S. and McCoy, E.C. (1979) Enhanced metabolism and mutagenesis of nitrosopyrrolidine in liver fractions isolated from chronic ethanol-consuming hamsters. Cancer Res., 39, 793-796.

Morgan, E.T., Koop, D.R. and Coon, M.J. (1982) Catalytic activity of cytochrome P-450 isozyme 3a isolated from liver microsomes of ethanol-treated rabbits. J. Biol. Chem., 257, 13951-13957.

Omura, T. and Sato, R. (1964) The carbon monoxide pigment of liver microsomes. I. Evidence for its haemoprotein nature. J. Biol. Chem., 239, 2370-2378.

Pelroy, R.A. and Gandolfi, A.J. (1980) Use of a mixed-function amine oxidase for metabolic activation in the Ames/Salmonella assay system. Mutation Res., 72, 329-334.

Phillipson, C.E. and Ioannides, C. (1983) Activation of aromatic amines to mutagens by various animal species including man. Mutat. Res., 124, 325-336.

Phillipson, C.E., Godden, P.M.M., Lum, P.Y., Ioannides, C. and Parke, D.V. (1984) Determination of cytochrome P-448 activity in biological tissues. Biochem. J., 221, 81-88.

Phillipson, C.E., Ioannides, C., Barrett, D.C.A. and Parke, D.V. (1985) The homogeneity of rat liver microsomal cytochrome P-448 activity and its role in the activation of benzo(a)pyrene to mutagens. Int. J. Biochem., 17, 37-42.

Seitz, H.K., Garro, A.J. and Lieber, C.S. (1978) Effect of chronic ethanol ingestion on intestinal metabolism and mutagenicity of benzo(a)pyrene. Biochem. Biophys. Res. Commun., 85, 1061-1066.

THE BINDING OF BENZO(A)PYRENE TO DNA AND THE ACTIVITY OF DRUG METABOLIZING ENZYMES OF OCTYL GALLATE TREATED RATS

W.Baer-Dubowska, J.Gnojkowski and D.Filikowska

Department of Biochemistry, Academy of Medicine
ul. Grunwaldzka 6, 60-780 Poznań, Poland

INTRODUCTION

Antioxidants such as butylated hydroxyanisole(BHA),butylated hydroxytoluene(BHT) and ethoxyquin can inhibit the induction of tumors in rodents by a variety of carcinogens including polycyclic aromatic hydrocarbons (Wattenberg L. 1972).Several investigations on the biochemical mechanism responsible for the protective effect of BHA against chemical carcinogenesis revealed that dietary BHA altered liver microsomal metabolism by reducing the production of reactive metabolites, markedly elevated conjugation enzyme activities and inhibited the formation of benzo(a)pyrene(BP) : DNA adducts (Dock L. et al. 1982). In a number of countries,gallic acid esters are used as antioxidants in food but their biological effects have not been well characterized.According to Depner et al.(1982) gallic acid esters affected the microsomal drug metabolizing enzymes of rat liver but the found effects were moderated.

In this study we have examined the effect of octyl gallate ip administration on the activity of nuclear drug-metabolizing enzymes of a rat liver,kidney and lung.We have also investigated the level of BP binding to DNA in the nuclear and micro - somal preparations from the liver of untreated and octyl gallate treated rats.

MATERIALS AND METHODS

G-^{3}H BP(20Ci/mmol) was purchased from International Amersham U.K. Octyl gallate from Fluka AG,Neu-Ulm. NADPH,unlabeled BP,cytochrome c, calf thymus DNA were obtained from Sigma Chemical Co(St.Louis MO,USA). BP-4,5-oxide was a gift of Dr P. Dansette(Paris). Others chemicals were of analytical grade.

Male Wistar rats received 2 ip injections per week for 2 weeks in the dose of 50 mg kg. Nuclei and microsomes were prepared by method of Alexandrov et al.(1976). The activity of aryl hydrocarbon hydroxylase(AHH) was measured by the method of Nebert and Gelboin(1968),the NADPH-cytochrome c reductase activity by the method of Phillips and Langdon(1962) and epoxide hydrolase by the fluorimetric procedure described by Dansette et al.(1979). Protein content was determined according to Lowry(1951).

Isolation and analysis of BP bound to DNA

Purified DNA_2or nuclei were incubated for 1 hour with 40 or 80 nmoles of ^{3}H-BP(2,7 Ci/mmole) and 4 mg of NADPH. Incubations contained nuclei(avarage 2 mg) or 5 mg of calf thymus DNA and 8 mg of microsomal protein in 4 ml of 0,25 sucrose Tris-KCl-MgCl$_2$ buffer. DNA was isolated and purified as previously described(Guenthner et al.1980) and determined spectrophotometrically. Aliquots were analyzed for radioactivity. The specific activity(pmol/mg) of the DNA solutions was deter mined.

RESULTS AND DISCUSSION

Table I summarizes the data on the influence of octyl gal late on the activities of nuclear AHH ,NADPH-cytochrome c reductase and epoxide hydrolase.

Table I
Effect of ip administration of octyl gallate on the nuclear drug -metabolizing enzymes

	AHH (pmol/min/mg)	NADPH-cyto-chrome c reductase (nmol/min/mg)	Epoxide hydrolase (pmol/min/mg)
Liver			
Control	37,80 ± 6,33	8,52± 1,38	53,55 ± 8,41
Octyl gallate	20,86 ± 7,36	7,24± 1,55	58,68 ±12,46
Kidney			
Control	1,26 ± 0,35	4,75± 0,35	24,70 ± 5,89
Octyl gallate	2,50 ± 0,43[a]	6,49± 0,18[a]	29,47 ± 1,24
Lung			
Control	5,10 ± 0,20	4,10± 1,30	ND
Octyl gallate	5,46 ± 0,58	4,80± 0,40	ND

All results are given as mean ± SD of 3-6 rats
[a]Statistically significant ($P < 0,05$) from control
ND, not determined

Intraperitoneal administration of octyl gallate increased the activity of kidney nuclear AHH and NADPH cytochrome c reductase and this increase was statistically significant. Epoxide hydrolase was determined only in liver and kidney and was slightly elevated. The effect of octyl gallate administration on the in vitro binding of ^{3}H-BP to DNA presents table II. Treatment of rats with octyl gallate inhibited the total binding level of BP to DNA of liver nuclei by 56%. The binding of BP to exogenous DNA in microsomal incubation mixtures was inhibited by 28%.

Table II

Effect of octyl gallate on the in vitro binding of BP to DNA

	Specific activity (pmol/mg DNA)	
	Control	Octyl gallate
Nuclei	$50,95 \pm 4,35$	$22,11 \pm 5,43$[a] $(43,39)$[b]
Microsomes with added DNA	$27,34 \pm 1,31$	$19,74 \pm 0,89$[a] $(72,20)$[b]

All results are given as mean $\pm$ SD of 2-3 determinations.
[a]Statistically significant ($P < 0,05$)
[b]Numbers in parentheses, percentage of control

The mechanism underlying the protective action of antioxidants against carcinogens has not been elucidated yet. Most authors have considered the possibility of a relationship between this action and the effects of antioxidants on enzymes involved in the metabolism of the carcinogens. Using 1% dietary levels of gallic esters Depner et al.(1982) have found inhibition of microsomal rat liver monooxygenase reactions by dodecyl and propyl gallate and induction of epoxide hydrolase activity. Octyl gallate did not affect the microsomal enzymes. The activation of carcinogens takes place not only in endoplasmic reticulum but also in the nucleus. It is suggested that in view of the proximity of the nuclear enzymes to DNA the nuclear monooxygenases may play an important role in carcinogenesis.

Our experiments have indicated that ip. administration of octyl gallate modified the activities of nuclear monooxygenases and reduced the binding of BP to DNA of the rat liver nuclei in a greater extent than to added DNA in microsomal preparations.

A similar effect was observed with liver nuclei and microsomes from BHA fed mice(Hennig et al.1983). It was suggested that BHA administration selectively modified the specific nuclear cytochromes catalyzing the metabolism of BP.

The data in this paper may indicate that modification of nuclear drug-metabolizing enzymes by octyl gallate may also

406

play a certain role in the biological effects of this anti -
oxidant.

REFERENCES

Alexandrov K.,Brooks P.,King H.W.S.,Osborne M.R.(1976)
Comparison of the metabolism of benzo(a)pyrene and binding to
DNA caused by rat liver nuclei and microsomes. Chem-Biol.
Interactions, 12 ,269-277
Dansette P.M.,Dubois S. and Jerina D.M.(1979) Continuos fluo-
rometric assay of epoxide hydrase activity.Anal.Biochem.,97,
340-345
Dock L.,Cha Y.N.,Jernstrom B. and Moldeus P.(1982) Effect of
2(3)-tert-butyl-4-hydroxyanisole on benzo(a)pyrene metabolism
and DNA-binding of benzo(a)pyrene metabolites in isolated mo-
use hepatocytes.,41,25-37
Depner M.,Kahl G.F. and Kahl R.(1982) Influence of gallic
acid esters on drug-metabolizing enzymes of rat liver.Fd.Chem.
Toxic.20,507-511
Guenthner T.M.,Jernstrom B. and Orrenius S.(1980) Carcinogenesis. 1, 407-417
Hennig E.E.,Demkowicz-Dobrzański K.K.,Sawicki J.T.,Mojska H.,
KujawaM.(1983) Effect of dietary butylated hydroxyanisole on
the mouse hepatic monooxygenase system of nuclear and microsomal fractions. Carcinogenesis. 4, 1243-1246
Lowry O.H.,Rosebrough N.J.,Farr A.L. and Randall R.J.(1951)
Protein measurement with the Folin phenol reagent.J.Biol.Chem.
193, 265-275
Nebert D.W. and Gelboin H.V.(1968) Substrate - inducible mi -
crosomal aryl hydroxylase in mammalian cell culture.J.Biol .
Chem. 243, 6242-6249
Phillips A.H. and Langdon R.G.(1962) Hepatic triphosphopyridine nucleotide-cytochrome c reductase : Isolation, characterization and kinetic studies. J.Biol.Chem.,237, 2652-2660
Wattenberg L.W.(1972) Inhibition of carcinogenic and toxic
effects of polycyclic hydrocarbons by phenolic antioxidants
and ethoxyquin. J.Natl.Cancer Inst.,48, 1425-1430

BIOTRANSFORMATION OF TUMOR-KILLING N^G-HYDROXYMETHYL-L--ARGININES IN EUCARYOTICS

Csiba,A.[1], Trézl,L.[2], Rusznák,I.[2] and Szarvas,T.[3]

[1]Municipal Hospital Péterfy, Department of Medicine and Clinical Pharmacology, H-1441 Budapest,
[2]Technical University of Budapest, Department of Organic Chemical Technology, H-1111 Budapest,
[3]Institute of Isotopes, Hungarian Academy of Sciences, H-1525 Budapest,Hungary

Reactions between amino acids and formaldehyde are usually explained according to the frequently quoted mechanism elaborated by Sörensen. Later methyl or hydroxymethyl derivatives of amino acids as well as Schiff-base formation could be detected among the reaction products /Tome and Naulet 1981/. Recently contribution of formaldehyde to the carcinogenic action of dimethyl nitoramine could be demonstrated /Swenberg et al. 1980/. Reaction of formaldehyde with both cyteine residues and cystine residues leads to thiazolidine derivatives /Nichols and Gruen 1970/. This reaction has been detected in vivo too /Neely 1964/. Cysteine is alleviating the toxicity of aldehyde in experimental animals also by the formation of a thiazolidine derivative /Spince et al. 1975/. Significant increase in the amount of alkali-hydrolysable urinary thiols occurred after the administration of ethanol, acetaldehyde or formaldehyde to experimental animals /Hemminki 1982/. L-lysine can be methylated and L-arginin can be hydroxymethylated spontaneously by formaldehyde /Trézl et al. 1981, 1983/. Stable hydroxymethyl-L-arginines could be prepared and isolated recently /Csiba et al. 1982a/. The reaction between formaldehyde and the guanidino group is rapid /k = 0.035 min^{-1}/. Its imino group is first and its primary amine group only second in this reaction. On the biological role of hydroxymethyl-L-arginine has already been reported in a previous paper /Csiba et al. 1982b/. Hydroxymethyl-L--arginines could be isolated from both of human blood and urine /Csiba et al. 1984/. A trimethyl-L-lysine promotes. cell proliferation, cleaning up the action of hydroxymethyl-L--arginine in this subject was the aim of the research work, reported here.

Materials and methods

Materials

The standard amino acids, L-arginine /Arg/, $N^G N^G$-dihyd-roxymethyl-L-arginine /DHMA/, N^G-monohydroxymethyl-L-arginine /MHMA/, N^G,N'^G-dihydroxymethyl-L-arginine /DHM'A/ and $N^G N^G,N'^G$ trihydroxymethyl-L-arginine /THMA/ were purchased from Reanal, Budapest, Hungary and were prepared according to our own method /Csiba et al. 1982a/. The eucaryotic baker's yeast was purchased from food-supply. All other materials were of analytical grade, Reanal, Budapest, Hungary.

The plate with dimensions of 20x20 cm was packed with Dowex 50x8 cation-exchange resin /Fixion 50x8, Chinoin, Budapest/. The combination of eluents was made of 50 g citric acid monohydrate, 30 g sodium hydroxyde, 7 ml cc.hydro-chloric acid and 450 ml distilled water. The ninhydrin spray reagent contained 0.2 g ninhydrin, 0.05 g cupric sulfate, 80 ml methanol and 20 ml 96 % acetic acid.

Methods

100 mg from mixture of hydroxymethyl-L-arginines and 5 g baker's yeast were dissolved in 100 ml of the nutritive solution according to Boas. The time and temperature of incubation were 48 hours and 30 ^{O}C, respectively. The blanks were: the blend of hydroxymethyl-L-arginines, the nutritive solution with yeast and/or the nutritive solution free from yeast. The incubated test mixture and the blanks were centrifuged and the supernatants and yeast cells were analysed one by one.

50 μl of the supernatants were dropped onto a Fixion 50x3 thin-layer sheet, respectively.

The yeast cells were washed three times with isotonic saline solution and then haemolysed with 50 ml of distilled water. The haemolysate was centrifuged and 100 μl of the supernatant was dropped onto the sheet.

The migration distance was 17 cm and the chromatographic analysis was 8 hours. The sheet was heated for development at 105 ^{O}C in 15 min. The hydroxymethyl-structures were identified on the basis of comparison with standard mono-, di- and trihydroxymethyl-L-arginines.

Results and discussion

R_f values of standard hydroxymethyl-L-arginines were 0.41 ± 0.02 /MHMA/, 0.29 ± 0.01 /Arg/, 0.20 ± 0.01 /DHMA/, 0.11 ± 0.01 /DHM'A/ and zero /THMA/, respectively /n = 3/. The di- and trihydroxymethyl spots were under the arginine standard spot. Other non-arginine amino acids were in position R_f 0.5. The retention of ornithin was: R_f = 0.61.

Conclusions

1./ In supernatants:
  /a/ the hydroxymethyl-L-arginines were retained;
  /b/ their quantity was reduced;
  /c/ the compound-ratio of hydroxymethyl-L-arginines
      to each other was unchanged;
  /d/ the amount of ornithin-pool has increased.

2./ In haemolysates:
  /a/ the proportion of L-arginine-pool was increased;
  /b/ the ornithin-pool was unchanged;
  /c/ the haemolysates were free of hydroxymethyl-L-
      -arginines.

After the incubation of haemolysated cells with hydroxymethyl-
-L-arginines, it could be observed chromatographically that
  /a/ the hydroxymethyl-L-arginines were eliminated;
  /b/ the ornithin-pool was unchanged;
  /c/ an unidentified ninhydrin-positive spot was found at
      position R_f = 0.13.

The ninhydrin-positive spot was no hydroxymethyl derivative
of L-arginine. It might be the spot of a decarboxylated
arginine derivative /agmatine/.

The inhibition of cell proliferation is assumed on the basis
of results, reported as follows:
/1/ The slow motion of cell proliferation might be due to
    inhibition of arginase /EC 3.5.3.1./, increase of
    arginine-pool and endogenous /non exogenous/ formaldehyde
    level.
/2/ The arginase is inhibited by hydroxymethylated guanidino
    group of L-arginine.
/3/ The hydroxymethyl-L-arginines cannot directly penetrate
    into the cell with passive transport.
/4/ The hydroxymethylated primary amino group of guanidino
    region can disintegrate in an instant to primary amino
    group and formaldehyde by unknown hydroxymethyltrans-
    ferase enzyme.
/5/ The dehydroxymethylated primary amino and hydroxymethylat-
    ed imino part of the arginine-guanidino group can hinder
    the function of arginase.
/6/ The arginine-pool increases.

It has already been reported that arginine and endogenous
formaldehyde /e.g. formaldehyde, generated from N-methyl
groups by demethylase enzymes/ can inhibit cell-proliferation
/Burns and Milner 1984, Cooksey et al. 1983, Wildavsky 1984/.

References

1. Burns, R.A., Milner, J.A. /1984/ Effect of arginine on the carcinogenicity of 7,12-dimethylbenzantracene and N-methyl--N-nitrosourea. Carcinogenesis 5/12/, 1539-1549.
2. Cooksey, P.G., Gate, E.N., Gescher, A. /1983/ The formation and metabolism of N-hydroxymethyl compounds. Biochem. Pharmacol. 32/20/, 3037-3043.
3. Csiba, A., Trézl, L., Tyihák, E., Graber, H., Vári, É., Téglás, G., Rusznák, I. /1982a/ Assumed role of L-arginine in mobilization of endogenous formaldehyde. Acta Phys. Acad.Sci.Hung. 59/1/, 35-43
4. Csiba, A., Trézl, L., Vári, É., Téglás, G. /1982b/ Possible role of the control of arginase and methionine adenosyl transferase in the transport-processes of endogenous formaldehyde and in hypermethylation. Medical Hypotheses 8, 209.
5. Csiba, A., Trézl, L., Tyihák, E., Szarvas, T., Rusznák, I. /1984/ N^G-hydroxymethyl-L-arginines: new serum and urine components and their characterisation by ion-exchange TLC. International Symposium on TLC, Szeged, Hungary, Abstr. No. 19.
6. Hemminki, K. /1982/ Urinary sulfur-containing metabolites after administration of ethanol, acetaldehyde and formaldehyde to rats. Toxicol. Letters 11, 1-6.
7. Neely, W.B. /1964/ The metabolic fate of formaldehyde^{14}C intraperitoneally administered to the rat. Biochem. Pharmacol. 13, 1134-1142.
8. Nichols, P.W., Gruen, L.C. /1970/ Thiazolidine-4-carboxylic acid. Aust. J.Chem. 23, 533-540.
9. Swenberg, J.A., Kerns, W.D., Mitchell, R.I., Gralla, E.J. /1980/ Induction of squamous cell carcinomas of the rat nasal cavity by inhalation exposure to formaldehyde vapor. Cancer Res. 40, 3398-3402.
10. Tome, D., Naulet, N. /1981/ Carbon 13 nuclear magnetic resonance studies on formaldehyde reactions with polyfunctional amino acids. Int. J. Peptide Prot. Res. 17, 501-508.
11. Trézl, L., Csiba, A., Rusznák, I., Tyihák, E., Szarvas, T. /1981/ L-arginine as an endogenous formaldehyde carrier in human blood and urine. Proc. Meet. Hung. Chem. Soc., Budapest, Hungary. Abstr. No. 205.
12. Trézl, L., Rusznák, I., Tyihák, E., Szarvas, T., Szende, B. /1983/ Spontaneous N-methylation and N-formylation reactions between L-lysine and formaldehyde. Biochem. J. 214, 289-292.
13. Wildavsky, A. /1984/ Formaldehyde regulation. Science 224, 550-555.

BIOCHEMICAL SIGNIFICANCE OF THE PROMOTION OF SINGLET OXYGEN
FORMATION AND CHEMILUMINESCENCE EMISSION BY HORSERADISH
PEROXIDASE IN SPONTANEOUS N^ε-METHYLATION AND N -FORMYLATION
REACTIONS IN A SYSTEM CONTAINING L-LYSINE, FORMALDEHYDE AND
HYDROGEN PEROXIDE

Trézl, L.[1], Rusznák, I.[1], Szarvas, T.[2] and Csiba, A.[3]

[1]Technical University of Budapest, Department of
Organic Chemical Technology, H-1521 Budapest
[2]Institute of Isotopes, Hungarian Academy of
Sciences, H-1525 Budapest
[3]Municipal Hospital Péterffy, Department of Medicine
and Clinical Pharmacology, H-1441 Budapest

The spontaneous formation of methylated and formylated
L-lysines and the inhibition of the reaction between L-lysine
and formaldehyde has been reported /Trézl et al. 1983/. The
promotion of the reaction by $NAD^+ + H^+$ coenzyme, and its inhibition by NAD^+ coenzyme could also be demonstrated /Csiba
et al. 1981/. Reductive methylation of L-lysine with formaldehyde is not difficult in the presence of $NaBH_4$ /Paik and
Kim 1980/. Spontaneous methylation of L-lysine could be made
easier with crown-ethers in the formaldehyde containing
system /Trézl et al. 1983/.
A great dela of liberated formaldehyde and a significant
amount of N -methylated L-lysines /in the proportion of 80 %
MML : 12 % DML : 6 % TML/ could be detected after the injection of dimethyl-nitrosamine into the liver of rats
/Tuberville and Craddock 1971/. The reaction between L-lysine
and activated formaldehyde leads to the formation of methylated lysines in very similar proportions, in the course of our
experiments.
Other precursors of formaldehyde /heroin, morphine,
caffeine, cycasin, aflatoxin-B_1, p-nitro anisole, dimethyl-
-hydrasine, N-methyl-N-nitroso urea, nicotine, vinyl-chloride
etc./ could also be detected in "in vivo" systems /Sawicki
and Sawicki 1978/.

Materials and methods

L-lysine /Reanal, Budapest, Hungary/, L-ascorbic acid
/Reanal/, Hydrogen peroxide /Reanal/, formaldehyde /Lachema,
Brno, Czechoslovakia/, N^α-acetyl-L-lysine methylamide
/CIBA-GEIGY, Basel, Switzerland/, Na-azide /NaN_3/ /Fluka,
Buchs, Switzerland/, 1-buthylamine /Fluka/, hydroxy urea
/Biogal Co. Debrecen, Hungary/, N^ε-monomethyl-L-lysine hydrochloride /MML//Sigma Chemical Co. St.Louis, Mo. USA/, N^ε-di-
methyl-L-lysine hydrochloride /DML//Vega, Columbia, USA/,
N^ε-trimethyl-L-lysine glutamate /TML//G.Richter, Budapest,
Hungary/, N^ε-formyl-L-lysine /F-LYS//Reanal/ were of
analytical grade and were used without purification. Horseradish peroxidase, R.Z. 3.0 Type VI. Activity 250-330 unit
per mg, solid, salt free powder /Sigma/ /^{14}C/ formaldehyde

/19.98 kBq/ml/ was prepared at the Institute of Isotopes of
the Hungarian Academy of Sciences.

TLC investigations

L-lysine /10 mM/ and formaldehyde /100 mM/ were allowed to
react in 0.1 M sodium phosphate buffer /pH 7/ at 37±1 °C.
Samples were taken from the reaction mixutre after 1, 10 and
60 minutes and the ninhydrine-positive products were detected
on FIXION 50x8 cation exchanging thin layer sheet. Under the
same conditions the following reactions were carried out:

a/ L-lysine /10 mM/, formaldehyde /100 mM/ and hydrogen
   peroxyde /10 and 50 mM/
b/ L-lysine /10 mM/, formaldehyde /100 mM/, hydrogen peroxide
   /10 mM/ and L-ascorbic acid /100 mM/
c/ L-lysine /10 mM/, formaldehyde /100 mM/, hydrogen peroxide
   /10 mM/ and Horseradish peroxidase /5 and 10 mg, respect-
   ively/

Analysis with labelled compounds

L-lysine /1 mM/ and /^{14}C/ formaldehyde /1 mM/ were allowed to
react in 0.1 M sodium phosphate buffer /pH 7/ at 37±1 °C. The
reaction mixture was analysed by TLC on a FIXION 50x8 thin
layer sheet after 1 and 60 minutes. The radioactivity distribution was determined by a thin layer chromatogram scanner.
Similar reactions and analyses were carried out also in the
presence of 1 mM and 5 mM hydrogen peroxide, respectively.

Chemiluminescence measurements

Chemiluminescence was monitored by photomultiplication method
with a Packard Tri Carb liquid scintillation counter /model
3390/ operated in the out of coincidence mode on the ^{3}H
channel. Scintillation counts were recorded in every 10 sec-
onds till the end of the tenth minute. The integrated results
of chemiluminescence were given in counts/min /c.p.m./. The
solution under study contained: L-lysine /1 mM/, formaldehyde
/1 mM/ and hydrogen peroxide /1 mM/ in 10 ml 0.1 M sodium
phosphate buffer /pH 7/. To this solution different promoting
and inhibiting compounds were added at 25 °C.
Reactions had been studied instead of L-lysine with L-arginine,
glycine, l-buthylamine and N$^{\alpha}$-acetyl-L-lysine methylamide,
respectively. All the reaction mixtures were tested with the
liquid scintillation counter.

Results

Great increase·in rate of the spontaneous reactions between
L-lysine and formaldehyde in a hydrogen peroxide containing
system could be demonstrated by means of TLC analysis.
Significant amount of MML, DML, TML and F-LYS could be detected by the end of the first minute of the reaction, but no
further remarkable change in their amount occurred within the
next nine minutes. Consequently, the excited state did not

last much longer in the system, than for a few minutes. Beside
the already known compounds also a new one /LYS-X/ was generated in the reaction activated by hydrogen peroxide. This
compound is N^{ε}-formyl-L-lysine. Its amount is in proportion
to the concentration of hydrogen peroxide in the system. The
more is the relative amount of LYS-X among the reaction products, the less is the amount of the N^{ε}-methylated lysines.
Approximately as much N^{ε}-methylated L-lysine was produced by
L-lysine and formaldehyde within one minute in the presence
of hydrogen peroxide, as in one hour in its absence. Under
the same compared conditions the amount of formyl compounds
was by one order of magnitude higher in the presence of hydrogen peroxide. It could be demonstrated that LYS-X was N^{α}-
-formyl-L-lysine. After boiling in 6 M hydrochloric acid solution it was hydrolyzed to L-lysine, like N^{ε}-formyl-L-lysine
and in thin layer chromatographic analysis it proved to be
identical with N^{α}-formyl-L-lysine, synthetized by Aizpurua's
method /Aizpurua and Palomo 1983/, in a mixture of L-lysine,
HCOOH /99 %/ and dimethylformamide, refluxing for four hours.
L-ascorbic acid inhibited the spontaneous formation of N^{ε}-
-methylated and formylated lysine also in the presence of
hydrogen peroxide. Horseradish peroxidase, on the other hand,
promoted the formation of these reaction products in the activated system. The rate of formation of N^{ε}-methylated-L-
-lysines in the reaction between L-lysine and activated formaldehyde increased by one order of magnitude, as demonstrated
in the presence of hydrogen peroxide of low concentration.
The relative amount of N-methylated-L-lysines and formylated-
-L-lysines is determined by the proportion of activated formaldehyde and formyl radical in the system. The higher is the
concentration of hydrogen peroxide in the system, the higher
is the relative amount of formyl-lysines among the reaction
products.
This explanation could be confirmed also by the measurement
of chemiluminiscence emission /Table 1/. c.p.m. occurring in
the system Lys+CH_2O can be considered as blank value. Notice-
able increase occurred in the $CH_2O+H_2O_2$ system, much higher,
however, in the system of Lys+$CH_2O+H_2O_2$. Even this high value
could be increased by the addition of horseradish peroxidase
to the last system. The addition of L-ascorbic acid to the
Lys-$CH_2O+H_2O_2$ system, however, decreased c.p.m. by one order
of magnitude. Combining L-ascorbic acid with glutathione in
the system, brought about a drop in c.p.m. close to the value
occurring in the $CH_2O+H_2O_2$ system. Other inhibitors /hydroxy
urea, Na-azide/ could also be found. The quenching effect of
Na-azide is the direct evidence of singlet oxygen but not of
peroxide anion. Lysine with blocked α-amino group /N^{α}-
acetyl-L-lysine-methylamide/ was more reactive than L-lysine
with $CH_2O+H_2O_2$. Consequently, ε-amino group of L-lysine in
peptides may equally be reactive with $CH_2O+H_2O_2$ as L-lysine
molecules are. No increase in c.p.m. was brought about in
$CH_2O+H_2O_2$ system by the addition of glycine or 1-buthyl-amine
instead of L-lysine. The addition of L-arginine to the
$CH_2O+H_2O_2$ system noticeably decreased the c.p.m. value. This
is in agreement with our earlier results /Csiba et al. 1982/,

that arginine cannot be N-methylated or formylated by form-aldehyde.

Table 1   Chemiluminescence of different systems after the reaction for one minute in 10 ml 0.1 M sodium phosphate buffer /pH 7/ at 25 ^{o}C

Reaction mixtures	counts/min/c.p.m./
1 mM L-lysine + 1 mM CH_2O	5×10^2
1 mM CH_2O + 1 mM H_2O_2	2.5×10^3
1 mM L-lysine + 1 mM CH_2O + 1 mM H_2O_2	1.05×10^5
1 mM L-lysine + 1 mM CH_2O + 1 mM H_2O_2 + + 5 mg horseradish peroxidase	1.25×10^5
1 mM L-lysine + 1 mM CH_2O + 1 mM H_2O_2 + + 1 mM L-ascorbic acid	3.5×10^4
1 mM L-lysine + 1 mM CH_2O + 1 mM H_2O_2 + + 3 mM L-ascorbic acid	1.1×10^4
1 mM L-lysine + 1 mM CH_2O + 1 mM H_2O_2 + + 1 mM L-ascorbic acid + 1 mM glutathione	6×10^3
1 mM L-lysine + 1 mM CH_2O + 1 mM H_2O_2 + + 1 mM hydroxy urea	5.2×10^4
1 mM L-lysine + 1 mM CH_2O + 1 mM H_2O_2 + + 1 mM Na-azide /NaN_3/	3.1×10^4
1 mM L-lysine + 1 mM CH_2O + 1 mM H_2O_2 + + 3 mM Na-azide /NaN_3/	8×10^3
1 mM N$^{\epsilon}$-acetyl-L-lysine methylamide + + 1 mM CH_2O + 1 mM H_2O_2	1.36×10^5
1 mM l-buthylamine + 1 mM CH_2O + 1 mM H_2O_2	1.2×10^3
1 mM L-arginine + 1 mM CH_2O + 1 mM H_2O_2	8×10^2
1 mM glycine + 1 mM CH_2O + 1 mM H_2O_2	1.4×10^3

Discussion

Rate and effectivity of spontaneous methylation and formyla-tion reactions between L-lysine and formaldehyde could sig-nificantly be increased by hydrogen peroxide. Proportion between methylated and formylated products depends on the concentration of hydrogen peroxide in the system. The lower is this concentration, the higher is the relative amount of the methylated derivatives. Activated reactions are accom-panied by chemiluminescence emission which can be promoted by the addition of horseradish peroxidase enzyme to the system. L-ascorbic acid /and a few other compounds/ may inhibit the spontaneous reactions even in activated systems. Based upon referred literature and our experiments the occurrence of the above chemical processes may be significant in biological systems, too. In biological systems free and bound lysines as well as peroxidase enzymes are included. The last ones act in the presence of hydrogen peroxide as their substrate. Formaldehyde may be liberated from a number of exogenous and endogenous precursors.

References

1. Csiba, A., Trézl, L., Rusznák, I., Tyihák, E. and Szarvas, T. /1981/ Regulation of spontaneous methylation of L-lysine by formaldehyde with NADH+H$^+$ coenzymes. Proc.Hung. Annu.Meet.Biochem. 21st Veszprém, Hungary. Abstr.No. 207-208.

2. Csiba, A. Trézl, L., Tyihák, E., Graber, H., Vári, É., Téglás, G. and Rusznák, I. /1982/ Assumed role of L-arginine in mobilization of endogenous formaldehyde. Acta Phys.Acad.Sci.Hung. 59/1/, 35-43.

3. Paik, W.K., Kim, S. /1980/ Protein Methylation. John Wiley and Sons, New York, p.44.

4. Sawicki, R. and Sawicki, C.R. /1978/ Aldehydes, Photometric analysis formaldehyde precursors. Academic Press, London.

5. Trézl., L. Ruszná, I., Tyihák, E., Szarvars, T., Szende,B. /1983/ Spontaneous N-methylation and N-formylation reactions between L-lysine and formaldehyde. Biochem.J. 214, 289-292.

6. Trézl, L., Bakó, P., Fenichel, L. and Rusznák, I. /1983/ Determination of methylated amino acids with crown-ethers. J. Chromatogr. 269, 40-43.

7. Tuberville, C. and Craddock, V.M. /1971/ Biochem.J. 124, 725-739.

EFFECT OF THE NITRO SUBSTITUENT ON THE MICROSOMAL METABOLISM OF BENZO[A]PYRENE

Fu, P.P. and M.W. Chou

National Center for Toxicological Research, Jefferson, Arkansas, USA

INTRODUCTION

Nitrated polycyclic aromatic hydrocarbons (nitro-PAHs) are mutagenic and carcinogenic environmental pollutants that require enzymatic activation to exert their biological effects (Beland et al., 1985). Metabolic studies on these compounds can yield information regarding their activation pathways and also on the effect of nitro substitution on the metabolism of the parent PAH. We have studied the aerobic metabolism of the model nitro-PAH, 1-nitrobenzo[a]pyrene (1-nitro-BaP) and its trans-9,10-dihydrodiol metabolite by liver microsomes of rats pretreated with 3-methylcholanthrene to determine the effect of 1-nitro substitution upon the metabolic pathways in comparison of those of the 3- and 6-nitro-BaP isomers and their respective parent hydrocarbons, benzo[a]pyrene (BaP) and benzo[a]pyrene trans-9,10-dihydrodiol.

MATERIALS AND METHODS

The 500 ml incubation mixture (pH 7.5) contained 25 mmol Tris-HCl, 1.5 mmol $MgCl_2$, 0.5 mmol $NADP^+$, 1 mmol glucose-6-phosphate, 50 units glucose-6-phosphate dehydrogenase, 500 mg microsomal protein and 40 μmol 1-nitro-BaP. The incubations were performed aerobically at 37°C for 1 hr, then quenched with acetone, and the metabolites were extracted with ethyl acetate. The metabolite, 1-nitro-BaP trans-9,10-dihydrodiol, was incubated under similar conditions except on a smaller scale.

[^{3}H]1-nitro-BaP (specific activity 114 mCi/mmol, from R. Roth, Midwest Research Inst.) was incubated for 10 min in a 1.0 ml incubation volume with 0.2 mg microsomal protein. [^{3}H]1-Nitro-BaP trans-9,10-dihydrodiol, obtained from metabolism of [3H]1-nitro-BaP (specific activity of 35 mCi/mmol), was similarly incubated.

The radioactive metabolites were separated by reversed-phase HPLC along with added non-radioactive metabolites as uv markers. Tritium was quantified with a FLO-ONE Model HP Radioactive Flow Detector. HPLC separations were conducted on a DuPont Zorbax ODS column (4.6 x 250 mm) eluted with a linear gradient of 70% methanol in water to 100% methanol over a period of 15 min at a flow rate of 1 ml/min. Metabolites were characterized by uv-visible, mass and 500 MHz proton NMR spectral analyses.

RESULTS

After incubation, the metabolites of 1-nitro-BaP were separated by reversed-phase HPLC (Figure 1A). The metabolites, identified as 1-nitro-BaP trans-9,10-dihydrodiol and 1-nitro-BaP trans-7,8-dihydrodiol, eluted at 9.1 and 18.4 min, respectively. 1-Nitro-BaP 7,8,9,10-tetrahydro-tetrol, which has a trans-cis-trans configuration between H_7-H_8, H_8-H_9, and H_9-H_{10}, respectively, was also identified as a minor metabolite (Chou and Fu, 1983). Oxidation of the 2,3- and 4,5-carbons was not detected. Quantitation of the metabolism by incubation of $[^3H]$1-nitro-BaP followed by HPLC separation (Figure 1B) indicated that more than 99% of the radioactivity was in the ethyl acetate extractable fraction, and that 1-nitro-BaP trans-9,10-dihydrodiol and 1-nitro-BaP trans-7,8-dihydrodiol accounted for 33 and 53%, respectively, of the total metabolites. A shorter incubation time and lower concentration of microsomal protein may be the reason that 1-nitro-BaP 7,8,9,10-tetrahydrotetrol was not formed.

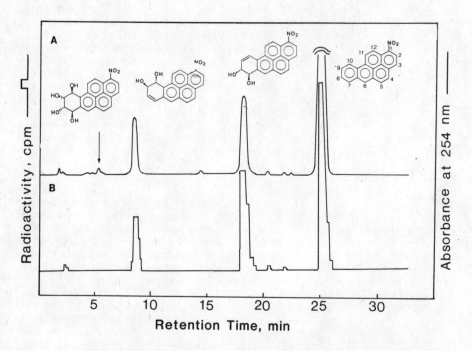

Figure 1. Reversed-phase HPLC profiles of ethyl acetate extractable metabolites obtained from incubation of 1-nitro-BaP; (a) UV-chromatogram and (b) radioactive chromatogram.

Further metabolism of 1-nitro-BaP trans-9,10-dihydrodiol afforded 1-nitro-BaP 7,8,9,10-tetrahydrotetrol as a predominant metabolite which accounted for 62% of the total. 3-Hydroxy-1-nitro-BaP trans-9,10-dihydrodiol was not identified.

DISCUSSION

The results indicate that 1-nitro-BaP is aerobically metabolized by liver microsomes of rats pretreated with 3-methylcholanthrene to yield 1-nitro-BaP trans-7,8-dihydrodiol and trans-9,10-dihydrodiol as the predominant metabolites. Metabolites derived from oxidation at the 2,3- or 4,5-double bond were not detected. Since metabolism of BaP (Holder et al., 1974) affords 3-hydroxy-BaP as the major metabolite and BaP trans-4,5-dihydrodiol as a minor metabolite, it appears that nitro substitution at the 1-position dramatically affects the regioselectivity of cytochrome P-450-catalyzed BaP oxidation. This is further supported by the observation that incubation of 1-ntiro-BaP trans-9,10-dihydrodiol resulted in formation of vicinal dihydrodiol epoxides while similar incubation with BaP trans-9,10-dihydrodiol yielded primarily the 3-phenol with only trace amounts of vicinal dihydrodiol epoxides being formed. Thus, these results indicate that the 1-nitro substitution can inhibit the cytochrome-P-450-mediated oxidation of the aromatic double bond ortho to the substituent. This finding is contrary to the previous report that 3-hydroxy-1-nitropyrene is a metabolite of 1-nitropyrene (El-Bayoumy and Hecht, 1983).

Similar to the metabolic pattern of 1-nitro-BaP, metabolism of 3-nitro-BaP also yielded the corresponding trans-7,8- and 9,10-dihydrodiol as the predominant metabolites (Chou and Fu, 1985). A trans-4,5-dihydrodiol was not detected, indicating that, contrary to the cases of 1-nitropyrene (Djuric et al., 1985), 7-nitrobenz[a]anthracene (Fu and Yang, 1983) and 9-nitroanthracene (Fu et al., 1985), a nitro substituent can inhibit the cytochrome-P-450-mediated oxidation at the aromatic double bond peri to the substituent.

We previously reported that metabolism of 6-nitro-BaP afforded 3-hydroxy-6-nitro-BaP as the major metabolite, and 1-hydroxy-6-nitro-BaP, 6-nitro-BaP-1,3-hydroquinone and 6-nitro-BaP-3,9-hydroquinone as minor metabolites (Fu et al., 1982; Fu and Chou, 1982). Both 3-hydroxy-6-nitro-BaP and 6-nitro-BaP-3,9-hydroquinone have been found to be formed through epoxide intermediates (Chou et al., 1983). Thus, comparison of the metabolites of 1-, 3-, and 6-nitro-BaP with those of BaP indicate that a nitro substituent at these three positions can drastically affect the regioselectivity of the cytochrome P-450 metabolizing enzyme. It is worth noting that the 7,8- and 9,10-dihydrodiols of 1- and 3-nitro-BaP, 1- and 3-hydroxy-6-nitro-BaP, 6-nitro-BaP-1,9-hydroquinone, and 6-nitro-BaP-3,9-hydroquinone are more mutagenic in bacteria than their corresponding substrates in the presence or in the absence of the S9 activation enzymes (Beland et al., 1985; Chou et al., 1985). However, presumably due to the orientation of the nitro substituent, 1- and 3-nitro-BaP are potent direct-acting mutagens while 6-nitro-BaP is not (Fu et al., 1985). Thus, an introduction of a nitro substituent to BaP increases the number of activation pathways and enhances its mutagenic activity in bacteria.

ACKNOWLEDGEMENT

We thank Cindy Hartwick for her assistance in the preparation of this manuscript.

420

REFERENCES

1. Beland, F.A., R.H. Heflich, P.C. Howard, and P.P. Fu. The in vitro metabolic activation of nitro polycyclic aromatic hydrocarbons, In: American Chemical Society Monograph, R.G. Harvey (ed), in press (1985).
2. Chou, M.W., F.E. Evans, S.K. Yang, and P.P. Fu. Evidence for a 2,3-epoxide as an intermediate in the microsomal metabolism of 6-nitrobenzo(a)pyrene. Carcinogenesis 4, 699-702 (1983).
3. Chou, M.W. and P.P. Fu. Evidence for the metabolic formation of a vicinal dihydrodiol-epoxide from the potent mutagen 1-nitro-benzo[a]pyrene. Biochem. Biophys. Res. Commun. 117, 541-548 (1983).
4. Chou, M.W., R.H. Heflich, and P.P. Fu. Multiple metabolic pathways for the mutagenic activation of 3-nitrobenzo[a]pyrene. Carcinogenesis, in press (1985).
5. Djuric, Z., E.K. Fifer, R.H. Heflich, P.C. Howard, and F.A. Beland, Oxidative microsomal metabolism of 1-nitropyrene to DNA-binding derivatives. Proc. Am. Assoc. Cancer Res. 26, 108 (1985).
6. El-Bayoumy, K. and S.S. Hecht, Identification and mutagenicity of metabolites of 1-nitropyrene formed by rat liver, Cancer Res. 43, 3132-3137 (1983).
7. Fu, P.P., M.W. Chou, S.K. Yang, F.A. Beland, F.F. Kadlubar, D.A. Casciano, R.H. Heflich, and F.E. Evans. Metabolism of the mutagenic environmental pollutant, 6-nitrobenzo[a]pyrene: metabolic activation via ring oxidation. Biochem. Biophys. Res. Commun. 105, 1037-1043 (1982).
8. Fu, P.P. and M.W. Chou. In vitro metabolism of 6-nitrobenzo(a)pyrene: effect of the nitro substituent in the regioselectivity of the cytochrome P-450-mediated drug metabolizing enzyme system, pp. 71-74, In: Cytochrome Biochemistry, Biophysics and Environmental Implications, E. Hietanen, M. Laitenen and O. Hanninen (eds). Elsevier Biomedical Press, Amsterdam, The Netherlands, (1982).
9. Fu, P.P. and S.K. Yang. Stereoselective metabolism of 7-nitrobenzo[a]anthracene to 3,4- and 8,9-trans-dihydrodiols. Biochem. Biphys. Res. Commun. 115, 123-129 (1983).
10. Fu, P.P., L.S. Von Tungeln, and M.W. Chou. Metabolism of 9-nitroanthracene by rat liver microsomes: identification and mutagenicity of metabolites. Carcinogenesis, in press (1985).
11. Fu, P.P., M.W. Chou, D.W. Miller, G.L. White, R.H. Heflich, and F.A. Beland. The orientation of the nitro-substituent predicts the direct-acting bacterial mutagenicity of nitrated polycyclic aromatic hydrocarbons. Mutat. Res., in press (1985).
12. Holder, G., H. Yagi, P. Dansette, D.M. Jerina, W. Levin, A.Y.H. Lu, and A.H. Conney. Effects of inducers and epoxide hydrase on the metabolism of benzo[a]pyrene by liver microsomes and a reconstituted system: Analysis by high pressure liquid chromatography. Proc. Natl. Acad. Sci. U.S.A. 71, 4356-4359 (1974).

PLANT AND MICROBIAL CYTOCHROME P-450s

YEAST AND PLANT CYTOCHROME P-450 ENZYMES: TARGETS FOR AZOLE-DERIVATIVES

Hugo Vanden Bossche, D. Bellens, J. Gorrens, P. Marichal,
H. Verhoeven and G. Willemsens

Laboratory of Comparative Biochemistry,
Janssen Pharmaceutica Research Laboratories, B-2340 Beerse, Belgium

INTRODUCTION

Azole derivatives are potent and selective inhibitors of several cytochrome P-450 (cyt. P-450) enzymes in mitochondrial and microsomal membranes. For example the oral active antifungal imidazole derivative, ketoconazole, is a potent inhibitor of the cyt. P-450 dependent 14α-demethylase in Saccharomyces cerevisiae (Vanden Bossche, 1985) and Candida albicans (Vanden Bossche et al., 1980). At higher concentrations ketoconazole also affects in decreasing order the microsomal 17,20-lyase system in rat testis (Vanden Bossche et al., 1985), the 11-hydroxylase system in bovine adrenal cortex mitochondria, the bovine adrenal and pig testis mitochondrial cholesterol side-chain cleavage (unpublished results) and the 14α-demethylase system in rat liver microsomes (Willemsens et al., 1980). The microsomal 21-hydroxylase of bovine adrenal cortex is not inhibited by ketoconazole concentrations as high as 10^{-5} M.

Fifty per cent inhibition of the bovine adrenal cortex 11-hydroxylase is achieved at 3×10^{-7} M etomidate, an imidazole derivative used for the induction and maintenance of anaesthesia (Vanden Bossche et al., 1984a). At a two times higher concentration etomidate affects cholesterol side-chain cleavage. Etomidate is devoid of antifungal activity and does not affect cyt. P-450 in yeast microsomes (unpublished results). L-etomidate, the less hypnotically active enantiomer is a much less potent inhibitor of the adrenal mitochondrial cyt. P-450 species (Vanden Bossche et al., 1984a).

Itraconazole, a triazole derivative that is orally active in the treatment of experimental aspergillosis, cryptococcosis, histoplasmosis, trichophytosis, candidosis and sporotrichosis (Van Cutsem et al., 1983) is a potent inhibitor of the 14α-demethylase in C. albicans (Vanden Bossche et al., 1984b) and Aspergillus fumigatus (Marichal et al., 1985). A fifty per cent decrease in the Soret peak height of the reduced carbon monoxide compound of cyt. P-450 of S. cerevisiae was achieved at 2×10^{-8} M. A similar effect on cyt. P-450 of dog testis microsomal cyt. P-450 is found at 10^{-5} M itraconazole only (Vanden Bossche et al., 1984b). This triazole derivative does not affect at concentrations as high as 10^{-5} M, testosterone biosynthesis in subcellular fractions of rat testis (Vanden Bossche et al., 1985). Itraconazole even at concentrations up to 10^{-5} M does not affect significantly the cyt. P-450 dependent cholesterol side-chain cleavage reaction and 11-hydroxylase from bovine adrenal mitochondria (unpublished results).

From this review it can be deduced that it is possible to develop compounds that interact selectively with cyt. P-450 species playing a role in important steps of sterol and steroid metabolisms. To evaluate better the selectivity of antifungal compounds, models are needed whose sensitivity is similar to that of pathogenic forms. It is the aim of this study to compare cyt. P-450 species of C. albicans with that of S. cerevisiae microsomes.

It has also been shown that multiple forms of cyt. P-450 with a dif ferent substrate specificity exist in plant microsomes (Reichhart et al., 1982). A number of imidazole and triazole derivatives, e.g. azaconazole, etaconazole, imazalil and propiconazole (the chemical structures are given in Fig. 1), are of use in plant protection. In this study it is proved that these compounds affect the yeast microsomal cyt. P-450 and are devoid of activity on the microsomal cyt. P-450 of Jerusalem artichoke tubers (Helianthus tuberosus L.) and of tulip bulbs (Tulipa gesnerana L.).

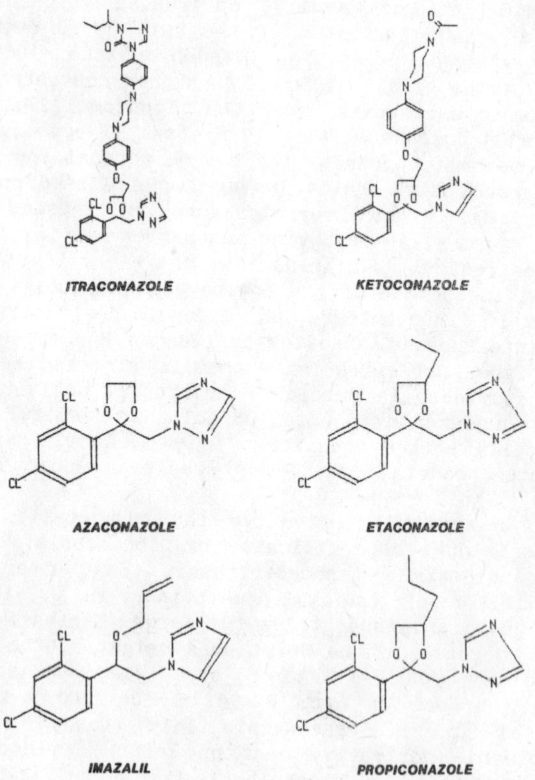

ITRACONAZOLE

KETOCONAZOLE

AZACONAZOLE

ETACONAZOLE

IMAZALIL

PROPICONAZOLE

Fig. 1.   Chemical structures of itraconazole, ketoconazole, azaconazole, etaconazole, imazalil, propiconazole.

MATERIAL AND METHODS

Strains, inocula and growth

Inocula of <u>Candida albicans</u> (ATCC 28516) were prepared as previously described (Vanden Bossche et al., 1975). One ml of a 24-h culture in casein-hydrolysate-yeast extract medium (CYG) was used to inoculate 200 ml of PYG_I-medium (polypeptone 5 g/l; yeast extract 10 g/l; glucose 10 g/l). Cells were grown for 24 h at 30° C and 30 ml (approx. 7 x 10^8 cells) were used to inoculate 5 l PYG_I-medium (polypeptone 10 g/l; yeast extract 10 g/l; glucose 40 g/l) as described by Ishidata et al. (1969) for <u>Saccharomyces cerevisiae</u>. Cells were grown for 14 h at 30° C on a magnetic stirrer.

Inocula of <u>S. cerevisiae</u> (B 19328-1) were prepared and cells were grown in PYG_I-medium as described by Vanden Bossche et al. (1984b). Twenty ml (approx. 5 x 10^8 cells) were used to inoculate 5 l of PYG_{II}-medium. Cells were grown for 15 h at 30° C on a magnetic stirrer.

Plants

Jerusalem artichoke tubers grown locally were stored in pot soil at 4° C in darkness. The preparation of tuber slices and induction of cyt. P-450 was as described by Reichhart et al. (1980). One-mm thick slices (300 g fresh weight) were inoculated in the dark at room temperature in 5-l Erlenmeyer flasks with 4 litres distilled water containing 25 mM $MnCl_2$ (pH 4.5). The incubation medium was vigorously bubbled with a stream of hydrated and filtered (Gelman filters: 0.2 μm) air.

Tulip bulbs were purchased locally. The swollen leaf part of the bulb was used.

Isolation of microsomes from S. cerevisiae and C. albicans

Microsomes of <u>S. cerevisiae</u> were isolated as described by Vanden Bossche et al. (1984b). Cells were harvested by centrifugation and suspended in 0.65 M sorbitol. They were broken in a Braun cell homogenizer and the homogenate centrifuged for 5 min at 1500 g. The supernatant was centrifuged for 20 min at 10,000 g and the supernatant recentrifuged as before. The microsomal pellet was obtained by centrifugation of the 10,000-g supernatant for 60 min at 105,000 g. The pellet was resuspended in 0.05 M potassium phosphate buffer containing 0.01 M EDTA (ph 7.4) and centrifuged as before. The final pellet was suspended in 0.1 M potassium phosphate buffer (pH 7.4) and stored under nitrogen at -70° C.

The isolation of microsomes from semi-anaerobically grown <u>C. albicans</u> was based on a method described by Aoyama et al. (1981). <u>C. albicans</u> was harvested, suspended in 0.65 M mannitol and broken in a Braun cell homogenizer. The homogenate was centrifuged for 5 min at 1500 g. The cell-free suspension was centrifuged at 10,000 g for 20 min. The microsomes were obtained by centrifugation at 100,000 g, washed with 0.1 M potassium phosphate buffer (pH 7.4) containing 10 mM EDTA and recentrifuged at 100,000 g. The pellet was resuspended in 0.1 M potassium phosphate buffer (pH 7.4) and centrifuged as before. The final pellet was suspended in the same buffer and stored under nitrogen at -70° C until used.

Isolation of a microsomal fraction from plant tissues

For the preparation of microsomal pellets from Jerusalem artichoke tubers the method of Reichhart et al. (1980) was used. Tuber tissue was homogenized for 2 min with a Moulinex mixer and 2 min with an Ultra-Turrax in 2 volumes of 0.1 M sodium phosphate buffer (pH 7.4) containing 15 mM 2-mercaptoethanol and 2 g insoluble polyvinylpyrrolidone (PVP, Janssen Chimica) per litre. The homogenate was filtered through 3 layers of cheese cloth and centrifuged for 15 min at 10,000 g. The resulting supernatant was centrifuged for 60 min at 100,000 g. The 100,000-g pellet was suspended in the extraction buffer containing 30 % glycerol and stored at -70° C until used.

The microsomal fraction of tulip bulbs was prepared as described by Rich et al. (1975). The washed swollen leaf parts (200 g) were suspended in 2 volumes of 0.3 M mannitol, 5 mM tricine, 0.1 % bovine serum albumin, 0.05 % cystine and 1 mM EDTA (pH 7.4), homogenized as described above and the homogenate filtered through cheese cloth. Its pH was adjusted to 7.2 with KOH 1 M. The cell debris, starch and nuclear, mitochondrial and intermediate fractions were removed by a 30-min centrifugation at 20,000 g. The supernatant was centrifuged at 100,000 g for 1 h. The resuspended pellet was centrifuged for 1 h at 100,000 g. The microsomal pellet was suspended in 0.3 M mannitol, 5 mM tricine and 1 mM EDTA (pH 7.2) and stored at -70° C until needed.

Spectral measurements

The cyt. P-450 content was determined by measuring the reduced carbon monoxide (CO) difference spectrum using 91 $cm^{-1}mM^{-1}$ as extinction coefficient (Omura and Sato, 1964). The absorbance increment between 448 nm (S. cerevisiae, C. albicans microsomes) or 450 nm (plant microsomes) and 490 nm was used for the calculation of the cyt. P-450 content. To measure the effects of drugs, the microsomal suspensions were diluted to 0.1 nmol/ml with 0.1 M potassium phosphate buffer of pH 7.4 (yeast and fungal microsomes) or with 0.1 M sodium phosphate buffer (pH 7.4) containing 15 mM 2-mercaptoethanol and 30 % glycerol (plant microsomes).

To trace the spectral transitions of the Soret band of cyt. P-450, associated with the addition of azole derivatives, microsomes were diluted 0.33 nmoles/ml. The suspension was then divided into the reference and sample cells. A base-line of equal light absorbance was established. The drugs were added to the sample and solvent (DMSO) to the reference cell and the resulting difference spectrum recorded. The spectral transitions of the Soret band and the reduced CO-difference spectra were traced with an Aminco DW-2C spectrophotometer.

Drugs were dissolved in dimethylsulfoxide (DMSO) for itraconazole 15 µl 6 M HCl per ml DMSO was added.

Protein

Protein was determined by the Bio-Rad protein assay (Bio-Rad Laboratories, Cat. No. 500.0006).

RESULTS AND DISCUSSION

As shown in Figure 2 and Table 1 the absorption maximum of the reduced CO-compound of a C. albicans microsomal fraction is at 448 nm. This is similar to that observed with the cyt. P-450 in S. cerevisiae (Yoshida and Kumaoka, 1975).

Immediately after the addition of the azole derivatives, ketoconale or itraconazole to the microsomal suspension, a dose-dependent decrease of the absorption difference between 448 and 490 nm was observed (Fig. 2).

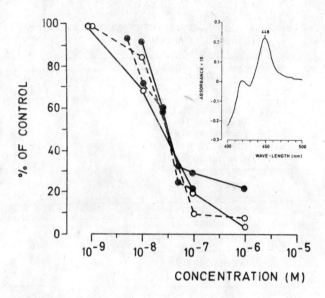

Fig. 2. Effects of ketoconazole (——) and itraconazole (---) on micro-somal cyt. P-450 obtained from S. cerevisiae (o) and C. albicans (●). The $\Delta A_{448-490\ nm}$ were determined and the per cent of control plotted against drug concentration. Insert shows a carbon-monoxide-difference spectrum of a microsomal preparation of C. albicans.

A 50 % decrease was obtained at 3.1×10^{-8} M of both ketoconazole and itraconazole (Table 2).

A similar effect on cyt. P-450 of S. cerevisiae was observed at 2.3×10^{-8} M ketoconazole and at 2.8×10^{-8} M itraconazole (Table 2). The sensitivity of the cyt. P-450 species of C. albicans and S. cerevisiae microsomes to imazalil sulfate, azaconazole, etaconazole and propiconazole was almost identical (Table 2). All these imidazole and triazole deriva-tives interfere at low concentrations with the cyt. P-450 species of S.

Table 1. Absorption maxima of the reduced CO-compound and
cyt. P-450 contents[a]

Species	Absorption maxima (nm)	Content nmoles/mg protein
C. albicans (ATCC 28516)	448	0.08 ± 0.04 (17)
S. cerevisiae (B 19328-1)	448	0.08 ± 0.04 (23)
Jerusalem artichoke tubers	450	0.30 ± 0.13 (25)
Tulip bulbs	450	0.14 ± 0.04 (5)

[a]The cyt. P-450 content was determined by measuring the dithionite re-duced CO-difference spectrum using an extinction coefficient of 91 $cm^{-1}mM^{-1}$. The absorbance increment between 448 (yeast microsomes) or 450 (plant microsomes) and 490 nm was used for the calculation of the cyt. P-450 content.

Table 2. Effects of nitrogen heterocycles on the microsomal
cyt. P-450 of yeast

Species	Drug	IC_{50}-value[a] $(x\ 10^{-8}\ M)$
C. albicans	ketoconazole	3.1
	itraconazole	3.1
	imazalil sulfate	5.0
	etaconazole	6.0
	propiconazole	6.1
	azaconazole	6.4
S. cerevisiae	ketoconazole	2.3
	itraconazole	2.8
	imazalil sulfate	7.5
	etaconazole	5.3
	propiconazole	6.9
	azaconazole	5.4

[a]By a weighed non-linear regression procedure, a sigmoidal dose-response
model was fitted to the individual observations and the corresponding
IC_{50}-values (50 % decrease in the peak height of the Soret band of the
reduced CO-compound) determined.

cerevisiae and C. albicans. On the contrary, they do not affect significantly the reduced CO spectrum of the cyt. P-450 forms of Jerusalem artichoke tubers or tulip bulbs. A 50 % decrease of the CO spectrum was not
achieved at 10^{-5} M.
     The oxidized cyt. P-450 in C. albicans membranes was found to give
typical type II binding spectra with all the azole derivatives tested.
Maxima were at about 430 and minima at 393 nm, thus similar to those obtained with cyt. P-450 in S. cerevisiae microsomes (Vanden Bossche et al.,
1984a). With most compounds maximum increase of Δ (430 nm - 393 nm) was
achieved at 5 x 10^{-7} M. This indicates that under the present conditions
the 6th binding place at the Fe^{3+}-atom is saturated at 5 x 10^{-7} M.
     Using Jerusalem artichoke tuber microsomes only imazalil sulfate in-
duces type II binding spectra with maxima at about 432 nm and minima at
402 nm. However, even at 10^{-4} M no maximum increase of Δ (432-402 nm) was
reached. These results further prove that these azole derivatives are
poor inhibitors of the cyt. P-450 species in these plant microsomes. This
suggests that plant cyt. P-450 species differ from those present in yeast
microsomes. Studies using other plant microsomes are in progress.
     The results presented here show that, at least for the evaluation of
the binding properties of azole derivatives to yeast microsomal cyt.
P-450, S. cerevisiae can be used instead of the yeast form of C. albicans.
Studies are going on to identify cyt. P-450 in the mycelium form.
     It is interesting to note that the azole derivatives do not differ
significantly (maximum a factor 3) in their effect on the reduced carbon
monoxide difference spectrum obtained with yeast microsomes. Therefore,
the differences in their antimycotic properties and effects on sterol
synthesis in intact cells should originate from differences in pharmacokinetic properties, e.g. uptake, distribution in the membranes and metabolism.

429

REFERENCES

Ishidate, K., Kawagushi, K., Tagawa, K., Hagihara, B. (1969). Hemoproteins in anaerobically grown yeast cells. J. Biochem. 65, 375-383.

Marichal, P., Gorrens, J., Vanden Bossche, H. (1985). The action of itraconazole and ketoconazole on growth and sterol synthesis in Aspergillus fumigatus and Aspergillus niger. Sabouraudia: J. Med. Vet. Mycol. 22, 13-21.

Omura, T., Sato, R. (1964). The carbon monoxide pigment of liver microsomes. I. Evidence for its hemoprotein nature. J. Biol. Chem. 239, 2370-2378.

Reichhart, D., Salaün, J.-P., Benveniste, I., Durst, F. (1980). Time course of induction of cytochrome P-450, NADPH - cytochrome c reductase, and cinnamic acid hydroxylase by phenobarbital, ethanol, herbicides, and manganese in higher plant microsomes. Plant Physiol. 66, 600-604.

Reichhart, D., Simon, A., Durst, F., Mathews, J.M., Ortiz de Montellano, R. (1982). Autocatalytic inactivation of plant cytochrome P-450 enzymes: selective inactivation of cinnamic acid 4-hydroxylase from Helianthus tuberosus by 1-aminobenzotriazole. Arch. Biochem. Biophys. 216, 522-529.

Rich, P.R., Cammack, R., Bendall, D.S. (1975). Electron paramagnetic resonance studies of cytochrome P-450 in plant microsomes. Eur. J. Biochem. 59, 281-286.

Van Cutsem, J., Van Gerven, F., Zaman, R., Heeres, J., Janssen, P.A.J. (1983). Pharmacological and preclinical results with a new oral and topical broad-spectrum antifungal, R 51 211. Proceedings 13th International Congress of Chemotherapy, Vienna, SS 4.8/1-11.

Vanden Bossche, H. (1985). Biochemical targets for antifungal azole derivatives. Hypothesis on the mode of action. In Current topics in medical mycology, M.R. McGinnis (Ed.). Springer Verlag, New York (in press).

Vanden Bossche, H., Willemsens, G., Van Cutsem, J.M. (1975). The action of miconazole on the growth of Candida albicans. Sabouraudia 13, 63-73.

Vanden Bossche, H., Willemsens, G., Cools, W., Cornelissen, F., Lauwers, W.F., Van Cutsem, J. (1980). In vitro and in vivo effects of the antimycotic drug ketoconazole on sterol synthesis. Antimicrob. Agents Chemother. 17, 922-928.

Vanden Bossche, H., Willemsens, G., Cools, W., Bellens, D. (1984a). Effects of etomidate on steroid biosynthesis in subcellular fractions of bovine adrenals. Biochem. Pharmacol. 33, 3861-3868.

Vanden Bossche, H., Willemsens, G., Marichal, P., Cools, W., Lauwers, W. (1984b). The molecular basis for the antifungal activities of N-substituted azole derivatives. Focus on R 51 211. In Mode of action of antifungal agents, A.P.J. Trinci, J.F. Ryley (Eds.). Cambridge University Press, Cambridge, pp. 321-341.

Vanden Bossche, H., Lauwers, W., Willemsens, G., Cools, W. (1985). The cytochrome P-450 dependent C17,20-lyase in subcellular fractions of the rat testis. Differences in sensitivity to ketoconazole and itraconazole. In Microsomes and drug oxidations, A.R. Boobis, J. Caldwell, F. de Matteis, C.R. Elcombe (Eds.). Taylor & Francis, Ltd., Basingstoke, U.K. (in press).

Willemsens, G., Cools, W., Vanden Bossche, H. (1980). Effects of miconazole and ketoconazole in sterol synthesis in a subcellular fraction of yeast and mammalian cells. In The host invader interplay, H. Vanden Bossche (Ed.). Elsevier Biomedical Press, Amsterdam, pp. 691-694.

Yoshida, Y., Kumaoka, H. (1975). Studies on the microsomal electron-transport system of anaerobically grown yeast. III. Spectral characteristics of cytochrome P-450. J. Biochem. 78, 785-794.

REGULATION OF CYTOCHROME P-450 FORMATION BY OXYGEN IN ALKANE-ASSIMILATING YEAST

W.-H. SCHUNCK, S. MAUERSBERGER, E. KÄRGEL, J. HUTH, and
H.-G. Müller

Central Institute of Molecular Biology, Academy of Sciences
of the GDR, Berlin-Buch, GDR

INTRODUCTION

In Candida maltosa as in several other yeast strains long-chain n-alkanes induce a highly specific cytochrome P-450 /P-450/ system, which catalyzes the first step of alkane degradation (Riege et al. 1981, Honeck et al. 1982, Schunck et al. 1983, Mauersberger et al. 1981, 1984).

By a so far unknown mechanism the cellular content of the alkane-hydroxylating P-450 is strongly increased in several Candida strains by oxygen limitation during cultivation on n-alkanes (Mauersberger et al. 1980, 1984, Gmünder et al. 1981, Sanglard et al. 1984). It is remarkable, that a similar response to oxygen limitation using glucose as carbon source was observed for another yeast P-450 system (Laurilla et al. 1984), which is involved in ergosterol biosynthesis, as proved for Saccharomyces cerevisiae by Aoyama et al. (1984). The aim of the present study was to find out the signal produced by oxygen limitation, which is directly responsible for enhanced P-450 formation in alkane-assimilating yeast.

MATERIAL AND METHODS

Most of the experiments were performed in a small-scale fermentor (working volume: 80-100 ml) equipped with automatic pH-titration, pO_2-electrode, high-speed stirring and supplies for air, nitrogen and carbon monoxide. General cultivation conditions wre as described by Mauersberger et al. (1984). Alkali consumption was used as an indirect growth parameter according to Verkooyen and Rietema (1980).

P-450 in intact cells was determined by means of CO-difference spectra. To avoid spectral interference of cyt oxidase its endogeneous reduction was prevented, if necessary, by addition of antimycin A.

Immunological quantitation of the P-450 protein was performed according to Kärgel et al. (1984).

RESULTS AND DISCUSSION

Cultivation of C. maltosa on n-hexadecane (maintained in the range of 0.5-1 %, v/v) results in the formation of a constant

specific P-450 content of about 0.05 nmoles per mg of cell
protein (Tab.1) under conditions of oxygen-saturation (pO_2 =
40-100 %). By adjusting a pO_2 of 3-5 % the growth is immediately reduced to 10 % and the P-450 content increases over a
period of 3 hours to a new about 5-times higher level. This
effect was fully reversible - if oxygen saturation is read-
justed both the growth rate and the specific P-450 content
return to their original values.

The enhanced P-450 formation during oxygen-limitation was
prevented by addition of cycloheximide to the culture medium.

Table 1   Influence of oxygen-limitation and selective CO-
           inhibition of alkane hydroxylation on the cellular
           cytochrome P-450 content

substrate[1]	pO_2 (%)	CO[2]	relative growth rate	cytochrome P-450 (nmoles/mg protein)
n-hexadecane	95-100	-	100	0.05
n-hexadecane	3- 5	-	10-15	0.27
n-hexadecane	70- 75	+	10	0.32
n-hexadecane + 1-hexadecanol	90-100	-	100	0.05
n-hexadecane + 1-hexadecanol	70	+	100	0.25
1-hexadecanol	95-100	-	100	0.03
1-hexadecanol	70	+	100	0.03

1) All experiments in the presence of 1.2 % (v/v) pristane
   (2,6,10,14-tetramethylpentadecane) as inert hydrocarbon
   phase to dissolve solid compounds
2) Compare Tab. 2

Among the oxygen-activating enzyme systems involved in the
metabolism of alkane-assimilating yeast the P-450 is distinguished by the lowest affinity towards oxygen (an apparent
Ko_2 of 26 µM for n-hexadecane utilization was found in contrast to 13 µM and $\leq$ 5 µM for 1-hexadecanol and glucose utilization, respectively; details will be published elsewhere).
Thus it can be assumed, that at low pO_2-values the P-450
system catalyzes the rate-limiting step of alkane degradation.

To simulate this situation we searched for conditions leading to a selective inhibition of the P-450 activity. As shown
in Tab. 2 the high CO-sensitivity of the P-450 provides an
appropriate opportunity.

By adjusting relatively low CO concentrations in the
culture medium (additional streaming with CO at such a rate
that the measured pO_2 decreases from 95 to 70 %) it was possible to reach complete inhibition of alkane utilization,
obviously without disturbing the basic metabolism of the yeast
cells. In particular, the uninhibited utilization of fatty
alcohols by a yeast culture blocked in alkane assimilation
indicates the high selectivity of CO-action.

Table 2   CO-sensitivity of utilization of different growth
          substrates by Candida maltosa

Substrate[1]	CO^{2}-inhibition (%)
n-alkanes ($>$ C_7)	100
1-tetradecene	100
1-hexadecanol	0
palmitic acid	0
2-hexadecanol	45
3-hexadecanon	45

1) All experiments in the presence of 1.2 % (v/v) pristane
   (see Tab. 1)

2) The pO_2 was decreased by additional streaming of the culture with carbon monoxide from 95-100 to 70 %; in control
   experiments with nitrogen the utilization rates of all
   substrates tested were not influenced by a pO_2 change down
   to 40 %

Using this method of selective CO-inhibition the same effect
was observed on the P-450 formation at a pO_2 of about 70 % as
in the case of oxygen limitation at a pO_2 of 3-5 % (Tab. 1).
To exclude the factor of growth-limitation besides hexadecane
a second non-repressive growth substrate (1-hexadecanol) was
added.
   In control experiments performed in absence of the n-alkane
substrate the specific P-450 content was not influenced by
CO. The low level of P-450 during cultivation on 1-hexadecanol
as the only carbon source was mainly due to the presence of
pristane (1.2 % v/v) which was used in all experiments as an
inert hydrocarbon phase to dissolve solid compounds like
fatty alcohols and fatty acids. Pristane represents a kind of
a gratuitous inducer which is not or only very slowly hydroxylated by the P-450 system.

CONCLUSIONS

Neither a low oxygen concentration itself nor factors arising
from growth-limitation were found to be direct signals that
trigger the enhanced P-450 formation which occurs in alkane-
assimilating yeast as a result of oxygen-limitation. The decisive point seems to be, however, the inhibition of the alkane hydroxylating activity in the presence of a compound
which acts both as an inducer and as a substrate of the P-450
system. Further experiments are in progress to prove the assumption that the P-450 content is mainly regulated by the
intracellular inducer concentration which depends on the ratio
of alkane transport rate into the cell and the P-450 activity.

434

REFERENCES

Aoyama, Y., Yoshida, Y. and Sato, R. (1984): J. Biol. Chem. 259, 1661

Gmünder, F.K., Käppeli, O. and Fiechter, A. (1981): Eur. J. Appl. Microbiol. Biotechnol. 12, 135

Honeck, H., Schunck, W.-H., Riege, P. and Müller, H.-G.(1982): Biochem. Biophys. Res. Commun. 106, 1318

Kärgel, E., Schmidt, H.E., Schunck, W.-H., Riege, P., Mauersberger, S., and Müller, H.-G. (1984): Anal. Lett. 17, 2011

Laurila, H., Käppeli, O., Fiechter, A. (1984): Arch. Microbiol. 140, 257

Mauersberger, S., Matyashova, R.N., Müller, H.-G., and Losinov, A.B. (1980): Eur. J. Appl. Microbiol.Biotechnol. 9, 285

Mauersberger, S., Schunck, W.-H. and Müller, H.-G. (1981): Z. Allgem. Mikrobiol. 21, 313

Mauersberger, S., Schunck, W.-H. and Müller, H.-G. (1984): Appl. Microbiol. Biotechnol. 19, 29

Riege, P., Schunck, W.-H., Honeck, H., and Müller, H.-G. (1981): Biochem. Biophys. Res. Commun. 98, 527

Sanglard, D., Käppeli, O. and Fiechter, A. (1984): J. Bacteriol. 157, 297

Schunck, W.-H., Riege, P., Honeck, H., and Müller, H.-G. (1983): Z. Allgem. Mikrobiol. 23, 653

Verkooyen, A.H.M. and Rietema, K. (1980): Biotechnol. Bioeng. 22, 615

THE YEAST ALKANE-HYDROXYLATING CYTOCHROME P-450 SYSTEM:
FURTHER CHARACTERIZATION AND REGULATION OF THE BIOSYNTHESIS

H.-G. MÜLLER, E. KÄRGEL, S. MAUERSBERGER, A. OTTO, P. RIEGE,
W.-H. SCHUNCK, B. WIEDMANN

Central Institute of Molecular Biology, Academy of Sciences
of the GDR, Berlin-Buch, GDR

Introduction

In the preseding conference on cytochrome P-450 the purification of an alkane-hydroxylating yeast cytochrome P-450
/P-450/ and of the corresponding NADPH-cytochrome P-450
/P-450 reductase/ to homogeneity as well as the reconstitution of the active enzyme system was reported (Müller et al.
1982). The P-450-linked terminal hydroxylation of n-alkanes
introduces a catabolic pathway, which allows the yeast to
grow on long-chain n-alkanes as the only sources of carbon
and energy. The P-450 is distinguished by high substrate
specificity and regioselectivity (Riege et al. 1981, Honeck
et al. 1982).

Material and Methods

The immunological techniques were outlined by Kärgel et al.
(1984). The N-terminal amino acid sequence was determined
by automated solid-phase Edman degradation (RANK HILGER
APS 240), except the first residue, which was detected after
dansylation. All other methods applied was described by
Mauersberger et al. (1984).

Results and Discussion

A comparative immunological study of the alkane-hydroxylating P-450 of the yeast Candida maltosa /P-450$_{C.m.}$/ demonstrates a substantial cross-reactivity of the anti(P-450$_{C.m.}$)
with yeast P-450 forms with the same physiological function.
A weak competition with the lanosterol-demethylating P-450

of Saccharomyces cerevisiae can be detected in a solid-phase
double-antibody radio-immuno-assay. P-450 forms of mammalia
(rabbit LM2 and LM4, bovine SSC) as well as of bacteria
($P-450_{CAM}$ of Pseudomonas putida, $P-450_{LIN}$ of Ps. incognita,
and also the alkane-hydroxylating P-450 of an Acinetobacter
strain) don't indicate any cross-reactivity in the highly sensitive RIA (Kärgel et al. 1985). The comparison of primary
structures from extensively studied mammalian P-450s and of
results of their immunological investigation has shown that
the latter reflect quite well the structural relatedness of
P-450 forms (Black and Coon 1985). Therefore, we conclude
from the results of the immunological study of yeast alkane-
hydroxylating P-450s that these forms should be very similar
to each other. However, there is a considerable distinctness
to all other P-450 species investigated.

This latter conclusion is supported by the comparison of the
NH_2-terminal sequences of $P-450_{C.m.}$ and of 27 P-450s (listed
by Black and Coon 1985), mostly of mammalian origin. The
occurence of 3 glutamic acid residues and 1 glutamine within
the foremost 8 positions (Table 1) is surprising for a membrane-bound P-450 and unique, so far known.

The main interest was directed to the regulation of the biosynthesis of the alkane monooxygenase system. As reported
earlier (Mauersberger et al. 1981) the cellular P-450 level
depends strongly on the carbon source utilized by the growing
yeast culture. n-Alkanes act as inducer, glucose as a re-
pressor and after growth on glycerol a low P-450 level indicates a non-induced state (Table 2). The other constituents
of the microsomal electron transport system and also the
enzymes of the further oxidation of the alkyl chain (alcohol
oxidase, aldehyde dehydrogenase, results here not presented)
demonstrate the same tendency (Mauersberger et al. 1984).
The good agreement of the results of both methods applied
for P-450 estimations (by CO-D only the intact P-450 is
estimated, by RIA P-450, P-420 and apoenzyme) proves that
no pool of P-450 apoenzyme exists (Kärgel et al. 1984).
Consequently, the assembly of prosthetic group and apoenzyme

Table 1

Comparison of the NH_2-terminal amino acid sequence of
$P-450_{C.m.}$ with $P-450_{LM2}$[+] and $P-450_{CAM}$[++]

	5	10
$P-450_{C.m.}$	A I E Q I I E E V L P Y	
$P-450_{LM2}$	M E F S L L L L L A G L	
$P-450_{CAM}$	T T E T I Q S N A N L A	

[+] Black et al. (1982)   [++] Haniu et al. (1982)

Table 2

Influence of the carbon source on the contents of the
constituents of the microsomal electron transport system
of the yeast Candida maltosa

carbon source	P-450[++] CO-D	RIA	cytochrome[++] b_5	P-450[+++] reductase	NADH-cyt c[+++] reductase
glucose	0	3	155	174	55
glycerol	110	100	367	234	131
n-alkanes[+]	465	410	621	746	246

[+] at air supply sufficient for unlimited growth
[++] pmoles $\times$ mg^{-1} protein
[+++] nmoles reduced cyt c $\times$ min^{-1} $\times$ mg protein^{-1}
CO-D: difference spectrometry of the reduced CO complex
RIA: radio-immuno-assay

to the holoenzyme should not be a regulation level for P-450
biosynthesis of any importance.

A quantitative analysis of the P-450 specific messenger was
carried out both from alkane-grown and from glucose-grown
C. maltosa cells by isolation of the Poly(A)-RNA fraction,
in-vitro translation in the wheat germ test system and subsequent immuno-precipitation with anti($P-450_{C.m.}$)IgG. Applied
to the SDS-PAGE the sediment of the P-450 specific immuno-
precipitation yields a single polypeptide with a migration
velocity being identical with P-450 (Wiedmann et al. (1985).
This result indicates that very likely no precursor with a

pro-sequence is synthesized, which is removed in a posttranslational process. Approx. 0.5-1 % of the whole translation product obtained with Poly(A)-RNA from alkane-grown cells
is P-450. Opposite to that the Poly(A)-RNA fraction of glucose-grown cells subjected to the identical procedure (wheat
germ translation, specific immunoprecipitation) yields only neg-
ligible traces of a polypeptide precipitable by anti($P-450_{c.m.}$)
at a comparable total translation activity. A comparison of
the contents of P-450-specific messenger and of P-450 in alkane-grown and glucose-grown cells demonstrates that the biosynthesis of P-450 is regulated on transcription level
(Wiedmann et al. 1985).

References

Black, S.D., Tarr, G.E., and Coon, M.J. (1982):
J. Biol. Chem. 257, 14616-14619

Black, S.D. and Coon, M.J. (1985) in Ortiz de Montellano,P.R.
(ed.) Cytochrome P-450: Structure, Mechanism, and Biochemistry,
Plenum Press New York

Haniu, M., Armes, L.G., Tanaka, M., Yasunobu, K.T., Shastry,
R.S., Wagner, G.C., and Gunsalus, I.C. (1982): Biochem.
Biophys. Res. Commun. 105, 889-894

Honeck, H., Schunck, W.-H., Riege, P., and Müller,H.-G. (1982)
Biochem. Biophys. Res. Commun. 106, 1318-1324

Kärgel,E., Schmidt,H.E., Schunck,W-H., Riege,P.,Mauersberger,S.,
and Müller, H.-G. (1984): Analytical Letters 17(B18),2011-2024

Kärgel, E. et al. (1985): in preparation)

Mauersberger, S., Schunck, W.-H., and Müller, H.-G. (1981):
Z. Allgem. Mikrobiol. 21, 313-321

Mauersberger, S., Schunck, W.-H., and Müller, H.-G. (1984):
Appl. Microbiol. Biotechnol. 19, 29.-35

Müller, H.-G., Schunck, W.-H., Riege, P., and Honeck, H.(1982)
in Hietanen, E. et al. (eds.) Cytochrome P-450: Biochemistry,.
Biophysics and Environmental Implications. Elsevier Amsterdam

Riege, P., Schunck, W.-H., Honeck, H., and Müller H.-G.(1981)
Biochem. Biophys. Res. Commun. 98, 527-534

Wiedmann, B. et al. (1985): in preparation

CHARACTERIZATION OF AN ALTERED CYTOCHROME P-450
FROM A MUTANT YEAST

Y. Yoshida*, Y. Aoyama*, T. Nishino**, H. Katsuki**,
U.S. Maitra***, V.P. Mohan***, D.B. Sprinson***

*Faculty of Pharmaceutical Sciences, Mukogawa University
Nishinomiya, Hyogo 663 Japan
**Faculty of Science, Kyoto University, Kyoto 606, Japan
***Columbia University, Roosevelt Hospital, New York, U.S.A.

A nystatin resistant yeast mutant (*Saccharomyces cerevisiae* SG1) is blocked in removal of the methyl group at C-14 of lanosterol (Trocha et al., 1977, and Aoyama et al., 1983). Lanosterol 14α-demethylation is catalyzed by a cytochrome P-450-containing monooxygenase system and the cytochrome responsible for this reaction (P-450/14DM) has been purified from a wild-type strain of *S. cerevisiae* (Yoshida and Aoyama, 1984, and Aoyama et al., 1984). By using the antibodies to P-450/14DM, we found the presence of some protein which was immunochemically indistinguishable from P-450/14DM in the microsomes of SG1 (Aoyama et al., 1983). Recently, we obtained a cytochrome P-450 preparation (P-450/SG1) from SG1 microsomes and found that the cytochrome was immunochemically the same as P-450/14DM. In this paper we describe purification and properties of P-450/SG1 and discuss the possibility that P-450/SG1 is an altered form of P-450/14DM.

EXPERIMENTAL PROCEDURE

Cultivation of yeasts and purification of P-450/14DM and P-450/SG1 were done according to the methods described in our recent paper (Yoshida and Aoyama, 1984). Antibodies to P-450/14DM were raised in rabbits as in (Aoyama et al., 1981). Lanosterol 14α-demethylase activity was determined as in (Aoyama et al., 1984). Peptide mapping was done according to the method of Cleveland et al. (1977). Buthiobate and ketoconazole were generous gifts from Dr. T. Kato (Sumitomo Chemical Co.) and Dr. H. Vanden Bossche (Janssen Pharmaceutica), respectively.

RESULTS AND DISCUSSION

P-450/SG1 could be purified from semi-anaerobically grown cells of SG1 by using the same method as that developed for the purification of P-450/14DM (Yoshida and Aoyama, 1984). Specific content of the purified P-450/SG1 was 10-12 nmol/mg protein and slight contaminants were detected by SDS-PAGE. Monomeric molecular weight of P-450/SG1 by SDS-PAGE was essentially the same as that of P-450/14DM (Mr=58,000) (Yoshida and Aoyama, 1984).

P-450/SG1 formed a single precipitin line with rabbit antibodies to P-450/14DM and the line was completely fused with that formed between the antibodies and P-450/14DM (Fig. 1). Peptide maps of P-450/SG1 after proteolysis with *Staphylococcus aureus* V8 proteinase or papain were super-imposable on those of P-450/14DM. These observations indicated that P-

450/SG1 was very similar to P-450/14DM and they could not be distinguished from each other by double immunodiffusion analysis and peptide mapping.

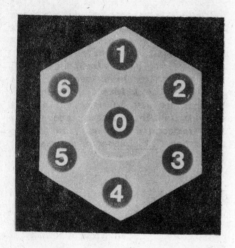

Fig. 1. Immunochemical identity of P-450/SG1 to P-450/14DM. Wells 1 and 4 contained 0.03 and 0.06 nmol, respectively, of P-450/14DM. Wells 2 and 3 contained 0.05 nmol of P-450/SG1. Wells 5 and 6 contained 0.1 nmol of P-450/SG1. Center well (0) contained 0.58 mg protein of anti-P-450/14DM.

P-450/14DM was slowly reduced by NADPH in the reconstituted system consisting of the cytochrome, NADPH-cytochrome P-450 reductase and DLPC, and the rate of reduction was greatly enhanced by the addition of lanosterol to the reconstituted system (Aoyama et al., 1984). In contrast, P-450/SG1 was reduced very slowly by the reductase system even in the presence of lanosterol.

The reconstituted P-450/14DM system converted lanosterol to 4,4-dimethylcholesta-8,14,24-trienol, the 14-demethylated product of lanosterol (Aoyama et al., 1984), but the P-450/SG1-containing system could not metabolize lanosterol. These facts indicated that P-450/SG1 could clearly be discriminated from P-450/14DM with respect to the catalytic activity.

Difference between P-450/SG1 and P-450/14DM was also observed on the reactivity with some antifungal agents that are known to interact specifically with P-450/14DM. Ketoconazole could interact with both of these P-450s but buthiobate bound only to P-450/14DM.

The most interesting difference between P-450/SG1 and P-450/14DM was observed on their optical absorption spectra. Ferric P-450/14DM and other usual low-spin P-450s show their Soret peaks at 416–418 nm and have the slightly higher α than β peaks (Yoshida et al., 1982, see Fig. 2). In contrast, Soret peak of ferric P-450/SG1 situated at 422 nm and the α band was observed as a weak shoulder on the β peak (Fig. 3). The slightly higher α than β peak of usual ferric low-spin P-450 is probably due to its unique ligand structure, O-Fe-S$^-$ (Yoshida et al., 1982, White and Coon, 1982, and Dawson et al., 1982). When the native 6th ligand trans to S$^-$ was replaced by an exogenously added nitrogenious one such as pyridine, imidazole and their derivatives, a few nm red-shift of the α, β and Soret bands and marked hypochromicity of the α peak were observed (Yoshida et al., 1982, and White and Coon, 1982). The absorption spectrum of ferric P-450/SG1 (Fig. 3) was superimposable on that of the 1-methylimidazole complex of P-450/14DM (Fig. 2). Based on these facts, it was strongly suggested that the 6th coordination position of ferric P-450/SG1 was occupied by a nitrogenous ligand. The ligand may be an imidazole of histidine in the apoprotein, because no imidazole or related compound was included in the P-450/SG1 preparation.

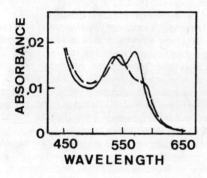

Fig. 2. Absorption spectra of ferric P-450/14DM and its 1-methylimidazole complex. ————— , Native ferric P-450/14DM; —— —— , 1-methylimidazole complex of ferric P-450/14DM.

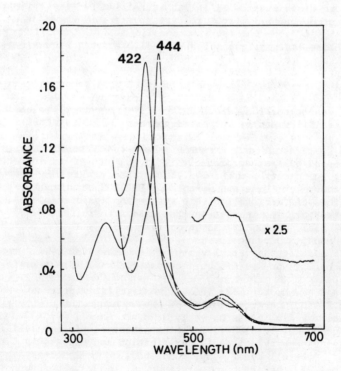

Fig. 3. Absorption spectra of P-450/SG1. ————— , Oxidized form; —— · —— , reduced form with $Na_2S_2O_4$; —— —— ,reduced CO-complex.

Upon reduction with hydrosulfite and binding to CO in the reduced state, P-450/SG1 showed the absorption spectra characteristic for P-450 though their Soret bands were slightly blue-shifted to 407 and 444 nm, respectively (Fig. 3). This fact indicated that the nitrogenous 6th ligand of P-450/SG1 was probably dissociated in the reduced state. P-450/SG1 is the only known P-450 species which shows the above-mentioned spectral properties and is considered to have N-Fe-S⁻ ligand structure.

Taken all observations together, it is highly likely that P-450/SG1 is an altered form of P-450/14DM. The alteration must occur at the activie site or some region affecting the active site. The alteration causes the binding of some internal nitrogenous ligand (probably His of the apoprotein) to the 6th coordination position of the heme iron and inactivates the cytochrome. Accordingly, P-450/SG1 is considered to be an interesting object of studies on structure-function relationship of cytochrome P-450.

ACKNOWLEDGEMENTS: We wish to thank Dr. T. Kato of Sumitomo Chemical Co. and Dr. H. Vanden Bossche of Janssen Pharmaceutica for their generous gifts of buthiobate and ketoconazole, respectively.

REFERENCES

Aoyama, Y., Okikawa, T. and Yoshida, Y. (1981). Evidence for the presence of cytochrome P-450 functional in lanosterol 14α-demethylation in microsomes of aerobically grown respiring yeast. Biochim. Biophys. Acta, 665, 596-601.

Aoyama, Y., Yoshida, Y., Hata, S., Nishino, T., Katsuki, H., Maitra, U.S., Mohan, V.P. and Sprinson, D.B. (1983). Altered cytochrome P-450 in a yeast mutant blocked in demethylating C-32 of lanosterol., J. Biol. Chem., 258, 9040-9042.

Aoyama, Y., Yoshida, Y. and Sato, R. (1984). Yeast cytochrome P-450 catalyzing lanosterol 14α-demethylation. II. Lanosterol metabolism by purified P-450/14DM and by intact microsomes., J. Biol. Chem., 259, 1661-1666.

Cleveland, D.W., Fischer, S.G., Kirschner, M.W. and Laemmli, U.K. (1977). Peptide mapping by limited proteolysis in sodium dodecyl sulfate and analysis by gel electrophoresis. J. Biol. Chem., 252, 1102-1106.

Dawson, J.H., Andersson, L.A. and Sono, M. (1982). Spectroscopic investigations of ferric cytochrome P-450-CAM ligand complexes. Identification of the ligand trans to cysteinate in the native enzyme., J. Biol. Chem., 257, 3606-3617.

Trocha, P.J., Jasne, S.J., and Sprinson, J.B. (1977). Yeast mutants blocked in removing the methyl group of lanosterol at C-14. Separation of sterols by high-pressure liquid chromatography., Biochemistry, 16, 4721-4726.

White, R.E. and Coon, M.J. (1982). Heme ligand replacement reactions of cytochrome P-450. Characterization of the bonding atom of the axial ligand trans to thiolate as oxygen., J. Biol. Chem., 257, 3073-3083.

Yoshida, Y. and Aoyama, Y. (1984). Yeast cytochrome P-450 catalyzing lanosterol 14α-demethylation. I. Purification and spectral properties., J. Biol. Chem., 259, 1655-1660.

Yoshida, Y., Imai, Y., and Hashimoto-Yutsudo, C. (1982). Spectrophotometric examination of exogenous-ligand complexes of ferric cytochrome P-450. Characterization of the axial ligand trans to thiolate in the native ferric low-spin form., J. Biochem., 91, 1651-1659.

CYTOCHROME P-450 OF YEASTS

Käppeli, O., Sanglard, D. and Laurila, H.O.

Department of Biotechnology, ETH-Hönggerberg,
8093 Zürich, Switzerland

Distribution of cytochrome P-450 in yeasts

Cytochrome P-450 was initially detected in whole cells of
yeasts of the genus Saccharomyces growing semi-anaerobically
with glucose as the carbon and energy source (Lindenmayer and
Smith, 1964) and in hydrocarbon degrading yeasts (Gallo et al.,
1971). Under fully aerobic conditions cytochrome P-450 was not
found in whole cells of glucose utilizing yeasts.

By using the subcellular fractionation method of Käppeli et al.
(1982) cytochrome P-450 was identified in the microsomal fraction of metabolically different yeasts (Table 1), which indicated its general distribution in these eucaryotes.

Table 1 : Metabolic characteristics of Trichosporon cutaneum,
Candida tropicalis and Saccharomyces uvarum, the
yeasts investigated

	T. cutaneum	C. tropicalis	S. uvarum
Ethanol formation from glucose : C-limitation	-	-	+
O-limitation	-	+	+
Growth on n-alkanes	n.d.*	+	-

* n.d., not determined

Oxygen limitation led to a derepression of cytochrome P-450
biosynthesis independent of the yeast strain and substrate
used (Table 2).

Table 2 : Influence of cultivation conditions on the microsomal cytochrome P-450 content of different yeasts

Yeast	Cytochrome P-450 content $(pmol\ mg^{-1}\ protein)$
Trichosporon cutaneum	
Glucose, C-limited	25
Glucose, O-limited	70
Candida tropicalis	
Glucose, C-limited	40
Glucose, O-limited	190
Alkane, C-limited	210
Alkane, O-limited	700
Saccharomyces uvarum	
Glucose, C-limited	35
Glucose, O-limited	330

Heterogeneity of cytochrome P-450 monooxygenases from yeasts

The monooxygenases of different yeasts exert different physiological functions. In hydrocarbon degrading yeasts it is involved in the primary hydroxylation of the substrate. In glucose
assimilating yeasts its role was related to sterol biosynthesis
(Aoyama et al., 1981). The question was whether there exist
different cytochrome P-450 species corresponding to the functions of monooxygenase systems.

By assessing several properties of the cytochrome P-450 monooxygenases originating from different yeasts or from yeasts
cultivated differently, two particular cytochrome P-450 forms
were distinguished (Table 3). The monooxygenase of hydrocarbon-
grown cells exhibited high aliphatic hydroxylation activity,
whereas that of glucose-grown cells was mainly active in the
14α-demethylation of lanosterol. The two types of activity were
sensitive to typical cytochrome P-450 inhibitors (carbon monoxide, propiconazole). These data indicate that different cytochrome P-450 forms are present in yeasts.

Heterogeneity of cytochrome P-450 monooxygenases from yeasts
was also manifest when whole cells of T. cutaneum and C. tropicalis were used in biphenyl hydroxylation. The resulting metabolite pattern from the two yeasts was clearly distinguishable

445

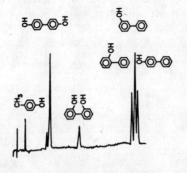

Trichosporon cutaneum

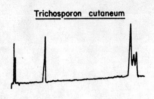

Candida tropicalis

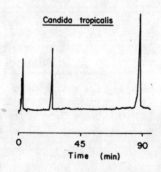

0   45   90
Time (min)

<u>Fig. 1</u> : Biphenyl hydroxylation products from <u>Trichosporon</u> <u>cutaneum</u> and <u>Candida tropicalis</u> as obtained by high pressure liquid chromatography. O-cresol was the internal standard.

446

Table 3 : Properties of cytochrome P-450 monooxygenases from
Candida tropicalis and Saccharomyces uvarum

Property	S. uvarum Glucose-grown	C. tropicalis Glucose-	Alkane-grown
Absorption maximum in the reduced CO-difference spectrum (nm)	447-448	447-448	450
Lauric acid hydroxylation activity (nmol/nmol P-450/min)	-	-	2.55
14α-demethylation of lanosterol (nmol/nmol P-450/min)	1.3	1.25	0.02
Propiconazole concentration for 50% inhibition of activity (µM)	0.2	0.1	75

(Fig. 1). With C. tropicalis 4-OH-biphenyl was predominantly
formed (over 95%) whereas with T. cutaneum 2-, 3- and 4-OH-
biphenyl constituted the metabolites. Since biphenyl hydroxylation was inhibited by propiconazole and penconazole, two
cytochrome P-450 inhibitors, it has to be assumed that multiple forms of cytochrome P-450 exist in yeasts.

References

Aoyama, Y., Okikawa, T. and Yoshida, Y. (1981) Evidence for
the presence of cytochrome P-450 functional in lanosterol 14α-
demethylation in microsomes of aerobically grown respiring
yeast. Biochim. Biophys. Acta, 665, 596-601.

Gallo, M., Bertrand, J.C. and Azoulay, E. (1971) Participation
du cytochrome P-450 dans l'oxidation des alcanes chez Candida
tropicalis. FEBS Letters 19, 45-49.

Käppeli, O., Sauer, M. and Fiechter, A. (1982) Convenient procedure for the isolation of highly enriched cytochrome P-450-
containing microsomal fraction from Candida tropicalis. Anal.
Biochem., 126, 179-182.

Lindenmayer, A. and Smith, L. (1964) Cytochromes and other
pigments of baker's yeast grown aerobically and anaerobically.
Biochim. Biophys. Acta 93, 445-461.

RECONSTITUTION OF n-ALKANE-OXIDIZING ACTIVITY OF A PURIFIED CYTOCHROME P-450 FROM Acinetobacter calcoaceticus strain EB 104

Otmar Asperger, Reinhard Müller and Hans-Peter Kleber
Sektion Biowissenschaften der Karl-Marx-Universität, Bereich
Biochemie, DDR 7010 Leipzig, G.D.R.

Recently, we reported that cytochrome P-450 (P-450) is induced in distinct strains of the bacterial genus Acinetobacter during growth on n-alkanes (Asperger et al. 1981, Asperger
et al.1984). Taking into account that only a restricted number of bacterial cytochromes P-450 is known and that rather
few data exist on bacterial enzymes oxidizing n-alkanes but
also favoured by the relatively high content of P-450 in our
strain (0.2 nmol/mg protein), we started with the purification of this P-450 and also with studies on its function and
on the composition of the whole system.

The organism used throughout these studies was Acinetobacter calcoaceticus strain EB 104 (Asperger et al. 1981). In
case, cells are grown on a liquid medium with n-hexadecane as
the sole source of carbon, the P-450 of cell-free extracts
obtained by EDTA/lysozyme treatment is mainly localized in
particulate fractions. It is predominantly found in membrane-
free supernatants if cells are grown on a complex medium and
n-hexane is added as the inducer. But in the latter case,
P-450 is contaminated with P-420 and for that reason P-450
was purified by us from n-hexadecane-grown materials. It was
extracted from the particulate fractions by addition of Triton X-100 to the generally used phosphate buffer (0.05 M,
pH 7.2, 100 g glycerole / l). P-450 revealed to be rather
stable in the presence of Triton X-100. Its purification was
achieved by chromatography of the "Triton extract" on DEAE-
cellulose followed by gel filtration on Sepharose 6B and

final chromatography on hydroxylapatite. The P-450 was homogeneous according to SDS-polyacrylamide gel electrophoresis and the yields amounted to about 20 %. Some of its properties are described in Fig. 1.

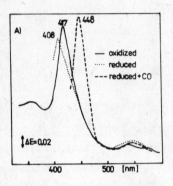

B)

Content of P-450 : 19 nmol/mg
Molecular weight : 52 000
Heme B   : 1 mol/mol
Isoelectric point: 4.7

Fig. 1. Absolute spectra (A) and molecular properties (B) of cytochrome P-450 purified from n-hexadecane-grown cells. The content of P-450 was determined according to Omura and Sato (1964).

With regard to the inducibility by n-alkanes (Asperger et al. 1984) the function of the P-450 as a n-alkane monooxygenase has been assumed. This is more stringently indicated by the fact that radiochemically measured oxidation of n-hexadecane by cell-free extracts is inhibited by CO, that this activity is distributed according to the same pattern as P-450 among subcellular fractions and that the activity is also detected in cells induced by the aromatic compound biphenyl. Induction of P-450 by biphenyl and also by several polycyclic aryl hydrocarbons has been detected more recently (Asperger et al. 1985).

However, to study the composition of the complete P-450 system, enzymatic reduction of P-450 was preliminary used as a method. For this, enzymatic reduction under aerobic conditions was followed by monitoring the kinetics and the degree of CO complex formation ($\Delta E_{448-490nm}$) in the presence of NADH (0.15 mM) and at a fixed concentration of P-450 (0.5 uM). The P-450 of cell-free extracts is very rapidly reduced

by NADH. After separation of ammonium sulfate-precipitated
cell-free extracts by gel filtration on Sephadex G-75, the
enzymatic reduction of P-450 that is eluted nearly in the
void volume requires the addition of two different components
(B and C) that are eluted from the same column (Table 1).

TABLE 1. REQUIREMENT OF TWO DIFFERENT PROTEINS FOR THE RE-
CONSTITUTION OF ENZYMATIC REDUCTION OF CYTOCHROME
P-450 AFTER THE SEPARATION OF THE SYSTEM ON
SEPHADEX G-75

Fractions	Enzymatic reduction	
	Degree[a] (%)	Rate[b] (nmol P-450/min/ml)
Before gel filtration	91	> 5.00[c]
After gel filtration:		
Cytochrome P-450	0	0.00
+ component B	0	0.00
+ component C	33	0.03
+ components B and C	77	1.03

a) % of dithionite-reduced P-450, b) initial rate
c) complete reduction within the deadtime of 6 sec

These results indicate that the P-450 system of Acinetobacter is composed of three components as it is found also
for other bacterial systems (Gunsalus et al. 1974). The spectral characteristics of the eluate containing component C are
quite similar to those of a ferredoxin isolated by us from
the same strain (Asperger et al. 1983). The functional properties of the component B correspond to a ferredoxin reductase. The highly purified P-450 is also reduced by these partially purified electron transfer components and we were also
able to reconstitute a functional active system that oxidizes
long-chain n-alkanes with these components (Fig. 2). Since
activity is only found in the presence of the highly purified
P-450, these results provide strong evidence for the function
of the cytochrome P-450 of A. calcoaceticus as a n-alkane
monooxygenase.

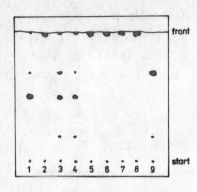

1) $1-^{14}C$-n-tetradecanol

2) complete assay,not incubated

3) complete assay

4) complete with cyanide

5) complete without P-450

6) complete without ferredoxin

7) complete without reductase

8) $1-^{14}C$-n-tetradecane

9) $1-^{14}C$-n-tetradecanoic acid

Fig. 2. Product formation from n-tetradecane by highly purified cytochrome P-450 supplemented with partially purified
ferredoxin and ferredoxin reductase from A. calcoaceticus.
(Schematic representation of the autoradiogram of incubation
mixtures separated by thin-layer chromatography).
1 ml assays containing 1 umol substrate ($15 \cdot 10^3$ Bq),
2 nmoles P-450, the reduction system and 1 umol NADH in Tri-
ton-containing phosphate buffer were incubated under vigo-
rous agitation for 30 min at $25^\circ C$, extracted with n-hexane
and the concentrated extracts were developed on silica gel
with n-hexane/diethyl ether/acetic acid (80/20/1).

REFERENCES

Asperger,O., Naumann,A., Kleber,H.-P. (1981) Occurrence of
cytochrome P-450 in Acinetobacter strains after growth on
n-hexadecane. FEMS Microbiol.Lett. 11 309-312

Asperger,O., Müller,R., Kleber, H.-P. (1983) Isolierung von
Cytochrom P-450 und des entsprechenden Reductasesystems
aus Acinetobacter calcoaceticus. Acta Biotechnol. 3
319-326

Asperger,O., Naumann, A., Kleber,H.-P. (1984) Inducibility of
cytochrome P-450 in Acinetobacter calcoaceticus by n-alkanes. Appl. Microbiol. Biotechnol. 19 398-403

Asperger,O., Stüwer,B., Kleber,H.-P. (1985) Aryl hydrocarbons
as inducers of cytochrome P-450 in Acinetobacter calcoaceticus. Appl. Microbiol. Biotechnol. in press

Gunsalus,I.C., Meeks,J.R., Lipscomb,J.D., Debrunner,P.,
Münck,E. (1974) Bacterial monooxygenases-the P-450 cytochrome system. In: Molecular mechanism of oxygen activation. (Ed. Hayaishi,O.) Academic press,New York.559-613

Omura,T., Sato,R. (1964) The carbon monoxide-binding pigment
of liver microsomes. II. Solubilisation, purification and
properties. J. Biol. Chem. 239 2379-2385

EFFECTS OF HERBICIDE ANTIDOTES ON REDUCED GLUTATHIONE CONTENT AND LEVELS OF CYTOCHROME P-450 AND GLUTATHIONE S-TRANSFERASE IN Zea mays L.

T. Kőmives, M. Balázs, V. Kőmives, F. Dutka

Central Research Institute for Chemistry of the
Hungarian Academy of Sciences, Budapest, Pusztaszeri ut 59
H-1525, Hungary

Chemical antidotes may protect maize (Zea mays L.) plants against thiocarbamate herbicide injury by facilitating the detoxication of the herbicide via glutathione conjugation by increasing the rate of sulfoxidation (Dutka and Kőmives, 1983; Kőmives and Dutka, 1980; Leavitt and Penner, 1979) and elevating the levels of reduced glutathione (GSH) and the enzyme cytosolic glutathione S-transferase (GST) (Kőmives et al.,1984; Lay and Casida, 1976; Mozer et al., 1983):

$$R^1-S-CO-NR^2R^3 \longrightarrow R^1-SO-CO-NR^2R^3 \xrightarrow[GST]{GSH} GS-CO-NR^2R^3$$

The aim of this work was to shed light on the possible involvement of the microsomal enzymes cytochrome P-450 and microsomal GST in the action of herbicide antidotes. For comparison, antidote effects on the levels of GSH and cytosolic GST were also measured.

MATERIALS AND METHODS

Naphthalic anhydride (NA) and other chemicals were commercial products. N,N-Diallyl dichloroacetamide (R-25788) was a generous gift from K. Fodor-Csorba. Seeds of maize (Pioneer 3906 hybrid, kindly provided by Dr. E. Széll) were treated with the antidotes according to the literature (Lay and Casida, 1976; Mozer et al., 1983). Determination of the content of GSH in the roots (Fedtke, 1981), cytosolic (Mozer et al., 1983) and microsomal GST (Morgenstern et al., 1980a) in the shoots using 1-chloro-2,4-dinitrobenzene as substrate, cytochrome P-450 (Omura and Sato, 1964) and protein (Lowry et al., 1951) after 48 h of treatment with the antidotes followed previous lines. Microsomes from shoots of etiolated maize seedlings were isolated by the method of Dr. U. Enkhart (private communication).

RESULTS

In accord with literature findings (Lay and Casida, 1976; Mozer et al., 1983) treatments of maize seedlings with R-25788

30*

led to increased GSH and cytosolic GST levels, while NA had no
significant effect on GSH content and induced a lower increase
in cytosolic GST activity (Table 1).

Microsomal GST levels were increased by both antidotes,
with NA being the less active (Table 1). No detectable microsomal GST was measured in maize seedlings when Triton X-100
was omitted from the assay mixture; and, *in vitro* treatments
of microsomal GST from control   and   antidote-treated plants
with 5,5'-dithiobis(2-nitrobenzoic acid) (DTNB) and N-ethylmaleimide (NEM) (1 mM, 30 sec, room temperature) did not result
in higher enzyme activity (data not shown). No significant
effects of the antidotes on cytochrome P-450 content was found.

Table 1. Effects of NA (0,5% w/w seed dressing) and R-25788
(0.1 mM in nutrient solution) on levels of GSH cytosolic and microsomal GST and cytochrome P-450 in
etiolated maize seedlings

	NA	R-25788	Control
GSH (μmole/g fresh weight)	0.43±0.04	0.69±0.11	0.45±0.06
Cytosolic GST (μmole/min/mg protein)	1.63±0.22	2.27±0.37	0.92±0.09
Microsomal GST (nmole/min/mg protein	95±12	147±18	67±9
Cytochrome P-450 (pmole/mg protein)	71±16	103±16	82±17

DISCUSSION

Metabolism of a number of organosulfur compounds in mammals
and plants involves oxidation at the sulfur atom (Hajjar and
Hodgson, 1982). In mammalian systems sulfoxidation reactions
are carried out primarily by microsomal cytochrome P-450 and/
or FAD-dependent monooxygenases, while in plants other enzymes
(peroxidases and lipoxygenases) may also be involved (Blee
et al., 1985).

Oxygenated sulfur compounds may be substrates for the
enzyme system GST (Lay and Casida, 1976). Cytosolic GST in
mammals is inducible: chemicals that increase the activity of
the cytochrome P-450 system lead to higher levels of cytosolic
GST (Morgenstern *et al.*, 1982). Interestingly, microsomal GST
is not induced by these treatments, though this enzyme is localized primarily in the same membrane as the cytochrome P-450
system (Morgenstern *et al.*, 1982). Microsomal GST of mammals,
however, can be activated *in vitro* by sulfhydryl reagents such
as DTNB and NEM (Morgenstern *et al.*, 1980b).

The data presented in this paper on microsomal GST show
that a) plants may have such a system,
b) this system in maize seedlings can not be activated
in vitro by DTNB and NEM, and
c) microsomal GST can be induced *in vivo* by appropriate
treatments.

Alleviation of thiocarbamate herbicide phytotoxicity to maize plants by NA and R-25788 may be due to increased activity of both cytosolic and microsomal GST system, as a result of the higher levels of the enzymes and, in case of R-25788, their primary substrate, GSH.

The finding that cytochrome P-450 levels in maize are not significantly influenced by NA and R-25788 does not rule out the possibility of increased sulfoxidation of the herbicide in antidote-treated plants. Rates of oxidation by this enzyme system in plants are not necessarily determined by cytochrome P-450 content (Higashi *et al.*, 1982; Hagmann *et al.*, 1983), or, alternatively, sulfoxidation of thiocarbamate herbicides in maize may be catalyzed by enzymes other than cytochrome P-450.

REFERENCES

Blee, E., Casida, J.E., and Durst, F. (1985) Sulfoxidation of methiocarb in higher plants. Biochem. Pharmacol. 34, 389-390

Dutka, F. and Kőmives, T. (1983) On the mode of action of EPTC and its antidotes. In "IUPAC Pesticide Chemistry: Human Welfare and the Environment", Ed. J. Miyamoto *et al.*, Pergamon Press, Oxford, pp. 213-218

Fedtke, C. (1981) Wirkung von Herbiziden und Herbizid-Antidots auf den Glutathion-Gehalt in Mais- und Soja-bohnenwurzeln. Z. Pflanzenkrankh. Pflanzenschutz, Sonderh. IX. 141-146

Hagmann, M.-L., Heller, W., and Grisebach, H. (1983) Induction and characterization of a microsomal flavonoid 3'-hydroxylase from parsley cell cultures. Eur. J. Biochem. 134, 547-554

Hajjar, N.P. and Hodgson, E. (1982) Sulfoxidation of thioether-containing pesticides by flavin-adenine dinucleotide-dependent monooxygenase of pig liver microsomes. Biochem. Pharmacol. 31, 745-752

Higashi, K., Karasaki, Y., and Nakashima, K. (1982) Possible multiplicity of microsomal cytochrome P-450 in higher plants. In "Microsomes, Drug Oxidations and Drug Toxicity", Eds. R. Sato and R. Kato, Japan Sci. Soc. Press, Tokyo, pp. 103-104

Kőmives, T. and Dutka, F. (198o) Mode of action of EPTC-antidotes on corn. Cereal Res. Commun. 8 627-633

Kőmives, V., Kőmives T., and Dutka F. (1984) Influence of herbicides on the glutathione S-transferase levels of maize seedlings. Acta Biochem. Biophys. Acad. Sci. Hung. 19, 130

Lay, M.-M. and Casida, J.E. (1976) Dichloroacetamide antidotes enhance thiocarbamate sulfoxide detoxification by elevating corn root glutathione content and glutathione S-transferase activity. Pestic. Biochem. Physiol. 6, 442-456

Leavitt, J.R.C. and Penner, D. (1979) In vitro conjugation of
glutathione and other thiols with acetanilide herbicides
and EPTC-sulfoxide and the action of the herbicide antidote R-25788. J. Agric. Food Chem. 27, 533-536

Lowry, O.H., Rosebrough, N.J., Farr, A.L., and Randall, R.J.
(1951) Protein measurement with the Folin phenol reagent.
J. Biol. Chem. 193, 265-275

Morgenstern, R., Meijer, J., DePierre, J.W., and Ernster, L.
(1980b) Reversible activation of rat liver microsomal
glutathione S-transferase activity. Eur. J. Biochem. 104,
167-174

Morgenstern, R., DePierre, J.W., and Ernster, L. (1980b)
Reversible activation of rat liver microsomal glutathione
S-transferase by 5,5'-dithiobis(2-nitrobenzoic acid) and
2,2'-dipyridyl disulfide. Acta Chem. Scand. B 34, 229-230

Morgenstern, R., Guthenberg, C., and DePierre, J.W. (1982)
Microsomal glutathione S-transferase. Purification, initial characterization and demonstration that is not
identical to the cytosolic glutathione S-transferases A,
B and C. Eur. J. Biochem. 128, 243-248

Mozer, T.J., Tiemeier, D.C., and Jaworski, E.G. (1983) Purification and characterization of corn glutathione S-transferase. Biochemistry 22, 1068-1072

Omura, T. and Sato, R. (1964) The carbon monoxide-binding
pigment of liver microsomes. II. Solubilization, purification and properties. J. Biol. Chem. 239, 2370-2378

THE INDUCTION OF CYTOCHROME P-450 TWO FORMS IN *CANDIDA* YEAST BY N-ALKANES OF DIFFERENT CHAIN LENGTH

Avetisova,S.M., Sokolov,Yu.I., Kozlov,V.I., Davydov,R.M., Davidov,E.R.
            Research Institute of Protein Biosynthesis, Moscow, USSR

Cytochrome P-450 is induced in yeast *Candida* by growth on n-alkanes. The properties of the alkane hydroxylation system of the yeast cultivated on n-alkanes or on n-tetradecane have been examined by many studies /Gallo et al.1976/, /Duppel et al.1973/, /Schunk et al.1983/. However, the similarity and differences in n-alkane hydroxylation by microrganisms between n-alkanes of different chain length is obvious /Fonken , Johnson 1972/. Therefore, the purpose of the present study was to investigate cytochromes P-450 synthesized by a cell during its cultivation on individual n-alkanes of different chain length.

MATERIALS AND METHODS

Cultivation

Candida maltosa yeast was obtained by periodic cultivation on n-alkanes (concentration 2% w/w): n-decane (C-10) and n-hexadecane (C-16). Cytochrome P-450 content was obtained as 18-20 and 30-35 nM of P-450/g W.W. for the process on C-16 and C-10,respectively.

Preparation of Microsomal Fraction

Yeast cells were collected by centrifugation (2500g×5 min) and then washed 3 times by cold deionized water and discrapted with glass beads in Dino-Mill (12% of dry weight, 3 1/h at 0-4°C) in 0.1 M potassium – phosphate buffer (pH 7,4) containing 0.25 M sucrose, 20% glycerol (w/w), 1 mM EDTA, 1 mM DTT. The microsomal fraction was obtained by centrifugation postmitochondrial supernatant at 105000g×120 min.

Isolation and Purification

Cytochrome P-450 was isolated according to the modified method of Azari and Wiseman /1982/. The modifications were:
- during solubilization the protein concentration was lowered to 20 mg of protein/ml and the duration of solubilization was increased up to 3 hs;
- centrifugation after dialysis was performed at 80000g×30 min as compared to 160000g;
- column with AO-Sepharose 4B (2.6×14 cm) was used for separation and the washing stage was introduced after the application of samples by a equilibration buffer containing 0.02% Emulgen 913.

Analytical Methods

Protein concentration was measured by the procedure of Lowry et al. /1951/. Cytochrome P-450 was measured from the CO-difference spectra as described by Omura and Sato /1964/. NADPH-cytochrome P-450 reductase was determined according to the method of Masters et al./1967/ and cytochrome b_5 - by the method of Ozols /1974/. Electrophoretic analysis was performed according to Laemmli /1970/ on gradient gels PAA 4/30 (Pharmacia Fine Chemicals, Sweden). Flash photolysis - by the method described by Imai /1982/.

RESULTS AND DISCUSSION

The procedure of cytochrome P-450 purification is shown in Table 1. Cytochrome P-450 was isolated from microsomes containing 0.5 and 1.0 nM P-450/ mg of protein for C-16 and C-10, respectively.

Table 1. Purification of cytochrome P-450

Purification stages	Protein mg		Cytochrome P-450					
			Total amount nM		Specific content nM/mg		Yield %	
	C-10	C-16	C-10	C-16	C-10	C-16	C-10	C-16
Microsomal fraction	1000	1200	1000	600	1	0.5	100	100
$(NH_4)_2SO_4$	410	600	805	480	1.9	0.8	80	80
AO-Sepharose 4B	35	47	620	390	17.7	8.2	62	65
HA 1(67kDa)	9	–	148	–	17.0	–	40	–
2(50kDa)	18	15	252	225	14.0	15.0		38

Solubilization of membranes by 1% sodium cholate with 0.1 % Triton X-100 caused a 100% (C-10) and 80% (C-16) yield of the enzyme.

Cytochrome P-450 was purified to an electrophoretically homogineous state. Introducing the washing with 0.02% Emulgen 913 at the stage of column chromatography (on AO-Sepharose 4B) helped to remove practically the all amount of cytochrome b_5 in the case of P-450$_{C-10}$ and the major amount in the case of cytochrome P-450$_{C-16}$.

Cytochrome P-450$_{C-10}$ preparations obtained after the procedure on AO-Sepharose 4B contained 17-18 nM/mg of protein. Electroforesis in SDS-PAGE revealed two bands of about the same intensity, corresponding to molecular weights 50 and 67 kDa. Separation of two forms of cytochrome P-450 was acieved by chromatography on hydroxylapatite.

Electrophoregrammes of proteins obtained at all stages of cytochrome P-450$_{C-16}$ purification revealed a cytochrome of molecular weight 50 kDa and showed an almost complete absence of the second form (67 kDa). The purified preparations did not contain cytochrome b_5 and NADPH-cytochrome P-450 reductase.

The difference in purification can be accounted for by the differece in the physical properties of the carbon substrates . Staining the samples according to Francis and Becker /1984/ showed the haem nature of the both purified proteins. Electrophoregrammes are shown in Fig.1.

The kinetic study of cytochrome P-450$_{C-10}$ (the fraction obtained after AO-Sepharose 4B) interaction with CO by the methods of flash photolysis showed the interaction to fit the two-stage kinetics: the initial phase made up about 40% of the reaction with the second order rate constant of $0.13 \times 10^7 M^{-1} sec^{-1}$. The second phase made up 60% of reaction with the second order rate constant of $0.15 \times 10^6 M^{-1} sec^{-1}$.

The reaction of cytochrome $P-450_{C-16}$ binding to CO had one stage with the second order rate constant of $0.27 \times 10^6 M^{-1} sec^{-1}$.

The rates of lauric acid hydroxylation *in vitro* were compared in a microsomal fraction and reconstituted system. The comparison showed (no data are presented) an increased hydroxylase activity of cytochrome $P-450_{C-10}$ towards fatty acids ω-oxidation which proved an order higher than that for cytochrome $P-450_{C-16}$.

The two-phase kinetics of cytochrome $P-450_{C-10}$ binding to CO, as well as the electrophoretical data and lauric acid oxidation *in vitro* indicate the existence of two forms of cytochrome in $P-450_{C-10}$ preparation and one form in $P-450_{C-16}$ preparation, the one with molecular weight 67 kDa being responsible for the fatty acids ω-oxidation and the other - for the n-alkanes hydroxylation.

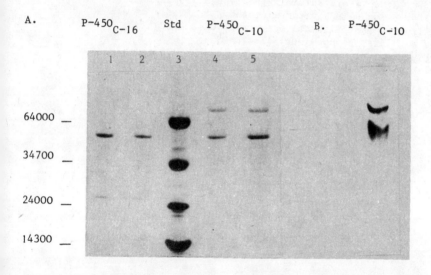

Figure 1. SDS-PAGE of purified cytochrome P-450 .
A. About 20 μg(2,4) and 30 μg(1,5) of proteins purified by affinity chromatography were electrophoresed in a system described by Laemmli /1970/.
B. About 80 μg of protein was electrophoresed and stained using the procedure of Francis and Becker /1984/.
Migration was from top to bottom.

REFERENCES

Azari,M.R., Wiseman,A., 1982, Purification and characterization of the cytochrome P-448 content of a benzo(a)purene hydroxylase from *Saccharomyces cerevisiae*, Anal.Biochem.129:129-138.
Duppel,W., Lebeault,J.M., and Coon,M.J., 1973, Properties of a yeast cytochrome P-450-containing enzyme system which catalyzes the hydroxylation of fatty acids, alkanes, and drugs, Eur.J.Biochem.36:583-592.
Fonken,G.S., Johnson,R.A., 1972, Chemical oxidation with microorganisms, chpter 1, p.40-59, Marcel Dekker, Inc., New York.

458

Francis,R.T., and Becker,R.R., 1984, Specific indication of hemoproteins in polyacrylamide gels using a double-staining process, Anal.Biochem.136: 509-514.

Gallo,M.,Poche,B., and Azoulay,E., 1976, Microsomal cytochromes of *Candida tropicalis* grown on alkanes, Biochim.Biophys.Acta 419:425-434.

Imai,I.,1982, Interaction of polycyclic hydrocarbons with cytochrome P-450, J.Biochem.92:67-77.

Laemmli,U.K., 1970, Cleavage of structural proteins during the assembly of the head of bacteriophage T4, Nature 227:680-685.

Lowry,O.H., Rosebrough,N.J., Farr,A.L., Randall,R.J., 1951, Protein measurement with the Folin phenol reagent, J.Biol.Chem.193:265-275.

Masters,B.S.S., Williams,C.H., Kamin,H., 1967, The preparation and properties of microsomal TPNH-cytochrome c reductase from pig liver, in Methods Enzymology, Section V11, Microsomal electron transport, v.10, p.565-573.

Omura,T., Sato,R., 1964, The CO-binding pigment of liver microsomes, 1. Evidens for its hemeprotein nature, J.Biol.Chem.239:2379-2385.

Ozols,L., 1974, Cytochrome b_5 from microsomal membranes of equine, bovine and porcine livers. Isolation and properties of preparations containing the membrane segments, Biochemistry 13:426-433.

Schunck,W.-H., Riege,P., Müller,H.-G., Scheler,W., 1983, Isolation and some molecular properties of cytochrome P-450 from the alkane assimilating yeast *Lodderomyces elongisporus*, Biochimiya 48:518-526.

Chemistry of Cytochrome P-450 Reactions

PROPERTIES OF A RECONSTITUTED AROMATASE CYTOCHROME P-450 SYSTEM FROM THE HUMAN PLACENTA: A COMMENT ON THE STEREOCHEMISTRY OF H-ELIMINATION AT C-1

L. Tan and N. Muto

Biochemistry Department, University of Sherbooke Medical Faculty,
Sherbrooke, Que., Canada, J1H 5N4

INTRODUCTION

The terminal steroid oxidase that catalyzes the multistep conversion of androgens to estrogens has been shown to be a cytochrome P-450 type monooxygenase (Thompson and Siiteri, 1974). This aromatase enzyme has thus far not been obtained in a homogeneous state from any biological source, although several groups have reported its partial purification (Zachariah et al. 1976; Thompson and Siiteri, 1976; Bellino and Osawa, 1978). The stoichiometry of the overall conversion is known, but the exact mechanism of the aromatization reaction is still the subject of controversy; in particular the last step of the sequence. Akhtar et al. (1982) proposed an enzyme-catalyzed, Baeyer-Villiger type third oxidation that also occurs at C-19. Fishman et al.(1977, 1981, 1982) believe that the third hydroxyl- ation takes place stereospecifically at the 2β-position, yielding a highly unstable ring A-intermediate as the endproduct of the enzymic reaction that then spontaneously collapses into the stable phenol. These mechanis- tic studies have been carried out either with crude microsomes, or at best with partially purified preparations. We have recently reported (Petit and Tan, 1985) that human placental aromatase undergoes irreversible inactivation when exposed to the epimeric 6-hydroperoxy derivatives of 4-androstene-3,17-dione (abbrev. 4-A). To understand how these suicide substrates interact with the enzyme, purification of the aromatase became mandatory. By a combination of bioaffinity-, and adsorption chromatography on AH-Sepharose, Con A-Sepharose and hydroxylapatite, we have been able to obtain the aromatase cytochrome P-450 in a high degree of purity, devoid of other cytochrome P-450 hydroxylases. Aromatase activity can be fully reconstituted by the addition of NADPH-cytochrome P-450 reductase. In this communication, we examine some of the properties of the reconstituted enzyme system.

RESULTS AND DISCUSSION.

Purification of aromatase cytochrome P-450. Briefly, frozen human placental microsomes were solubilized in 50 mM potassium phosphate buffer (pH 7.5) containing 20% glycerol, 1 mM DTT, 0.2 mM EDTA, 10 μM 4-A, and 1% CHAPS. After centrifugation, the supernatant was removed and purified by chromatography on AH-Sepharose, Con A-Sepharose, hydroxylapatite, and once again AH-Sepharose in this sequence. The thus purified enzyme showed a single, major band of mol. wt. 55,000 Da on SDS-PAGE. It had a specific cytochrome P-450 content of 4.15 nmol/mg protein, and a specific aromatase activity of 43.24 nmol/min/mg protein. This highly purified enzyme was used for the studies described in the following.

Inhibitory effect of various steroids. A reconstituted enzyme system was prepared by adding a homogeneous preparation of bovine hepatic cytochrome P-450 reductase (0.0048 μmol/min) to purified, human placental aromatase (0.383 μg= 1.6 pmol cytochrome P-450). The reconstituted aromatase was then incubated at 37 °C for 10 min in phosphate buffer in the presence of 0.1 μM [1,2-^{3}H]4-A, 0.1 μM steroid inhibitor, and a NADPH-generating system. After addition of 20 mg activated charcoal, the mixture was shaken, then centrifuged. The amount of radioactivity, released as tritiated-H_2O was then counted by removing an aliquot of the clear supernatant^2 and mixing with Aquasol. In the controls, no inhibitor was added. The results are shown in Table 1. At the relatively low steroid concentration of 0.1 μM tested, 1,6-dehydro-4-A showed the strongest effect with 57% inhibition. 19-Nor-4-A was also a strong aromatase inhibitor. Especially noteworthy is the difference between the two ring-A saturated androstanediones; the relatively planar 5α-androstanedione is eight times more inhibitory than the bulkier, 5β-epimer. It seems, therefore, that the inhibition is the strongest, when the 3-dimensional structure of the steroid molecule is also the most planar. C-6-oxygenated androstenedione derivatives were also found to be strong aromatase inhibitors. On the other hand, C_{21}-steroids such as progesterone, corticosterone and desoxycorticosterone did not cause any inhibition at all. These preliminary results with the reconstituted enzyme do not allow us to delineate the type of inhibition observed; in particular, if these steroid inhibitors compete with the natural aromatase substrate for the same active site.

TABLE 1. INHIBITORY EFFECT OF VARIOUS STEROIDS ON AROMATASE ACTIVITY OF RECONSTITUTED ENZYME SYSTEM.

Steroid (0.1 μM)	Inhibition (%)	Steroid (0.1 μM)	Inhibition (%)
None (control)	0	1-Dehydro-4-A	37
19-Nor-4-A*	43	6-Dehydro-4-A	11
5α-Androstane-3,17-dione	42	1,6-Dehydro-4-A	57
5β-Androstane-3,17-dione	5	6-Oxo-4-A	40
5-A*	23	6β-OOH-4-A	54

*4-A = 4-Androstene-3,17-dione; 5-A = 5-Androstene-3,17-dione.

Stereochemistry of H-elimination at C-1. Previous publications on the mechanism of estrogen biosynthesis have always maintained that the two H-atoms at C-1 and C-2 that are eliminated in the course of the aromatization reaction, are those that are positioned, respectively, in the 1β-equatorial and the 2β-axial configuration. However, we would like to emphasize that these studies were not only carried out with impure aromatase preparations, but that the absolute distribution of the tritium label at C-1 and C-2 of the aromatase substrate was often not established beyond any doubt. If we assume that Fishman's mechanism is the correct one and that in the third enzymatic step the 2β-H is indeed substituted by -OH with retention of configuration, then, based on chemical considerations governing conformational analysis, we may expect that in the fourth and final non-enzymatic step, trans-diaxial 1α-H,2β-OH elimination should be thermodynamically favored over the corresponding 1β(equatorial),2β(axial)-cis elimination reaction. To verify our hypothesis, we determined the formation of estrone with the reconstituted aromatase system by three methods: 1) by RIA; 2) by the formation of tritiated H_2O after removal of labeled steroids by chloroform extraction, or 3) by^2 charcoal adsorption of the [^{3}H]steroids. Two differently labeled substrates were incubated: namely [1,2-^{3}H]4-A, and [19-C^{3}H$_3$]4-A. As explicitly specified to us by the manufacturer (N.E.N., Boston, MA), the absolute distribution of the tritium label in the [1,2-^{3}H]4-A used in this study, is as follows: C-1α (13%), C-1β (44%), C-2α (10%), and C-2β (33%).

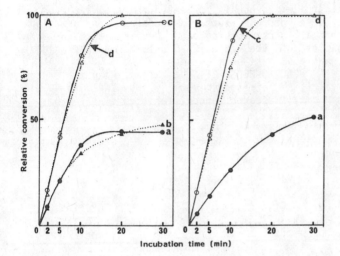

Fig.1 .

RIA AND RADIOMETRIC ANALYSIS OF ESTRONE FORMATION WITH RE-CONSTITUTED AROMATASE ENZYME SYSTEM.

Curves :
(a) charcoal-adsorption method,
(b) CHCl$_3$-extraction method,
(d) radioimmunoassay,
(c) corrected curves-a.

When [1,2-^{3}H]4-A was used as substrate (Fig.1A) and the aromatization was measured as the amount of tritium exchanged to [^{3}H]water, a plateau was reached after 20 min with 46±2% of tritium released; both with the char-coal-(curve-a), and with the CHCl$_3$-extraction method (curve-b). We empha-size that this value does not accurately reflect true conversion of 4-A to estrone, but represents only the total tritium-percentage that is lost; either from the C-1β/C-2β-, or from the C-1α/C-2β-pair. The radioimmuno-assay (curve-d) does indicate the actual formation of estrone. This RIA-curve also reaches a plateau at 20 min with 100% conversion. If aromatization indeed entails the loss of the 1β- and 2β-protons, the maximum value of 77% [^{3}H]water should have been found. Instead, the experimentally found values from two experiments were between 44-48%. This is very close to the theoretically expected value of 46% when the 1α and 2β protons are, in fact, the ones that are lost during the aromatization reaction. By multiplying each experimental point on the Y-axis of curve-a by the factor 2.17(=100:46) and replotting them, we obtain curve-d. As can be seen, the experimental RIA-curve (d) and the thus calculated curve-c (based on loss of the 1α,2β-protons), become congruent. These results clearly point to the loss of the 1α- and 2β-protons, and not that of the 1β and 2β protons, during aromatization. For further confirmation, we used [^{3}H]4-A as a second substrate, but now containing all of the [^{3}H]label at the C-19 methyl group (cf. Fig.1B). With this substrate, for each molecule of 4-A that is aromatized, 2 molecules of [^{3}H]water and 1 molecule of [^{3}H]HCOOH are released into the incubation medium. In this experiment, an isotope effect was found to slow down the reaction rate. Consequently, curve-a was still increasing after 30 min. Curve-c represents curve-a that was replotted by applying a correction factor of 3.2 (cf. Miyairi and Fishman), due to the isotope effect.

In résumé, with two differently labeled substrates, the reconstituted aromatase enzyme gave RIA-, and [³H]-exchange curves (c and d of Figs.1A and 1B) that were nearly identical. These results can only be adequately explained by invoking a reaction mechanism that proceeds via the combined loss of the axial 2β-proton and the axial 1-alpha-proton. Before non-enzymatic stabilization to estrone of the labile 2β-OH-19,19-di-OH(or -19-oxo)-androstenedione intermediate can occur, the chemical groups or bonds involved must first be favorably oriented and assume the proper conformation. In Fig.2, we propose such a preferred conformational transition-state in which all participating bonds are staggered, allowing for spontaneous elimination of the axial 2β-OH, 1α-H, and the oxidized 19-CH$_3$-group in an all-trans, and anti-parallel manner.

ACKNOWLEDGEMENTS

We thank Dr. S. Miyairi, Rockefeller University, New York, for a gift of [19-C ³H₃]4-A and the Medical Research Council for financial support.

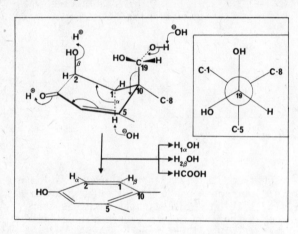

Fig. 2.

PROPOSED MECHANISM OF NON-ENZYMATIC STEP IN THE AROMATIZATION SEQUENCE.

Inset: Newman projection viewed along the C-19/C-10 bond.

REFERENCES

Akhtar, M., Calder, M.R., Corina, D.L. and Wright, J.N. (1982). Mechanistic studies on C-19 demethylation in oestrogen biosynthesis. Biochem. J. 201 , 569–580.

Bellino, F.L. and Osawa, Y. (1978). Solubilization of estrogen synthetase from human term placental microsomes using detergents. J. Steroid Biochem. 9 , 219–228.

Fishman, J. and Goto, J. (1981) Mechanism of estrogen biosynthesis. Participation of multiple enzyme sites in placental aromatase hydroxylations. J. Biol. Chem. 256 , 4466–4471.

Fishman, J. (1982) Biochemical mechanism of aromatization. Cancer Res. (Suppl.) 42 , 3277s–3280s.

Goto, J. and Fishman, J. (1977). Participation of a nonenzymatic transformation in the biosynthesis of estrogens from androgens. Science, 195 , 80–81.

Miyairi, S. and Fishman, J. (1985) Radiometric analysis of oxidative reactions in aromatization by placental microsomes. J. Biol. Chem. 260 , 320–325.

Petit, A. and Tan, L. Inactivation of human placental aromatase by the epimeric 6-hydroperoxyandrostenediones. J. Steroid Biochem. (in press).

Thompson, E.A. Jr. and Siiteri, P.K. (1974). Utilization of oxygen and reduced NADP by human placental microsomes during aromatization of androstenedione. J. Biol. Chem. 249 , 5364–5372.

Thompson, E.A. Jr. and Siiteri, P.K. (1976). Partial resolution of the placental microsomal aromatase complex. J. Steroid Biochem. 7 , 635–639.

Zachariah, P.K., Lee, Q.P., Symms, K.G. and Juchau, M.R. (1976). Further studies on the properties of human placental cytochrome P-450. Biochem. Pharmacol. 25 , 793–800.

THE EFFECT OF 3-/4-METHOXYPHENYL/-1-PHENYL-2-AMINOPROPANE ON THE METABOLISM OF 1,3-DIPHENYL-2-AMINO PROPANE IN RAT LIVER MICROSOMES, ISOLATED PERFUSED RAT LIVER AND IN VIVO BILIARY ELIMINATION EXPERIMENTS

K.JEMNITZ, J.SZAMMER, G.DÉNES, L.ÖTVÖS

CENTRAL RESEARCH INSTITUTE FOR CHEMISTRY OF THE HUNGARIAN ACADEMY OF SCIENCES, BUDAPEST, HUNGARY

The importance of microsomal cytochrome P-450 monooxygenase in the metabolism of many endogenous and foreign compouns, including steroids, drugs, carcinogens and pesticides is well documented /1./ The rate and the route of the metabolism of a drug can be highly dependent on other compounds present in the organism at the same time As the number of the potential substrates is much higher than the number of different P-450-s, this enzyme may be the site of many drug interactions.

In our present study 1,3-diphenyl-2-amino propane /DIPA/ was used as a model compound for microsomal monooxygenation and the effect of one of its derivatives 3-/4-methoxyphenyl/1-phenyl-2-aminopropane /4-MeO-DIPA/ on DIPA metabolism was investigated. We chose this compound because it's inhibitory properties on DIPA metabolism seemed to be the most interesting among the other derivatives, that had been investigated in earlier experiments. /2./ First the interaction of the two compounds was studied in liver microsomes, but it was also interesting to see how this effect was changing choosing more complex levels of the organization as isolated perfused liver and in vivo biliary excretion.

In vitro experiments with rat liver microsomes

Liver microsomes were prepared from male CFY rats in 0.15 M KCl, 0.01 M phosphate buffer, pH 7.4. Protein concentration was determined by the procedure of Lowry et al/3./ Cytochrome P-450 was detemined by the method of Omura and Sato /4./. MI complex formation was carried out according to Lindeke et al /5./, substrates were used in 0.5mM. The rate of metabolism was determined by NADPH consumption of the substrates. NADPH utilization was determined by dual-wavelength spectroscopy, against microsomes incubated in the absence of any substrate at 340 nm. The reactions were initiated by simultaneous addition of NADPH /2 μM/ to both cuvettes. The extinction coefficient used for determination of the NADPH concentration was $6.22 \cdot 10^{-3} M^{-1} cm^{-1}$.

	MI-complex formation[a]	NADPH consumption[b]
DIPA	1.89 ± 0.05	1.10 ± 0.15
4-MeO-DIPA	0.07 ± 0.03	0.37 ± 0.06
DIPA + 4-MeO-DIPA	1.14 ± 0.03	

a. $\Delta A_{/455-490/} \cdot cm^{-1} \cdot \text{P-450 mM}^{-1} \cdot \min^{-1}$

b. $\text{mM NADPH} \cdot \text{mM P-450}^{-1} \cdot \min^{-1}$

Table.I. The effect of 4-MeO-DIPA on the in vitro DIPA metabolism

Experiments with isolated perfused rat livers

Male CFY rats were anesthetized with urethane. The operative procedure and perfusion apparatus were the same as described by Miller /6./. Liver was perfused at 37°C with 100 ml of Krebs-Hensleit salt solution containing 100 mg glucose. After the initial 10 minutes of perfusion /2-^{14}C-DIPA and 4-MeO-DIPA with /2-^{14}C/-DIPA together respectively were added to the perfusate in 0.4 mM. The separation technics of the metabolites was described earlier /7./, the quantitative measurements were performed by radioactivity determinations.

4-OH-DIPA / 10^{-6} mmol/ml perfusate /

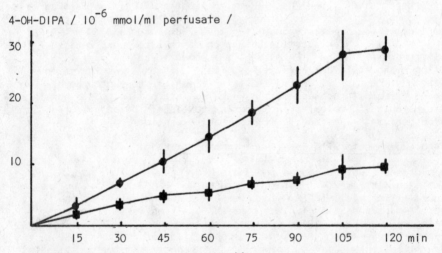

● 4-OH-DIPA in experiments, when /2-^{14}C/-DIPA was the only substrate
◗ 4-OH-DIPA in case when both substrates were added together

Fig.I. The effect of 4-MeO-DIPA on DIPA metabolism in
isolated perfused rat liver

In vivo experiments

Male CFY rats were anesthetized by urethane, and the biliary
ducts were cannulated. Bile was collected over 30 minutes intervals
for 2 hours. 0.5 mM /2-^{14}C/-DIPA and /2-^{14}C/-4-MeO-DIPA were
administered to the animals respectively. For the study of the
effect of 4-MeO-DIPA on DIPA biliary excretion 4-MeO-DIPA was
administered to the animals 60 minutes before the /2-^{14}C/-DIPA
administration. Animals were treated i.v. The quantitative
determinations of the metabolites were described above.

Fig.2.   Fig.3.

radioactivity /.10^{-3} mM/   excreted metabolism in % of
                                        the total

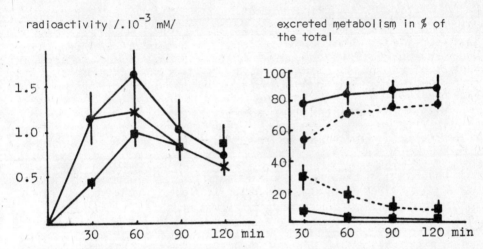

• The elimination profile of DIPA; x The elimination profile of 4-MeO-
DIPA; ■ The elimination profile of DIPA in the presence of 4-MeO-DIPA
Fig.2. Biliary excretion of DIPA, 4-MeO-DIPA and the
       effect of 4-MeO-DIPA on DIPA biliary elimination

  •   4-OH-DIPA;   ■ DIPA
———— Rate of excretion of metabolites, when only DIPA was administered
----- Rate of excretion of metabolites, when both compounds were
      administered
Fig.3. The effect of 4-MeO-DIPA on the ratio of DIPA
       metabolites excreted in the bile

In conclusion, the results of the present study show, that 4-MeO-
DIPA inhibits the metabolism of DIPA, and this effect can be well
observed in all levels of organization that were investigated in
our present experiments. These data are not sufficient to decide
the nature of this inhibition. One may conclude, that the rate of
metabolism of 4-MeO-DIPA is much slower than that in case of DIPA
/Table and Fig 2./, so the competition for NADPH cannot explain

31*

468

the inhibitory effect, and this refers to the inhibition by
forming MI complex too.
Biliary excreation data show, that 4-MeO-DIPA does not only
change the rate of DIPA elimination with the bile, but also the
distribution of metabolites. When only DIPA is administered to
the animals the amount of excreated DIPA just conjugated with β-
glucuronic acid was below 5 % of all metabolites. When both
compounds were present in the organism the percentage of this
metabolite was much higher, almost 30 % during the first half of
an hour.

References

1. Gillette, J.R. Annu.Rev.Pharmacol. 14 271-292 /1974/

2. Jemnitz, K. PH.D. Thesis, Eötvös Lóránd University, Budapest
    /1981/

3. Lowry, O.H. Rosebrough, N.J., Farr, A.L., Randall, R.J.
    J. Biol. Chem. 193 265-275 /1951/

4. Omura, T. Sato, R., J. Biol. Chem. 239 2370-2378 /1964/

5. Lindeke, B., Paulsen, U. Anderson, E., Biochem. Pharmac. 28
    3629 /1979/

6. Miller, L.L. Technique of Isolated Rat Liver Perfusion 11-53
    /Isolate Liver Perfusion and Its Applications, I.Bartoseh,
    A. Quaitani, L.L. Miller eds/, Raven Press, New York /1973/

7. Juvancz, Z., Ledniczky, M. Imre L., Bihary M, and Vajda, M.
    J. Chromatogr 337 121-125 /1985/

STRUCTURE-ACTIVITY RULES FOR PHENOBARBITAL-TYPE INDUCERS OF MICROSOMAL MONOOXYGENASES

Ilya B. TSYRLOV, Konstantin E. GERASIMOV

Laboratory of Xenobiochemistry, Institute of Experimental Medicine, Academy of Medical Sciences of the U.S.S.R., Siberian Division, 630091 Novosibirsk (U.S.S.R.)

INTRODUCTION

The study of the mechanisms of monooxygenase induction and elucidation of factors determining the specificity of inductive effect of a xenobiotic (PB-,MC-,PCN-type,etc) are the problem of biochemistry of drug-metabolizing enzymes and applied branches of knowledge. The present paper shows that the major factor in structure-activity correlations of PB-type inducers is localization and state of the position which undergoes monooxygenation at cytochrome P-450 active centre. Preservation of this position intact or its blockage ensures the pronounced PB-like induction properties for a xenobiotic; localization of the position in molecule of PB-type inducer results in the specificity of enzyme de novo synthesized.

MATERIALS AND METHODS

Inbred C57B/6 mice (20 g) were injected i.p. by amobarbital and ω-1 methylamobarbital once in a dose of 100 mg per kg. During 4 days Wistar rats (50 g) were treated i.p. by phenobarbital (100 mg per kg), 3-methylcholanthrene (40 mg per kg), aminopyrine and isopropylaminoantipyrine (IPAAP)(80 mg per kg). Isolation of liver microsomal fraction, determination of cytochrome P-450 content, NADPH oxidation rate, activity of aminopyrine N-demethylase were performed as described (Tsyrlov et al.1976). Velocities of microsomal hydroxylation of cyclohexane and tetralin were measured by liquid chromatography (Anshits et al.1979). Quantities of immunochemically determined cytochromes in microsomes were evaluated by rocket immunoelectrophoresis (Pickett et al.1981). In radiological investigations inducers were administered 16 h, and L-(1-^{14}C) leucine - 8 and 4 h before animals were killed. Fluorography of gel plates after SDS-PAAG electrophoresis of microsomal proteins was carried out using Mark-III counter.

RESULTS AND DISCUSSION

As it was shown by us previously (Tsyrlov et al.1976), PB-type inducers are the most potent in their parent (non-metabolizing) form. In fact, hydroxylated PB metabolites lose entirely their induction properties (Gesteil et al.1980). In experiments with chlorinated naphthalene and biphenyl derivatives this hypothesis was confirmed (see also: Goldstein et al. 1977; Ahotupa,Aitio,1980). Thus, chlorinated naphthalene (the site of hydroxylation at 1-position is blocked) was able to stimulate activity of microsomal demethylase though naphthalene itself did not have any inductive properties. PB-cytochrome P-450 is known to hydroxylate biphenyl molecule at 2- and 4-positions. Biphenyl isomers chlorinated at these positions are not practically metabolized in microsomes (Shulte,Acker, 1974) and as a result of this one can observe PB-type induction of monooxygenase system.

That is why it seems possible to convert typical substrates of monooxygenase into genetic inducers of the system by direct chemical modification of the substrate's position that undergoes monooxygenation. An evidence has been obtained in experiments with typical cytochrome P-450 substrates - amobarbital and aminopypine. The main data are presented in Table 1.

Table 1

	Control	Methylamo-barbital	4-IPAAP	PB
Mouse liver microsomes:				
P-450 content* **	0.66	1.12	–	2.3
NADPH oxidation** **	10.8	17.6	–	20.0
Aminopyrine demeth.	2.64	5.0	–	9.8
Rat liver microsomes:				
P-450 content* **	0.55	–	1.17	2.2
Aminopyrine demeth. ***	1.7	–	3.03	5.1
P-450 specific cont.	3-5%		32%	40%

*nmol/mg protein. **nmol/min·mg protein. ***determined by rocket immunoelectrophoresis using antibodies against PB-P-450

Amobarbital and aminopyrine themselves are not characterized as inducers of liver monooxygenase. Contrary, methylamobarbital (the hydroxylating position at ω-1 carbon atom in alyphatic side chain is blocked) causes more than 2-fold prolongation of sleeping time for mice as compared with amobarbital (see also Table 1). As to aminopyrine, substitution of two methyl groups in its molecule at $-N(CH_3)_2$ position for isopropyl group (i.e. 4-IPAAP) is accompanied by following effects. Firstly, fluorography of plates after electrophoresis revealed de novo synthesis of microsomal hemoprotein with m.w. 51 000. Microsomes are characterized by increase in the total and specific cytochrome P-450 content and in the rate of aminopyrine N-demethylation (Table 1). Thus, typical monooxygenase substrates were converted to inducers by chemical modification of position subjected to oxidation at P-450.

The data obtained lead to revision of two known rules formulated by H.Remmer (1972) for PB-type inducers. First, their induction effects appears to depend not so much on concentration of durable P-450-substrate complex, as on concentration of free xenobiotic in cytoplasm. Second, our results as well as the data on a greater induction activity of chlorinated PB derivatives (Poland et al.1980) testify to the fact that blockage of specific position in the substrate is more important than relative hydrophobicity of its molecule.

Table 2

	oxidation rate, 10^{-10} mol/min·mg protein		
	cyclohexane into cyclohexanol	tetralin into α-tetralol	β-tetralol
Control microsomes	4.0	4.3	0.3
PB-microsomes	34.0	163.0	5.1
MC-microsomes	1.0	3.8	-

According to immunochemical data, a few monooxygenase molecular forms are known to be present in PB-microsomes. However in any case the number of substrates metabolized by these forms is at least two order larger. These data stipulate the necessity in revealing the factors determining substrate specificity of PB-type monooxygenases. In this sense the data presented in Table 2 are of interest. Thus, after PB administration the rate of hydroxylation at the secondary carbon atom in molecules of cyclohexane and tetralin rises drastically. It is known that treatment of animals by PB results in accelerated rate of hydroxylation at the secondary carbon atom in hexane, N-methylcarbazole and laurate (Frommer et al.1974; Coon et al.1980; Okita,Masters,1980). PB induction is followed by the rise of hydroxylation of a number of barbiturates (Freudenthal,Carroll,1973) due to increase of hydroxylation at secondary carbon atom in alyphatic derivatives (see the data on amobarbital). It is important that metabolism of PB itself proceeds by hydroxylation at the secondary carbon atom in ethyl side chain.

From the data obtained we can conclude that after PB administration the microsomal monooxygenase de novo synthesized catalyses metabolism of substrates by the way characteristic for the inducer itself. The hypothesis seems to be valid: PB-type inducer effects both the genome and the synthesized enzyme during formation of its active centre. This ensures the appointed level of positional specificity of induced monooxygenase. In fact, formation of the enzyme active structure occurs in the medium containing the inducer which is the potential substrate for the enzyme. According to the Koshland's theory (1963) concerning enzyme's active centre plasticity, the centre undergoes conformational changes due to binding of PB-type inducer. This leads to formation of a relatively rigid template suitable for metabolism of the given inducer and other xenobiotics metabolizing by a similar way. Thus, geometric and electronic tune of the active centre to PB-type inducer provides a high catalytic activity and positional specificity of de novo synthesized monooxygenase. The data obtained by other authors also confirm the fact that when rats are treated by PB the posttranslational stage is the rate-limiting step in biosynthesis of catalytically active cytochrome P-450 (Dubois et al.1980; Lechner,Sinogas,1980).

REFERENCES

Ahotupa M.,Aitio A. (1980) Effect of chlorinated naphthalenes and terphenyls on the activities of drug metabolizing enzymes in rat liver. Biochem.Biophys.Res.Commun.,93,250-257
Anshits A.,Sokolovsky B.,Burylin S.,Lyakhovich V.,Tsyr-

472

lov I. (1979) Influence of substrate and inducer structure upon catalytic activity of microsomal enzymes. Kinetics and Catalysis (USSR),20,865-869

Coon M.,Koop P.,Persson A.,Morgan E. (1980) Inducible and constitutive isozymes of microsomal cytochrome P-450. In: Biochemistry,Biophysics and Regulation of Cytochrome P-450 (J.-A. Gustafsson et al.,eds),Elsevier,Amsterdam-N.Y.-Oxford,7-15

Gresteil T.,Manu J.,Dansette P.,Leroux J. (1980) In vivo administration of hydroxylated phenobarbital metabolites: effect on rat hepatic cytochrome P-450, epoxide hydrotase and UDP-glucuronosyltransferase. Biochem.Pharmacol.,29,1127-1133

Dubois R.,John M.,Waterman M. et al. (1980) Induction of hepatic cytochrome P-450: study of the molecular events following phenobarbital administration to rats. In: Biochemistry, Biophysics and Regulation of Cytochrome P-450 (J.-A.Gustafsson et al.,eds),Elsevier, Amsterdam-N.Y.-Oxford,397-404

Freudenthal R.,Carroll F. (1973) Metabolism of certain commonly used barbiturates. Drug Met.Revs.,2,2655-2678

Frommer U.,Ullrich V.,Orrenius S. (1974) Influence of inducers and inhibitors on the hydroxylation pattern of n-hexane in rat liver microsomes. FEBS Lett.,41,14-16

Goldstein J.,Hickman P.,Bergman H. et al. (1977) Separation of pure polychlorinated biphenyl isomers into two types of inducers on the basis of induction of cytochrome P-450 or P-448. Chem.-Biol. Interactions,17,69-87

Koshland D. (1963) Correlation of structure and function of enzyme action. Science,142,1533-1541

Lechner M.,Sinogas C. (1980) Changes in gene expression during liver microsomal enzyme induction by phenobarbital. In: Biochemistry,Biophysics and Regulation of Cytochrome P-450 (J.-A.Gustafsson et al.,eds), Elsevier, Amsterdam-N.Y.-Oxford,404-414

Okita R.,Masters B. (1980) Effect of phenobarbital treatment and cytochrome P-450 inhibitors on the laurate ω- and (ω-1)-hydroxylase activities of rat liver microsomes. Drug Metab.Disposition,8,147-151

Pickett C.,Jeter R.,Morin J.,Lu A. (1981) Electroimmunochemical quantitation of cytochrome P-450, cytochrome P-448 and epoxide hydrolase in rat liver microsomes. J.Biol.Chem., 256,8815-8820

Poland A.,Mak I.,Glover E. et al. (1980) 1,4-bis/2-(3,5-dichloropyridiloxy)/benzene, a potent phenobarbital-like inducer of microsomal monooxygenase activity. Mol.Pharmac.,18, 571-580

Remmer H. (1972) Induction of drug metabolizing enzyme system in the liver. Eur.J.Clin.Pharmac.,5,116-136

Shulte E.,Acker L. (1974) Identifizierung und metabolisiebarkeit von polychlorierten biphenylen. Naturwissenschaften,61,79-80

Tsyrlov I.,Zakharova N.,Gromova O.,Lyakhovich V. (1976) Possible mechanism of induction of liver microsomal monooxygenases by phenobarbital. Biochim.Biophys.Acta,421,44-56

THEORETICAL ASPECTS OF CYTOCHROME P-450

IMPLICATIONS ABOUT THE ORIGIN OF CYTOCHROME P-450$_{LM}$ ISOZYME MULTIPLICITY FROM THE AMINO ACID SEQUENCES

Dus, K.M.

The E.A. Doisy Department of Biochemistry, St. Louis University
School of Medicine, St. Louis, MO   63104, U.S.A.

CONCEPTUAL OUTLINE

The recent technical advances in biochemistry and molecular biology have
provided us with an enormous number of complete amino acid sequences of
proteins, some of them even exceeding 500 residues and/or being in rather
short supply. And the pace of sequence determinations is still increasing. Considering this wealth of new data and the intensity of effort
behind the explosive development it seems timely to define an organiza-
tional framework and to reach an appreciation of the whole information
content of amino acid sequences.

There are many types of information provided by the precise sequence of
amino acid residues of a protein. For instance, an investigator may
inquire about the rules of protein folding that dictate how the primary
structure translates into the secondary and tertiary structure of the
protein. But in the following pages P-450 hemeproteins and related monoxygenase components are employed as examples to illustrate additional
information inherent in amino acid sequences. This information may be
used to trace sequence homology and to reveal structure-function correlations. Since all the proteins compared in this presentation are related
to each other by sequence homology, regardless of their function, it is
concluded that they originated from the same ancestral polypeptide. Each
protein in this set has attained its own characteristic sequence and
function, yet suitable alignment of selected segments still reveals in
each the presence of conserved regions of internal sequence repitition
with concomittant homology to corresponding segments of the other members
of this protein family (1). This observation is compatible with the
assumption that their current size and individuality is the result of
progressive gene duplications and independent, function-directed mutational adjustments. A very similar process, yet limited to partial gene
duplications of but a few selected areas of P-450$_{LM}$ is suggested as the
cause of isozyme multiplicity. It is assumed that the location of these
sites in the linear sequence has been selected such that after folding,
the 3D structure of the protein will bring them into contact with the
heme environment. In this manner the duplicated sequences may increase
the size and flexibility of the substrate binding site and enhance its
hydrophobic character. This simple concept, if it applies, can accommo-

date many potential P-450$_{LM}$ isozymes, in addition to those which have
already been found and characterized. Since the original P-450$_{LM}$ is
unknown, the sequence of the shorter and presumably older bacterial
P-450$_{CAM}$ is employed for alignment to pinpoint the major areas of
sequence repitition leading to drug hydroxylating and aryl hydrocarbon
hydroxylating P-450$_{LM}$ isozymes. Since P-450$_{CAM}$ is the only P-450 currently available in crystalline form suitable for X-ray diffraction (2),
this comparison carries additional implications concerning the 3D structure of P-450$_{LM}$ isozymes.

RESULTS AND DISCUSSION

Table I is organized into 6 sections to demonstrate that cytochromes
P-450$_{CAM}$ (3), P-450$_{LM}$ (4,5) and b$_5$ (6), as well as NADPH cytochrome P-450
reductase (7), and PdX (8), and AdX (9), all satisfy the requirement of
internal sequence repitions (1,10). These repeated sequences show a
theme in variation which occurs, somewhat modified, in each of these
proteins, thus making their amino acid sequences partially homologous to
each other. The scoring method records identity as 0, one and two base
changes as 1 and 2, respectively, and a gap as 3. The sum of these
weights is then divided by the number of positions in the sequence to
reach a value (expressed in minimal base changes) for the relative fit
attained by alignment of these sequences. The smaller the value, the
better the fit. To qualify, sequences must be at least 6 residues long
without gap while extended sequences may contain gaps (not included in
the score) provided coincidences occur at regular intervals. Both types
of comparison are illustrated in the alignment of sequences of P-450$_{LM-2}$
and its reductase. The extended comparison of the 4 sequences (Section
5) illustrates both the common origin and the process of gradual modification. Moreover, the fact that the similarity in character of these
sequences was conserved by the constraints of evolutionary selection may
tell us something important bout a common purpose or function of these
sequences within their respective proteins. Since they all have to
interact with closely related reductases this structural feature should
not go unnoticed or unexplored. In fact, the last section provides a
further clue along this line by comparing the putative binding site of
PdX to P-450$_{CAM}$, with its sole and C-terminal Trp, to a portion of the
membrane embedded foot of b$_5$ (6,11). Both proteins are able to bind to
P-450$_{CAM}$ and compete for the same site. Thus, sequence homology can
point to structure-function correlations.

The understanding of the general process of progressive gene duplications
by which the monoxygenase enzyme systems might have originated, is of
great importance for the consideration of the process by which the
P-450$_{LM}$ isozymes could have been generated. As shown in Table II, and in
subsequent tables, the very same concepts of gene duplication and subsequent independent mutational modification may also apply to the problem
of P-450$_{LM}$ isozyme multiplicity and its evolutionary origin. The only
new aspect which seems characteristic for P-450$_{LM}$ and has not previously
been observed with other protein families, is the occurrence of "hot
spots" in the sequence where partial gene duplications may arise easily
and repeatedly while the rest of the polypeptide chain remains essentially unaffected. The mechanism by which this selection is made is not
clear at this time but a simple rationale can be offered which opens the
problem to experimental testing. Perhaps, the duplicated areas are

TABLE I

PARTIAL SEQUENCE ALIGNMENT OF MONOXYGENASE COMPONENTS

EXAMPLES OF INTERNAL HOMOLOGY IN P-450 CAM

BASE CHANGES

```
          128
     ArgIleGlnLeuAlaCysSerLeuIleGlu   138    1.0
157  ProIleArgIlePheMetLeuLeuAlaGlyLeuProGlu
       1  1  0  1  1  2  0  0  2  1               0.9
     AlaLysArgMetCysGlyLeuLeuLeuVal   245

     356  LeuGlyGlnSerLeuAlaAla   362
            0  2  1  1  0  1  0          0.7
     369  LeuLysGluTrpLeuThrArg   375
```

P-450 CAM AND P-450 LMb SEGMENTS

```
CAM   141
LeuGluAsnArgIleGlnGluGlnAlaCysSerLeuIleGluSerLeuArg
  1  0  2  0  0  0  0  2  0  2  1  0  1  0  2  0   0.76
ValGluArgIleGlnGluGlnAlaGlnCysLeuValGlnGluLeuArg
LM   158
```

```
       151                160
CAM   AspTyrAlaGluProPhePheProIleArgIle
        0  0  1  1  2  1  0  1               0.90
LM    AspTyrThrAspArgGlnPheLeuArgLeu
       189                198
```

```
      348                               362
CAM   PheGlyHisGlySerHisLeuCysLeuGlyGlnSerLeuAlaArg
        0  1  2  0  2  1  1  0  0  0  1  1  0  0      0.67
LM    PheSerThrGlyLysArgGlyCysLeuGlyGluGlyIleAlaArg
      429                               443
```

NADPH-CYTOCHROME P-450 REDUCTASE COMPARED TO CYTOCHROME P-450 LM-2

```
                  180              190                  200
REDUCTASE  TyrGluHisPhe - AsnMetGlyLysTyr - ValAspGlnArgLeuGluGlnLeuGlyAlaGlnArg IlePheGluLeuGly
LM-2       TyrGlyIlePheAlaAsnGlyLysTrpLysTrpArgAlaLeuArgArgPheSer - LeuAlaThrMetArgAspPheGlyMetGly
           111                                    120                   130                       137
```

```
           120    AlaLeuArgArgPheSer - LeuAla
P-450 LM-2         1  1  0  0  0  3  0  1       0.78       LM-2   90   GlyGlnAlaGluAspPhe   95
      353    GluIleGlnArgPheSerAspLeuVal                                0  2  0  0  1  0      0.50
              1  2  0  0  1  1  1  0  1      0.78       REDUCTASE  89   GlyThrAlaGluGluPhe   0.67
      196    GlyAlaGlnArgIlePheGluLeuGly                               1  1  0  1  1  0      0.67
REDUCTASE  187  ValAspGlnArgLeuGluGlnLeuGly   152       REDUCTASE  147  AspAsnAlaGlnAspPhe  152
```

HOMOLOGY OF P-450 CAM AND PUTIDAREDOXIN

```
                                              24
PdX   12   ArgTyrAlaGluAspValAlaAspGlyValSerLeuMet
             0  0  2  2  1  1  0  0  1  1  0  0  1  0   0.69
CAM   209  ArgArgGlnIleLysProGlyThrAspAlaIleSerLeuVal
                                              221
```

```
PdX   45   CysAlaThrCysHisVal   50
             0  1  1  2  1  0       0.83
CAM   240  CysGlyLeuLeuLeuVal   245
```

COSERVED SEQUENCES IN 4 RELATED PROTEINS

```
P-450 LMb    316   LysTyrProHisValAla - GluLys - ValGlnLysIle
b5            58    PheGluAsp - ValGlyHisSerThrAspAlaArgGluLeuSerLysThr
AdX           64    PheGlu - LysLeuGluAlaIleThrAsnGln - GlyAsnAsnMetLeu
PdX           56    PheThrAspLysValProIle - AlaAsnGlnArgGlu IleGlyMetLeu
```

```
P-450 LMb    328   AspGlnValIleGlySer - His   341    AspAspArgSerLysMet
b5            74    - PheIleGluGluLeuHis     82      AspAspArgSerLysIle
AdX           79    AspLeuAlaTyrGly - Leu    85      ThrAspArgSerArgLeu
PdX           72    GluCysValThrAlaGluLeuLys  81     - Asn - SerArgArgLeu
```

HOMOLOGY AT PUTATIVE   PdX   99   ValAsp - ValProAspArgGlnTrp-OH 106
BINDING SITE TO P-450 CAM   b5   98   IleAspSerAsnProSer - TrpTrp 105

TABLE II

CONSERVED PROLINE CLUSTER

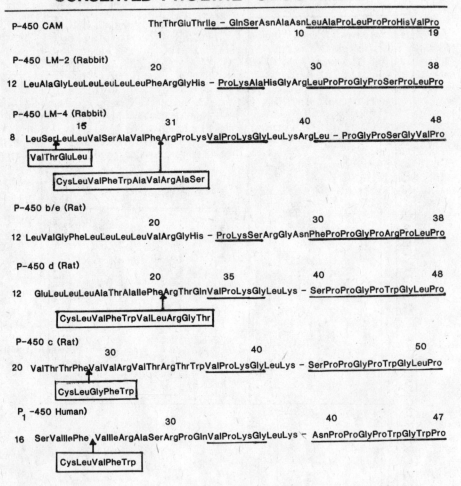

479

TABLE III

ALIGNMENT OF SEQUENCES SURROUNDING CONSERVED CYSTEINES

P-450 CAM   125 LeuGluAsnArgIleGlnGluLeuAlaCysSerLeuIleGluSerLeuArg 141

P-450 LM-2(Rabbit) 142 ValGluGluArgIleGlnGluGluAlaGlnCysLeuValGluGluLeuArgLys 159

P-450 LM-4 (Rabbit) 150 SerAsnProAlaSerSer - SerSerCysTyrLeuGluGluHisValSerLys 166

P-450 b/e (Rat) 142 ValGluGluArgIleGlnGluGluAlaGlnCysLeuValGluGluLeuArgLys 159

P-450 d (Rat) 148 SerAspProThrSer - ValSerSerCysTyrLeuGluGluHisValSerLys 164

P-450 c (Rat) 153 SerAspProThrLeu - AlaSerSerCysTyrLeuGluGluHisValSerLys 169

P_1-450 (Human) 149 SerAspProAlaSerSer - ThrSerCysTyrLeuGluGluHisValSerLys 165

P-450 CAM   348 PheGlyHisGlySerHisLeuCysLeuGlyGlnSerLeuAlaArgArgGluIleIleValThrLeuLeuGlu 371

P-450 LM-2 (Rabbit) 429 PheSerLeuGlyLysArgIleCysLeu GlyGluGlyIleAlaArgThrGluLeuPhe - ThrIleMetGlu 455
                                                                         LeuPhePheThr

P-450 LM-4 (Rabbit) 429 PheGlyLeuGlyLysArgArgCysIleGlyGluThrLeuAlaArgTrpGluValPhe - IleLeuLeuGln 455
                                                                         LeuPheLeuAla

P-450 b/e (Rat) 429 PheSerThrGlyLysArgIleCysLeuGlyGlyIleAlaArgAsnGluLeuPhe - ThrIleLeuThr 455
                                                                         LeuPhePheThr

P-450 d (Rat) 449 PheGlyLeuGlyLysArgArgCys Ile GlyGluIleProAlaLysTrpGluValPhe - IleLeuPheSer 475
                                                                         LeuPheLeuAla

P-450 c (Rat) 454 PheGlyLeuGlyLysArgLysCysIleGlyGluThrIleGlyArgLeuGluValPhe - IleLeuPheSer 480
                                                                         LeuPheLeuAla

P_1-450 (Human) 450 PheGlyMetGlyLysArgLysCysIleGlyGluThrValAlaArgTrpGluValPhe - IleLeuLeuGln 476
                                                                         LeuPheLeuAla

exposed on the surface where insertion of additional sequences may not cause any significant distortion of the overall architectural design of the protein. We must remember that all isozymes are able to interact with the same reductase. Finally, there must be a great selective advantage associated with these inserted sequence repeats. Since the isozymes have altered substrate preferances and are inducible by substrate it seem reasonable to invoke selection of partial gene duplications such that only portions on the surface are involved which in the 3D structure contribute to the substrate binding site. The alignment of sequences close to the amino termini of 7 P-450 hemeproteins in <u>Table II</u> focuses on a conserved cluster of proline residues. This feature is most peculiar and has been noted for some time (10). Due to the kink in the polypeptide chain caused by proline it may imply the start of a structural domain which is common to bacterial and membrane-inserted P-450s. Of practical significance for our sequence alignment in <u>Table II</u> is the fact that the position of this proline cluster in the early part of all P-450 sequences constitutes a readily visible point of correspondence regardless of the presence and uneven length of the signal peptide in some of them. No signal peptide is present in P-450$_{CAM}$. P-450$_{LM-4}$ of rabbit (12), as well as P-450d and P-450c of rat (13), and P$_1$-450 of human origin (14), show a new characteristic feature, the proposed partial sequence repeats, called out in boxes and aligned below the main sequence at certain "hot spots". One area where repititions seem to occur frequently is seen in <u>Table II</u> as a highly hydrophobic decapeptide is inserted after the Phe in position 20 of P-450$_{LM-4}$ and P-450d, while P-450c and P$_1$-450 have only a 5-residue repeat. P-450$_{LM-4}$ has an additional 4-residue repeat after Ser in position 9. It is important to note that excision of these repeats from the main sequence improves the alignment significantly.

<u>Table III</u> presents the alignment of sequences surrounding conserved cysteines thought to be involved in chelation to the heme iron of P-450. In accordance with this assignment we see good homology for all P-450s entered into the alignment, reinforcing the rationale that the structural integrity of the catalytic site must be maintained. In the lower half of this comparison appears another 4-residue insert of very hydrophobic residues in P-450$_{LM}$ isozymes relative to P-450$_{CAM}$.

The main emphasis of this presentation is on <u>Table IV</u> which features the alignment of sequences in the suggested area of multiple sequence repititions of P-450$_{LM}$ isozymes. The concept of "hot spots" is best demonstrated by this segment of the P-450$_{LM}$ isozyme sequences. The homology to the P-450$_{CAM}$ sequence is significantly improved by "looping off" the inserted repeats. It is apparent that P-450$_{LM-2}$ of rabbit and P-450b/e of rat have very comparable sequence repeats, which in turn are inserted into very similar positions within the respective sequences, while P-450c and P-450d of rat have more extended inserts which appear to be in part the result of a second repitition of the first repeats, and/or possibly of a third duplication process. Again, the positions of insertion are very comparable for this pair of sequences. Clearly, the repeats represent over two thirds of all the hydrophobic residues in this section of the P-450$_{LM}$ sequences and thus account for the general increase in hydrophobic character. Another interesting aspect is the fact that the repeats nevertheless still align quite well to the main sequence and to the sequence of P-450$_{CAM}$. The increase in the number and hydrophobicity

481

TABLE IV

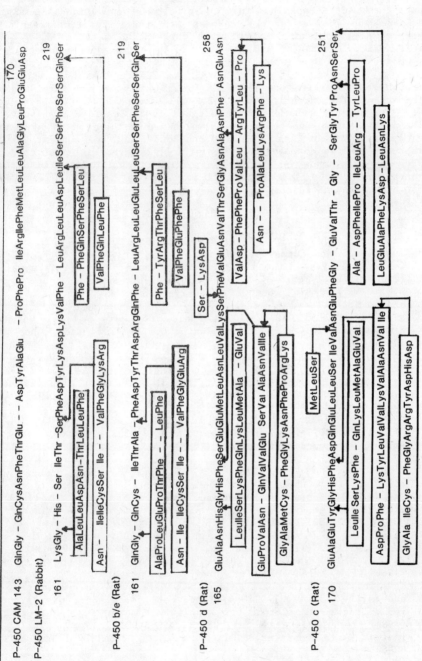

SUGGESTED AREA OF MULTIPLE SEQUENCE REPITITIONS IN P–450 LM ISOZYMES

P-450 CAM 143   GlnGly – GlnCysAsnPheThrGlu - – AspTyrAlaGlu   – ProPhePro IleArgIlePheMetLeuAlaGlyLeuProGluGluAsp

170

P-450 LM-2 (Rabbit)

219

161   LysGly – His – Ser IleThr –SerPheAspTyrLysAspLysValPhe – LeuArgLeuLeuAspLeuIleSerSerPheSerSerGlnSer

AlaLeuLeuAspAsn–ThrLeuLeuPhe

Asn –   IleIleCysSer Ile – – ValPheGlyLysArg

Phe – PheGlnSerPheSerLeu

ValPheGlnLeuPhe

P-450 b/e (Rat)

219

161   GlnGly – GlnCys – IleThrAla –PheAspTyrThrAspArgGlnPhe – LeuArgLeuLeuGluLeuSerSerPheSerSerGlnSer

AlaProLeuGluProThrPhe – – LeuPhe

Asn – Ile IleCysSer Ile – – ValPheGlyGluArg

Phe – TyrArgThrPheSerLeu

ValPheGluPhePhe

Ser – LysAsp

258

P-450 d (Rat)

165   GluAlaAsnHisGlyHisPheSerGluGluMetLeuAsnLeuValLysSerPheValGluAsnValThrSerGlyAsnAlaAsnPhe – AsnGluAsn

LeuIleSerLysPheGlnLysLeuMetAla – GluVal

GluProValAsn – GlnValValGlu SerVal AlaAsnVallle

GlyAlaMetCys – PheGlyLysAsnPheProArgLys

ValAsp – PhePheProValLeu – ArgTyrLeu – Pro

Asn – – ProAlaLeuLysArgPhe – Lys

MetLeuSer

P-450 c (Rat)

251

170   GluAlaGluTyrGlyHisPheAspGlnGluLeuGluSer IleValAsnGluPheGly – GluValThr – Gly – SerGlyTyr ProAsnSerSer

LeuIle SerLysPhe – GlnLysLeuMetAlaGluVal

AspProPhe – LysTyrLeuValValLysValAlaAsnVal Ile

GlyAla IleCys – PheGlyArgArgTyrAspHisAsp

Ala – AspPheIlePro IleLeuArg – TyrLeuPro

LeuGluAlaPheLysAsp – LeuAsnLys

32 Vereczkey

of the repeats in P-450$_{LM}$ isozymes seems to correlate roughly with the increase in size and hydrophobicity of their preferred substrates which induce them. It is possible that the process of induction also controls the size and nature of the repeats and their position within the sequence. But if there is a direct correlation between the sequence portion which is repeated and the point of insertion of the repeat, it is not understood at this time.

Although the comparison to P-450$_{CAM}$ is arbitrary it has furnished much information. The extension to other P-450 sequences will be very important. Furthermore, the sequences aligned in Tables I-IV represent only those parts of the selected proteins which are best suited to demonstrate the application of evolutionary concepts and to suggest certain structure-function correlations. It will be of great interest to derive a "consensus sequence" of P-450 hemeproteins which, together with the 3D structure of P-450$_{CAM}$, might reveal more of the intricate features of P-450 structure and function.

REFERENCES

1. Dus, K.M. (1984), Proc. Natl. Acad. Sci. USA 81, 1664-1668.
2. Yu, C.-A., and Gunsalus, I.C. (1974) J. Biol. Chem. 249, 102; Yu, C.-A., Gunsalus, I.C., Katageiri, M. Suhara, K., and Takemori, S. (1974) J. Biol. Chem. 249, 94; Poulos, T.L., Perez, M., and Wagner, G.C. (1982) J. Biol. Chem. 257, 10427-10429.
3. Haniu, M., Tanaka, M., Yasunobu, K.T., and Gunsalus, I.C. (1982) J. Biol. Chem. 257, 12657-12663; Haniu, M., Armes. L.G., Yasunobu, K.T., Shastry, B.A., and Gunsalus, I.C. (1982) J. Biol. Chem. 257, 12664-12671.
4. Fujii-Kuriyama, Y., Mizukami, Y., Kawajiri, K., Sogawa, K., and Muramatsu, M. (1982) Proc. Natl. Acad. Sci. USA 79, 2793-2797; Heinemann, F.S., and Ozols, J. (1983) J. Biol. Chem. 258, 4195-4201.
5. Tarr, G.E., Black, S.D., Fujita, V.S. and Coon, M.J. (1983) Proc. Natl. Acad. Sci. USA 80, 6552-6556.
6. Ozols, J. (1972) J. Biol. Chem. 247, 2242-2245; Ozols, J., and Gerard, C. (1977) J. Biol. Chem. 252, 8549-8553.
7. Porter, T.D., and Kasper, C.B. (1985) Proc. Natl. Acad. Sci. USA 82, 973-977.
8. Tanaka, M., Haniu, M., Yasunobu, K.T., Dus, K., and Gunsalus, I.C. (1974) J. Biol. Chem. 249, 3689-3701.
9. Tanaka, M., Haniu, M., Yasunobu, K.T., and Kimura, T. (1973) J. Biol. Chem. 248, 1141-1157.
10. Dus, K.M. (1982) Xenobiotica 12, 745-772.
11. Sligar, S.G., Debrunner, P.G., Lipscomb, J.D., Namtvedt, M.J., and Gunsalus, I.C. (1974) Proc. Natl. Acad. Sci. USA 71, 3906-3910.
12. Fujita, V.A., Black, S.D., Tarr, G.E., Koop, D.R., and Coon, M.J. (1984) Proc. Natl. Acad. Sci. USA 81, 4260-4264.
13. Kawajiri, K., Gotoh, O., Sogawa, K., Tagashira, Y., Muramatsu, M., and Fujii-Kuriyama, Y. (1984) Proc. Natl. Acad. Sci. USA 81, 1649-1653; Sogawa, K., Gotoh, O., Kawajiri, K., and Fujii-Kuriyama, Y. (1984) Proc. Natl. Acad. Sci, USA 81, 5066-5070.
14. Tsaiwal, A.K., Gonzalez, F.J., and Nebert, D.W. (1985) Science 228, 80-82.

STRUCTURE-AFFINITY RELATIONSHIPS FOR 2,3,7,8-TETRACHLORODIBENZO-p-DIOXIN (TCDD) RECEPTOR BINDING

Mikael Gillner[1], Jan Bergman[2], Christian Cambillau[3], Birgitta Fernström[1], and Jan-Åke Gustafsson[1]

[1]Department of Medical Nutrition, Karolinska Institute, Huddinge University hospital F69, S-141 86 Huddinge, Sweden, and [2]Department of Organic Chemistry, Royal Institute of Technolology, S-100 44 Stockholm 70, Sweden, and [3]Department of Chemistry and Molecular Biology, Swedish University of Agricultural Sciences, S-750 07 Uppsala, Sweden

INTRODUCTION

The apparently structurally dissimilar polycyclic aromatic hydrocarbons (PAH) and chlorinated dioxins are well-known inducers of specific forms of cytochrome P-450. These compounds bind with high affinity and specificity to a protein with low capacity that is conceived to regulate the induction, the TCDD receptor (Poland and Knutson 1982). In order to elucidate common structural requirements for TCDD receptor ligands we have studied such compounds by means of molecular structure studies and receptor binding assays.

MATERIALS AND METHODS

2,3,7,8-Tetrachlorodibenzofuran (TCDBF) and [1,6-^{3}H]TCDD were kind gifts form Drs. C. Rappe, University of Umeå (Umeå, Sweden) and A. Poland, McArdle Laboratory for Cancer Research (Madison, WI), respectively. 3,3',4,4'-Tetrachlorobiphenyl was generously provided by Dr. Åke Bergman, Wallenberg Laboratory, University of Stockholm (Stockholm, Sweden). Indolo[3,2-b]carbazole, indolo[2,3-a]carbazole, 5H-benzo[b]carbazole, 6H-indolo[2,3-b]quinoxaline, compound IV in Fig 1., and 3,3'-diindolylmethane were synthesized as described in the literature. 6,12-Dihydroxyindolo[3,2-b]carbazole was made by reduction of the corresponding quinone. Other chemicals were obtained from commercial sources.

Rat liver cytosol was used as a source of TCDD receptor. Competitors were dissolved in DMSO. Binding assays by electrofocusing in polyacrylamide gel and determinations of IC$_{50}$ values by log-logit plots were carried out essentially as described before (Johansson et al. 1982).

Molecular structure studies were performed with a vector display unit driven by a VAX 11/750 computer running an interactive program. Crystallographic data from the literature were used as input data when available. If not, models of molecules were built from standard bond and angle values. Van der Waals radii of the atoms were added and images of the molecules plotted so that the main plane of the molecules coincided with the plotting plane. The van der Waals radii values for the bowls were (in Å): H, 1.2; O, 1.4; N, 1.5; Cl, 1.8; C, 1.7.

RESULTS

The capacity of different compounds to compete with [^{3}H]TCDD for its specific binding sites in rat liver cytosol are listed as IC$_{50}$ values in Table 1. The structural formulas of some of these are shown in Fig. 1.

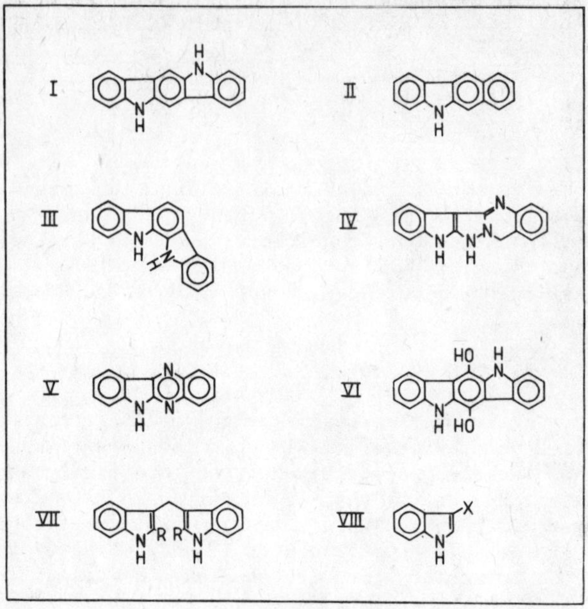

Fig 1. Structural formulas for some TCDD receptor-ligands.

Competitor	IC_{50} (nM)
Dibenz[a,h]anthracene	2.5 ± 0.7
TCDBF	3.6 ± 0.2
Indolo[3,2-b]carbazole (A)	3.6 ± 2.6
5.6-Benzoflavone (D)	25.9 ± 12.6
Benzo[a]pyrene (B)	42.4 ± 9.6
5H-Benzo[2,3-a]carbazole (II)	69.6 ± 26.2
3,3',4,4'-Tetrachlorobiphenyl (G, H)	> 150
Indolo[2,3-a]carbazole (E) (III)	> 150
Compound IV (IV)	> 150
7,8-Benzoflavone (F)	> 150
6H-Indolo[2,3-b]quinoxaline (V)	> 150
6,12-Dihydroxyindolo[3,2-b]carbazole (VI)	> 1500
3,3'-Diindolylmethane (VII; R=H)	> 1500
Indole-3-carbinol (VIII; X=CH_2OH)	> 1500

Table 1. Binding affinities of various competitors for the specific TCDD binding sites in rat liver cytosol estimated as described in Methods.

The most potent ligands found were dibenz[a,h]anthracene, TCDBF, and indolo[3,2-b]carbazole (I). The latter compound may be chemically derived from 3,3'-diindolylmethane, one of the most potent inducers of aryl hydrocarbon hydroxylase (AHH) occuring in Brussels sprouts (Loub et al. 1975).

The angular indolo[2,3-a]carbazole (III) was a much weaker ligand than the linear indolo[3,2-b]carbazole, which may indicate that a relatively linear ring system is essential for high-affinity TCDD receptor binding of indoles. A corresponding difference in receptor affinity was observed when the linear 5,6-benzoflavone was compared to the angular 7,8-benzoflavone.

An opening of the ring system of the indolocarbazoles, leading to 3,3'-diindolylmethane (VII, R=H) greatly reduced receptor affinity, perhaps due to loss of the rigid flat structure of the indolocarbazoles in these analogues. Similarly, chlorinated biphenyls, which have quite a flexible structure bind to the TCDD receptor with considerably lower affinity than the corresponding chlorinated dioxins with their rigid flat structure.

The low affinity of 6,12-dihydroxyindolo[3,2-b]carbazole (VI), as compared to indolo[3,2-b]carbazole, for the TCDD receptor may indicate that metabolism (e. g. hydroxylation) is not required for binding.

The fact that indoles smaller than indolo[3,2-b]carbazole, such as 5H-benzo[b]carbazole (II), 6H-indolo[2,3-b]quinoxaline (V) have lower affinity for the TCDD receptor than indolo[3,2-b]carbazole may indicate that a minimal molecular size is required for TCDD receptor binding. The low receptor affinity of compound IV may be due to its non-linearity. Indole-3-carbinol (VII, X=CH$_2$OH), another well-known AHH-inducer from Brussels sprouts failed to significantly displace [^{3}H]TCDD from its receptor.

As noted earlier, (Poland and Knutson 1982) the concept of a 3 x 10 Å rectangle with the centers of the halogen atoms in its corners as a structural requirement for TCDD receptor-ligands fails to account for the relatively high affinity of some PAH (e. g. benzoflavones). Since the chlorine atoms have relatively large van der Waals radii, we hypothesized that if the van der Waals radii are considered, a more general structure-affinity relationship could be established.

When the van der Waals radii of all atoms of TCDD are included, the molecule fits into a smallest rectangular envelope of 6.8 x 13.7 Å. In Fig. 2 such envelopes are drawn around computer-generated plots of some high-affinity TCDD receptor ligands. It is apparent that TCDD (A) and indolo[3,2-b]carbazole (C) are approximately isosteric when the van der Waals radii are considered. Dibenz[a,h]anthracene (not shown) and 5,6-benzoflavone (D) also fit quite well into the envelope. On the other hand, the angular indolo[2,3-a]carbazole (E) and 7,8-benzoflavone (F) do not fit into the envelope when the phenanthrene or carbazole parts of the molecules are given the same orientation as their corresponding linear analogues. Benzo[a]pyrene (B) is too short and broad for the envelope. These findings indicate that a good fit into the 6.8 x 13.7 Å envelope is correlated with high TCDD receptor-affinity.

Despite the reasonably good fit of 3,3',4,4'-tetrachlorobiphenyl (G and H) into the 6.8 x 13.7 Å envelope its receptor affinity is quite low. This may be due to that the phenyl rings are not coplanar.

DISCUSSION

The results seem to corrobate earlier findings for chlorinated hydrocarbons, namely that a critical size and planarity of a ligand is required for high affinity-binding (IC$_{50}$ < 150 nM) to the TCDD receptor. A limitation of the model of a 3 x 10 Å rectangle as a generalized structure-affinity relationship for the binding of ligands to the TCDD receptor is that it does not account for the binding of PAH and indoles to the receptor. The alternative model we have proposed here, a 6.8 x 13.7 Å envelope including the atomic van der Waals radii of the TCDD receptor-ligand, does not suffer from this limitation, since the 6.8 x 13.7 Å envelope seems to give a good fit for the unhalogenated high-affinity binding compounds investigated here, as well as for chlorinated ligands.

We hope to be able to extend and refine our new concept so that it can be used to predict binding affinity of new molecules. For prediction

486

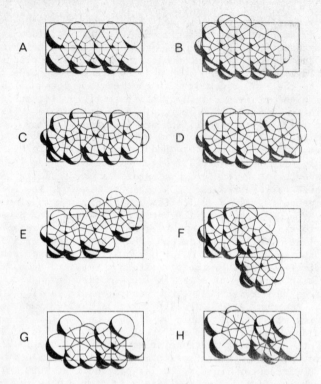

Fig. 2. Images of some TCDD receptor ligands plotted with the van der Waals radii included by means of a computer as described in Methods.

of AHH-inducing capacity of TCDD receptor ligands, the widely differing solubility, and rates and routes of metabolism between halogenated and unhalogenated ligands will also have to be taken into account.

ACKNOWLEDGEMENTS

This study was supported by the Bank of Sweden Tercentary Foundation, the Ekhaga foundation, the Funds of the Karolinska Institute, the Swedish Cancer Society, and the Swedish Council for Planning and Coordination of Research.

REFERENCES

Johansson, G., M. Gillner, B. Högberg, and J.-Å. Gustavsson (1982). The TCDD receptor in intestinal mucosa and its possible dietary ligands. Nutr. Cancer. 3:134-144.

Loub, W. D., L. W. Wattenberg, and D. W. Davis (1975). Aryl hydrocarbon hydroxylase induction in rat tissues by naturally occuring indoles of cruciferous plants. J. Natl. Cancer Inst. 54:985-988.

Poland, A., and J.C. Knutson (1982). 2,3,7,8-Tetrachloro-p-dibenzodioxin and related halogenated aromatic hydrocarbons: examination of the mechanism of toxicity. Ann. Rev. Pharmacol. Toxicol. 22:517-554.

SECONDARY STRUCTURE PREDICTION FOR CYTOCHROME P-450

Jiří Hudeček and Pavel Anzenbacher

Department of Biochemistry, Charles University, Albertov 2030
128 40 Prague 2, Czechoslovakia

INTRODUCTION

When direct information from an absolute method (X-ray crystallography) is lacking, reasonable estimates of the secondary structure is necessary. This may be achieved by two approaches: (i) by indirect methods which give only percentual figures of individual types of secondary structure; and (ii) by prediction methods, based on statistical analysis of known protein structures or on theoretical models of interactions governing formation of secondary structure. With integral membrane proteins, classical prediction methods based on statistical analysis (developed for soluble globular proteins) were questioned because of nonpolar membrane interior. The aim of this work is to predict the cytochrome P-450 (P-450) secondary structure including an approach to P-450 as to an integral membrane protein. Prediction methods used for P-450 were tested with bacteriorhodopsin known to be typical membrane protein.

METHODS

First type of prediction method used in this paper evaluates hydrophobicity of segments of polypeptide chain twenty one or five residues long (hydrophobicity profiles H_{21} or H_5). In this method, a sum of individual hydrophobicities G of amino acids is computed (G is free energy of transfer of given side chain from polar to nonpolar environment obtained from n-octanol/water partition coefficients; Putyera et al., manuscript in preparation). Analogous approach was applied also for membrane preferences (M_{21} and M_5) computed as sum of preference parameters for membranes P_m introduced by Argos et al.(1982). In both these methods, maxima of functions H_{21} and M_{21} indicate possible location of transmembrane segments[1] (average length of a transmembrane segment should be ~ 3nm corresponding to 20 residues in ⍺-helix). Minima of H_5 and M_5 indicate possible appearance of loops exposed to aqueous environment.

Finally, third method was based on joint prediction by three independent methods developed for globular proteins (Fasman

488

(1980), Garnier et al.(1978), Barkovsky and Bandarin(1979)).

RESULTS AND DISCUSSION

Hydrophobicity profiles H_{21}, H_5 and membrane preferences M_{21} and M_5 as well as joint predictions were calculated for rat liver microsomal P-450, forms b (sequence from Fujii-Kuriyama et al.(1982)), d (Kawajiri et al.(1984)) and e (Yuan et al.(1983)) and for rabbit liver microsomal P-450 LM 2(Tarr et al.(1983)) and for P-450$_{CAM}$ from Pseudomonas putida (Haniu et al.(1982)). Results for P-450b and P-450$_{CAM}$ are displayed in (illustrative) Fig. 1. First part from the N-terminus (~20 residues) is composed of hydrophobic amino acids. This part is completely lacking in the P-450$_{CAM}$, which is not a membrane protein. Hence, it may be speculated that this region is a signal sequence responsible for anchoring in the membrane. Next region (~150 residues) is more hydrophilic and is probably exposed to the solution. Third part involves at least three regions with high hydrophobicity and may be interpreted as composed of α-helices spanning the membrane. On the contrary, P-450$_{CAM}$ (with missing first hydrophobic part) has only two segments with remarkable preference for membrane and hydrophobicity. Location of α-helices found by joint prediction (with exception of the part at the N-terminus) differs markedly from maxima of membrane preference and hydrophobicity. It reflects the unsuitability of methods based on properties of globular proteins for prediction of membrane protein topology.

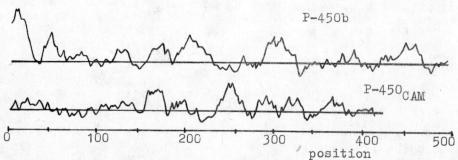

Figure 1. Membrane preferences (M_{21}) for P-450b and P-450$_{CAM}$

Earlier attempts to predict the secondary structure and membrane topology of P-450 were published by Gotoh et al. (1983) (for P-450b) and Tarr et al.(1983) (for P-450 LM 2). The approach of Gotoh et al.(1983) is completely based on the methods developed for globular proteins. Main objection to this prediction is that it did not take into account the membrane character of P-450b, thus, not giving the picture of P-450 membrane topology. Moreover, several regions predicted as β-strands are more probably buried in the membrane and involved in α-helices.

The second report(Tarr et al.(1983))deals with prediction

methods used are not given. The proposed spatial arrangement consists of eight spiral segments spanning the membrane. The length of hydrophobic transmembrane helices was assumed to be 15 to 17 residues which may be too short to span the lipid bilayer (usually, about 20 residues are taken).

Detailed analysis revealed similar character of topology of all five microsomal P-450s studied. Among them, forms b and e can be regarded as one very closely related group. Forms c and d are also alike but having distinctly altered region in the second quarter of the sequence. Results for P-450b can be expressed by two alternative arrangements (A and B, Fig. 2). Scheme A is based on the assumption that all seven hydrophobic ⊥-helices are embedded into membrane. As a result, two more hydrophilic globular domains of considerable size are formed at both sides of the membrane (Arg21-Pro167; Glu 322-Glu 434). Both these domains provide potential heme-binding sites with sequences conserved among various P-450s. Although this scheme is consistent with models of Tarr et al. (1983) and Heinemann and Ozols (1983), necessary consequence (for maintaining the length of 20 residues per transmembrane segment) is that several highly polar residues (Glu 218,281, 286; Arg 232,308; Lys 225) are forced to be in the membrane. This situation is energetically unfavorable. Second model (Scheme B) leaves these parts out of the membrane forming only one globular part anchored by four helices into the membrane.

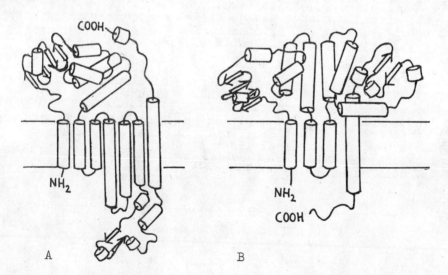

Figure 2. Scheme of proposed P-450b structure.

REFERENCES

Argos P., Rao J.K.M., Hargrave P.A. (1982) Structural prediction of membrane-bound proteins. Eur.J.Biochem. 128, 565-575

Barkovsky E.V., Bandarin V.A. (1979) Prediction of structural parts of globular proteins from their primary structure. Bioorg. chim.(Moscow) 5, 24-34 (in Russian)

Fasman G.D. (1980) Prediction of protein conformation from the primary structure. Ann.N.Y.Acad.Sci. 348, 147-159

Fujii-Kuriyama Y., Mizukami Y., Kawajiri K., Sogawa K., Muramatsu M. (1982) Primary structure of a cytochrome P-450: Coding nucleotide sequence of phenobarbital-inducible cytochrome P-450 cDNA from rat liver. Proc.Natl.Acad.Sci. U.S.A. 79, 2793-2797

Garnier J., Osguthorpe D.J., Robson B. (1978) Analysis of the accuracy and implications of simple methods for predicting the secondary structure of globular proteins. J.Mol. Biol. 120, 97-120

Gotoh O., Tagashira Y., Izuka T., Fujii-Kuriyama Y. (1983) Structural characteristics of cytochrome P-450. J. Biochem.(Tokyo) 93, 807-817

Haniu M., Armes L.G., Yasunobu K.T., Shastry G.A., Gunsalus I.C. (1982) Amino acid sequence of the Pseudomonas putida cytochrome P-450. J.Biol.Chem. 257, 12664-12671

Heinemann F.S., Ozols J. (1983) The complete amino acid sequence of rabbit phenobarbital-induced liver microsomal cytochrome P-450. J.Biol.Chem. 258, 4195-4201

Kawajiri K., Gotoh O., Sogawa K., Tagashira Y., Muramatsu M., Fujii-Kuriyama Y. (1984) Coding nucleotide sequence of 3-methylcholanthrene inducible cytochrome P-450d cDNA from rat liver. Proc.Natl.Acad.Sci.U.S.A. 81, 1649-1653

Tarr G.E., Black S.D., Fujita V.S., Coon M.J. (1983) Complete amino acid sequence and predicted membrane topology of phenobarbital-induced cytochrome P-450 (isozyme 2) from rabbit liver microsomes. Proc.Natl.Acad.Sci.U.S.A. 80, 6552-6556

Yabusaki Y., Shimizu M., Muramaki H., Nakamura K., Ojeda K., Ohkawa H. (1984) Nucleotide sequence of a full length cDNA coding for 3-methylcholanthrene induced rat liver cytochrome P-450MC. Nucl.Acid Res. 12, 2929-2938

Yuan P.-M., Ryan D.E., Levin W., Shively J.E. (1983) Identification and localization of amino acid substitutions between two phenobarbital inducible rat hepatic microsomal cytochromes P-450 by microsequence analysis. Proc. Natl.Acad.Sci.U.S.A. 80, 1169-1173

FREE COMMUNICATIONS

IMMOBILIZATION OF MICROSOMAL FRACTION AND ITS APPLICATION FOR CARRYING OUT CYTOCHROME P-450 DEPENDENT REACTIONS

S.A.Andronati, T.I.Davidenko, O.V.Sevastyanov

Physico-Chemical Institute after A.V.Bogatsky,
Academy of Sciences of the Ukrainian SSR,
Odessa, U S S R

Interest towards the problems of immobilization of sub-cellular structure is due to the fact that the majority of metabolic processes are realized using multienzymic complexes catalyzing chemical transformations coordinated in time and space. Being isolated these complexes are not stable and lose quickly their catalytic activity. One of the above complexes is cytochrome P-450 dependent system of liver microsomes containing cytochrome P-450, NADPH-dependent cytochrome P-450, reductase and phosphatidylcholine. Realization of the mentioned systems immobilization would allow to obtain liver and kidney models, novel biological catalysts for oxidation and reduction processes, new medicinal preparations, enzymic electrodes, objects for studying some metabolic and toxicologic aspects and correlation of immunologic and monooxygenase systems.

Existing techniques for immobilization of liver microsomes on inorganic matrix (silochrome) using bifunctional cross-linking reagents (cyanuric chloride and 2,4-toluenediisocyanite) allow to obtain preparations possessing only esterase activity, showing, however, no monooxygenase one (Bogatsky et al., 1979). At covalent binding of liver microsome with BrCN-activated Sepharose (Tänig et al., 1975) it is necessary to separate monooxygenase system of liver microsomes into the components: cytochrome P-450, NADPH-dependent reductase and phosphatidylcholine. At the immobilization of the system of microsomes there is no monooxygenase activity observed. However, in previous publication (Tänig et al., 1977) it was shown that at immobilizing liver microsomes on BrCN-Sepharose 15% of the initial activity is preserved of cytochrome P-450 determined vs. aminophenazone and amidopyrine. The increase of the stability of liver microsome monooxygenase system is also observed at immobilization into polyacrylamide substituted partially by acylhydrazide groups with the transversal cross-linking of glyoxylems: activity was completely preserved at 37°C for 5 hours (Tawetz et al., 1984). In this system 0-de-methylation was observed of p-nitroanisole. However, the activity of NADPH-cytochrome c-reductase in immobilized microsomes was 7 times lower than in the fresh suspension. The work of Schubert et al. should be noted describing introduction of rabbit liver microsomes into 5% gelatin gel which resulted in 60% preservation of amidopyrine N-demethylating activity. On the base of the above preparation authors have proposed enzymic electrode for determination of NADPH, glucoso-6-phosphate and aniline.

The present work is aimed at the revealing of the possibility to obtain stable preparations of rat liver microsomes at the immobilization on the implanted polyethylenes.

Polyethylenes with 6-10% of the implanted polyacrylic acid and 5.6% of the implanted polyallyl alcohol were used. Grafting was realized by gas-phase polymerization initiated by ionizing radiation. Microsomes were isolated from white rat liver (Eriksson et al., 1978), monooxygenase activity was determined according to the reported procedure (Karuzina et al., 1977). Characteristics of native microsomes are as follows: microsomal protein yield - 24.5 mg of protein per 1 g of liver, activity by aniline 3.9 nmol of substrate per 1 nmol of cytochrome P-450 per 1 min at 37°C.

Below the scheme of immobilization is given:

$$\text{$\int$-COO}^{\ominus} + \overset{\overset{\displaystyle N-R'}{\|}}{\underset{\underset{\displaystyle H-N-R''}{|}}{C}}{}^{\oplus} \longrightarrow \text{$\int$-CO-}\overset{\overset{\displaystyle O}{\|}}{\underset{\underset{\displaystyle NHR''}{|}}{C}}\overset{N-R'}{} \xrightarrow{H_2N-E} \text{$\int$-CONH-E} + \overset{\overset{\displaystyle HN-R'}{}}{\underset{\underset{\displaystyle HN-R''}{}}{C=O}} ;$$

$$\text{$\int$-CH=CH-CH}_2\text{ OH} + \text{[quinone]} \longrightarrow \text{$\int$-CH=CH-CH}_2\text{-O-[phenol OH]} \longrightarrow$$

$$\longrightarrow \text{$\int$-CH=CH-CH}_2\text{O-[quinone]} \xrightarrow{H_2N-E} \text{$\int$-CH=CH-CH}_2\text{-O-[catechol-NH-E]} .$$

At the immobilization on polyethylene with 6-10% of the implanted polyacrylic acid in the presence of p-1-cyclohexyl-3(2-morpholinoethyl)carbodiimide m-p-toluene sulphonate at the ratio carbodiimide:polyethylene with the implanted polyacrylic acid 1:5-1:18, pH 7.2-7.6 preparations were obtained characterized by the considerable preservation of N-dealkylating activity by dimethylaniline, amidopyrine, ethylmorphine (116, 197.8, 160%, respectively) and 25% of monooxygenase activity determined by aniline in the presence of cumene hydroperoxide. Activity of native microsomes is designated as 100%. 0.1M K-phosphate buffer with 0.1M KCl, 1 mM EDTA and 1mM dithiotreitol on 20% glycerol by the volume within the range of pH 7.2-7.6 since these pH values provide the maximal preservation of monooxygenase activity (MA) of immobilized microsomes (substrate dimethylaniline):

pH	6.0	6.5	7.0	7.2	7.4	7.8	8.0
MA,% of initial	60	81	104	116	116	104	91

Mass ratio soluble carbodiimide/polyethylene with the implanted polyacrylic acid (CDI/PA) may be varied from 1:8 to

1:12, its change resulting in the decrease of monooxygenase activity of immobilized microsomes (substrate dimethylaniline)

CDI/PA	1:5	1:6	1:8	1:10	1:12	1:15	1:20
MA,% of initial	66	83.5	110	116	110	46	36

Immobilization was carried out at 5°C for 24 hours. Liver microsomes immobilized as above proved stable for 4 weeks at 5°C. Temperatural optimum of immobilized microsomes is 50°C (by aniline) and 37°C (by dimethylaniline), optimal pH - 7.6, the increase being observed of the stability in acid and alkaline regions.

At immobilization of microsomes on polyethylene with the implanted polyallyl alcohol cross-linking agent - n-benzoquinone the preparations were obtained with 147.6% preservation of monooxygenase activity determined by aniline.in the presence of cumene hydroperoxide, 214.6% (in the presence of NADH-generating system and O_2) and 266.5% (in the presence of NADPH and O_2).

Microsomes immobilized by this procedure proved stable for 6 months at the temperature of storage 4-5°C. During this period monooxygenase activity reduced by 20%. Temperatural optimum for immobilized microsomes was 37°C (substrate dimethylaniline) and pH optimum 7.6. Ratio microsomal protein/polyethylene with the implanted polyallyl alcohol was 0.3-1:10. Monooxygenase activity within the above mentioned interval was maximal, at the ratio 1.25:10 the value of activity somewhat changes, however, percentage of the binding of microsomal protein reduces:

Microsomal protein/carrier	0.2:10	0.3:10	0.5:10	0.75:10	1.0:10	1.25:10
MA,% of initial	70	96.5	98.6	96.4	100	99.3

Thus, at the immobilization of microsomal fraction on polyethylene with the implanted polyacrylic acid and polyallyl alcohol the preparations are formed characterized by the increased stability with the high percentage of binding and preservation of the initial monooxygenase activity. The priority of the given type of carriers and the given technique of immobilization is possibly due to the increase of the distance from the active microsomal centres owing to the modification of polyethylene by the implanted polyallyl alcohol and polyacryl acid parallel with the specific "friable" packing as well as with the hydrophobic character of the mentioned carriers.

The possibility was considered of multiple using of microsomes immobilized on the implanted polyethylenes which showed the necessity of carrying out the reactions of aniline hydroxylation and N,N-dimethylaniline N-demethylation in the presence of glycerol. Here, for 5% glycerol the activity is preserved for 15 cycles (See Table).

Table

Effect of glycerol on cytochrome P-450 dependent
activity of microsomal fraction immobilized on
polyethylene with the implanted polyacryl acid

Cycle No	p-hydroxylation of aniline, % of maximal			N-demethylation of N,N'-dimethylaniline, % of maximal				
	Volume percent of glycerol			Volume percent of glycerol				
	0	5	10	0	5	10	15	20
1	100	100	100	100	100	100	100	100
2	65.3	55.8	90.6	66.7	55.7	74.6	71.3	50.0
3	28.6	47.9	78.2	63.3	52.9	73.7	71.3	38.9
4	26.6	37.3	68.8	35.9	51.7	69.3	65.3	24.1
6	20.4	34.5	50.1	29.1	39.1	65.8	60.4	0
10	16.3	23.9	33.1	21.4	27.6	42.1	47.5	0
13	0	18.6	21.8	13.7	24.1	37.7	32.7	0
15	0	15.9	0	9.4	21.8	22.8	23.8	0

In column regime immobilized microsomal preparations (37°C) catalyzed N,N-dimethylaniline N-demethylation and hydroxylation of aniline and phenazepam (7-bromo-5-(o-chlorophenyl)-1,2-dihydro-3H-1,4-benzodiazepine-2-one) for 9-11 hours.

REFERENCES

A.V.Bogatsky, O.V.Sevastyanov, T.I.Davidenko et al.(1979). Immobilization of Microsomal Fraction of White Rat Liver on Silochrome. Dokl. AN Ukr.SSR, Ser.B, N 8, 654-656.
G.-R.Tänig, D.Baess, K.Ruckpaul. Verfahren zur Tragerfixierung des Monooxygenasesystems der Lebermikrosomen.(1975).Pat. N 113013.
G.-R.Tänig, D.Baess, K.Pommerening, K.Ruckpaul. (1977).Eigenschaften der Zytochrom P-450-abhangigen Monooxygenase bei Fixierung an unlosliches Tragermaterial. Ergebn. Exp. Med.,24, N 1, 99-101.
Lennart C. Eriksson (1978). Preparation of liver microsomes with high recovery of endoplasmic reticulum and a low grade of contamination. Biochim. et biophys. acta, 508, N 1, 155-164.
I.I.Karuzina, A.I.Archakov.1977. Modern Methods in Biochemistry. M.: Meditsina, 49-62.
F.Schubert, D.Kirstein, F.Scheller, P.Mohr. Immobilization of cytochrome P-450 and its application in enzyme electrodes. (1980). Anal. Lett., B13, N 13, 1167-1178.
F.Schubert, F.Scheller, P.Mohr, W.Scheler. (1982). Microsomal NADPH-oxidase and hybrid electrodes. Anal. Lett., 15, N B8. 681-698.
A.Jawetz, A.S.Perry, A.Freeman, E.Katchalski-Katzir. (1984). Monooxygenase activity of rat liver microsomes immobilized by entrapment in a crosslinked prepolymerized polyacrylamide hydrazide. Biochim. et biophys. acta. Gen.subj., 798, N 2, 204-209.

496

BIOSYNTHESIS OF MINERALOCORTICOIDS BY CYTOCHROME P-450$_{11\beta}$-SYSTEM

OKAMOTO, M., OHTA, M., WADA, A., OHNISHI, T., SUGIYAMA, T. & YAMANO, T.

Department of Biochemistry, Osaka University Medical School,
4-3-57 Nakanoshima, Kita-ku, Osaka, 530 Japan

INTRODUCTION

Cytochrome P-450$_{11\beta}$ of adrenocortical mitochondria has been known for
several years to catalyze 11β-hydroxylation and 18-hydroxylation of
11-deoxycorticosterone (DOC) (Watanuki et al., 1977; Suhara et al.,
1978; Watanuki et al., 1978). Recent studies in our laboratory showed
further catalytic activities performed by this cytochrome. Thus
P-450$_{11\beta}$ catalyzes both 11β-hydroxylation and 19-hydroxylation of
18-hydroxy-11-deoxycorticosterone (18(OH)DOC) (Momoi et al., 1983; Fujii
et al., 1984), and P-450$_{11\beta}$ also catalyzes the formation of aldosterone
(ALDO) from corticosterone (B) or 18-hydroxycorticosterone (18(OH)B)
(Wada et al., 1984; Ohnishi et al., 1984; Wada et al., 1985). These
findings strongly suggest a pivotal role of this cytochrome in the
biosynthesis of mineralocorticoids. In this paper our attention will be
directed to (1) a factor regulating the various hydroxylation activities
of P-450$_{11\beta}$, and (2) 19-hydroxylation of DOC catalyzed by P-450$_{11\beta}$.

MATERIALS AND METHODS

P-450$_{11\beta}$ was purified to homogeneity from bovine adrenocortical mitochondria according to the method described by Wada et al. (1985). The
methods of the purification of NADPH-adrenodoxin reductase and adrenodoxin were described previously (Sugiyama & Yamano, 1975; Suhara et al.,
1972). For the usual enzyme assay, a steroid (100 μM) was incubated
with a reconstituted system of P-450$_{11\beta}$ in the presence of NADPH-re-
generating system under aerobic conditions for 4 min. The products were
extracted with CH_2Cl_2, and analyzed by HPLC. The columns and the
solvents used for the HPLC were TSK-Gel LS 310 in 98% CH_2Cl_2/1.8% EtOH/
0.2% H_2O for normal phase (NP)-HPLC and ODS in 60% MeOH/40% H_2O for
reverse phase (RP)-HPLC, respectively. The retention time of a steroid
was expressed as a relative value to that of the internal standard
(cortisol). 19-Oxo-11-deoxycorticosterone (19-oxo-DOC) was synthesized
by Dr. T. Terasawa of Shionogi Research Laboratory.

RESULTS AND DISCUSSION

The pathways established previously When DOC was incubated with a reconstituted system of P-450$_{11\beta}$, the hydroxylations took place at 11β-

position (production of B) with a turnover number of 85 nmol/min/nmol
P-450 and 18-position (production of 18(OH)DOC) with 7.6 nmol/min/nmol
P-450. As Momoi et al. (1983) demonstrated, the hydroxylation at either
11β- or 19-position of 18(OH)DOC was catalyzed by P-450$_{11\beta}$. Because of
a large Km for the substrate, these reactions did not seem to reach the
maximal velocities even in the initial presence of 100 μM 18(OH)DOC.
Turnover numbers obtained under these conditions were 8.1 for 11β-hy-
droxylation (production of 18(OH)B) and 4.6 for 19-hydroxylation
(production of 18,19-dihydroxy-11-deoxycorticosterone (18,19(OH)$_2$DOC)).

When B was incubated with P-450$_{11\beta}$, we found in the incubation mixture a product of 18-hydroxylation, 18(OH)B, and a further-oxidized product, ALDO (Wada et al., 1984). Turnover numbers for the two reactions
were 1.1 for 18(OH)B-formation and 0.23 for ALDO-formation.

Effect of phospholipids on the P-450$_{11\beta}$-catalyzed reactions As we
reported previously, the P-450$_{11\beta}$-catalyzed production of ALDO from B
was markedly enhanced by the addition of phospholipids to the reaction
system (Wada et al., 1984). Thus the turnover number of ALDO-formation
in the presence of phospholipids was 2.0; about 9-fold the number in the
absence of the lipids. In contrast to the ALDO-formation, the extent of
stimulation observed on 18(OH)B-formation was only 4-fold (the turnover
number in the presence of the lipids was 4.6). We also examined the
effect of phospholipids on the 11β-hydroxylation of DOC catalyzed by
P-450$_{11\beta}$. As shown in Fig. 1, B-formation from DOC was markedly inhibited in the presence of the lipids. The turnover number was depressed
from about 60 in the absence of the lipids to about 10 in the presence
of 100 nmol phospholipids under these experimental conditions.

The electron-transfer components of P-450 reside in the lipid bilayer
of biomembrane. Therefore it is reasonable that phospholipids influence
the efficiency of the electron transport of P-450-system, thus affecting
the rate of catalysis conducted by P-450. However, the effect of the
phospholipids on the reactions catalyzed by P-450$_{11\beta}$ cannot be simply
explained in this way, because the single enzyme catalyzes the hydroxylations at the various sites of the steroid substrate and the extent of the
effect of the phospholi-pids differs in each reac-tion. Therefore it is
conceivable that the active site of P-450$_{11\beta}$ itself is finely modulated
by an environmental factor such as membrane phospho-lipid. When a steroid
substrate binds to the active site of P-450$_{11\beta}$, the topological interrela-
tionship between the substrate and the cataly-tic site of the enzyme
would be influenced by the phospholipid present in the vicinity of the active

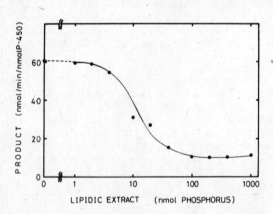

Fig. 1. Effect of lipidic extracts on 11β-
hydroxylase activity of P-450$_{11\beta}$: Phospholipid extracts of bovine adrenocortical
mitochondria were added to the reaction
mixture of DOC with P-450$_{11\beta}$, and the rate
of B-formation was plotted against the
amount of the lipids added.

site. In other words, the lipid might determine the site of hydroxylation in the steroid molecule.

<u>19-Hydroxylation of DOC catalyzed by P-450$_{11\beta}$</u> The discovery of 18,19-(OH)$_2$DOC in the reaction products of 18(OH)DOC with P-450$_{11\beta}$ prompted us to search the reaction mixture of DOC with the cytochrome for 19-hydroxy-11-deoxycorticosterone (19(OH)DOC). When DOC (100 μM) was incubated with P-450$_{11\beta}$ (30 pmol) for 4 min and the reaction products were analyzed by HPLC, we found a steroid product having a retention time identical to that of the authentic 19(OH)DOC (the relative retention time was 0.40 in NP-HPLC and 0.78 in RP-HPLC). The turnover number of 19(OH)DOC-formation under these conditions was 7.0. When a larger amount of P-450$_{11\beta}$ was used for the reaction and the products were analyzed, the peak of 19(OH)DOC disappeared in the chromatogram and the new unidentified peaks appeared instead. These results suggest that 19(OH)DOC produced was further metabolized to the other steroids by P-450$_{11\beta}$. Therefore we incubated 19(OH)DOC with P-450$_{11\beta}$ and analyzed the reaction products. The above-mentioned unidentified peaks appeared again in the chromatogram. One of the peaks had the relative retention time of 0.08 in NP-HPLC and 1.08 in RP-HPLC, which coincided with those of 19-oxo-DOC. To further confirm the identity of this peak substance, it was collected and its mass spectrum was analyzed. The FD/MS spectrum of the unidentified substance gave a M$^+$ peak of m/e 344, and so did the mass spectrum of the authentic 19-oxo-DOC. These results suggest that P-450$_{11\beta}$ can catalyze 19-oxidation of 19(OH)DOC to produce 19-oxo-DOC.

<u>General discussion</u> The P-450$_{11\beta}$-catalyzed reactions discussed in this paper are illustrated in Fig. 2. ALDO is the most potent physiologically-occurring mineralocorticoid, and all the reactions in the biosynthetic metabolism of ALDO starting from DOC are catalyzed by a single enzyme, P-450$_{11\beta}$. That the oxidative metabolism of the angular 19-methyl group of DOC is also catalyzed by P-450$_{11\beta}$ has an additional very important implication in the role of this cytochrome in the biosynthesis of mineralocorticoids. 19-Nor-11-deoxycorticosterone (19-nor-DOC) is known to be a mineralocorticoid as potent as ALDO, and Gomez-Sanchez et al. (1979) found that this steroid was excreted in the urine of adrenal regeneration hypertension rat. They also suggested 19(OH)DOC and 19-oxo-DOC as the precursors of 19-nor-DOC (Gomez-Sanchez et al., 1982). We confirmed the mineralocorticoid activity of 18-hydroxy-19-nor-11-deoxycorticosterone (Okamoto et al., 1983). These findings suggest that

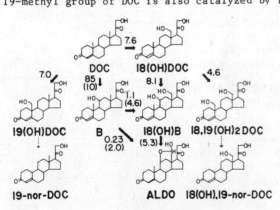

Fig. 2. P-450$_{11\beta}$-catalyzed reactions: The reactions catalyzed by P-450$_{11\beta}$ are shown by arrows with their turnover numbers (nmol/min/nmol P-450). The numbers in parentheses are the turnover numbers in the presence of phospholipids. The enzymatic nature of the reactions shown by dotted arrows remains to be established.

$P-450_{11\beta}$ is deeply involved in the biosynthesis of another series of potent mineralocorticoids, 19-nor-steroids.

Acknowledgments This work was supported by a research grant from Science and Technology Agency of Japan.

REFERENCES

Fujii, S., Momoi, K., Okamoto, M., Yamano, T., Okada, T. & Terasawa, T. (1984): 18,19-Dihydroxydeoxycorticosterone, a new metabolite produced from 18-hydroxydeoxycorticosterone by cytochrome $P-450_{11\beta}$. Chemical synthesis and structural analysis by ^{1}H NMR. Biochemistry, 23 2558-2564.

Gomez-Sanchez, C.E., Holland, O.B., Murry, B.A., Lloyd, H.A. & Milewich, L. (1979): 19-Nor-deoxycorticosterone: A potent mineralocorticoid isolated from the urine of rats with regenerating adrenals. Endocrinology, 105 708-711.

Gomez-Sanchez, C.E., Gomez-Sanchez, E.P., Shackleton, C.H.L. & Milewich, L. (1982): Identification of 19-hydroxydeoxycorticosterone, 19-oxo-deoxycorticosterone and 19-oic deoxycorticosterone as products of deoxycorticosterone metabolism by rat adrenals. Endocrinology, 110 384-389.

Momoi, K., Okamoto, M., Fujii, S., Kim, C.Y., Miyake, Y. & Yamano, T. (1983): 19-Hydroxylation of 18-hydroxy-11-deoxycorticosterone catalyzed by cytochrome $P-450_{11\beta}$ of bovine adrenocortex. J. Biol. Chem., 258 8855-8860.

Ohnishi, T., Wada, A., Nonaka, Y., Okamoto, M. & Yamano, T. (1984): Effect of phospholipid on aldosterone biosynthesis by a cytochrome $P-450_{11\beta}$-reconstituted system. Biochem. Inter., 9 715-723.

Okamoto, M., Momoi, K., Yamano, T., Nakamura, M., Odaguchi, K., Shimizu, T., Okada, T. & Terasawa, T. (1983): Affinity of 18,19-dihydroxydeoxy-corticosterone and 18-hydroxy-19-nor-deoxycorticosterone to aldosterone receptor and their mineralocorticoid activity. Biochem. Inter., 7 687-694.

Sugiyama, T. & Yamano, T. (1975): Purification and crystallization of NADPH-adrenodoxin reductase from adrenocortical mitochondria. FEBS Lett., 52 145-148.

Suhara, K., Takemori, S. & Katagiri, M. (1972): Improved purification of bovine adrenal iron-sulfur protein. Biochim. Biophys. Acta, 253 272-278.

Suhara, K., Gomi, T., Sato, H., Itagaki, E., Takemori, S. & Katagiri, M. (1978): Purification and immunochemical characterization of the two adrenal cortex mitochondrial cytochrome P-450-proteins. Arch. Biochem. Biophys., 190 290-299.

Wada, A., Okamoto, M., Nonaka, Y. & Yamano, T. (1984): Aldosterone biosynthesis by a reconstituted cytochrome $P-450_{11\beta}$ system. Biochem. Biophys. Res. Commun., 119 365-371.

Wada, A., Ohnishi, T., Nonaka, Y., Okamoto, M. & Yamano, T. (1985): Synthesis of aldosterone by a reconstituted system of cytochrome $P-450_{11\beta}$ from bovine adrenocortical mitochondria. J. Biochem., in the press.

Watanuki, M., Tilley, B.E. & Hall, P.F. (1977): Purification and properties of cytochrome P-450 (11β- and 18-hydroxylase) from bovine adrenocortical mitochondria. Biochim. Biophys. Acta, 483 236-247.

Watanuki, M., Tilley, B.E. & Hall, P.F. (1978): Cytochrome P-450 for 11β- and 18-hydroxylase activities of bovine adrenocortical mitochondria: One enzyme or two? Biochemistry, 17 127-130.

SEPARATION AND CHARACTERIZATION OF CYTOCHROME P-450 FROM RABBIT COLON MICROSOMES

EMI KUSUNOSE, MASATOSHI KAKU, SATORU YAMAMOTO, KOSUKE ICHIHARA* AND MASAMICHI KUSUNOSE

Toneyama Research Institute, Osaka City University Medical School and Toneyama National Hospital, Toyonaka, and *Department of Biochemistry, Kawasaki Medical School, Kurashiki (Japan)

INTRODUCTION

Little is known regarding colon cytochrome P-450 (P-450) except for the work of Strobel's group. Fang and Strobel (1978) prepared microsomes from rat colon mucosa, which contained P-450 and catalyzed the metabolism of various drugs as well as benzo(a)pyrene. Recently, Strobel et al. (1983) found that human colon microsomes also possessed a P-450-dependent drug metabolizing system, and adapted the electroblotting-technique with antibodies raised against rat liver P-450 for the identification of particular forms of P-450 in human and rat colon.

Recently, we have found that microsomes from rabbit small intestine and colon mucosa catalyze prostaglandin A_1 (PGA1) ω-hydroxylation (Kusunose et al. 1984). When this microsomal PGA1 ω-hydroxylase activity is expressed in terms of the turnover rate of P-450, the values of the small intestine and colon microsomes are much higher than those of other tissue microsomes, suggesting the occurrence of a P-450 that is very active in PGA1 ω-hydroxylation. A specific form of P-450 capable of hydroxylating PGA1 has been isolated and highly purified from the small intestine microsomes (Kaku et al. 1984). Its turnover rate is the highest among those reported for purified P-450. These findings prompted us to determine whether the colon microsomes also contain such a P-450. The present paper describes the separation and partial purification of three distinct P-450 fractions from rabbit colon microsomes (Kaku et al. 1985). PGA1 is effectively hydroxylated at the ω-position by a particular form of P-450.

RESULTS AND DISCUSSION

P-450 was solubilized with 0.7% cholate from rabbit colon mucosa microsomes. Three P-450 fractions, designated as "P-450ca", "P-450cb", and

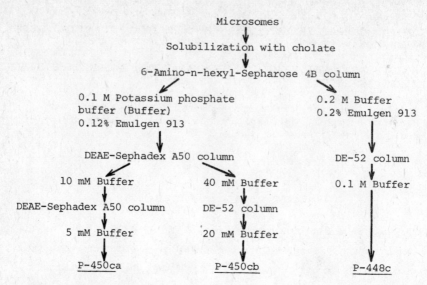

Microsomes
↓
Solubilization with cholate
↓
6-Amino-n-hexyl-Sepharose 4B column

0.1 M Potassium phosphate buffer (Buffer) 0.12% Emulgen 913 →

0.2 M Buffer 0.2% Emulgen 913 →

DEAE-Sephadex A50 column

DE-52 column

10 mM Buffer ↓   40 mM Buffer ↓   0.1 M Buffer ↓

DEAE-Sephadex A50 column   DE-52 column

5 mM Buffer ↓   20 mM Buffer ↓

P-450ca   P-450cb   P-448c

Scheme 1. Separation and partial purification of P-450ca, P-450cb and P-448c from rabbit colon microsomes.

"P-448c", were separated and partially purified to specific contents of 1.2, 2.6, and 1.5 nmol P-450/mg protein, respectively (Scheme 1). In spite of the low content of P-450, P-450ca migrated as the single major polypeptide band with a molecular weight of about 53,000 on SDS-PAGE (Fig. 1). This band was stained with tetramethylbenzidine. On the other hand, P-450cb exhibited two major bands, but only one of these was stained with tetramethylbenzidine, and its molecular weight was about 57,000. P-448c showed one major band with a molecular weight of about 56,000; this band was also stained with tetramethylbenzidine. However, when P-448c was subjected to high-performance liquid chromatography (HPLC), two P-450 peaks were observed. This fraction presumably contains at least two forms of P-450 with similar mobilities on SDS-PAGE.

The absorption maxima in the CO-reduced difference spectra of P-450ca, P-450cb, and P-448c were observed at 451, 450, and 449 nm, respectively.

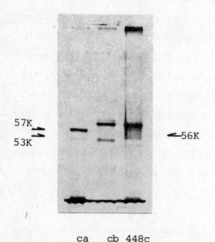

57K
53K
56K

ca   cb 448c

Fig. 1. SDS-PAGE of P-450 from rabbit colon microsomes. P-450ca, 8 µg; P-450cb, 22 µg; P-448c, 13 µg.

Tabele I.   Catalytic activities of P-450ca, P-450cb and P448c.   Turnover rates for each substrate are expressed as nmol of products formed/min/nmol of P-450.

Substrate	P-450ca Cytochrome b_5 With	Without	P-450cb Cytochrome b_5 With	Without	P-448c Cytochrome b_5 With	Without
PGA_1	20.5	2.74	0	0	0	0
PGE_1	0.2	-	-	-	-	-
Caprate	6.34	-	-	-	-	-
Laurate	22.4	2.11	0	-	0	-
Myristate	9.44	-	-	-	-	-
Aminopyrine	0	0	7.46	3.23	2.74	0.99
Benzphetamine	6.81	4.40	9.97	6.52	4.48	8.81
Benzo(a)pyrene	0	-	0.1	-	1.74	1.20
7-Ethoxycoumarin	0	-	0	-	0.01	-

Table I summarizes the catalytic activities of these P-450.   P-450ca effectively hydroxylated PGA_1 and fatty acids in a reconstituted system. When P-450ca, NADPH-cytochrome P-450 reductase, phosphatidylcholine, or NADPH was omitted, no activity was observed.   Cytochrome b_5 had a marked stimulatory effect.   The prostaglandin B_1 (PGB_1) derivative obtained from the PGA_1 hydroxylation product coincided with authentic 20-hydroxy-PGB_1 in its retention time on HPLC.   19-Hydroxy-PGB_1 was not detected in the HPLC profile.   Among the fatty acids tested, laurate was the most effective substrate.   The apparent Km for laurate was 2.63 µM.   11-Hydroxy- and 12-hydroxylaurate were identified as the major products of laurate hydroxylation. In contrast to P-450ca, neither P-450cb nor P-448c had detectable activity toward prostaglandins and fatty acids.   These P-450 catalyzed the N-demethylation of aminopyrine and benzphetamine.   Moreover, P-448c also had benzo-(a)pyrene hydroxylase activity.

Four monooxygenase activities were compared with microsomes from different regions of colon in order to examine the localization of these P-450 (Fig. 2).   When the activities were expressed in terms of the turnover rate of P-450, PGA_1 ω-hydroxylase activity showed a high degree of correlation with laurate ω- and (ω-1)-hydroxylase activities.   On the other hand, no correlation was found among the activities toward PGA_1, aminopyrine and benzo(a)pyrene.   The results suggest that P-450ca preferentially occurs in

504

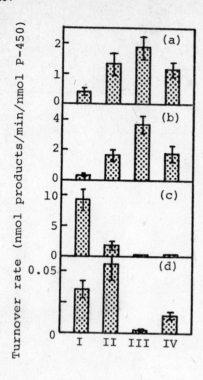

Fig. 2. Comparison of four monooxygenase activities (expressed in terms of
the turnover rate of P-450) in microsomes from four different regions of
colon mucosa. a, PGA_1; b, laurate;
c, aminopyrine; d, benzo(a)pyrene. I,
appendix; II, cecum; III, ascending
colon; IV, descending colon.

ascending colon, while P-450cb and P-448c are preferentially present in
appendix and cecum, respectively.

In conclusion, rabbit colon mucosa microsomes contain at least three
different forms of P-450, one of which is specialized for the ω-oxidation
of prostaglandins, the other being involved in the metabolism of xenobiotics such as drugs and polycyclic hydrocarbons.

REFERENCES

Fang, W. and Strobel, H.W. (1978) The drug and carcinogen metabolism system of rat colon microsomes. Arch. Biochem. Biophys. 186, 128-138.
Kaku, M., Ichihara, K., Kusunose, E., Ogita, K., Yamamoto, S., Yano, I.,
and Kusunose, M. (1984) Purification and characterization of cytochrome
P-450 specific for prostaglandin and fatty acid hydroxylase activities
from the microsomes of rabbit small intestine mucosa. J. Biochem. 96,
1883-1891.
Kaku, M., Kusunose, E., Yamamoto, S., Ichihara, K., and Kusunose M. (1985)
Multiple forms of cytochrome P-450 in rabbit colon microsomes. J. Bio-
chem. 97, 663-670.
Kusunose, E., Kaku, M., Ichihara, K., Yamamoto, S., Yano, I., and Kusunose,
M. (1984) Hydroxylation of prostaglandin A1 by the microsomes of rabbit
intestinal mucosa. J. Biochem. 95, 1733-1739.
Strobel, H.W., Newaz, S.N., Fang, W.F., Lau, P.P., Oshinsky, R.J., Stralka,
D.J., and Salley, F.F. (1983) Evidence for the presence and activity of
multiple forms of cytochrome P-450 in colonic microsomes from rats and
humans. Extrahepatic Drug Metabolism and Chemical Carcinogenesis,
Rydstrom, J., Montelius, J., and Bengtsson, M. eds., Elsevier Science
Publishers, Amsterdam, pp. 57-66.

Rat liver cytochrome P-450 metabolism of Noscapine

Mats O. Karlsson, Bengt Dahlström & Sten W. Jakobsson

ACO Läkemedel AB, S-171 03 Solna and Department of
Forensic Medicine, Karolinska Institutet, S-104 01 Stockholm
Sweden

Noscapine is an isoquinoline alkaloid obtained from opium,
but its pharmacological properties are quite distinct from
the phenanthrene derivatives or narcotic alkaloids. It is
a non-narcotic centrally active antitussive agent with few
side-effects and no addiction liability.
Noscapine is extensively metabolised and only small amounts
of unchanged noscapine are excreted in the urine in several
species including man (1).
The present study was undertaken in order to characterize
the cytochrome P-450 mediated metabolism in the liver.
Methods. Microsomes were prepared from starved male Spraque-
Dawley rat livers , and incubated for 20 minutes, in 0.16 M
phosphate buffer, pH 7.4 at 37°. NADPH was generated from
NADP by using isocitrate dehydrogenase , isocitrate and
essential ions. Subfractions from mitochondria, endoplasmic
reiculum and cytosol were obtained by differential centrifugation.
Extent of metabolism of noscapine was estimated as disappe-
rance of the substrate. This was analyzed by HPLC.

Results. Metabolism of Noscapine was found to be significantly localized only in the microsomal fraction. The metabolism
was dependent of O_2 and NADPH; the reaction velocity was at
maximum at about 37° and at pH 7.3 - 7.4. Noscapine metabolism was completely inhibited by a gas mixture containing
40% CO, 4% O_2 and 56% N_2 (Table). Thus it could be concluded
that Noscapine in the rat liver is metabolized by the cytochrome P-450 containing monooxygenase system.

Table
NOSCAPINE METABOLISM
ug metabol./min/mg prot.

Complete system	1.45
- NADPH	0
+ CO, 40%	0.015
boiled microsomes	0

<u>Conclusion</u>. Noscapine is metabolized in the rat liver via a cytochrome P-450 coupled enzyme reaction localized in the endoplasmic reiculum.

<u>REFRENCE</u>

1. TSUNODA, N, and Yoshimurh, H, Metabolic rate of Noscapine, Xenobiotica 1981, vol. 11, no 1, 23-32.

PAPERS RECEIVED AFTER JUNE 1

CARCINOGENESIS AND CYTOCHROME P-450

G. Feuer [1,2]

Departments of Clinical Biochemistry and
Pharmacology,
University of Toronto, Toronto, Canada

The effect of carcinogenesis on cytochrome P-450 and related cell functions was studied in albino rats. Tumor was produced by two systems: (a) hepatocarcinoma by DEN and 2-acetylaminofluorene (Solt-Farber model), and (b) R3230AC implanted mammary adenocarcinoma. In these models the effect of the tumor on the function of the hepatic endoplasmic reticulum was studied. In 2-AAF model nodules showed decreased Phase I. components (cytochrome P-450, mixed function oxidase), lower membrane progesterone content, receptor binding and a metabolic shift to the reductive pathway. Phase II. components (glutathione, GSH-acyl transferase) were increased. In hepatoma these parameters were further changed. In R3230 mammary tumor Phase I. components of hepatic endoplasmic reticulum were reduced, Phase II components increased; progesterone content and receptor binding were also raised. These studies indicate an association between tumor development and cytochrome P-450 and metabolically-linked enzymes.

It has been established that cytochrome P-450 represents the site of interaction between the cell and various xenobiotics including carcinogens (1). The function of cytochrome P-450 and metabolically-linked enzyme systems can be beneficial or detrimental e.g. conversion of chemicals to reactive forms with carcinogenic activity (1). In this study we view this from another side: What is the effect of carcinogenesis on cytochrome P-450 and related cell functions?

Footnotes:

[1] Supported by the Medical Research Council of Canada.

[2] I would like to acknowledge the assistance of Dr. R. Cameron Dr. J. Kellen, Dr. W. Roomi and Mr. C. DiFonzo in preparing this manuscript.

The biologic events of liver carcinogenesis have been studied thoroughly using sequential models (2,3). Applying such analyzable systems, a series of tissue changes is evident in the liver in response to chemical carcinogens, namely: foci, hepatocyte nodules-remodelling or persistent nodules, hepatocellular cancer and metastasizing liver cancer (2,3). The focal proliferations, foci and nodules, show a characteritic pattern of changes in drug metabolizing enzymes involving profound decreases in phase I components and increases in phase II components whereas surrounding liver tissue shows the normal pattern (2). This same pattern of changes is seen in liver cancers (2-5).

Biochemical measurements of nodule tissue compared to surrounding or normal tissue in a number of different laboratories have consistently shown that nodules and cancers have decreased (4-10) drug metabolizing functions. In the present experiments using the resistant-hepatocyte (Solt-Farber) model in Fischer rats (6) cytochrome P-450 content, aminopyrine-N-demethylase activity (APDM) and aryl hydrocarbon hydroxylase activity were decreased, and glutathione S-acyl-transferases (GST), glutathione (GSH), epoxide hydrolase (EH) and UDP-dlucuronyl transrerase (UDP-GT) were increased, despite their depressed basal levels (Fig. 1). Cytochrome P-450 content and APDM activity in nodules were found to be responsive to induction by phenobarbital (PB) (7-9).

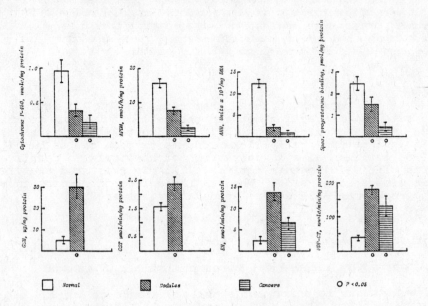

Fig. 1: *Levels of cytochrome P-450, Phase I and Phase II components of the drug metabolising enzyme system, and specific progesterone binding in microsomes of normal liver, nodules and cancers brought about by DEN-SAAF treatments of male Fischer 344 rats (6).*

Histochemical and immunohistochemical studies have confirmed that the majority of individual nodules show decreased levels of four different cytochrome P-450's (10), and increased levels of GSH (11), GST's including: the B and C forms (10), Ya,Yb (12) and Yp subunits (13), EH (10, 12, 14) and UDP-GT (15). Furthermore, the majority of individual hepatocytes within each nodule show these increases or decreases and at a similar level relative to each other and the surrounding cells (10, 12, 14). When foci are studied in this way, a similar pattern to the nodules was seen but only about 20% of foci differed from the normal liver enzyme pattern and some foci showed increases in the PB-1 type of cytochrome P-450 protein immunohistochemically (10).

In the Solt-Farber model there is a synchrony of growth and development of nodules (2,3,6), two types of nodules with very different biological behaviour can be identified and these different nodules have different phase II enzyme patterns (14, 16). The emergence and growth of numerous nodules in response to selection in this model is followed by redifferentiation or remodelling of the majority of nodules to normal liver phenotype (16, 17). Two enzymes studied histochemically in these remodelling nodules, EH and gamma-glutamyl transferase, both are seen to decrease towards normal liver levels during remodelling (21). In persistent nodules, the phase II enzymes remain elevated and are increased in cancers (12-14). This "material continuity" of phenotypic characteristics (18) between persistent nodules and cancers is paralleled by a biologic continuity, namely, persistent nodules are one site at which cancers are seen to arise (2, 6).

Nodules generated by four different protocols (19) showed this similar pattern of decreased phase I (cytochrome P-450, APDM) enzyme activity and increased phase II (GSH, GST) components. In these protocols, initiation was similar in that a chemical carcinogen was used: either diethylnitrosamine (3 protocols) or 2-acetylaminofluorene (2-AAF) in one protocol. Selection or promotion was by very different means. PB diet, choline-methionine deficient diet, orotic acid diet or 2-AAF plus carbon tetrachloride (Fig. 2). This study suggests that the distinct pattern of enzyme changes was dependent on the initial carcinogen exposure and that the expression of these changes in the nodules was idenpendent of the means to cause the differential growth of these nodules. Another interesting finding was that nodules implanted in the spleen and growing for up to 70 weeks in the spleen retained this original nodule phenotype of decreased phase I and increased phase II components (19).

512

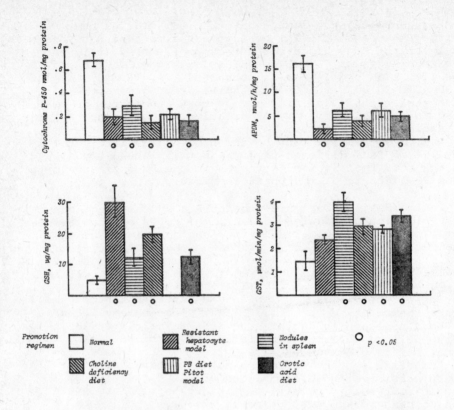

Fig. 2: *Levels of cytochrome P-450, Phase I and Phase II components of the drug metabolizing enzyme system in microsomes from the liver of normal male Fischer 344 rats and microsomes from nodules isolated from the liver of rats treated with various diets as promoters of cancer, and nodules grown in spleen (19).*

A similar pattern of enzyme changes in the liver was observed in response to cancer growing at a distant site (subcutaneous tissue). In this study, a transplantable R3230 AC mammary adenocarcinoma was grown subcutaneously in the rat (20) and its remote effects on the normal host liver enzymes studied (21). The development of the breast cancer was associated with a decrease of Phase I components (cytochrome P-450, APDM and an increase of Phase II components (GSH, GST and DT-diaplorase) and membrane-bound specific progesterone binding (Fig. 3).

Recent studies on persistent nodules have suggested that the underlying modification responsible for these enzyme changes in the nodules may be at the transcriptional level. Nodule mRNA's were compared to those in surrounding liver or normal liver. It was found that GST mRNA's specific for the

Ya/Yc and Yb subunits were increased 3-fold and 5-fold respectively, and DT-diaphorase mRNA was 5-7 fold increased over
controls (22). In addition, the DT-diaphorase gene in nodules
showed hypomethylation (R. Cameron, personal communication).
The nodule DT-diaphorase and GST-Yb mRNA's still retain their
capacity for regulation (induction) by 3-methylcholanthrene
(3-MC) showing cytochrome P448 dependency, whereas nodule GST-
Yz/Yc mRNA's did not respond to 3-MC regulation. This latter
data suggested that within the nodules there is a specific
defect in the regulatory mechanism(s) that leads to an induction of the Ya/Yc mRNA's in normal tissue by xenobiotics (22).

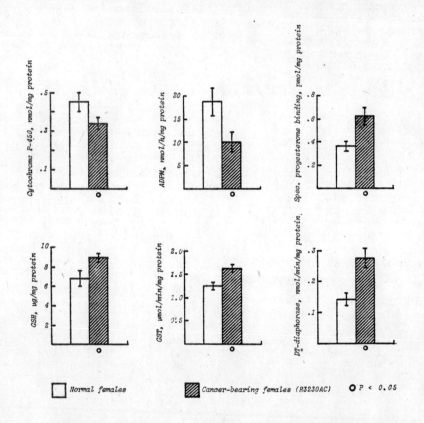

Fig. 3: *Levels of cytochrome P-450, Phase I and Phase II components of
the drug metabolizing enzyme system, and specific progesterone binding in
microsomes from liver of normal female Fischer 344 rats and from the liver
of rats with implanted R3230AC adenocarcinoma (20).*

Liver carcinogenesis therefore does have striking effects on cytochrome P-450 or P-448 and related cell functions. The pattern of changes in these enzymes is consistently found in an characteristic of both the focal proliferations and liver cancers arising during the process. This pattern may be the basis of the "resistance" of nodule hepatocytes and "initiated" hepatocytes to the cytotoxic effects of xenobiotics (including chemical carcinogens), mechanism for promotion and selective growth of nodules (2,3). Moreover, resulting from the effect of the tumor on distant tissue sites a decreased differentiation may exist in the liver, causing changes in some enzyme patterns related to cytochrome P-450 and metabolically-linked hepatocyte parameters. Recent evidence for mRNA and DNA changes in some of these enzymes (22) opens the way for investigations into the genetic basis for the carcinogenic process, using cytochrome P-450's and related proteins and their genes as genetic probes.

REFERENCES

1. Coon, M.J., Conney, A.H., Estabrook, R.W., Gelboin, H.V., Gillette, J.R. and O'Brien, P.J., Eds. Microsomes, Drug Oxidations, and Chemical Corcinogenesis, Academic Press, New York, (1980).

2. Farber, E., Biochim, Biophys. Acta 738, 171 (1984).

3. Farber, E. and Cameron, R., Adv. Cancer Res. 31, 125 (1980).

4. Gravela, E., Feo, F., Canuto, R.A., Garcia, R. and Gabriel, L., Cancer Res. 35, 3041 (1975).

5. Cameron, R., Sweeney, G.D., Jones, K., Lu, G., and Farber, E., Cancer Res. 36, 3888 (1976).

6. Solt, D.B., Medline, A. and Farber, E., Am. J. Pathol. 88, 595 (1977).

7. Okita, K., Noda, K., Fukumoto, Y. and Takemoto, T., Gann, G, 7, 899 (1976).

8. Feo, F., Canuto, R. A., Garcea, R., Brosso, O. and Caselli, G. C., Cancer Lett. 5, 25 (1978).

9. Kaneko, A., Dempo, K., Kaku, T., Yokoyama, S., Satoh, M., Mori, M. and Onoe, T., Cancer Res., 40, 1658 (1980).

10. Buchmann, A., Kuhlmann, W., Schwarz, M., Kunz, W., Wolf, C. R., Moll, E., Friedberg, T. and Oesch, F., Carcinogenesis 6, 513, (1985).

11. Deml, E. and Oesterle, D., Cancer Res., 40, 490, (1980).

12. Cameron, R. G., Ekern, K., Pickett, C. B. and Farber, E., VII Int. Congress of Histochemistry and Cytochemistry, Helsinki, Finland, (1984).

13. Sato, K., Kitahara, A., Satoh, K., Ishikawa, T., Tatemasu, M. and Ito, N., Gann. 75, 199, (1984).

14. Enomoto, K., Ying, T. S., Griffin, M. J. and Farber, E., Cancer Res. 41, 3281, (1981).

15. Fischer, G., Ullrich, D. and Bock, K. W., Carcinogenesis 6, 605, (1985).

16. Enomoto, K. and Farber, E., Cancer Res. 42, 2330, (1982).

17. Tatematsu, M., Nagamine, Y. and Farber, E., Cancer Res. 43, 5049, (1983).

18. Foulds, L., Neoplastic Development, Vols. 1 and 2, Academic Press, London (1969 and 1975).

19. Roomi, M. W., Ho. R. K., Sarma, D. S. R. and Farber, E., Cancer Res. 45, 564 (1985).

20. Feuer, G. and Kellen, J. A., Clin. Science 64, 303 (1983).

21. Shapot, V. S., Adv. Cancer Res. 30, 89, (1979).

EVIDENCE FOR VARIATIONS IN MEMBRANE PHOSPHOLIPIDS FOLLOWING THE INDUCTION OF CYTOCHROME P-450 AND P-448

M.S.I. DHAMI[1,2], D.V. PARKE[2] AND G. FEUER[3]

Department of Clinical Biochemistry and Pharmacology, University of Toronto, Toronto, Ontario, Canada,[3] Department of Biochemistry, University of Surrey, Guildford, Surrey, England,[2] Canadian Memorial Chiropractic College, Toronto, Ontario, Canada.[1,2]

In view of the dependence of mixed function oxidase system on phospholipids the effect of aroclor-1254 (PCB) and 3-methylcholanthrene was studied (MC) on hepatic microsomal system in the male rat. Aminopyrine N-demethylase representing cytochrome P-450, 7-ethoxyresorufin and 7-ethoxycoumarin deethylase representing cytochrome P-448 were significantly raised by PCB. MC only increased 7-ethoxyresorufin and 7-ethoxycoumarin deethylase. PCB induction raised microsomal phosphatidylcholine (PC), while MC induced phosphatidylethanolamine (PE). Acyl components of phospholipids were also modified. The saturated: unsaturated fatty acid ratio was altered from 43/57 in controls to 37/63 with PCB and 44/56 with MC. MC affected only fatty acids in PE fraction, whereas PCB altered the ratio in both PC and PE indicating that induction of cytochrome P-450 involves unsaturated fatty acids whereas P-448 changes are related to saturated components of membrane phospholipids.

The phospholipid components of membrane enzymes constitute an integral part of their catalytic activity and regulatory action (1). Since many drugs modify the synthesis of phospholipids (2-5), it was of interest to determine their action on molecular composition of these species.

Male Wistar rats were treated with 4 daily i.p. doses of PCB and MC, 80 mg/kg body weight, dissolved in arachis oil. Various methods applied for the isolation of hepatic microsomes, separation and quantitation of microsomal phospholipid and measurement of enzyme activity are described in previous publications (6,7).

PCB and MC treatments significantly increased aminopyrine N-demethylase activity (APDM). MC also increased 7-ethoxyresorufin deethylase (EROD) activity while PCB caused an induction to a lesser extent. Similarly, PCB and MC treatments both equally induced 7-ethoxycoumarin deethylase (ECD) activity. EROD is an apparently more specific and sensitive assay for cytochrome P-448 induction. PCB and MC treatments increased cytochrome P-450 or cytochrome P-448, respectively (Table 1).

Table 1: Effect of foreign compounds on the liver microsomal enzyme activities and cytochrome P-448 or P-450 in male rats.*

Treatment	APDM	ECD	EROD	P-450 or P-448
	nmol/h/mg protein		pmol/min/mg protein	nmol/g liver
Control	28.3 ± 1.6	2.6 ± 0.4	28.0 ± 3.0	20.4 ± 1.6
PCB	$69.7 \pm 4.2^{\dagger\dagger}$	$31.7 \pm 4.8^{\dagger\dagger}$	$51.0 \pm 9.0^{\dagger\dagger}$	$34.6 \pm 1.9^{\dagger\dagger}$
MC	$40.5 \pm 3.1^{\dagger}$	$39.9 \pm 5.1^{\dagger\dagger}$	$320.0 \pm 50.0^{\dagger\dagger}$	$32.9 \pm 1.4^{\dagger\dagger}$

*value are expressed as mean $\pm$ S.E.M. of four rats in each group.
Values differ significantly from control group; $\dagger p < 0.05$; $\dagger\dagger p < 0.005$.

The effects of microsomal phospholipids are seen in Table 2. PCB treatment increased membrane PC, PE and LPC but decreased the SM content. In contrast, MC induction only increased PE. Fatty acid content of microsomal phospholipids showed variations (Table 3). PCB significantly increased palmitic, palmitoleic, stearic and oleic acids and caused a three-fold increase in linoleic and arachidonic acids. MC did not affect these acyl derivatives. The ratio of saturated to unsaturated fatty acids was modified from 43/57 in controls to 37/63 with PCB and 44/56 with MC treatments.

Table 2: Effect of foreign compounds on phospholipid composition of liver microsomal preparation in the male rats.*

Phospholipid Fraction	Phospholipid Content (μmol P/g liver)		
	Control	PCB	MC
Total	8.0	12.5	8.3
Phosphatidylcholine	3.9 ± 0.22	$7.1 \pm 0.40^{\dagger\dagger}$	3.6 ± 0.25
Phosphatidylethanolamine	1.9 ± 0.08	3.2 ± 0.50	$2.5 \pm 0.10^{\dagger}$
Lysophosphatidylcholine	0.4 ± 0.02	$0.6 \pm 0.03^{\dagger}$	0.4 ± 0.02
Phosphatidylinositol	0.5 ± 0.01	0.6 ± 0.05	0.5 ± 0.04
Sphingomyelin	0.4 ± 0.02	$0.2 \pm 0.02^{\dagger\dagger}$	0.3 ± 0.02
Phosphatidylserine	0.5 ± 0.02	0.4 ± 0.04	0.6 ± 0.04
Phosphatidic acid	0.4 ± 0.01	0.4 ± 0.02	0.4 ± 0.02
Recovery %	94	93	91

*Values are expressed as means $\pm$ S.E.M. of four rats in each group,
Values differ significantly from control; $\dagger p < 0.05$; $\dagger\dagger p < 0.005$.

In PC fractions of microsomes, PCB treatment increased stearic, oleic, arachidonic and docosahexaenoic acids and three- and two-fold palmitoleic and linoleic acids, respectively (Table 3). MC treatment did not affect these acyl components. PCB administration increased the total amount of fatty acids in this fraction. The relative amounts of saturated and unsaturated fatty acids were control 43/57; PCB 34/66; MC 41/69.

PCB caused a two-fold increase in palmitic, oleic and docosahexaenoic acids and a three-fold increase in palmitoleic and linoleic acids in PE fraction of microsomes (Table 3). MC treatment mainly affected palmitic, palmitoleic, stearic, oleic and docosahexaenoic acids. The ratios of saturated to unsaturated fatty acids were control, 46/54; PCB, 42/58; MC, 49/51.

Table 3: Effect of foreign compounds on fatty acid composition of liver microsomal preparation in the male rats.[a]

Fraction	Treatment	Fatty Acid (µg/g Liver)								Ratio Sat./Unsat
		16:0	16:1	18:0	18:1	18:2	20:4	22:6	Total	
Whole Microsomes	Control	920 ± 20	100 ± 6	860 ± 20	550 ± 20	670 ± 30	680 ± 30	330 ± 20	4980 ± 150	43/57
	PCB	1320 ± 120[†]	190 ± 20[†]	1400 ± 140[†]	700 ± 60[†]	1990 ± 140[††]	860 ± 180[††]	360 ± 30	9610 ± 740[††]	37/63
	MC	1010 ± 90	90 ± 10	840 ± 70	650 ± 50	650 ± 60	570 ± 50	350 ± 30	4980 ± 460	44/56
Microsomal-Phosphatidyl-choline	Control	290 ± 10	20 ± 1	240 ± 20	110 ± 5	180 ± 6	240 ± 20	80 ± 3	1240 ± 30	43/57
	PCB	310 ± 30	60 ± 10[††]	320 ± 10[†]	140 ± 10[†]	370 ± 30[††]	390 ± 30[†]	160 ± 15[††]	1970 ± 140[††]	34/66
	MC	280 ± 20	20 ± 1	220 ± 20	120 ± 10	170 ± 15	260 ± 20	90 ± 5	1230 ± 80	41/59
Microsomal-Phosphatidyl-ethanolamine	Control	120 ± 2	6 ± 1	150 ± 3	70 ± 10	50 ±	140 ± 3	40 ± 1	590	46/54
	PCB	240 ± 20[††]	20 ± 2[††]	240 ± 20[†]	150 ± 10[††]	170 ± 10[††]	190 ± 10[†]	80 ± 3[††]	1120 ± 90[††]	42/58
	MC	200 ± 20[††]	10 ±	210 ± 10[†]	110 ± 10[†]	70 ± 5	180 ± 15	50 ± 3[†]	840 ± 60[†]	49/51

[a]Values are expressed as means ± S.E.M. of four rats in each group. Values differ significantly from control group:
† p <0.05; †† p <0.005.
(Sat.Unsat. indicates saturated and unsaturated fatty acids; µg/g indicates micrograms per g liver).

It has been postulated that cytochrome P-450 pocesses a broad specificity substrate binding sites which accomodate a diversity of substrates whereas cytochrome P-448 has a different binding site with narrow specificity (8). The performance of cytochrome-pigment related function of the endoplasmic reticulum is connected with a characteristic phos-

pholipid composition. Various treatments modify microsomal
phospholipids (2-5). The present experiments revealed that
PCB or MC altered microsomal phospholipids and their acyl components. PCB modified the saturated/unsaturated fatty acid
ratio mainly by affecting unsaturations of PC fractions. On
the other hand, MC only altered PE saturated acyl components.
This is in agreement with earlier investigations (9). The
difference between the effects of PCB and MC on microsomal
phospholipid acyl components may indicate that the induction
of cytochrome P-450 involves the incorporation of unsaturated
fatty acids into PC and PE moieties of this complex, while
that of cytochrome P-448 is connected with the incorporation
of saturated fatty acids into PE fractions.

References

1. Harada, N. and Omura, T. (1981), Biochem. J., 89, 237-248.

2. Davison, S. C. and Wills, E. (1974), Biochem. J., 140,
   461-468.

3. Ishidate, K. and Nakazawa, Y., (1976), Biochem. Pharmacol.,
   25, 1255-1260.

4. Ilyas, M. S., De La Iglesia, F. A. and Feuer, G., (1978),
   Toxicol. Appl. Pharmacol., 45, 69-77.

5. Sastry, B. V. R., Statham, C. N., Meeks, R. G. and Axelrod,
   J., (1981), Pharmacology, 23, 211-222.

6. Feuer, G., (1979), Drug Metab. Rev., 9, 147-169.

7. Dhami, M. S. I., (1985), Ph.D. Thesis, University of Surrey,
   Guildford, England.

8. Becker, J. F., Meeham, T. and Bartholomew, J. C., (1978),
   Biochim. Biophys. Acta., 512, 136-146.

EFFECT OF SEX HORMONES ON MICROSOMAL PROGESTERONE RECEPTOR BINDING AND ITS RELATION TO THE CYTOCHROME P-450 SYSTEM

R.G. Cameron[1], M.W. Roomi, L. Stuhne-Sekalec, and G. Feuer[2]

Departments of Clinical Biochemistry, Pharmacology and Pathology, University of Toronto, Toronto, Canada

The identification of specific progesterone binding to liver microsomes and its association with cytochrome P-450 led to studies on the effect of sex hormones in the rat on progesterone binding and cytochrome P-450 content. Estradiol exposure decreased cytochrome P-450 content and progesterone binding in female rats. Ovariectomy also decreased specific progesterone binding. These parameters were relatively unaffected in male rats by castration or testosterone treatment. It seems that in the female rat specific progesterone binding may be involved in the modification of cytochrome P-450 linked drug metabolism by hormonal manipulation.

Specific progesterone binding to rat liver microsomes has now been demonstrated for both females (1,2) and males (3). A relationship of this progesterone binding and cytochrome P-450 content and function has been recently shown in male rats by the demonstration of parallel response to cobalt-heme exposure (4), lead exposure (5), and in preneoplastic liver nodules and cancers (6) as illustrated in Figure 1. Specific progesterone binding and cytochrome P-450 are both decreased in nodules and cancers, and in response to lead. Cobalt-heme increased specific progesterone binding and decreased cytochrome P-450 at similar doses and times after exposure (4). These results suggested that progesterone may have some modulating role for drug metabolizing enzyme function.

Footnotes:

1. Scholar of the Canadian Liver Foundation.
2. Supported by the Medical Research Council of Canada.

522

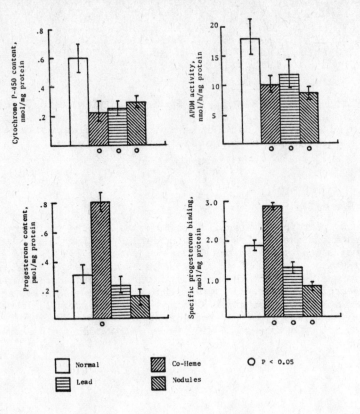

Fig. 1: Cytochrome P-450 and progesterone content, aminopyrine
N-demethylase activity and specific progesterone binding in liver microsomes
of normal male Fischer 344 rats; rats treated with Co-heme, 50 μmol/kg body
weight s.c. single dose; lead nitrate, 50 μmol/kg body weight i.v., single
dose; and in nodules of liver microsomes brought about by DEN-2AAF treatment
(7).

   Manipulations of the sex hormonal levels in male and
female rats using castration and/or testosterone exposure,
ovariectomy and/or estradiol exposure were combined with measures of the cytochrome P-450 content and specific progesterone
binding to microsomes to test this hypothesis. Adult Fischer-
344 rats, male and female, were used. Cytochrome P-450
and progesterone binding studies and cobalt-heme exposures
were performed as described previously (1,2,4). Estrogen
and testosterone were given as subcutaneously implanted 1 cm
silastic tubes each containing 100 mg (7). It was found that
in females, estradiol decreased cytochrome P-450 content by
50% in normal rats and by 70% in cobalt-heme treated rats
(Figure 2). Specific progesterone binding to liver microsomes

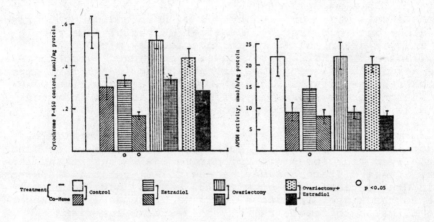

Fig. 2: *Cytochrome P-450 content and aminopyrine N-demethylase activity of female Fischer 344 rats; effects of estradiol, ovariectomy, ovariectomy + estradiol and Co-heme treatment (50 μmol/kg body weight, s.c., single dose).*

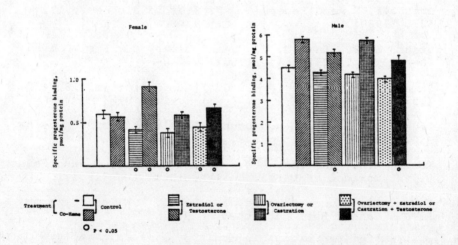

Fig. 3: *Effect of hormonal manipulations on specific progesterone binding in hepatic microsomes from Fischer 344 rats. Manipulations include females: estradiol, ovariectomy, ovariectomy + estradiol; males: testosterone, castration, castration + testosterone and Co-heme treatment (50 μmol/kg body weight, s.c., single dose).*

524

in female rats was decreased by 30% with estradiol or ovariectomy alone but increased by 70% after estradiol plus cobalt-
heme and 30% after estradiol and ovariectomy plus cobalt-
heme (Figure 3). Males also showed a decreased cytochrome
P-450 and an increased progesterone binding in all rats exposed
to cobalt-heme (Figures 1 and 3). Only the castration plus
testosterone group showed any significant effects in males
with a 9% decrease in progesterone binding.

Hormonal manipulations were very effective in females
but not in males in altering both specific progesterone binding to liver microsomes and cytochrome P-450 content and function. These differences may be due to sex-related differences in drug metabolizing enzymes (8) and may have some relevance to the higher incidence in males of human liver cancer
(9 to 1, males to female ratio) in (Western) low-risk
countries (9).

References

1. Drangova, R. and Feuer, G., J. Steroid Biochem., 13,
   629, (1980).

2. Drangova, R., Law, M. H. R. and Feuer, G., Res. Commun.
   Chem. Pathol. 29, 183, (1980).

3. Yamada, M. and Miyagi, H., J. Steroid Biochem. 16, 437,
   (1982).

4. Feuer, G., Roomi, M.W., Stuhne-Sekalec, L. and Cameron,
   R.G., Xenobiotica in press, (1985).

5. Roomi, M.W., Cameron, R.G., Stuhne-Sekalec, L. and Feuer,
   G., J. Toxicol. Environ. Health, submitted (1985).

6. Feuer, G., Stuhne-Sekalec, L., Roomi, M.W. and Cameron,
   R. G., Cancer Res. Submitted, (1985).

7. Mishkin, S. Y., Farber, E., Ho, R. K., Mulay, S. and
   Mishkin, S., Hepatology, 3, 308, (1983).

8. Kato, R., Drug Metab. Rev. 3, 1, (1974).

9. Sumithran, E. and MacSween, R.N.M., Histopathology, 3,
   447, (1979).

EFFECT OF BILE ACIDS ON MICROSOMAL PROGESTERONE RECEPTOR BINDING AND CYTOCHROME P-450 CONTENT

GHOSHAL, A., GRACON, S., ROOMI, M.W., CAMERON, R.G., STURGESS, J.M., DE LA IGLESIA, F.A. AND FEUER, G.

DEPARTMENTS OF CLINICAL BIOCHEMISTRY, PHARMACOLOGY AND PATHOLOGY, UNIVERSITY OF TORONTO, TORONTO, CANADA AND WARNER-ALMBERT/PARKE-DAVIS PHARMACEUTICAL RESEARCH ANN ARBOR, MICHIGAN, USA

Since progesterone causes an increased biliary cholesterol output, it may be involved in the onset of bile acid-induced cholestasis. The effect of bile acid accumulation was studied on ^{3}H-progesterone binding in hepatic microsomes. In competition experiments in vitro lithocholic, glycolithocholic, taurolithocholic, deoxycholic, glycodeoxycholic, chenodeoxycholic, hyodeoxycholic, cholic, glycocholic and taurocholic acid could not displace ^{3}H-progesterone from microsomal binding. Monohydroxy derivatives however, increased progesterone binding. In microsomes isolated from the liver of female rats treated with cholestatic doses of taurolithocholic, lithocholic, chenodeoxycholic or cholic acid in vivo only taurolithocholic acid increased progesterone binding sites and reduced cytochrome P-450 content and aminopyrine N-demethylase activity in a dose dependent manner. Taurolithocholic acid also raised microsomal cholesterol content suggesting a conformational change which results in a greater capacity for progesterone and cholesterol binding.

Steroids, particularly oral contraceptives, are capable of inducing intrahepatic cholestasis (1-3). Other steroids causing cholestatic response include bile acids, particularly the monohydroxy derivatives (4). Bile acids can also interact with components of the cytochrome P-450 system resulting in a "hypoactive hypertrophic endoplasmic reticulum (ER)", one of the morphologic changes seen during the onset of intrahepatic cholestasis (5). Progesterone increases total biliary cholesterol output by the hepatocyte. Since an increased cholesterol content in the bile canalicular membranes is prominent in bile acid-induced cholestasis, the possibility arises that the actions of progesterone and bile acids on the hepatocyte may be related through the specific binding sites present in liver microsomes. This study explores this relationship by examining the effect of various bile acids in vitro and in vivo on progesterone binding, cholesterol and cytochrome P-450 content of rat liver microsomes.

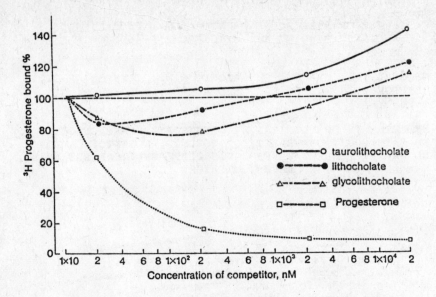

Fig. 1: *In vitro effects of monohydroxyl bile acids on hepatic*
microsomal progesterone binding in female Wistar rats.

Wistar albino female rats were exposed to cholestatic
doses in vivo, liver microsomes prepared and progesterone
binding, cholesterol content and cytochrome P-450 content
measured as previously reported (6,7).

In vitro studies showed that only the monohydroxy
bile acids increased progesterone binding to liver microsomes
(Fig. 1). The effect of the monohydroxy derivatives, particularly taurolithocholic acid (TLA) may be due to the deficiency
of hydroxyl groups, allowing these compounds to more readily
insinuate themselves into the membrane and inducing a conformational change in the progesterone binding site.

In the in vivo experiments only TLA increased the number
of binding sites in a dose dependent manner (Fig. 2). TLA
also raised microsomal cholesterol level (Fig. 3). Taurocholic acid (TCA) decreased cytochrome P-450 and aminopyrine demethylase activity (Fig. 3). Cholesterol content was slightly
raised by TCA and this compound competed with TLA. The effect

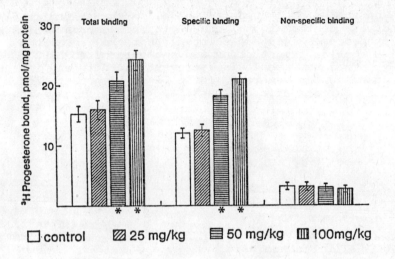

Fig. 2: In vivo effects of taurolithocholate on the hepatic microsomal binding of progesterone in female Wistar rats.

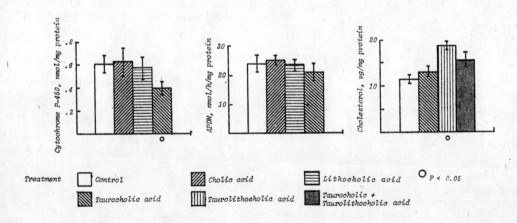

Fig. 3: Effect of bile acid treatment of rats on microsomal cytochrome P-450 and cholesterol content and aminopyrine N-demethylase activity. Treatments: cholic acid, lithocholic acid taurocholic acid, 200 mg/kg body weight 7 daily i.v. doses; taurolithocholic acid, 100 mg/kg body weight single i.v. dose, given to female Wistar rats.

of TLA on cholesterol supports the claim that taurolithocholate
may induce a conformational change in the membrane which mod-
ifies cytochrome P-450 and related functions and allows
steroids such as progesterone and cholesterol to bind more
readily. This may result in decreased fluidity and if applied
to the canalicular membrane, could be an important factor in
initiating intrahepatic cholestasis.

References

1. Einarsson, K., Ericsson, J.L.E., Gustafson, J.S. and
    Zietz, E., Biochim. Biophys. Acta 369, 279, (1974).

2. Feuer, G. and Dhami, M.S.I., Can. Fam. Phys. 28, 1299,
    (1982).

3. Boelsteria, U.A., Rakhit, G. and Balazs, T., Hepatology
    3, 12, (1983).

4. Palmer, R.H., Bile salts and the liver. In: Progress in
    Liver Disease. (Popper, H. and Schaffner, F., eds.), Vol.
    VII, pp. 221-242, Grune and Stratton, New York, USA,
    (1982).

5. Schaffner, F. and Popper, H., Lancet 2, 355, (1969).

6. Ghoshal, A.K., M.Sc. Thesis, University of Toronto,
    Toronto, Canada, (1984).

7. Feuer, G., Roomi, M.W., Stuhne-Sekalec, L. and Cameron,
    R.G., Xenobiotica, in press, (1985).

THE EFFECT OF PARTIAL HEPATECTOMY ON CYTOCHROME P-450, MICROSOMAL HEME CONTENT AND PHASE I AND PHASE II ENZYMES

ROOMI, M.W., CAMERON, R.G. AND FEUER, G.[1]

DEPARTMENTS OF PATHOLOGY AND CLINICAL BIOCHEMISTRY,
UNIVERSITY OF TORONTO, TORONTO, CANADA

The effect of partial hepatectomy (PH) on microsomal cytochrome P-450, aminopyrine N-demethylase (APDM), glutathione S-acyltransferase (GST), DT-diaphorase and glutathione (GSH) and heme content has been examined in male Fischer rats. At 12 and 24 hrs there was no change in cytochrome P-450 content and APDM activity, however it showed a decrease two days after PH persisting until the end of 7 days. The recovery of cytochrome P-450 was gradual and restored to normal levels completely by two weeks. APDM activity remained low throughout with only a partial recovery at the end of four weeks. Microsomal heme content was similar to cytochrome P-450. GSH and DT-diaphorase increased, GST decreased from day 1 until 4 weeks. These results suggested that one of the mechanisms for the decrease in cytochrome P-450 could be the decrease in heme synthesis which in turn may be associated with a decreased synthesis or an increased degradation of heme.

The initiation of carcinogenesis with chemicals requires cell (hepatocyte) proliferation within a few days of the chemical exposure (1-3). Many biochemical changes are occurring during liver regeneration (4,5) and these changes may be important for the initiation process such as DNA repair (3,6). In order to examine whether alterations in the balance of activiation-inactivation of chemical carcinogenes may plan a role in this enhanced initiation response to chemicals during hepatocyte proliferation, phase I and II components were measured sequentially for hours to days after partial hepatectomy (PH).

Footnotes:

[1]Supported by the Medical Research Council of Canada.

530

Partial hepatectomies were performed on male Fischer
rats and the levels of cytochrome P-450, microsomal heme, aminopyrine-N-demethylase (APDM), glutathione (GSH), glutathione S-
acyl-transferase (GST) and DT-diaphorase were measured as described previously (7) at 6 hrs., 12 hrs., 1, 2, 3, 7, 14 and
28 days after PH (Figures 1 and 2).

At 6 and 12 hrs. after PH, GST decreased and DT-
diaphorase increased slightly and these levels remained slightly
altered even at 4 weeks post PH (Figure 2). APDM was decreased
40% and GSH increased 2-fold at 1 day until 4 weeks (Figures 1
and 2). Cytochrome P-450 and microsomal heme content both decreased in parallel by day 2 and returned to normal levels by
2 weeks after PH (Figure 1).

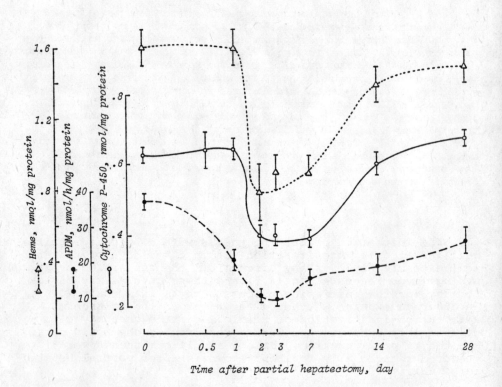

Fig. 1: *Effect of partial hepatectomy on cytochrome P-450 and heme
contents and aminopyrine N-demethylase activity in male Fischer 344 rats.*

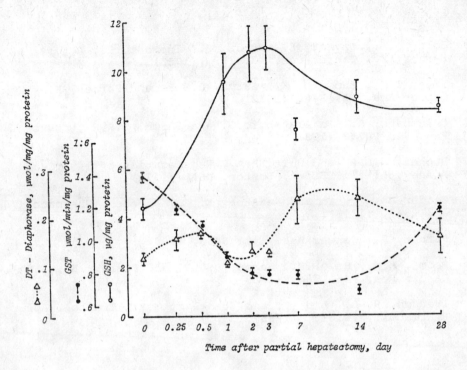

Fig. 2: Effect of partial hepatectomy of Phase II components of the drug metabolising enzyme system in male Fischer 344 rats.

Cell proliferation after PH is therefore associated with a number of changes in phase I and II components. The decreases in cytochrome P-450 content could be related directly to the decreased microsomal heme which may in turn relate to alterations in synthesis and/or catabolism of heme. Recent work by Iversen et al. (8) has shown similar changes after PH to ours with cytochrome P-450 decreasing at 2 days post PH and a Phase II component UDP-glucuronyl transferase fluctuating in activity over the first week post PH. The relative significance of these changes for the initiation by chemical carcinogens is now under study in our laboratories.

35*

References

1. Solt, D.B., Cayama, E., Tsuda, H., Enomoto, K., Lee, G. and Farber, E., Cancer Res. 43, 188, (1983).

2. Ying, T.S., Enomoto, K., Sarma, D.S.R. and Farber, E., Cancer Res. 42, 876, (1982).

3. Columbano, A., Rajalakshmi, S. and Sarma, D.S.R., Cancer Res. 41, 2079, (1981).

4. Bucher, N.L.R. and Malt, R.A., Regeneration of liver and Kidney, Little, Brown, Boston, USA, (1971).

5. Hendersen, P., Th and Kersten, K. J., Biochem. Pharmacol., 19, 2343, (1970).

6. Farber, E., Cancer Res., 44, 5463, (1984).

7. Roomi, M.W., Ho, P.K., Sarma, D.S.R. and Farber, E., Cancer Res., 45, 564, (1985).

8. Ivensen, P. L., Liu, Z. and Franklin, M.R., Toxicol. Appl. Pharmacol., 78, 10, (1985).

ALTERNATE PATHWAYS FOR DIOXYGEN AND PEROXIDE ACTIVATION AND REDUCTION

M.J. Coon, A.D.N. Vaz, L.D. Gorsky, and D. Pompon

Department of Biological Chemistry, Medical School
The University of Michigan, Ann Arbor, Michigan 48109-0010, U.S.A.

The versatility of cytochrome P-450-containing enzyme systems in catalyzing a myriad of reactions with organic substrates, including oxygenation, reduction, desulfuration, dehalogenation, and so forth, is now widely recognized. Indeed, the substrates are almost unlimited in number considering that many newly synthesized organic compounds, such as drugs, serve as substrates although they have not previously been presented to biological systems. P-450 is also nonselective with respect to agents that serve as oxygen donors. Although molecular oxygen is the donor of primary physiological importance, various peroxy compounds and other oxidants can be substituted (1-3). The present paper is concerned with alternate pathways by which the oxygen donors are reduced or "activated" and with the possibility that substrates and other components of the system, such as cytochrome $\underline{b}_5$, may influence the relative rates of the pathways and thus modify the action of individual P-450 isozymes.

OXYGEN-UTLILIZING REACTIONS CATALYZED BY P-450

Determination of the stoichiometry of hydroxylation reactions catalyzed by liver microsomal cytochrome P-450 (P-450$_{LM}$) is complicated by the occurrence of competing reactions. In the reconstituted system free of catalase, the initial rates of NADPH oxidation, O_2 consumption, and product formation were found (4) to correspond to the sum of the monooxygenase activity (Eq. 1, where RH represents the substrate and ROH the corresponding product) and the oxidase activity (Eq. 2):

$$RH + O_2 + NADPH + H^+ \rightarrow ROH + H_2O + NADP^+ \qquad (1)$$

$$O_2 + NADPH + H^+ \rightarrow HOOH + NADP^+ \qquad (2)$$

Recently, however, a study of the oxidation of ethanol to acetaldehyde, catalyzed by P-450 isozyme 3a isolated from liver microsomes of rabbits chronically administered ethanol, showed that the sum of acetaldehyde and hydrogen peroxide formed was considerably less than predicted by either oxygen consumption or NADPH oxidation (5). This observation led to further study, which revealed that under certain conditions the various isozymes of P-450 catalyze an additional reaction in which 2 molecules of NADPH are consumed per molecule of O_2 (6). This oxidase reaction, unlike that involving the 2-electron reduction to give H_2O_2, involves a

4-electron transfer and is presumed to yield water as shown in Eq. 3:

$$O_2 + 2 \ NADPH + 2H^+ \rightarrow 2 \ H_2O + 2 \ NADP^+$$

This will be referred to as the 4-electron oxidase reaction to distinguish it from monooxygenase-derived H_2O formation (Eq. 1). It should be emphasized that final proof that water is the product requires confirmation by experiments with $^{18}O_2$.

The stoichiometry of P-450-catalyzed reactions has been studied extensively, but previous reports have not identified a 4-electron oxidase reaction either in the absence of added compounds or in the presence of hydroxylatable substrates. On the other hand, Staudt et al. (7) have reported that with the uncoupler perfluoro-n-hexane no product was formed in liver microsomes and 2 mol of NADPH were oxidized per mol of oxygen consumed. The NADPH-dependent production of hydrogen peroxide by liver microsomes (8-11) is influenced by the other components present, including substrates and products as shown in the reconstituted system (4). Thus, the demonstration of a 1:1:1 molar stoichiometry for NADPH uptake, O_2 uptake, and product formation with various substrates in microsomes and mitochondria from different tissues requires that the rate of the endogenous oxidase activity (Eq. 2) be unaltered during substrate hydroxylation. Zhukov and Archakov (12) have examined the stoichiometry of NADPH oxidation in liver microsomes and concluded that oxygen may be directly reduced to water in such preparations.

The influence of substrates on the stoichiometry of P-450$_{LM}$-catalyzed reactions is shown in Table I, which includes selected data from Gorsky et al. (6).

Table I. Stoichiometry of P-450$_{LM}$-catalyzed reactions. P-450 isozyme 2 was used in the reconstituted system.

Substrate	Molar ratio, NADPH/O_2/ H_2O + organic products	Molar ratio, NADPH unaccounted for/O_2 unaccounted for
None	1.30 : 1.0 : 0.70	2.0
Cyclohexane	1.05 : 1.0 : 0.99	-
N,N-Dimethylaniline	1.04 : 1.0 : 0.94	-
Benzene	1.24 : 1.0 : 0.71	1.8
n-Pentanol	1.38 : 1.0 : 0.57	1.9
Perfluorocyclohexane	1.74 : 1.0 : 0.19	1.9

In those instances in which the H_2O_2 and organic products formed are inadequate to account for the O_2 and NADPH consumed, the ratio of about 2 calculated for extra NADPH to extra O_2 utilized is compatible with water formation. The results show that cyclohexane and dimethylaniline shut off the 4-electron oxidase pathway, whereas benzene and n-pentanol do not. Perfluorocyclohexane, a non-hydroxylatable pseudosubstrate, gave even greater uncoupling. To generalize from these and other experiments, of the substrates examined some have no effect on the oxidase reaction yielding hydrogen peroxide or the 4-electron oxidase reaction, some are inhibitory, and some are stimulatory, but the same substrate does not necessarily have the same effect on the two reactions.

As shown earlier (13), purified P-450 isozyme 2 catalyzes oxygen transfer from peroxy compounds to substrates in the absence of NADPH, NADPH-cytochrome P-450 reductase, and molecular oxygen, and the reaction has the following stoichiometry:

$$RH \quad + \quad XOOH \quad \rightarrow \quad ROH \quad + \quad XOH,$$

(4)

where RH and ROH again represent the substrate and product, respectively, and XOOH represents the peroxy compound (for example, cumene hydroperoxide). In another paper at the present symposium, evidence is presented for a pathway as yet incompletely understood in which peroxides are degraded with net oxidation of NADPH (14).

EFFECTS OF CYTOCHROME b_5 ON P-450 - CATALYZED REACTIONS

No evidence was obtained with the reconstituted system containing purified P-450, reductase, and phosphatidylcholine of an obligatory requirement for additional components in the hydroxylation of laurate, testosterone, or drugs and other foreign compounds in the reconstituted system (15). Studies in several laboratories have shown, however, that cytochrome b_5 (b_5), which functions as an electron carrier between NADH-cytochrome b_5 reductase and the terminal desaturase in the fatty acyl-CoA desaturase system, may play a role in P-450-dependent reactions in microsomal membranes. The involvement of the NADH-dependent pathway was indicated by the synergistic effect on the oxidative N-demethylation of substrates when NADH as well as NADPH was added to rat (16) or rabbit microsomal suspensions (17). In addition, studies on the steady state levels of reduced b_5 and oxyferrous P-450 in microsomes showed that b_5 may be involved in transfer of the second electron in the catalytic cycle, the b_5-dependent pathway being more important when NADH is also available as an electron donor (18,19). This conclusion has been supported by immunochemical studies in which antibody to b_5 (20-22) or to NADPH-cytochrome c reductase (21) was used as a probe. On the other hand, the complexity of the system has been shown by the finding (23,24) that substrates of the different oxidative enzyme systems alter the electron flow patterns in microsomes and also by the conclusion of Staudt et al. (7) that the synergistic effect of NADH can be explained as an electron-sparing effect for NADPH rather than electron transfer from NADH to oxygenated P-450.

As reviewed elsewhere (25), the addition of purified b_5 to reconstituted microsomal enzyme systems results in a diversity of effects which suggest that it plays a direct, though complex, role in P-450-catalyzed reactions. As noted by several laboratories, b_5 has been found to stimulate, inhibit, or have no effect on various reactions in the reconstituted system depending on the substrate and isozyme of P-450 being studied and the conditions employed. For example, Lu et al. (26) reported that b_5 is not an obligatory component in the NADPH-dependent demethylation of benzphetamine or hydroxylation of benzpyrene by purified rat P-450 but appears to be involved in the hydroxylation of testosterone at the 16α-position and of chlorobenzene. However, in several instances an obligatory b_5 requirement has been demonstrated: for p-nitroanisole demethylation by a form of rabbit P-450$_{LM}$ isolated by b_5-Sepharose chromatography and assayed with Triton X-100 in place of phospholipid (27), for prostaglandin A, E_1, and E_2 hydroxylation by isozyme 2 of rabbit P-450$_{LM}$ in the usual reconstituted system with phospholipid present (28), and for dehalogenation of an anesthetic (29). The stoichiometry of reduced pyridine nucleotide consumption during benz-phetamine demethylation by a system containing $\underline{b}_5$ and NADH-cytochrome $\underline{b}_5$ reductase as well as P-450 and its reductase indicated that electrons are transferred to P-450 via $\underline{b}_5$ (30). More recently, direct evidence has been presented for the binding of $\underline{b}_5$ to P-450 (31,32) as well as for the transfer of electrons from reduced $\underline{b}_5$ to P-450 (31,33).

This laboratory has recently compared the effects of cytochrome $\underline{b}_5$ with manganese-protoporphyrin IX substituted for heme with those of native cytochrome $\underline{b}_5$ and the apoenzyme on the oxygenation of substrates in the reconstituted system containing liver microsomal cytochrome P-450, NADPH-cytochrome P-450 reductase, and phosphatidylcholine (25). Mn-$\underline{b}_5$, unlike $\underline{b}_5$, remains essentially fully oxidized in the presence of NADPH and NADPH-cytochrome P-450 reductase under aerobic conditions. The effects of various concentrations of $\underline{b}_5$ and its derivatives were deter-mined at constant P-450 and reductase concentrations. The general conclusion was that almost all of the effects of $\underline{b}_5$ on the oxygenase system are accounted for by its role as an electron carrier.

The detailed effects of cytochrome $\underline{b}_5$ on the decay of the ferrous dioxygen complexes of P-450$_{LM2}$ and P-450$_{LM4}$ from rabbit liver microsomes were then studied by stopped-flow spectrophotometry (34). The P-450 ($Fe^{II}O_2$) complexes accept an electron from reduced cytochrome $\underline{b}_5$ and, in a reaction not previously described, donate an electron to oxidized cytochrome $\underline{b}_5$ to give ferric P-450. A comparison with the electron-transferring properties of ferrous P-450 under anaerobic conditions allowed determination of the limiting steps of the two reactions involv-ing the oxygenated complex. The rate of decay of the dioxygen complex was increased in all cases with $\underline{b}_5$ present; however, with oxidized $\underline{b}_5$ a large increase in the rate was observed with P-450 isozyme 4 but not with isozyme 2, whereas the opposite situation was found when reduced $\underline{b}_5$ was used.

From the results obtained, a scheme was proposed in which the ferrous dioxygen complex decomposes rapidly into another species differ-ing from ferric P-450 in its spectral properties and from the starting complex in its electron-transferring properties. The scheme, which indicates how competition among spontaneous decay, cytochrome $\underline{b}_5$ oxida-tion, and cytochrome $\underline{b}_5$ reduction by the ferrous O_2 complex may influence substrate hydroxylation (34), is given in Fig. 1.

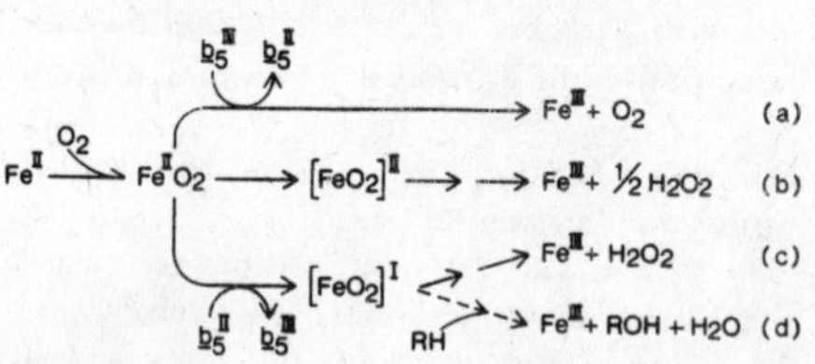

Fig. 1. Proposed reactions of the Ferrous dioxygen complex of P-450$_{LM}$ with ferric or ferrous cytochrome $\underline{b}_5$. RH represents a typical substrate and ROH represent the corresponding product.

The extent of $\underline{b}_5$ reduction or oxidation in single turnover experi-ments was limited by the decomposition of the ferrous dioxygen complex into another species by some internal electronic or possibly even confor-mational rearrangement, as discussed earlier (34). The absorption spectrum of this species is probably too close to that of the ferrous dioxygen complex to allow direct observation with P-450$_{LM4}$. With P-450$_{LM2}$, the first and most rapid phase of the spectroscopic decay correlates well with the calculated course for the ferrous dioxygen complex. A similar triphasic decay was also found recently for P-450 isozyme 3b (35). Thus, it may be assumed that with both enzymes the first phase of the apparent ferrous dioxygen decay is associated with an internal electron rearrangement. The remaining two phases may be associated with a bimolecular dismutation reaction giving back the fully oxidized enzyme and some hydrogen peroxide. This hypothesis is supported by the lack of superoxide anion formation during autoxidation of P-450 isozyme 4 (35), in contrast to the case with P-450$_{cam}$ (36). Furthermore, recent experiments have shown that the kinetics of H_2O_2 production during the autoxidation of P-450$_{LM4}$ is identical to that of the apparent biphasic decay of the $[FeO_2]^{II}$ intermediate (37).

The nature of the proposed rapid electron rearrangement within the ferrous dioxygen species is not evident. We know that the overall redox state is unchanged and that the species formed is unable to reduce b_5. As the P-450 ferrous dioxygen complex contains four redox centers, oxygen, iron, porphyrin, and thiolate ligand, several routes may lead to a more thermodynamically stable electron distribution. The simplest is an electron transfer between Fe^{II} and O_2 to give $Fe^{III}O_2^-$, a structure similar to that of oxyhemoglobin (38-40) or the stable O_2 complex of P-450$_{cam}$ (41,42). The formation of a thiyl radical or a porphyrin cation radical are other possibilities. With either radical, O_2 may be reduced to the level of hydrogen peroxide in the absence of an external electron donor and then immediately dissociate. This would be in accord with the absence of a lag in hydrogen peroxide formation (37) and the absence of an effect of the P-450 concentration on the rate of decay. On the other hand, formation of such a radical would be expected to induce a large spectral change, similar to that observed with peroxidases (43), but no such effect has been observed with P-450.

The various reactions of the P-450 dioxygen complex all lead to the generation of ferric P-450, as pictured in Fig. 1. Pathway a, involving electron transfer to ferric b_5, and pathway b, involving spontaneous autoxidation with hydrogen peroxide formation (35), would both be expected to interfere with substrate hydroxylation by removing the precursor of the "active oxygen" species (3). The third route, electron transfer from ferrous b_5 to the O_2 complex, would be expected to yield H_2O_2 in the absence of substrate (pathway c) or, in the presence of substrate (RH), to generate the active oxygen species upon cleavage of the oxygen-oxygen bond with eventual formation of water and hydroxylated substrate (ROH) according to pathway d.

The relevance of these pathways to the observed stimulatory and inhibitory effects when cytochrome b_5 is added to the reconstituted system in the presence of various P-450 isozymes and substrates is currently being investigated by steady-state kinetics. Preliminary results (44) indicate that the method of reconstitution of the P-450-containing liver microsomal enzyme system with b_5, including the order of addition of the components, the concentration of the b_5, and the length of incubation prior to initiation of the reaction by NADPH, governs the steady state catalytic activity obtained. For example, the addition to cytochrome P-450 isozyme 2, NADPH-cytochrome P-450 reductase, and phosphatidylcholine of concentrated b_5 results in extensive inhibition of benzphetamine demethylation and NADPH oxidation, whereas the addition of dilute b_5 results in extensive stimulation of the demthylation reaction. The inhibition is partly relieved by prolonged incubation, suggesting that the mixed aggregate originally formed may slowly rearrange to form a more active aggregate, presumably the same as that formed with dilute b_5. P-450 isozyme 4-catalyzed aminopyrine demethylation and aniline p-hydroxylation are not stimulated by b_5, as predicted from the model described above.   End-point stoichiometry measurements were carried out with P-450 isozyme 2 in the absence of $\underline{b}_5$ or in the presence of $\underline{b}_5$ under optimal conditions.   With dimethylaniline, the utilization of NADPH and O_2 was completely accounted for by the sum of H_2O_2 and formaldehyde produced, and when $\underline{b}_5$ was present a large increase in formaldehyde and decrease in H_2O_2 were seen.   With cyclohexanol, a substrate known not to interfere with the 4-electron NADPH oxidase reaction, the H_2O_2 and cyclohexanone formed did not account for all of the NADPH and O_2 utilized; when $\underline{b}_5$ was added, a large increase in cyclohexanone and a nearly stoichiometric decrease in H_2O_2 were obtained.

Previous work from two other laboratories has shown a relationship between $\underline{b}_5$ and H_2O_2 production.   Ingelman-Sundberg and Johansson (45) found that $\underline{b}_5$ stimulated P-450 isozyme 2-catalyzed $\underline{p}$-nitroanisole deme-thylation with a concomitant decrease in hydrogen peroxide formation when these proteins were incorporated along with NADPH-cytochrome P-450 reductase into phospholipid vesicles.   However, they found no effect of $\underline{b}_5$ on this reaction or on 7-ethoxycoumarin deethylation in the usual micellar reconstituted system.   In view of our present findings, their method of reconstitution may have accounted for the different effects seen in the two systems.   Imai (46) reported that the addition of $\underline{b}_5$ to a reconstituted system resulted in a stimulation of 7-ethoxycoumarin deethylation and a decrease in the rate of H_2O_2 formation.   Taken to-gether, these and other more recent data indicate that when $\underline{b}_5$ is recon-stituted with the P-450 isozyme 2 system under optimal conditions, substrate monooxygenation is enhanced, NADPH oxidation is unaffected, and hydrogen peroxide formation is decreased.

OVERALL SCHEME AND SUMMARY

The pathways discussed in the present paper are shown in Fig. 2 in relation to the basic mechanistic cycle.   The reactions involved in the O_2- dependent and peroxide-dependent substrate hydroxylations have been reviewed (3).   The routes for hydrogen peroxide formation, either direct-ly from protonation of a 2-electron-reduced dioxygen species or indirect-ly from dismutation of superoxide, are indicated.   As shown by the dashed arrow, the ferric-O_2 (or ferrous-O_2) complex of P-450 isozyme 4 yields H_2O_2 without producing superoxide as a detectable intermediate (35); possibly a dismutation reaction is involved.   Assuming that the 4-electron NADPH oxidase reaction yields H_2O (Eq. 3), several lines of evidence indicate that it is not formed via the reduction of H_2O_2 (6). We postulate, instead, that the "active oxygen", presumably $(Fe-O)^{3+}$, is directly reduced to H_2O.   Finally, the H_2O_2 liberated is shown as serving as a potential oxygen donor for substrate hydroxyation.   However, such high levels of H_2O_2 are required for this reaction that it is believed to be of doubtful physiological significance.   As discussed above, sub-strates and $\underline{b}_5$ may control the action of P-450 by influencing the rela-tive rates of the monooxygenase, 2-electron oxidase, and 4-electron oxidase activities.

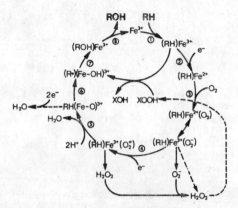

Fig. 2. <u>Proposed scheme for mechanism of action of cytochrome P-450 in hydroxylation reactions</u>. The mechanistic cycle presented earlier (3). Reations involving the pathways discussed in the present paper have been added to. RH represents the substrate and ROH the corresponding product.

Acknowledgment: This research was supported by Grant AM-10339 from the National Institutes of Health.

REFERENCES

1. Kadlubar, F.F., Morton, K.C., and Ziegler, D.M. (1973) Biochem. Biophys. Res. Commun. 54, 1255-1261.
2. Rahimtula, A.D., and O'Brien, P.J. (1974) Biochem. Biophys. Res. Commun. 60, 440-447.
3. White, R. E., and Coon, M. J. (1980) Annu. Rev. Biochem. 49, 315-356.
4. Nordblom, G. D., and Coon, M. J. (1977) Arch. Biochem. Biophys. 180, 343-347.
5. Morgan, E. T., Koop, D. R., and Coon, M. J. (1982) J. Biol. Chem. 257, 13951-13957.
6. Gorsky, L. D., Koop, D. R., and Coon, M. J. (1984) J. Biol. Chem. 259, 6812-6817.
7. Staudt, H., Lichtenberger, F., and Ullrich, V. (1974) Eur. J. Biochem. 46, 99-106.
8. Gillette, J.R., Brodie, B.B., and LaDu, B.N. (1957) J. Pharmacol. Exp. Ther. 119, 532-540.
9. Thurman, R.G., Ley, H.G., and Scholz, R. (1972) Eur. J. Biochem. 25, 420-430.
10. Hildebrandt, A.G., and Roots, I. (1975) Arch. Biochem. Biophys. 171, 385-397.
11. Hildebrandt, A.G., Heinemeyer, G., and Roots, I. (1982) Arch. Biochem. Biophys. 216, 455-465.
12. Zhukov, A.A., and Archakov, A.I. (1982) Biochem. Biophys. Res. Commun. 109, 813-818.
13. Nordblom, G. D., White, R. E., and Coon, M. J. (1976) Arch. Biochem. Biophys. 175, 524-533.
14. Vaz, A.D.N., and Coon, M.J., paper presented at this meeting.
15. Haugen, D. A., van der Hoeven, T. A., and Coon, M. J. (1975) J. Biol. Chem. 250, 3567-3570.
16. A.H. Conney, R.R. Brown, J.A. Miller, and E.C. Miller: Cancer Res. 17, 628-633 (1957).

540

17. Cohen, B.S., and Estabrook, R.W. (1971) Arch. Biochem. Biophys. 143, 46-53.
18. Hildebrandt, D.A., and Estabrook, R.W. Arch. Biochem. Biophys. 143, 66-79 (1971).
19. Noshiro, M., Ullrich, V., and Omura, T. (1981) Eur. J. Biochem. 116, 521-526.
20. Mannering, G.J., Kuwahara, S., and Omura, T. (1976) Biochem. Biophys. Res. Commun. 70, 951-956.
21. Sasame, H.A., Thorgeirsson, S.S., Mitchell, J.R., and Gillette, J.R. (1974) Life Sci. 14, 35-46.
22. Noshiro, M., Harada, N., and Omura, T. (1979) Biochem. Biophys. Res. Commun. 91, 207-213.
23. Jansson, I., and Schenkamn, J.B. (1977) Arch. Biochem. Biophys. 178, 89-107.
24. Jansson, I., and Schenkman, J.B. (1978) Arch. Biochem. Biophys. 185, 251-261.
25. Morgan, E. T., and Coon, M. J. (1984) Drug. Metab. Dispos. 12, 358-364.
26. Lu, A.Y.H., Levin, W., West, S.B., Vore, M., Ryan, D., Kuntzman, R., and Conney, A.H., (1975). In Cytochromes P-450 and b_5: Structure, Function, and Interaction (Cooper, D.Y., Rosenthal, O., Snyder, R., and Witmer, C., Eds), pp.447-466, Plenum Press, New York.
27. Sugiyama, T., Miki, N., and Yamano T., (1980) J. Biochem. (Tokyo) 87, 1457-1467.
28. Vatsis, K.P., Theoharides, A.D., Kupfer, D., and Coon, M.J. (1982) J. Biol. Chem. 257, 11221-11229.
29. Canova-Davis, E., and Waskell, L. (1984) J. Biol. Chem. 259, 2541-2546.
30. Imai, Y., and Sato R., (1977) Biochem. Biophys. Res. Commun. 75, 420-426.
31. Bonfils, C., Balny, C., and Maurel, P. (1981) J. Biol. Chem. 256, 9457-9465
32. Chiang, J.Y.L (1981) Arch. Biochem. Biophys. 211, 662-673.
33. Werringloer, J., and Kawano, S. (1980) In Microsomes, Drug Oxidations, and Chemical Carcinogenesis (Coon, M.J., Conney, A.H., Estabrook, R.W., Gelboin, H.V., Gillette, J.R., and O'Brien, P.J., Eds), pp. 469-478, Academic Press, New York.
34. Pompon, D., and Coon, M. J. (1984) J. Biol. Chem. 259, 15377-15385.
35. Oprian, D. D., Gorsky, L. D., and Coon, M. J. (1983) J. Biol. Chem. 258, 8684-8691.
36. Sligar, S.G., Lipscomb, J.D., Debrunner, P.G., and Gunsalus, I. C. (1974) Biochem. Biophys. Res. Commun. 61, 290-296.
37. Gorsky, L.D. (1984) Doctoral thesis, The University of Michigan
38. Weiss, J.J. (1964) Nature (Lond.) 202, 83-84.
39. Peisach, J., Blumberg, W.E., Wittenberg, B.A., and Wittenberg, J.B. (1968) J. Biol. Chem. 243, 1871-1880.
40. Yamamoto, T., Palmer, G., Gill, D., Salmeen, I.T., and Rimai, L. (1973) J. Biol. Chem. 248, 5211-5213.
41. Sharrock, M., Münck, E., Debrunner, P.G., Marshall, V., Lipscomb, J.D., and Gunsalus, I.C. (1973) Biochemistry 12, 258-265.
42. Sharrock, M., Debrunner, P.G., Schulz, C., Lipscomb, J.D., Marshall, V., and Gunsalus, I.C. (1976) Biochim. Biophys. Acta 420, 8-26.
43. Dolphin, D., Forman, A., Borg, D.C., Fajer, J., and Felton, R.H. (1971) Proc. Natl. Acad. Sci. U.S.A. 68, 614-618.
44. Gorsky, L.D., and Coon, M.J. (1985) Fed. Proc. 44, 1611.
45. Ingelman-Sundberg, M., and Johansson, I. (1980) Biochem. Biophys. Res. Commun. 97, 582-589.
46. Imai, Y. (1981) J. Biochem. (Tokyo) 89, 351-362.

CHARACTERIZATION OF RABBIT LIVER MICROSOMAL P-450 ISOZYME
2 BY FLUORESCENCE PROPERTIES AND STATE OF AGGREGATION
AS AFFECTED BY A DETERGENT

K. Inouye and M.J. Coon

Department of Biological Chemistry, Medical School
The University of Michigan, Ann Arbor, Michigan 48109-0010, U.S.A.

Rabbit liver microsomal P-450 isozyme 2 (P-450LM$_2$) (1-3) has been characterized by its immunochemical properties (4-6), EPR, absorption, and CD spectra (3,7), and interactions with other components of the enzyme system (8-10). In addition, the complete amino acid sequence has been established by the Edman method (11,12). Much remains to be learned, however, about the function of specific amino acid residues in substrate binding and in the catalysis of various reactions, as well as about the states of aggregation of this and other P-450 cytochromes isolated from membranes. Indeed, it is not known to what extent individual residues participate in the aggregation process or whether it is a result of general hydrophobic attractions. The present paper is concerned with the fluorescence properties of P-450 isozyme 2 and with the influence of a zwitterionic detergent, 3-[(3-chloramidopropyl)-dimethylammonio]-1-propanesulfonate (CHAPS), on the state of aggregation of this purified cytochrome.

PROPERTIES OF THE TRYPTOPHAN RESIDUE IN P-450 ISOZYME 2 AS DETERMINED BY FLUORESCENCE

At the present time the only residue in P-450 isozyme 2 with a function known with some assurance is Cys$_{436}$. As reported recently (13), the reaction of purified rabbit liver microsomal P-450 isozyme 2 with 4,4'-dithiobis(2-nitrobenzoate) exhibits first order kinetics and results in the modification of a single thiol, but causes no net loss of the native ferrous-carbonyl spectrum. The site of isozyme 2 rapidly labeled by this compound and by monobromobimane, a fluorescent reagent for thiol groups, was shown to be Cys$_{152}$. The results obtained strongly suggest that Cys$_{152}$ does not provide the proximal thiolate ligand to the heme iron atom. Since Cys$_{152}$ represents one of the two highly conserved cysteine-containing regions in the P-450 cytochromes, it appears likely that the other region, containing Cys$_{436}$ in this rabbit cytochrome (corresponding to Cys$_{355}$ in bacterial P-450cam, Cys$_{436}$ in rat P-450 b or e, Cys$_{461}$ in rat P-450 c, Cys$_{456}$ in rat P-450 d or mouse isozyme 3, and Cys$_{458}$ in mouse isozyme 1) (14) is the source of the thiolate ligand to the heme.

Since similar attempts at chemical modification of the single tryptophyl residue (Trp$_{121}$) in P-450 isozyme 2 were unsucessful, we have turned to fluorescence analysis (15). Hemeproteins are generally non-fluorescent, which may be explained by the action of the heme as a sink to accept energy from excited aromatic amino acid residues in the protein (16). However, some hemeproteins have been reported to emit weak fluorescence, including hemoglobin (17) and rabbit P-450 isozymes 2 and 4 (18), which indicates that in these proteins the distance between the heme and certain tryptophyl residues might be greater than the Förster critical distance for energy transfer (17). The observation that action spectra for the photodissociation of carbon monoxide-heme complexes of respiratory enzymes show a peak in the ultraviolet absorption region of the protein led to the proposal (19) that the energy of the excited amino acid residues was transferred to the heme and thus caused the dissociation (20).

A typical fluorescence emission spectrum for P-450 isozyme 2 upon exposure to light at 295 nm, which is known to excite only Trp residues in proteins, is shown in the <u>inset</u> to Fig. 1. The peak has a maximum at 335 nm. Electrophoretically homogeneous preparations of this cytochrome varying in specific content from 6.7 to 16.4 nmol of P-450 per mg of protein, and thus differing in the amount of apoenzyme present, were examined. In all instances the emission spectra had the same shape with a maximum at 335 nm. Thus, the environmental state of the tryptophyl residue appears to be independent of the heme content of the preparation. As indicated, the relative fluorescence at 335 nm decreases in a linear fashion with respect to increasing specific P-450 content of the preparations. According to a linear least squares regression analysis, the relative fluorescence of 100% apoenzyme would be 13.9 ± 0.4 and that of 100% holoenzyme (with an apparent specific content of 19.1 ± 0.6 nmol of P-450 per mg of protein) would be zero. At the maximal specific content of 18.0, calculated from the molecular weight of 55,700 as determined from the amino acid sequence (11), the F_{335} value was estimated to be 6% that of the apoprotein. Thus, almost all of the observed fluorescence of the cytochrome must come from the apoenzyme in the preparations; until methods

(taken from Ref. 15.)

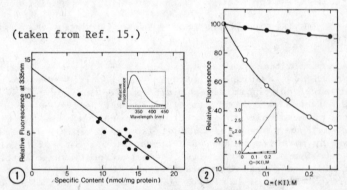

Fig. 1. Inset, fluorescence emission spectrum of P-450 isozyme 2 upon excitation at 295 nm. The P-450 concentration was 9.1×10^{-7} M and the specific content was 13.5 nmol per mg of protein.

Main figure, relationship between relative fluorescence intensity at 335 nm and the specific content of different P-450 isozyme 2 preparations. For comparison, all results are expressed per mg of protein per ml.

Fig. 2. Quenching by potassium iodide of fluorescence of P-450 isozyme 2 (●) with excitation at 295 nm and emission at 335 nm and of N-acetyltryptophanamide (o) with excitation at 295 nm and emission at 355 nm. KI was varied up to 0.25 M in the usual buffer mixture containing 5.4×10^{-7} M P-450 (specific content, 13.5 nmol per mg protein), or of 1.1×10^{-5} M acetyltryptophanamide. KCl, which has no quenching effect, was added to bring the sum of KI and KCl to 0.25 M. Inset, Fo/F vs. the iodide concentration.

become available for the isolation of the apo- and holoenzymes completely free of each other, the value of 6% for the contribution of the latter should be considered provisional.

The effects of ions which are known to quench the fluorescence of other proteins (21,22) were tested with P-450 isozyme 2. Figure 2 shows that the relative fluorescence intensity of the cytochrome decreased slightly with increasing KI concentrations. No P-420 was formed, even at the highest KI level used. In contrast, much more extensive quenching was observed with the model compound, N-acetyltryptophanamide, reaching about 50% at 0.13 M and 70% at 0.25 M iodide. As shown in the inset to Fig. 2, Fo/F, the ratio of the relative fluorescence in the absence of KI to that in the presence, was plotted against the iodide concentration. According to the Stern-Volmer equation (21), Fo/F equals $1 + K_Q[Q]$ where [Q] is the concentration of the quencher and K_Q is the dissociation constant for the complex between the quencher and the tryptophyl residue. The K_Q values of iodide for the cytochrome and the model compound were estimated to be 0.38 and 8.0 M^{-1}, respectively. In experiments not shown, CsCl was shown to quench the fluorescence emission at 335 μm of acetyltryptophanamide strongly and that of the trp residue of the cytochrome only slightly. In addition, the effects of methanol and ethanol indicated that these alcohols have greater accessibility to the Trp_{121} environment than do iodide and cesium ions.

P-450 isozyme 2 is known to be largely aggregated in the hexameric to octameric state in aqueous solution (23). If the distance between Trp_{121} of apoprotein and the heme(s) of the neighboring protein(s) in the aggregate is within the Förster's critical distance (30 to 40 Å) (24,25), the fluorescence of the tryptophyl residue should be quenched almost completely even in the mixed aggregates of holo- and apoproteins. The fact that F_{335} is quenched in a linear and almost stoichiometric manner with respect to the increase in specific content suggests that the fluorescence of Trp_{121} is quenched by a single heme, presumably in the same protein molecule. The most reasonable estimate for the distance between Trp_{121} and the heme in the aggregated form is less than 40 Å for the intramolecular case and greater than 40 Å for the intermolecular case.

Whether the tryptophyl residue facilitates electron transfer from NADPH via the reductase flavins to the heme of P-450 isozyme 2 remains to be established. Recently, Schwarze et al. (26) estimated the distance between the NH_2-terminal methionine of this cytochrome and its heme group to be 26.1 Å by measuring the energy transfer from the fluorescein isothiocyanate-modified methionine to the heme. They concluded that this distance is too large to be surmounted by a thermally activated tunneling mechanism, and that the NH_2-terminus does not participate in the electron transfer pathway from reductase to the heme.

EFFECTS OF A ZWITTERIONIC DETERGENT THAT DISSOCIATES RABBIT LIVER MICROSOMAL CYTOCHROME P-450 ISOZYME 2 TO THE MONOMERIC FORM

In view of the usefulness of CHAPS in converting P-450 isozyme 2 to a monomer retaining some of the catalytic properties of the aggregated form, other effects of this detergent on the cytochrome have been examined. CHAPS gives a difference spectrum with isozyme 2 similar to that with "type II" substrates and decreases CO binding to the reduced cytochrome. The concentration required for 50% maximal change is 0.6 and 0.7% for these two effects, measured as $A_{424-390}$ and $A_{452-490}$, respectively. The detergent causes an increase in the fluorescence of the P-450, measured by an

544

increase in the intensity at 335 nm upon excitation at 295 nm. Gel chromatography indicated that about half of the cytochrome is converted to the monomer in 1.0% detergent and that almost complete conversion occurs in 1.5% detergent. These and other experiments suggest that with increasing levels of CHAPS the substrate-binding site is modified and an amino group in the protein becomes a ligand to the iron, followed by eventual conversion of the protein aggregates to P-450 monomers.

Acknowledgement: This research was supported by Grant AM-10339 from the National Institutes of Health.

REFERENCES

1. van der Hoeven, T.A., Haugen, D.A., and Coon, M.J. (1974) Biochem. Biophys. Res. Commun. 60, 569-575.
2. Imai, Y., and Sato, R. (1974) Biochem. Biophys. Res. Commun. 60, 8-14.
3. Haugen, D.A., and Coon, M.J. (1976) J. Biol. Chem. 251, 7929-7939.
4. Dean, W.L., and Coon, M.J. (1977) J. Biol. Chem. 252, 3255-3261.
5. Park, S.S., Persson, A.V., Coon, M.J., and Gelboin, H.V. (1980) FEBS Lett. 116, 231-235.
6. Park, S.S., Cha, S.-J., Miller, H., Persson, A.V., Coon, M.J., and Gelboin, H.V. (1982) Mol. Pharmacol. 21, 248-258.
7. Chiang, Y.L., and Coon, M.J. (1979) Arch. Biochem. Biophys. 195, 178-187.
8. French, J.S., Guengerich, F.P., and Coon, M.J. (1980) J. Biol. Chem. 255, 4112-4119.
9. Ruckpaul, K., Rein, H., Ballou, D.P., and Coon, M.J. (1980) Biochim. Biophys. Acta 626, 41-56.
10. Ruckpaul, K., Rein, H., Blanck, J., Ristau, O., and Coon, M.J. (1982) Acta Biol. Med. Ger. 41, 193-203.
11. Tarr, G.E., Black, S.D., Fujita, V.S., and Coon, M.J. (1983) Proc. Natl. Acad. Sci. USA 80, 6552-6556.
12. Heinemann, F.S., and Ozols, J. (1983) J. Biol. Chem. 258, 4195-4201.
13. Black, S.D., and Coon, M.J. (1985) Biochem. Biophys. Res. Commun. 128, 82-89.
14. Black, S.D., and Coon, M.J. (1985) in Cytochrome P-450: Structure, Mechanism, and Biochemistry (Ortiz de Montellano, P., Ed) Plenum Publishing Corp., in press.
15. Inouye, K., and Coon, M.J. (1985) Biochem. Biophys. Res. Commun. 128, 676-682.
16. Udenfriend, S. (1962) Fluorescence Assay in Biology and Medicine, Academic Press, New York.
17. Weber, G., and Teale, F.J.W. (1959) Faraday Discuss. Chem. Soc. 27, 134-141.
18. Chiang, Y.-L., and Coon, M.J. (1979) Arch. Biochem. Biophys. 195, 178-187.
19. Warburg, O. (1932) Angew. Chem. 45, 1-6.
20. Bücher, T., and Kaspers, J. (1947) Biochim. Biophys. Acta 1, 21-34.
21. Lehrer, S.S. (1971) Biochemistry 10, 3254-3263.
22. Ricchelli, F., Rossi, E., Salvato, B., Jori, G., Bannister, J.V., and Bannister, W.H. (1983) in Oxy Radicals and Their Scavenger Systems. Vol 1. Molecular Aspects (Cohen, G., and Greenwald, R.A., Eds), pp.320-323, Elsevier, New York.
23. French, J.S., Guengerich, F.P., and Coon, M.J. (1980) J. Biol. Chem. 255, 4112-4119.
24. Stryer, L. (1960) Radiation Res. Suppl. 2, 432-451.
25. Förster, T. (1960) Radiation Res. Suppl. 2, 326-339.
26. Schwarze, W., Bernhardt, R., Jänig, G.-R., and Ruckpaul, K. (1983) Biochim. Biophys. Res. Commun. 113,353-360.

NADPH OXIDATION BY HYDROPEROXIDES: CATALYSIS BY CYTOCHROME P-450 AND EVIDENCE FOR HOMOLYTIC CLEAVAGE OF THE PEROXIDE BOND

A.D.N. Vaz and M.J. Coon

Department of Biological Chemistry, The University of Michigan, Ann Arbor, Michigan 48109, U.S.A.

INTRODUCTION

The extensive role of the cytochrome P-450 family of isozymes in the oxidation of endobiotic and xenobiotic compounds and in the detoxification and generation of reactive intermediates capable of cell toxicity and carcinogenicity is well documented (Microsomes and Drug Oxidations, 1984). By contrast, reductive reactions involving peroxides and cytochrome P-450 have received relatively little attention (Mc Lane et al., 1983). The oxidation of substrates by these isoenzymes is effected by a net transfer of two electrons from NADPH to dioxygen as shown in Eq. 1. Concomitant with the monooxidation reaction is the reduction of dioxygen to hydrogen peroxide shown in Eq. 2. Recently this laboratory (Gorsky et al., 1984) has reported a four-electron reduction of dioxygen, presumably to water as in Eq. 3.

$$NADPH + RH + O_2 + H^+ \rightarrow NADP^+ + ROH + H_2O \qquad (1)$$
$$NADPH + O_2 + H^+ \rightarrow NADP^+ + H_2O_2 \qquad (2)$$
$$2\ NADPH + O_2 + 2H^+ \rightarrow 2\ NADP^+ + [2H_2O] \qquad (3)$$

The mechanism of reductive dioxygen activation by this class of hemeproteins in the monooxygenation process has been the subject of much investigation (White and Coon, 1980). Cytochrome P-450 can also serve as a peroxygenase in catalyzing the reaction,

$$RH + YOOH \rightarrow ROH + YOH \qquad (4)$$

where RH represents a substrate and YOOH a peroxy compound such as a hydroperoxide or peracid serving as the oxygen donor (Nordblom et al., 1976). A recent report by Lee and Bruice (1985) on the reaction of various peroxy compounds with porphyrins suggests that the activation of the peroxide bond may be homolytic or heterolytic, with homolysis being the likely route for compounds where the pK_a of the resultant alcohol is above 11.0. Earlier studies in this laboratory (Coon and Blake, 1982) indicate that no distinct intermediate similar to horseradish peroxidase Compound I is formed by the reaction between the hydroperoxides and cytochrome P-450. Furthermore, the substrate hydroxylation rate constant with hydroperoxides was shown to be sensitive to alterations in both reactants, a result compatible with a homolytic mechanism for oxygen-oxygen bond cleavage but not with heterolytic formation of a common iron-oxo intermediate from various hydroperoxides.

In the present study we have investigated the reduction of hydroperoxides as a possible model for the 4-electron reduction of dioxygen by cytochrome P-450, in a reconstituted system consisting of rabbit liver microsomal cytochrome P-450 isozyme 2, NADPH-cytochrome P-450 reductase, and dilauroylglyceryl-3-phosphoryl choline. Under anaerobic conditions the oxidation of NADPH by hydroperoxides is dependent on the presence of both the reductase and P-450. Analysis of the stoichiometry of the reaction and of the nature of the products formed in the case of benzyl and cumene hydroperoxides indicates that this reductive process may be described by the reaction,

$$NADPH + YOOH + H^+ \rightarrow NADP^+ + YX + H_2O \qquad (5)$$

where YX indicates that reduction of the hydroperoxide does not follow a simple two-electron transfer from NADPH to the hydroperoxide to yield the corresponding alcohol, but possibly involves stepwise one-electron reductions resulting in homolytic cleavage of the peroxide bond and rearrangement of the alkoxy radical to yield the product.

EXPERIMENTAL PROCEDURES

Cytochrome P-450 isozyme 2 and NADPH-cytochrome P-450 reductase were purified by published procedures (Haugen and Coon, 1976; French and Coon, 1979). In all experiments with the reconstituted system, a 1:1 ratio of reductase to P-450 was used, while phosphatidylcholine (PC) was at a concentration of 30 μg/ml. Reconstitution of the enzyme system with PC was routinely done by combining appropriate volumes of the reductase (14-16 μM)and the P-450 (25-42 μM) with a freshly sonicated PC dispersion (1 mg/ml). All studies were done at 25°C in 50 mM potassium phosphate buffer, pH 7.4, containing 10 mM EDTA. NADPH oxidation was monitored at 340 nm (ϵ_{340} = 6.22 mM^{-1}cm^{-1}). Enzymes and the hydroperoxide were added to the NADPH in buffer from separate arms of a cuvette after the system was made anaerobic under an atmosphere of nitrogen. After addition of the enzymes, reactions were monitored at 340 nm for 3 to 5 min to ascertain anaerobic conditions. The hydroperoxide was then added, and the reaction was monitored for 2 to 3 min. Reaction products of benzyl and cumene hydroperoxides were analyzed by high pressure liquid chromatography on an IBM LC/9533 liquid chromatographic system with an IBM LC/9540 data integrator. An IBM C-18 reverse-phase column was used, and the eluate was monitored at 254 nm. For products of cumene hydroperoxide, an isocratic solvent system (40% methanol:60% water) was used at a flow rate of 1 ml/min. Acetophenone, cumenol, and cumene hydroperoxide had retention times of 11.6, 12.8 and 14.6 min, respectively. Detection limits for cumenol and cumene hydroperoxide were 2.5 nmol, while that for acetophenone was 10 pmol. Separation of benzyl alcohol and benzyl hydroperoxide was obtained with 25% methanol:75% water with retention times of 6.9 and 7.3 min, respectively.

RESULTS AND DISCUSSION

That the oxidation of NADPH is P-450-dependent is clearly indicated in Table 1. A maximal rate of NADPH oxidation was obtained only in the presence of native P-450 and reductase (Expt. 1). Thermal denaturation of either enzyme resulted in a significant decrease in NADPH oxidation (Expts 2 and 3), and no oxidation was observed in the absence of both enzymes (Expt. 6). Of interest is the small but significant oxidation of NADPH in the presence of P-450 alone (Expt. 4). While this rate is small relative to the complete system, it is significant since NADPH alone is not capable of electron transfer to P-450. This suggests that NADPH oxidation may be occuring by a reactive species generated from the hydroperoxide by the cytochrome.

Expt.	Components omitted	NADPH oxidation nmol/min/nmol P-450
1	None	30.0
2	None, except P-450 boiled	0.8
3	None, except reductase boiled	2.0
4	P-450	0.05
5	Reductase	3.6
6	P-450 and reductase	0.08

Table 1. Effect of omission of various components of the complete reconstituted system consisting of P-450 LM$_2$ (0.05 μM), reductase (0.05 μM), and PC (30 μg/ml), in 50 mM potassium phosphate buffer, pH 7.4, with 10 mM EDTA. In Expts. 2 and 3 the respective enzymes were boiled for 5 min prior to reconstitution.

In experiments not presented here, the effect on the rate, and with some reagents the extent, of NADPH oxidation by cumene hydroperoxide of a substrate (benzphetamine), or a heme ligand (CO), or free radical scavengers such as phenol, dithiothreitol (DTT), or hydroquinone was examined. While benzphetamine (1 mM) was without effect, CO at a saturating concentration inhibited the rate of NADPH oxidation by 64%. While this inhibition suggests that heme iron ligation by the hydroperoxide is essential for a significant rate of NADPH oxidation, the lack of complete inhibition suggests that porphyrin peripheral electron transfer may also be occurring, or that this transfer becomes significant when heme ligation is unavailable. The inhibition of the rate of NADPH oxidation by phenol, DTT, or hydroquinone was 50% or less, and could not be increased at higher concentrations. The molar ratio of NADPH oxidized to cumene hydroperoxide consumed was found to be one to one. This stoichiometry was reduced to 0.6:1 and 0.49:1 in the presence of phenol (10 mM) and DTT (5 mM), respectively. These results strongly support the idea that part of the rate and extent of NADPH oxidation by cumene hydroperoxide is due to a non-enzymatic route.

Product analysis of reaction mixtures containing benzyl and cumene hydroperoxides indicated that the corresponding alcohols are not products of the enzymatic reduction of these hydroperoxides. Acetophenone has been identified as one product of the reaction with cumene hydroperoxide. This indicates that reduction of cumene hydroperoxide by P-450 must involve the intermediate formation of the cumyloxy radical in a Haber-Weiss type reaction, which then undergoes elimination, presumably of a methyl radical, to yield acetophenone. In this connection it is noteworthy that methyl radicals have been identified in reactions of cumene hydroperoxide with rat liver microsomes (Griffin, 1980).

ACKNOWLEDGMENTS: We thank Robert Clark for the preparation of P-450 isozyme 2 and NADPH-cytochrome P-450 reductase. This research was supported by NIH Grant AM-10339.

REFERENCES

Coon, M.J., and Blake II, R.C. (1982) In Oxygenases and Oxygen Metabolism (Nozaki, M., Yamamoto, S., Ishimura, Y., Coon, M.J., Ernster, L., and Estabrook, R.W., Eds.) pp. 485-495, Academic Press, New York.

French, J.S., and Coon, M.J. (1979) Arch. Biochem. Biophys. 195, 565-577.

Gorsky, L.D., Koop, D.R., and Coon, M.J. (1984) J. Biol. Chem. 259, 6812-6817.

Griffin, B.W. (1980) In Microsomes, Drug Oxidations, and Chemical Carcinogenesis (Coon, M.J., Conney, A.H., Estabrook, R.W., Gelboin, H.V., Gillette, J.R., and O'Brien, P.J., Eds.) pp. 319-322, Academic Press, New York.

Haugen, D.A., and Coon, M.J. (1976) J. Biol. Chem. 251, 7929-7939.

Lee, W.A., and Bruice, T.C. (1985) J. Amer. Chem. Soc. 107, 513-514.

Mc Lane, K. E., Fisher, J., and Ramakrishnan, K. (1983). Drug Metabolism Reviews 14, 741-799.

Microsomes and Drug Oxidations (1984). Boobis, A.R., Caldwell, J., De Matteis, F., and Elcombe, C.R., Eds., Taylor & Francis, London.

Nordblom, G.D., White, R.E., and Coon, M.J. (1976) Arch. Biochem. Biophys. 175, 524-533.

White, R.E., and Coon, M.J. (1980) Annu. Rev. Biochem. 49, 315-356.

CYTOCHROME P-450 AND ITS IMPLICATION IN ENVIRONMENTAL PROBLEMS

HÄNNINEN, O., LINDSTRÖM-SEPPÄ, P., PESONEN, M. AND HARRI, M.

Department of Physiology, University of Kuopio, P.O.Box 6, 70211 Kuopio, Finland.

INTRODUCTION

The chemical loading of the environment has awakened wide concern. The use of chemicals for industrial, agri- as well as silvicultural and for communal purposes is extensive and growing. Furthermore the provision on energy by burning for the industry, transport as well as for housing contributes considerably to the emission of chemicals to the nature due to pyrolytic and pyrosynthetic reactions. The quality of the living environment has changed in large areas in the world. The health of plants, animals and also of man is endangered in several locations of the globe. Some species show diminished population numbers and reduced reproduction rates as well increased signs of disease.

For the control of the environment the levels of several compounds have been monitored with chemical analysis. The samples have been collected from air, water and soil as well as from the living organisms. If the chemicals are acutely toxic, such analyses give results which are easily understandable. Most of the chemicals released to the nature are, however, without such acute effects. Furthermore their concentrations are low. The determination of low concentrations require specialized well equipped laboratories. Furthermore one must remember that only those chemicals are monitored which are known and considered with the present status of knowledge to have significance. Furthermore such analyses seldom give answers on the interactions between the chemicals. The interactive resposes can be seen only in the living organisms themselves. Most of the living organisms appear to have at least some ability to metabolize foreign compounds. This ability can enhance the toxicity through a release of reactive intermediates. Metabolism to water soluble end products is the final fate, since up to this the chemicals are retained by the cellular lipids of the organism. The chemical loading very often increases the rate of metabolism of foreign compounds in organisms. Cytochrome P-450 has a key position in the metabolism of a great variety of compounds. It has also a central role in the formation of reactive intermediates, which cause genotoxic and other delayed harmful effects. Cytochrome P-450 cored enzyme activities are under extensive endogenous and exogenous control. Thus there are a number of good reasons to examine cytochrome P-450 in organisms, when environmental problems are studied. Furthermore monooxygenase systems can be used as reagents in analysis of the nature of chemical loaders. In the following, a few examples of such studies are described.

ACID RAIN

One of the most serious current environmental problems in the highly industrialized countries is the acid rain (OECD 1981). It affects directly and indirectly the vegetation trough the mobilization of the soil components. Furthermore acidic water flow to rivers and lakes lowers their pH and hamper the living of the aquatic organisms (Muniz 1981). The effects of acidification have been dramatic e.g. in Sweden, where thousands of lakes have been affected (Dickson 1975). The acidification finally kills the fish. There are few studies available on the effects of short term exposure to low pH on the biotransformation in fish. Laitinen et al. (1982) have shown that in the common white fish (Coregonus lavaretus) the hepatic 7-ethoxycoumarin O-deethylase increased after an exposure to pH 3. On the other hand in the splake (Salvelinus fontinalis x Salvelinus namaycush) the aryl hydrocarbon hydroxylase was lower in acid exposed than in control fish. In both fish species UDPglucuronosyltransferase activity increased in acidified water. The acid water affects also the biotransformation in extrahepatic tissues (Laitinen et al. 1984). Thus in the common white fish the renal 7-ethoxycoumarin O-deethylase increased, but the renal aryl hydrocarbon hydroxylase decreased. In the duodenal mucosa 7-ethoxycoumarin O-deethylase showed an increasing activity. There is an obvious need for further studies on the early effects of the acidification on the aquatic species and on the interactions of the acid rain and pollutants.

BIOMONITORING WITH AQUATIC SPECIES OF THE WATER POLLUTANTS

Aquatic organisms can be considered as lipid drops in water. They concentrate lipophilic compounds to their tissues. The chemicals are retained over long periods of time if not metabolized into water soluble form. Since fish are also important sources of protein, their tissues have been used for the analysis of environmental pollutants. The occurrence of the pollutants and their levels are, however, affected by the rate of metabolism. Thus chlordane levels are low in the muscle and liver of the rainbow trout (Salmo gairdneri) compared to those e.g. of vendace (Coregonus albula) living in the same waters. Vendace livers show considerably lower rate of various monoxygenase and other biotransformation activities as well as lower cytochrome P-450 contents than seen in the rainbow trout (Pyysalo et al. 1981). Several, perhaps most fish and other aquatic animals have biotransformation ability (Dewaide 1971, Koivusaari et al. 1980, Ade et al 1982). The biotransformation ability at least in some of these species appears to respond to the environmental pollutants. E.g. in Salmonid fish polyaromatic hydrocarbons and polychlorinated compounds cause an induction of the hepatic monoxoygenase system (Lindström-Seppä et al. 1985). The exposure of e.g. rainbow trout to crude oil can enhance ten fold or even more some of the monoxoygease activities in rainbow trout within few days (Lindström-Seppä et al. 1985). In nature elevated biotransformation activities have been measured after oil spills from ships (Payne and Penrose 1975, Stegeman 1981). High monoxygenase activities in fish liver have been seen over long periods of time in oceans (Stegeman 1978). In brakish water in boreal region the response to oil spill may have rather short (few months) duration (Lindström-Seppä et al 1985 in this volume). Thus the monitoring of the monoxoygenase activities and cytochrome P-450 provide useful information on the environmental loading. It should, however, be kept in mind not all chemicals known as inducers in mammalians cause an induction in the fish (Lindström-Seppä et al. 1985). In other aquatic genera the response may be even smaller than in fish. Furthermore the physical factors of the environment as well as physiological functions like spawning affect extensively the biotransformation ability in fish in nature (Koivusaari et al. 1981). The level of chemical loading can be so low that it is not able to elevate the monoxoygenase activities above the control level, although increased levels of pollutants can be detected in chemical analysis (Lindström-Seppä et al 1985). It should be remembered that the genetic variation in the wild polulations is considerable, which also hampers the evaluation of the data obtained.

The occurrence of the malignant tumors in the fish like Northern pike is high (Papas et al. 1976). This species shows considerable biotransfromation ability i.e. ability also to activate pollutants to reactive intermediates (Balk et al. 1980). It is possible that the occurrence of tumors in fish tissues correlates with the pollution.

BIOTRANSFORMATION IN WILD RODENTS AND ENVIRONMENTAL QUALITY

There is lots of information available on the cytochrome P-450 and mono-oxygenase acitivites as well as on their control in common laboratory rodents. These species use in nature when living wild as feed rubbish heaps and other waste material. They live in drainages and other areas considered dirty. The wild rodents are thus supposedly exposed to inducers present in the environment. Thus they would serve useful biomonitors of their surroundings. Furthermore the wild animals are also exposed to extreme climatic conditions, and presumably they are also forced to a high locomotor acitivity. Both these have been reported to increase the metabolism of xenobiotics (Gayathri et al. 1973, Frenkl et al. 1980).

We have made an attempt to compare the biotransformation ability in laboratory and wild rats of the same age of both sexes. The wild rats were trapped from a mucipal dumping ground (Juankoski) having a permanent rich stock of wild rodents. The dumping ground contained all kinds of wastes including food refuges. The laboratory rats of Wistar, F344 and BN strains were kept in Macrolon gages, and they were fed on pelleted chow (Hankkija, Finland). In addition to the hepatic biotransformation enzyme activities few enzymes were measured from the muscle homogenates (M. gastrocnemius) to indicate the physical fitness of the animals studied.

The activities of the oxidative marker enzymes from the muscle were 38-50 per cent higher in the wild rats (Fig. 1). Such levels can be obtained in laboratory rodents, if they are forced to run for 1 h daily (Harri et al. 1982). These results clearly show that the wild rats had a better physical fitness than the laboratory rats. No sex differences were observed.

Between the wild and laboratory rats there were no significant differences in the liver-to-body weight ratios. Both the content of cytochrome P-450 as well as the level of aryl hydrocarbon hydroxylase were, however, signifi-cantly lower in the wild than in the laboratory rats. In both groups males showed higher values. (Fig.1)

If we assume that the cytochrome P-450 and monoxygenase activities indicate the chemical loading in the living environment, the results obtained suggest that the rats in municipal dumping ground live in a cleaner environment than in the laboratory. It is natural that the animals can make their choices in the dumping grounds - which in fact are not possible in animal laboratories. Furthermore our results showed that wild rats, in

552

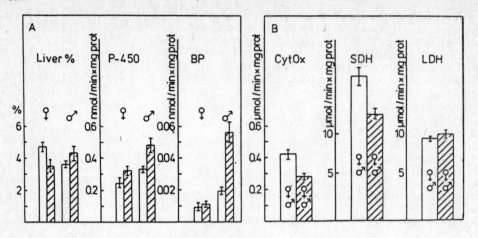

Fig. 1.A. Relative liver weight (%), content of cytochrome P-450 and activity of benzo(a)pyrene hydroxylase (BP) in liver microsomes from wild (open columns) and laboratory rats (shaded columns).
 B. Activity of mitochondrial cytochrome oxidase (CytOx), succinate dehydrogenase (SDH) and cytosolic lactate dehydrogenase (LDH) as representatives of muscular energy metabolism in wild and laboratory rats.

spite of their superior physical fitness, had lower biotransformation ability which is also suggest for a possibility of avoidance of inducers in dump grounds.

EARTHWORMS AND BIOTRANSFORMATION

Earthworm is an important species in the decomposition of organic litter such as leaves, straws and animal manure. They have also significance for the aeration of the soil. The earthworms contain cytochrome P-450 (Liimatainen and Hänninen 1982). They are also able to metabolize at least some of the chemicals used in agriculture (Stenersen 1984). Unfortunately very little is known on the control of the monooxygenase systems in these animals. Therefore it is not know, if the content of environmetal pollutants and the level of monoxoygenase activities could be used an indicator of the chemical loading in the soil. Since the earthworm is important for the ecology of the soil, further studies would be recommendable.

CYTOCHROME P-450 AS A REAGENT IN THE ANALYSIS OF AIR AND WATER POLLUTION

Chemical analysis of the pollutants has its limitations. Therefore the screening of mutagenic materials during the processing of e.g. the raw water for human use has been applied to provide additional information. In such studies bacterial tests have been used. Since these bacteria are incapable to activate indirect mutagens, induced mammalian liver preparations containing high amounts of cytochrome P-450 must be used as additional factors (Ames et al. 1975, Ames 1979).

The most important sources of air pollution are the exhausts of burning processes in the energy provision of industry, transport and housing. These emissions are mutagenic. (Natusch 1978). Especially peat burning appears to release such compounds, which can be detected Salmonella strain TA98 in the presence of cytochrome P-450 containing system (Kivelä-Ikonen et al. 1985). In air a number of reactions take place. At first the emission products show high direct mutagenicity, but later their nature changes and the addition of S-9 mix increases the mutagenicity of the material (Jantunen et al. 1984).

It has been demonstrated that the chlorination of the organic material present in the surface waters generate mutagenic derivatives. The industrial discharges and humus from nature are the major sources of organic compounds. E.g. in Finland most of the surface waters contain high amounts of humus. The pulp industry uses high volumes of water. The bleaching process in pulp production releases lots of chlorinated compounds into the nature. Thus it is not a wonder that in Finland mutagens occur in high concentrations in top water obtainder from processed surface water (Vartiainen et al. 1985). The highest levels have been seen during the winter and the lowest during the spring and autumn, when the water layers are efficiently mixed by the up and down currents. In studies of the nature of these mutagens the Ames tests with an activation in the presence of S-9 mix has been used. It appears, however, that the mutagens are mostly directly acting. The addition of the mammalian tissue preparation only diminishes the mutagenic activity. This effect cannot be explained by the presence of protein, since similar amounts of bovine serum albumin is not so effective.

CONCLUSION

The biomonitoring of the environment is necessary in the followup of the effects of chemical loading. The wide occurrence of cytochrome P-450 in nature and its increase in many organisms by several common pollutants encourage the measurement of cytochrome P-450 and monooxygenase activities to complement direct chemical analyses. Since the cytochrome P-450 is envolved in the production of reactive intermediates and thus contributes to the carcinogenicity of environmental pollutants even in wild species, studies on cytochrome P-450 and related activities in different species are recomendable. Such studies provide additional general information on the enzyme system itself and promote our understanding of its multisided functions.

REFERENCES

Ade,P., Banchelli Soldaini, M.G., Castelli, M.G., Chiesara, E., Clementi, F., Fanelli, R., Funari, E., Ignesti, G., Marababini, A., Oronesu, M., Palmero, S., Prisino, R., Ramundo Oralndo, A., Silano, V., Viarengo, A. and Vittozzi, L. (1982) Comparative Biochemical and Morphological Characterization of Microsomal Preparations of Rat, Quail, Trout, Mussel and Daphnia Magna. in Cytochrome P-450, Biochemistry, Biophysics and Environmental Implications (Eds. E. Hietanen, M. Laitinen, O. Hänninen) Elsevier, Amsterdam, 387-390.

Ames, B.N. (1979) Identifying Environmental Chemicals Causing Mutations and Cancer. Science 204, 587-593.

Ames, B.N., McCann, J. and Yamasaki, E. (1975) Methods for Detecting Carcinogens and Mutagens with the Samonella/Mammalian-microsome

Mutagenicity Test. Mutation Research 31, 347-364.

Balk, L., Meier, J., Seidegård, J., Morgenstern, R. and Depierre, J.W. (1980) Initial Characterization of Drug-metabolizing Systems in the Liver of the Northern Pike, Esox lucius. Drug Metab. Disp. 8, 98-103.

Dewaide,J.H. (1971) Metabolism of Xenobiotics, Comparative and Kinetic Studies as a Basis for Environmental Pharmacology. Dissertation. University of Nijmegen, The Netherlands, 163 pages.

Dickson, W. (1975) The Acidification of Swedish Lakes. Rep.Inst.Freshw. Drottningholm 54, 8-20.

Frenkl, R. Györe, A. and Szeberenyi, Sz. (1980) The Effect of Muscular Exercise on the Microsomal Enzyme System of the Rat Liver. Eur.J.Appl.Physiol. 44, 135-140.

Gayathri, A.K., Rao, M.R.S. and Padmanaban, G. (1973) Drug Metabolism in Cold Exposed Animals. Ind.J.Biochem.Biophysics 10, 31-33.

Harri, M. Kuusela, P. and Oksanen-Rossi, R. (1982) Modification of Training-induced Responses by Repeated Norepinephrine Injections in Rats. J. Appl.Physiol. Respirat.Environ.Exercise Physiol. 53, 665-670.

Jantunen, M., Liimatainen, A., Ramdahl, T. and Itkonen, A. (1984) Rapid Changes in Peat Combustion Product Mutagenicity after Release into Atmosphere. XIV Annual Meeting of the European Environmental Mutagen Society, Moscow, Abstracts p. 225.

Kivelä-Ikonen, P., Lihtamo, H., Hänninen, O., Jantunen, M. and Itkonen, A. (1985) The Mutagenicity of Peat-fired Power Plant Flue Gases in Salmonella/ Mutagenicity Test. To be published. (in Finnish available in Publications of the Ministry of Commerce and Industry, Energy Department, Series D:41, Helsinki (1983) pp 25-27.

Koivusaari, U., Harri, M. and Hänninen, O. (1981) Seasonal Variation of Hepatic Biotransformation in Female and Male Rainbow Trout (Salmo gairdneri) Comp.Biochem.Biophys. 70 C, 149-157.

Koivusaari, U., Lindström-Seppä, P., Lang, M., Harri, M. and Hänninen, O. (1980) Occurrence of Cytochrome P-450 in Certain Fresh Water Species in Northern Europe. Biochemistry, Biophysics and Regulation of Cytochrome P-450 (eds J-Å Gustafsson et al) Elsevier, Amsterdam, 455-458.

Laitinen, M. Hietanen, E. Nieminen, M. and Pasanen, P. (1984) Acidification of Water and Extrahepatic Biotransformation in Fish. Environmental Pollution (Ser. A) 35, 271-278.

Laitinen, M. Nieminen, M. and Hietanen, E.(1982) The Effect of pH Changes of Water on the Hepatic Metabolism of Xenobiotics in Fish. Acta pharmacol. et toxicol 51, 24-29.

Muniz, I.P. (1981) Acidification - Effects on Aquatic Organisms. Third International Conference on Energy Use Management, Berlin (West), (Eds R.A. Fazzolore, and C.B. Smith). Vol. IV A 101-A 123.

Liimatainen, A. and Hänninen, O. (1982) Occurrence of cytochrome P-450 in the Earthworms Lumbricus terrestis. In Cytochrome P-450. Biochemistry, Biophysics and Environmental Implication (Eds. E. Hietanen, M. Laitinen and O. Hänninen) Elsevier, Amsterdam, pp 255-258.

Lindström-Seppä, P., Koivusaari, U., Hänninen, O. and Pyysalo, H. (1985) Cytochrome P-450 and Monooxygenase Activities in the Biomonitoring of Aquatic Environment. Pharmazie 40, 232-234.

Natusch, D.S.F. (1978) Potentially Carcinogenic Species Emitted to the Atmosphere by Fossil-fuel Plants. Environ. Health Perspect. 32, 79-90.

OECD (1981) The Cost and Benefits of Sulphur Oxide Control, Publications of the Organization for Economic Cooperation and Development.

Papas, T.S., Dahlberg, J.E. and Sonstergard (1976). Type C-Virus in Lymphosarcoma in Northern Pike (Esox lucius). Nature, 261, 506-508.

Payne, J.F. and Penrose, W.R. (1975) Induction of Hydrocarbon (benzo(a)pyrene) Hydroxylase in Fish by Petroleum. Bull. Env. Contam. Toxicol. 14, 112-116.

Pyysalo, H., Wickström, K., Litmanen, R., Lindström-Seppä, P., Koivusaari, U. and Hänninen, O. (1981) Contents of Chlordane-, PCB- and DDT Compounds and the Biotransformation Capacity of Fishes in the Lake Area of Eastern Finland. Chemosphere 10, 865-876.

Stegeman, J.J. (1978) Influence of Environmental Contamination on Cytochrome P-450 Mixed-function Oxygenases in Fish: Implications for Recovery in the Wild Harbor Marsh. J. Fish. Res. Board Can. 35, 668-674.

Stegeman, J.J. (1981) Polynuclear Aromatic Hydrocarbons and Their Meabolism in the Marine Environment. In Polycylic Hydrocarbons and Cancer, Acad.Press, New York, 1-60.

Stenersen, J. (1984) Detoxication of Xenobiotics by Earthworms. Comp.Biochem.Physiol. 78C, 249-252.

Vartiainen, T., Liimatainen, A., Jääskeläinen, S., Kauranen, P. and Kalliokoski, P. (1985) High Mutagenic Activities in Chlorinated Drinking Water in Finland. Fourth International Conference on Environmental Mutagens. Stockholm, 1985. Abstracts.

THE REACTION OF 15-HYDROPEROXYEICOSATETRAENOIC
ACID WITH MICROSOMAL CYTOCHROME P-450

Randy H. Weiss
Ronald W. Estabrook

Department of Biochemistry
University of Texas Health Science Center
Dallas, Texas   USA 75235

Hydroperoxides can be substituted for NADPH and molecular oxygen in the cytochrome P-450 catalyzed oxidation of some substrates (White and Coon, 1980). As shown by product studies, the mechanism of hydroperoxide-dependent oxidations is different from that of NADPH-supported reactions (Capdevila, et al., 1980; Hlavica, et al., 1983). Cytochrome P-450 can also catalyze hydroperoxide-dependent lipid peroxidation (O'Brien and Rahimtula, 1975). Cytochrome P-450 has been proposed as a propagating agent in the formation of reactive intermediates from fatty acid hydroperoxides in NADPH-dependent lipid peroxidation (Svingen, et al., 1979). Wheeler (1983) has reported a cytochrome P-450 catalyzed reaction of linoleic acid hydroperoxide with rat liver microsomes in the absence of NADPH in which the only reactants were apparently the hydroperoxide and molecular oxygen. Due to the importance of arachidonic acid metabolism and the ability of cytochrome P-450 to support hydroperoxide reactions (Capdevila, et al., 1984), we have initiated a study of the reaction of 15-hydroperoxyeicosa-5,8,11,13-tetraenoic acid (15-HPETE) with rat liver microsomes. This reaction is catalyzed by cytochrome P-450 and does not require NADPH.

METHODS

Synthesis of 15-HPETE. The procedure of Funk, et al. (1976), was used to synthesize ^{14}C 15-HPETE from ^{14}C arachidonic acid. The crude 15-HPETE was purified by silicic acid column chromatography, eluting with 95/5 hexane/ethyl ether to remove unreacted arachidonic acid and then 85/15 hexane/ethyl ether to elute 15-HPETE. The 15-HPETE was further purified by HPLC (using conditions given in Figure 3). Yields of 60-70% were obtained of purified 15-HPETE.

Microsomal fractions were prepared according to the procedure of Remmer, et al. (1976), using male Sprague-Dawley rats. Procedures for other experiments are given in the Figure and Table legends.

RESULTS AND DISCUSSION

Oxygen uptake. The addition of 15-HPETE to liver microsomes prepared from phenobarbital (PB)-treated rats results in a rapid uptake of molecular oxygen. Little oxygen uptake is observed with boiled microsomes treated with 15-HPETE. As shown in the superimposed oxygen electrode tracings of Figure 1, the extent of oxygen uptake is dependent

on the concentration of 15–HPETE. A similar pattern of oxygen
utilization has been observed following cumene hydroperoxide addition to
rat liver microsomes (Weiss and Estabrook, unpublished results).
Determination of the stoichiometry for the 15–HPETE reaction indicates
that about 1 nmole of oxygen is used for each nmole of 15–HPETE added.

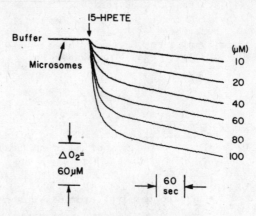

Figure 1. Oxygen uptake as a function of 15–HPETE
concentration. Liver microsomes prepared from PB-treated
rats were suspended at a concentration of 2 mg/ml in 50 mM
tris–HCl, pH 7.5, 150 mM KCl and 10 mM $MgCl_2$. Oxygen uptake
was measured using an oxygen electrode.

The initial rate of oxygen uptake is dependent on the 15–HPETE
concentration.
    The extent of oxygen uptake is independent of the protein
concentration (Table I). This supports the observation of a fixed
stoichiometry. The initial rate of oxygen uptake is also dependent on
the concentration of cytochrome P–450. N,N-Dimethylaniline, a substrate
for cytochrome P–450 (Nordblom, et al., 1976), inhibits the oxygen
uptake, with concomitant production of formaldehyde. This shows that
15–HPETE and microsomal cytochrome P–450 can support the N-dealkylation
of N,N-dimethylaniline. At a concentration of 2 mM, N,N-dimethylaniline
inhibits the extent of oxygen uptake 47% and the initial rate 37% when
100 µM 15–HPETE is added to liver microsomes (2 mg/ml) prepared from
PB-treated rats.
    The extent of oxygen uptake is also dependent on the type of
microsomes (cytochrome P–450) used. The extent and initial rate of
oxygen uptake observed following 15–HPETE treatment of liver microsomes
from PB- >ciprofibrate (CIP)- ∿ 3-methylcholanthrene (3–MC)- > un-treated
rats.

Table I. Oxygen Uptake as a Function of Protein
Concentration

Concentration of protein (mg/ml)	Oxygen uptake	
	extent (μM)	initial rate (μM/min)
0.25	119	104
0.50	125	143
1.0	124	211
2.0	135	456

The concentration of 15-HPETE added was 100 μM. Oxygen
uptake was measured as in Figure 1.

15-HPETE metabolites. In addition to oxygen uptake, the treatment of rat
liver microsomes with 15-HPETE in the absence of NADPH leads to the
formation of at least three polar metabolites (A-C) as shown in Figure 2.
Greater than 70% of the 15-HPETE is metabolized in 2 min. The major
metabolites A and B have no absorbance at 254 nm indicating the loss of
the conjugated diene structure of 15-HPETE. Using the reaction
conditions given in Figure 3, metabolite(s) A is formed in 16-23% yield;
B, 49-58%; C, 5-9%; and recovered 15-HPETE in 14-26% yield with PB-,
3-MC-, CIP- and un-treated rats. The relative constancy of the pattern
of products formed, as seen using liver microsomes with different
compositions of cytochromes P-450, indicates that the heme moiety rather
than the protein is an important feature in the formation of 15-HPETE
metabolites. The identity of metabolites A and B remain to be
determined. 15-Hydroxyeicosa-5,8,11,13-tetraenoic acid (15-HPETE)
appears to be metabolite C since it absorbs light at 254 nm and has a
retention time on HPLC similar to that of metabolite C.

Anaerobic experiments show that oxygen is not required for the
formation of the metabolites. This suggests that cytochrome P-450 may be
catalyzing a rearrangement of 15-HPETE. The observed oxygen uptake may
be due in part to lipid peroxidation. The formation of thiobarbituric
acid reactive products support this explanation. That lipid peroxidation
is catalyzed by cytochrome P-450 is shown by the different extents and
initial rates of oxygen uptake observed with different types of
microsomes.

The corresponding alcohol derived from the hydroperoxide is an
expected product of the hydroperoxide-dependent oxidation of a substrate
by cytochrome P-450 (Nordblom, et al., 1976). However, the presence of
N,N-dimethylaniline does not change the metabolite profile observed in
the reaction of 15-HPETE with rat liver microsomes, i.e. an increase in
the yield of the expected alcohol, 15-HETE, derived from 15-HPETE is not
observed. This suggests that a mechanism different from that of other
cytochrome P-450 catalyzed hydroperoxide-supported reactions is operative
with 15-HPETE.

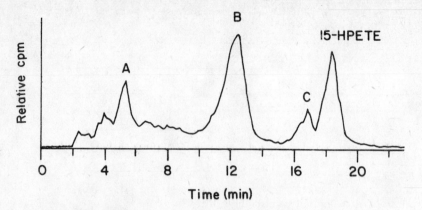

Figure 2. HPLC chromatogram of the metabolites derived from 15-HPETE in the reaction of 15-HPETE with liver microsomes prepared from untreated rats. Microsomes (2 mg/ml) were treated with 100 μM ^{14}C 15-HPETE for 2 minutes at 25°C. The reaction was quenched with EA and HCl and extracted with EA. The products (in ethanol) were analyzed by reverse-phase (C_{18}) HPLC using a linear gradient of 50% CH_3CN/50% H_2O/0.1% acetic acid to 100% CH_3CN/0.1% acetic acid for 40 minutes at a flow rate of 1 ml/min. The radioactivity eluting off of the column was monitored by liquid scintillation counting.

ACKNOWLEDGEMENTS

This work was supported in part by a grant from the USPHS (NIGMS 16488) and The Robert A. Welch Foundation (I-959).

REFERENCES

Capdevila, J., Estabrook, R.W., Prough, R.A. (1980) Arch. Biochem. Biophys. 200, 186-195.

Capdevila, J., Saeki, Y., Falck, J.R. (1984) Xenobiotica 14, 105-118.

Funk, M.O., Isaac, R., Porter, N.A. (1976) Lipids 11, 113-117.

Hlavica, P., Golly, I., Mietaschk, J. (1983) Biochem. J. 212, 539-547.

Nordblom, G.D., White, R.E., Coon, M.J. (1976) Arch. Biochem. Biophys. 175, 524-533.

O'Brien, P.J., Rahimtula, A. (1975) J. Agric. Food Chem. 23, 154-158.

Remmer, H., Greim, H., Schenkman, J.B., Estabrook, R.W. (1976) Methods Enzymol. 10, 703-708.

Svingen, B.A., Buege, J.A., O'Neal, F.O., Aust, S.D. (1979) J. Biol. Chem. 254, 5892-5899.

Wheeler, E.L. (1983) Biochem. Biophys. Res. Commun. 110, 646-653.

White, R.E., Coon, M.J. (1980) Ann. Rev. Biochem. 49, 315-356.

Author Index

564

SUBJECT INDEX

Dimethylaminoazobenzene 385
Dimethylaniline-N-demethylation
   105
Dimyristoylphosphatidylcholine
   181
Dioxygen activation 83
Dioxygen complex 41
Dioxygen splitting 41
1,3-diphenyl-2-amino-propane 465
DNA binding 375

Electron density 33
Electron distribution 11
Electron spin resonance 41
Electron transfer 121
Electronic structure 41
Energy transport 15
Enthalpy 63
Entropy 63
Enzymatic competition 285
Epoxide hydrolase 139, 361, 365,
   403
Ergosterol 431
ESR spectroscopy 91
Estramustine 391
Estrogen synthetase 461
Estrogene 219
Etaconazole 423
Ethanol 99, 303, 307
Ethanol oxidation 95
7-Ethoxycoumarin 191
7-Ethoxycoumarin-o-deethylase 211,
   345, 353
7-Ethoxyresorufin 191
7-Ethoxyresorufin-o-deethylase
   247, 339, 345, 395
7-Ethoxyresorufin-o-deethylation
   335
Ethylmorphine 191, 295
Ethylmorphyne-N-demethylase 281
Ethylxanthogenate 87
Evolutionary divergence 475
5-Exo-hydroxycamphor 49

Fatty acid (ω-1)-hydroxylase 501
Fatty acid ω-hydroxylase 501
Fe(III)EDTA 99
Fe-CO vibration 11
Fe-EDTA 95
Feminization 231
Fish 357
Flash photolysis 455
Flumecinol 253, 261, 269, 291
Fluorene-9-one 299
Formaldehyde 411
Free energy correlation 49

Free radicals 67

Gel permeation chromatography
   391
Gene duplication 475
Genetic induction 469
Gerbils 323
Glucose-6-phosphatase 361
Glucose oxidase 63
Glutathione 451
Glutathione-S-transferase 319,
   361, 365, 451

Haem 3
Haloperidol 285
Helical motions 3
Heme oxygenase 281
Hemoproteins 57, 79
Hepatocytes 299
Hepatoprotection 67
Hepatoprotection by scavangers
   67
Hepatoredoxin 113, 185
Hepatoredoxon reductase 113, 185
Hepatotoxicity 67, 323
Herbicide 345
Herbicide antidote 451
Heterologous reconstitution 159
Heterooligomers 139
n-Hexadecane 431, 455
Hexobarbital 281
Hexobarbitone 277
Homology 475
Hydrocarbons 87
Hydrogen peroxide 75
Hydroperoxide 57
Hydrophobic amino acids 487
Hydrophobic transmembrane
   helices 487
19-hydroxydeoxycorticosterone
   497
Hypophysectomy 219

Imazalil 423
Imidazole 95
Imipramine 285
Imipramine demethylase 285
Immobilization 493
Immunoinhibition 247
Indocyanin green 273
Inducers 291
Inhibiting oxidation 57
Internal repeats 475
Ionic stabilizations 49
Isotope effects 95
Isozyme selectivity 163